D0138822

The physiological ecology of seaweeds

The physiological ecology of seaweeds

Christopher S. Lobban
University of New Brunswick
Saint John
New Brunswick, Canada

Paul J. Harrison
University of British Columbia
Vancouver
British Columbia, Canada

Mary Jo Duncan
University of British Columbia
Vancouver
British Columbia, Canada

Cambridge University Press
Cambridge
London New York New Rochelle
Melbourne Sydney

Published by the Press Syndicate of the University of Cambridge
The Pitt Building, Trumpington Street, Cambridge CB2 1RP
32 East 57th Street, New York, NY 10022, USA
10 Stamford Road, Oakleigh, Melbourne 3166, Australia

First published 1985

Printed in the United States of America

Library of Congress Cataloging in Publication Data

Lobban, Christopher S.

The physiological ecology of seaweeds.

1. Marine algae – Physiology. 2. Marine algae –
Ecology. I. Harrison, Paul J. (Paul James), 1941– .
II. Duncan, Mary Jo. III. Title.
QK570.2.L63 1985 589.4′5041 84–9584
ISBN 0 521 26508 8

Contents

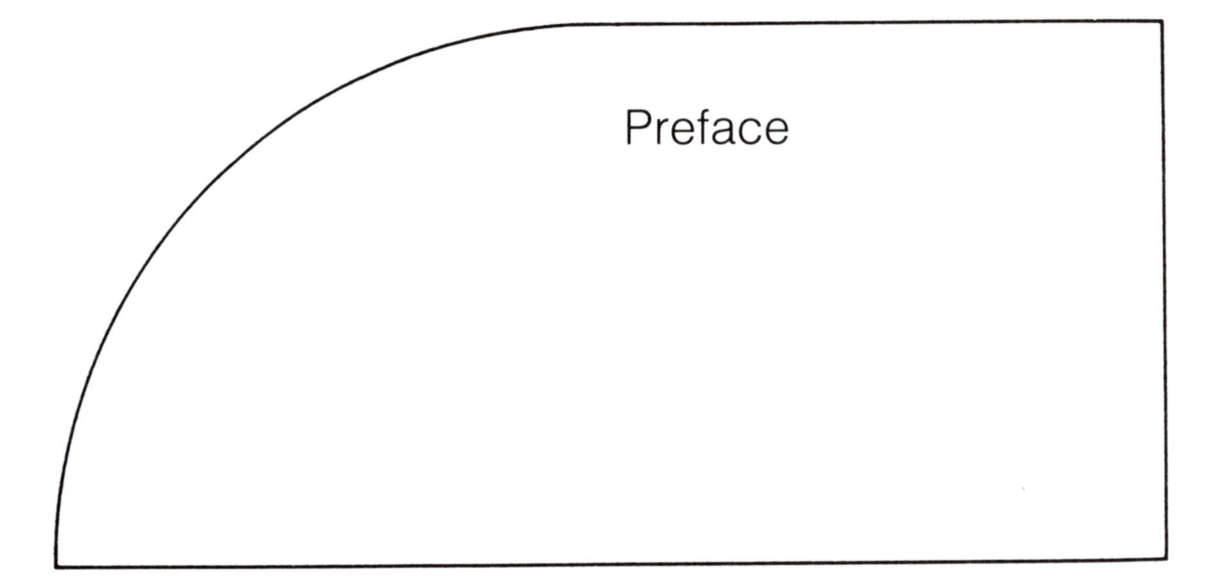

Preface

We feel a sense of wonder at how beings live. The beauty of nature is not only in the forms and colors but also in the interplay between environment and the organism. The classic phycology text deals very largely with structure, reproduction, and systematics. Our object in writing this book has been to provide a discussion of the physiology and ecology of seaweeds that is short enough, and written in an appropriate style, to be a textbook, yet covers the subject matter in sufficient depth to be of use for a one-semester course at both upper undergraduate and postgraduate levels.

The book grew out of lecture materials for a course at the University of New Brunswick, Saint John, and is intended as a sequel to a course in seaweed structure and reproduction. We have assumed that the student has taken such a course and is familiar with elementary cell biology, but we do not presuppose courses in plant physiology, plant ecology, or biochemistry. We have chosen to omit the phytoplankton and microbenthos because of the very different habitats and algal divisions of these populations.

We have all edited the entire work, and each of us took primary responsibility for writing certain chapters. M. J. D. wrote Chapter 11, she and I together wrote Chapter 2, P. J. H. wrote Chapters 6 and 8, and I wrote the rest.

Sections of the text have been critically read by N. J. Antia, W. P. Cochlan, H. Kauss, C. A. Kennedy, N. M. Price, C. B. Schom, T. E. Thomas, and W. N. Wheeler. The bulk of the work has been reviewed by R. G. S. Bidwell and J. R. Waaland, whose constructive criticisms were very helpful. Last, by no means least, we thank P. C. Silva for assistance in ridding the text of obsolete nomenclature. The manuscript was completed while C. S. L. was on sabbatical leave at Scripps Institution of Oceanography.

C. S. L.

1 Introduction

1.1 The plants and their environment

Seaweeds are marine algae of the classes Chlorophyceae, Phaeophyceae, and Rhodophyceae. They live attached to the seabed between the top of the intertidal zone and the maximum depth to which adequate light for growth can penetrate. They interact with other marine organisms and all interact with their physicochemical environment. Among the major environmental factors affecting seaweeds are light, temperature, salinity, water motion, and nutrient availability. Among the biological interactions are relations between seaweeds and their epiphytic bacteria, fungi, algae, and sessile animals; interactions between herbivores and plants (both macroalgae and epiflora); and the roles of herbivore predators. Finally, there are internal factors that regulate seaweed growth and development. Dissolved and particulate organic carbon from seaweeds provides food for detritivores and filter feeders; moreover, seaweeds are used as food by humans. Together these topics make up the physiological ecology of seaweeds.

Ocean vegetation is dominated by evolutionarily primitive plants: the algae. No mosses, ferns, or gymnosperms are found in the oceans, and only a few diverse angiosperms (the seagrasses) and even fewer lichens occur in marine habitats (though the latter are scarcely known). The water column, of course, is the domain of the phytoplankton, unicellular or colonial plants. The benthos of rocky shores is dominated by the seaweeds, though there are numerous and abundant microalgae beneath them and on their surfaces. Muddy and sandy areas have fewer seaweeds, since most species cannot anchor there (genera such as *Udotea* produce penetrating, rootlike holdfasts). In such areas the seagrasses become an important part of the vegetation, particularly in tropical and subtropical areas (Helfferich & McRoy 1980, Ferguson et al. 1980, Dawes 1981). Salt marshes have rather more diverse and dominant angiosperm floras. The paucity of fully marine angiosperms is balanced by a paucity of freshwater macroalgae; Rhodophyceae and Phaeophyceae are particularly scarce. That there are relatively few marine angiosperms may reflect the recent origin of the phylum and the problems of readaption to the sea, including the physiological problems imposed by the osmotic strength of seawater and its very different ion composition compared with soil (Queen 1974). The higher plants evolved the means to colonize the land; all their advances can be seen as adaptations to those drier habitats. The fundamental division of the vascular plant body into roots and shoots, vascular tissue itself, the waxy cuticle, pollen and seeds, are all useful on land and have contributed to the dominance of the seed plants there. What, then, are the features of seaweeds that make them so well suited to their environment?

One of the important features of most seaweeds, in contrast to phytoplankters, is that they are multicellular. Multicellularity confers the advantage of allowing extensive development in the third dimension of the water column. (The siphonous green algae form large noncellular thalli supported by trabeculae, ingrowths of the siphon wall.) Seaweeds usually grow vertically away from the substratum; this habit brings them closer to the light, enables them to grow large without extreme competition for space, and enables them to harvest nutrients from a greater volume of water. Support tissue is not usually necessary for this upward growth, since most small seaweeds are slightly buoyant and the water provides support. Support tissue is metabolically expensive, since it is nonphotosynthetic. However, strength and resilience are required to withstand water motion. Some of the larger seaweeds (e.g., *Pterygophora*) do have stiff, massive stipes; but others (e.g., *Hormosira*) employ flotation to keep them upright. Many of the kelps and fucoids have special gas-filled structures, pneumatocysts, but in other seaweeds (e.g., erect species of *Codium*; Dromgoole, 1982) gas trapped among the filaments achieves the same effect. (*Codium fragile* has become a nuisance in New England because large plants become buoyant enough to carry off cultivated shellfish to which they are attached – Wassman & Ramus 1973.)

Nutrient (and water) uptake and photosynthesis take place over virtually the entire surface of the seaweed thallus, in contrast to vascular land plants. However, a second important feature of multicellularity is that it allows division of labor between tissues, which is developed to various degrees in seaweeds. Differentiation and specialization among cells of algal thalli range from virtually nil, as in *Ulva*, where all cells except the rhizoids serve both vegetative and reproductive functions, to differentiated photosynthetic, storage, and translocation tissues in a variety of organs including stipe, blades, and pneumatocysts, as in certain kelps. Physiological differentiation between organs of large brown algae has recently been demonstrated, although, of course, no seaweed shows the differentiation seen in vascular plants. Even in vascular plants, the cells are biochemically more general than are animal cells: organs of vascular plants – stems, leaves, roots, flowers – all contain much the same mix of cells, whereas animal organs each contain only a few specialized cell types. The small diversity of cells in an algal thallus means that each cell is physiologically and biochemically even more general than vascular plant cells. For example, cortical cells of seaweeds function not only in photosynthesis but also in nutrient uptake. There are specialized cells in some algae, however, including sieve elements and gland cells.

A consequence of multicellularity is reduced surface-to-volume ratio of the organism. The reduction is small in uniseriate filaments, where only the end walls adjoin other cells, but is great in massive parenchymatous forms. Thus seaweeds are potentially somewhat less efficient in gas and mineral exchange than are phytoplankton. This is particularly true for thick seaweeds in which a long distance separates internal cells from the edge of the thallus (or, more correctly, the outer edge of the boundary layer where water movement begins). This is balanced to some extent by the potential for nutrient storage in medullary cells. Luxury uptake and storage of nitrogen and phosphorus have been demonstrated in a number of seaweeds, as has carbohydrate storage. Such mechanisms are important in habitats where there are pronounced seasonal changes in nutrient and light availability.

Seaweeds lack resting stages, in contrast to land plants and freshwater algae. Many seaweeds adapt to changing climatic seasons through their life cycles, and their reproduction can be cued by the environment: photoperiod and water temperature. In some of the morphologically less complex species, reproductive or developmental alternatives are determined by the environment; in others there is an obligatory cycle of generations cued by the seasons. Many of these seaweeds have two markedly dissimilar stages, which seem particularly suited to widely different seasons, with the cryptic stage active in seasons the other stage cannot survive. In such alternation of heteromorphic generations in seaweeds the large stage may be the gametophyte or the sporophyte, or both stages may have the same ploidy level. Moreover, it is important for us not to assume that winter is a poor season for seaweed growth: the large stage may be found in winter and the cryptic stage in summer (e.g., in many *Porphyra* species). Many seaweeds in the same habitats have isomorphic stages or grow year round. In the marine environment the nutrient season (which we cannot sense directly) is as important as daylength or temperature. Grazing must also be considered. Seaweed life histories often include such alternative means of reproduction as parthenogenesis and apomeiosis, which permit the repetition of one stage when syngamy fails to occur.

Algal classes are noted for their wide diversity of pigments, but whether this is a factor in their success in the variable light environment of the sea is a subject of much debate. As detailed in Section 2.7.4, there is poor correlation between the class of the predominant algae and the depth of the water. More likely, the different pigment compositions (as well as other biochemical differences) represent various equally successful symbioses in the evolution of eukaryotes.

1.2 The study of seaweed physiology and ecology

In Chapters 2 through 8, we examine the major physical factors and how each affects the plants from the biochemical level through to the population level. These factors – light, temperature, salinity, water motion, nutrients, and pollution – are followed by a chapter on seashore communities, which introduces the biological interactions and completes and integrates the discussion of the physical environments of seaweeds. Chapters 10 and 11 extend the integrated approach, first in an examination of the development and life histories of seaweeds, finally in the practical application of seaweed physiological ecology to mariculture.

Ecology, physiology, and biochemistry are part of a continuum of study to answer the question, How do organisms live? One cannot study this question without first knowing the structure and life histories of the organisms (textbooks on this include those by Bold & Wynne 1978 and Dawes 1981; classification summaries are given by Wynne & Kraft 1981 and Round 1981). Equally, one should not study only the anatomy, for this is like trying to understand human biology merely through dissection of the body. Ecology can be defined as the study of organisms in their natural environment, but it usually focuses on levels beyond the individual organism. Physiology is the study of the responses of individuals or their parts. Biochemistry is the study at the molecular level. (The continuity is especially clear in unicellular individuals.) These categories – ecology, physiology, and biochemistry – are artificial inasmuch as they treat the same phenomena at a smaller and smaller scale.

Seaweed biology has some aspects in common with other disciplines, including marine zoology and vascular plant physiology. These connections can provide useful suggestions for work on seaweeds, but comparisons, although easily made, can be misleading. Moreover, there is a great danger of speculative comparisons being transformed into "fact." The marine habitat of seaweeds is shared by benthic animals and by phytoplankton. Seaweeds are set apart from marine animals by a host of differences, including their nutrition. The differences from phytoplankton are much less marked, but include differences in structure and biochemistry between classes, multicellularity, and attachment to the seabed. Seaweed physiology has some parallels in higher plant physiology, and aspects of basic eukaryote cell biochemistry are common. The marine environment differs in several important ways from the terrestrial, and might be expected to cause significant differences in physiology between marine and terrestrial plants. Yet the physiology of marine angiosperms is not greatly different from that of their dry-land counterparts; as mentioned, the differences are principally in relation to the high osmotic pressure of seawater, the substantial differences in ionic composition between seawater and soil, and the hydrodynamic forces. The much greater differences between seaweeds and angiosperms stem from the great evolutionary divergence of the two groups, rather than from the habitat.

The use of taxonomic species to define ecological entities is a handicap, as Harper (1982) has emphasized in a thought-provoking essay. This is because "the criteria used by the taxonomist for the delineation of taxa are chosen deliberately from the conservative and stable features of morphology that are not subject to marked genetic variation, polymorphism or phenotypic change. These same criteria . . . may be quite inappropriate for describing the ecologically relevant differences between individuals, populations and communities . . . The failure of taxonomic categories to fit as ecological categories is not surprising . . . yet it may be just the taxonomically useless characters that are mainly responsible for determining the precise ecologies of organisms."

We close this brief introductory chapter with a caveat: Let the reader beware: In trying to make a cohesive story out of fragmentary literature, especially in view of our wish (and need) to select examples rather than citing every relevant paper, we may give an impression of uniformity or completeness. Students are cautioned that this is a *false* impression: In most cases only one or a few studies have been done on a particular topic, and a phenomenon demonstrated in a particular alga under certain conditions will not necessarily turn out to be the same in other algae or under other conditions. Lewin (1974) cautioned thus: "There is still a tendency . . . to over-generalize on the basis of investigations on no more than one or two examples. Some workers even today tend to propound such statements as 'Lysine synthesis in the Euglenophyceae goes thus and so' although they may have examined only one strain of one species of one genus grown under only one set of known conditions; or, worse, they may have analysed only a single tuft of seaweed, grown under unspecified and probably unknown conditions, and identified merely to genus by whichever passing botanist was rash enough to do so . . . Only the future will show how wrong they may have been." Very much more work needs to be done to fill in what is so far little more than a framework, but we hope in the following pages to present the essentials of that framework.

2 Light and photosynthesis

2.1 **Introduction**

Light is without doubt the most important factor affecting plants; it is also one of the most complex. This complexity arises in the first place from the nature of light itself, and secondly from the great number of effects it has on plants. Seaweeds grow in an exceptionally dynamic and diverse light environment. The continuous ebb and flood of tides have a profound effect on the quantity and quality of the sun's energy reaching seaweeds and add greatly to the variation already present in irradiance at the earth's (or ocean's) surface. The primary importance of light to seaweeds is in providing the initial energy for photosynthesis, and ultimately for all biological processes. It is the signal for many events throughout the life cycles of the algae, including reproduction, growth, and distribution. It is involved in synthetic processes in ways still not understood. The quality and quantity of light depend on such factors as depth and the number of particles in the water. Most physiological experiments involving light are initially conducted in the laboratory, under artificial light. Knowledge of the emissive properties of the light sources is essential if the results are to have ecological significance.

In this chapter the nature of light in general, and underwater in particular, is presented first, together with discussion of the methods and units used in measuring light quantity and quality. Then algal pigments are described and put in their structural context, the thylakoids of the chloroplast. The remainder of the chapter covers the light reactions of photosynthesis in general and field studies of photosynthesis in marine algae.

2.2 **Radiant energy principles and measurement**

2.2.1 Properties of radiant energy including light

Radiant energy from the sun, propagated through space in the form of waves, ranges across the electromagnetic spectrum from the long-wave, low-energy quanta of the radio region to the short-wave, high-energy cosmic rays (Fig. 2.1). Wavelength (λ) and frequency (ν) are related to the speed of light by the equation:

$$\lambda(\text{nm}) \times \nu\ (\text{s}^{-1}) = c = 3 \times 10^8\ \text{m s}^{-1}$$

Since the speed is constant (in a given medium), that is, the speed of light, frequency is inversely proportional to the wavelength. Paradoxically, energy travels not only as a series of waves but also in indivisible packets or quanta. The energy of a quantum ($\mathcal{E}$) is directly proportional to the frequency and inversely proportional to the wavelength:

$$\mathcal{E} = h\nu = hc/\lambda$$

where h is Planck's constant, 6.6×10^{-34} J-s. The significance of these equations to photosynthesis lies in the relation between wavelength and energy. As the wavelength increases from the blue region (400 nm) toward the red (700 nm), the amount of energy per photon decreases.

"Light" refers to the narrow region of the spectrum that is visible to the human eye, plus the ultraviolet and infrared wavelengths. There are important differences between the spectral sensitivities of the human eye and plant photosynthetic pigments (Figs. 2.2, 2.3). Rhodopsin, the visual pigment, has one major peak, at 556 nm, with absorption decreasing on either side, whereas the chlorophylls and other photosynthetic pigments have major absorption peaks across a broad region. The term "photosynthetically active radiation" (PAR) has come into use as more descriptive of the spectral properties of photosynthetic pigments. PAR is defined as wavelengths of 400 to 700 nm. There is, however, some evidence that photosynthetic absorbance extends down to 300 nm in the green alga *Ulva lactuca* and the tetrasporangial stage (*Trailliella intricata*) of the red alga *Bonnemaisonia hamifera* (Halldal 1964).

Figure 2.1. Electromagnetic spectrum showing position of visible, UV, and IR wavelengths. (After Withrow & Withrow 1956)

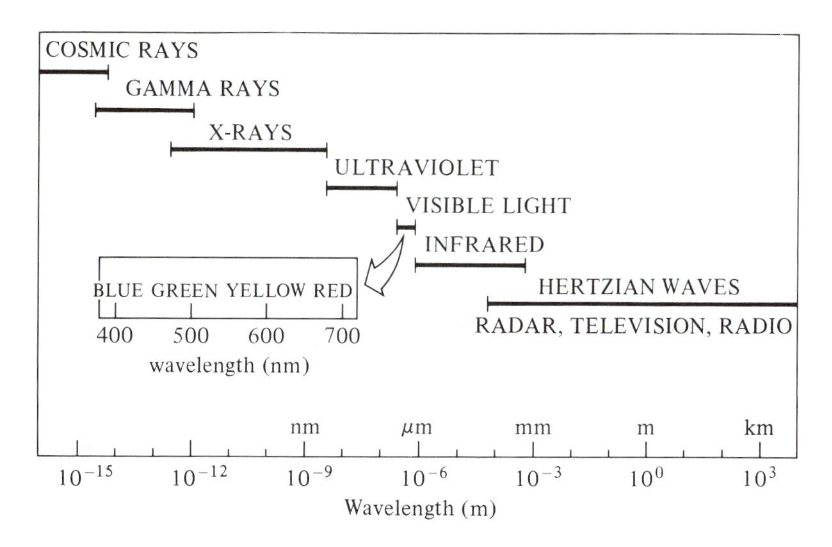

The red wavelengths, between 650 nm and 750 nm, are separated into two portions: near red (visible), 650 to 700 nm; and far red, from 700 nm to 750–800 nm. The distinction between these two regions is important in the activity of the reversible pigment phytochrome, which is involved in photomorphogenesis (Sec. 10.4.1). Short wavelengths, below 400 nm, are the ultraviolet, with the near UV extending to about 320 nm and the far UV to about 180 nm, where quartz and air begin to absorb strongly (Withrow & Withrow 1956). The shortest UV rays from the sun to reach the earth have wavelengths of about 290 nm. (Shorter UV wavelengths can be obtained from germicidal lamps.)

Specific terms are used to distinguish between energy emanating from a source, whether it is the sun or a lamp, and energy being intercepted by an object such as a seaweed. ''Radiance'' is the flux (rate of flow) radiated by a source; ''irradiance'' is the flux intercepted per unit area. Typically, plant physiologists are concerned with irradiance. With plant cultures in the laboratory, irradiance may be increased or decreased by placing the plants closer to or further from the lamps, or by using filters to decrease the quantity of light. Radiance may be altered by choosing different light sources (see below, and Fig. 2.7), or by using a different number of lamps. The end result will, of course, be a change in irradiance, that is, the amount of energy reaching the plants. Often the term ''light intensity'' is used loosely and incorrectly instead of irradiance. This is especially true in older biological literature.

All the terms mentioned above apply to energy. The energy of a photon is inversely proportional to the wavelength. Thus as the wavelength increases from the blue region (400 nm) to the red region (700 nm), the amount of energy per photon decreases. Pigments absorbing photons at different wavelengths are receiving different amounts of energy, but the energy available for photosynthesis is about the same in all cases (for reasons explained in Sec. 2.5.1); thus the number of photons received (photon flux density) is a useful measure of irradiance.

Figure 2.2. Sensitivity of the human eye to visible light, showing the peak in the green region. (After IES Handbook 1952)

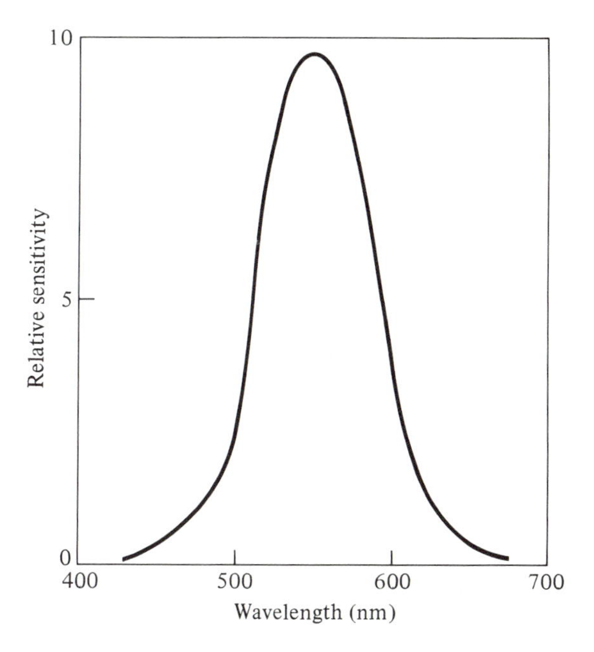

The unit most commonly used to measure photon flux density is the einstein (E) or microeinstein (μE), which refers to a mole of photons. Some authors would use simply mole (or micromole) because the einstein is not an SI unit. However, as Lüning (1981a) has pointed

Figure 2.3. Spectrum of solar energy at the earth's suface (upper dotted curve), and absorption spectra of algal pigments. (From Gantt 1975 *BioScience* 25: 781–788. Copyright © 1975 by the American Institute of Biological Sciences)

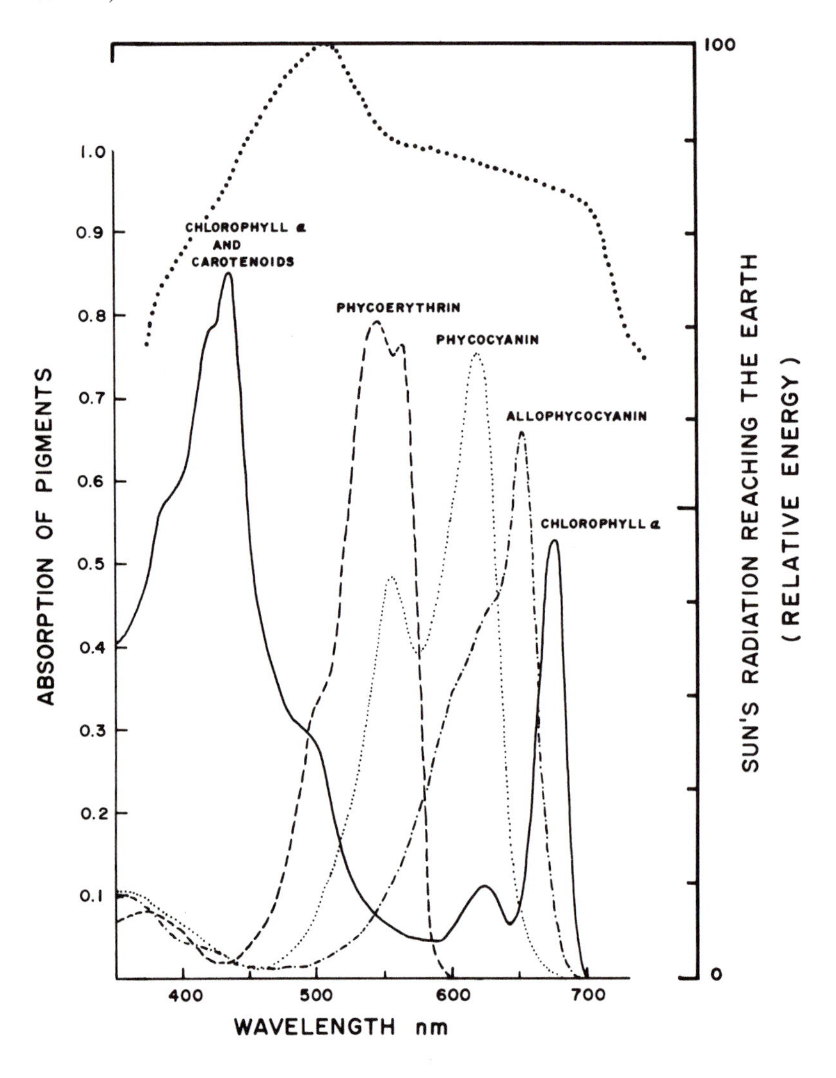

out, that would remove "the last verbal hint. . .that a quantity having to do with light is being measured." The appropriate SI units for measuring energy are joules (J) and watts (W) (see Leedale 1968). The joule is used to measure the total *quantity* of energy transferred in the quanta. The watt is used to measure radiant flux (Smith & Tyler 1974). Thus, $1\ W\,m^{-2} = 1\ J\ m^{-2}\ s^{-1}$. Besides the accepted international units, many others are found in the literature.

Perhaps the most common in older biological literature, and unfortunately still in use, are the illuminance units for radiant flux, foot-candles and lux. They refer to the spectrum as perceived by the human eye. Illuminance units cannot be precisely converted into irradiance units except for specified light sources, but Table 2.1 gives some approximate values.

2.2.2 Measurement

The reason why much of the biological light data are given in lux or foot-candles is that photometers, or light meters, are easy to use and inexpensive; they were for a long time the only instruments available. These data can be misleading, however, because the instruments are designed to measure light absorbed by the human eye, not by plant pigments. The ideal instrument for plant physiologists would measure the exact number of quanta absorbed by plant pigments. Such an ideal is unattainable, but it is possible to measure the energy or quanta relatively uniformly across the spectrum, and thus obtain a better measure of the light available to plants than is obtained from an illuminance meter. Instruments are presently available that can measure total irradiance and/or total quanta across the PAR spectrum, as shown

Table 2.1. *Approximate conversion factors for irradiance units (1 W m^{-2}) to quantum irradiance (in μE m^{-2} s^{-1}) and illuminance (in lux)*

Light source	Quantum irradiance (μE m^{-2} s^{-1})	Approximate illuminance (lux)
Daylight above water surface	4.6	250
Daylight below water surface	4.2	—
White fluorescent tubes		
"Daylight 5000 de Luxe"[a]	4.6	250
"Cool white"[b]	4.6	360
"Warm white"[c]	4.7	390
Quartz iodine lamp[d]	5.0	250
Tungsten (100 W)	—	240

[a]Osram 40 W/19, as measured with spectroradiometer.
[b]Sylvania FR96T12-CW-VHO-135.
[c]Westinghouse F96T12/WW/SHO.
[d]Sylvania 250W-T-4Q-CL.
Source: Lüning (1981a), with permission of Blackwell Scientific Publications.

Figure 2.4. Energy sensitivity of an ideal irradiance meter and an ideal quanta meter compared with two commercial quanta meters (broken lines). The ideal quanta meter has a lower energy response to compensate for the greater energy of quanta at shorter wavelengths. (From Lüning 1981a with permission of Blackwell Scientific Publications)

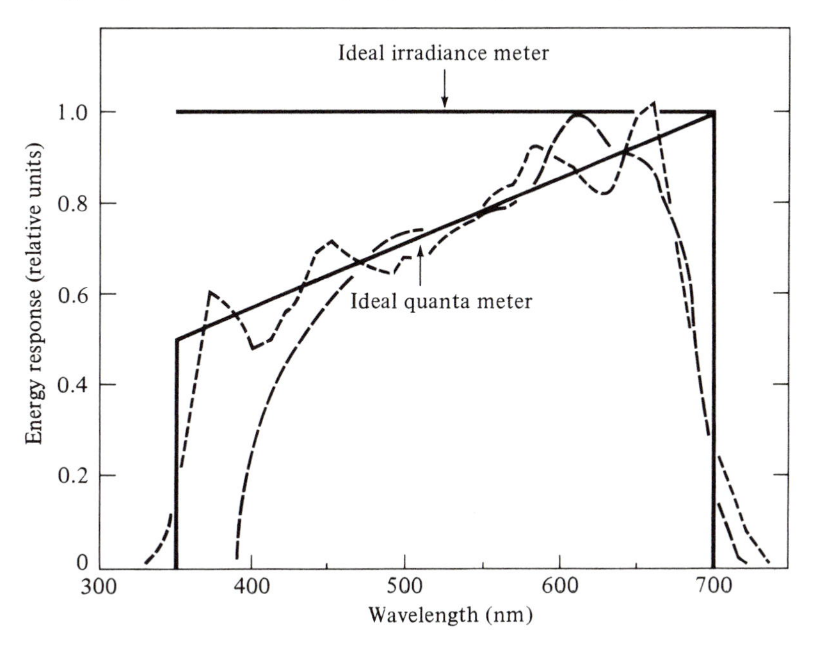

in Figure 2.4. Spectroradiometers and spectroquanta meters permit measurement of energy or quanta, respectively, available in specific wave bands (e.g., Tyler & Smith 1970, Burr & Duncan 1972). Such instruments are usually equipped with filters to separate and transmit discrete wavelengths to the detector, at intervals from 5 nm to 25 nm apart. This is important for measurement in the underwater environment, where light quality changes rapidly with depth (see Sec. 2.3.1). Such instruments can be installed in the sea to give continual readings of irradiance at a given spot. Whether one uses a total irradiance meter, a spectroradiometer, or both depends on the information required. For example, Lüning & Dring (1979) measured underwater irradiance at 20-min intervals for a year in a subtidal kelp zone off Helgoland (North Sea). These data provided a good estimate of the total amount of energy per square meter available annually to the plants. Measurements with a spectral irradiance meter gave the amount of energy or number of quanta at discrete wavebands, thereby defining the transmissive quality of the water.

An important part of a light-measuring instrument, especially one used underwater, is the collector. This is the window through which the light beams pass on their way to the detector. The window may be a flat surface of translucent diffusing material, such as opal glass, known as a 2π or cosine collector, which measures all rays coming from the hemisphere above it. Alternatively, it may be a 4π collector, that is, a sphere collecting rays from all directions. Both types are useful for biological measurements. Phytoplankton are more or less spherical, as are their plastids, so they are essentially 4π collectors. Many seaweeds present a more or less horizontal surface to collect radiant energy, and might be better represented by a 2π collector. (Although many seaweed chloroplasts are spherical, other factors restrict the directions of light harvesting; these include thallus thickness, the position and numbers of plastids, the habit of the alga, and its position in the water column and on the substratum.) In most cases oceanographers measure downwelling irradiance, the "flux incident per unit area measured on a horizontally oriented cosine collector facing upward" (Smith 1969, 1974). Upwelling light could be measured by inverting the collector, but downwelling irradiance is the quantity most commonly measured by oceanographers to optically classify natural waters. Scalar irradiance, however, is considered by some to be the better measurement of radiant energy available in the sea for biological processes. Scalar irradiance, measured with a 4π collector, is defined as "the integral of irradiance distribution over all directions about a point at a given depth underwater" (Smith & Tyler 1976).

2.3 Sources of radiant energy

2.3.1 Solar energy

Radiation reaching the surface of the earth consists of two components, direct sunlight and "skylight." The latter may come from a clear sky or from a sky partly or completely cloudy (Szeicz 1974). Although solar irradiance just outside the earth's atmosphere varies little (Laue & Drummond 1968), that reaching the earth's surface shows large variations as a result of scattering and absorption by the atmosphere. Molecules of atmospheric gases and fine dust particles scatter much of the incoming energy. Because these particles are small, relative to the wavelengths of PAR, the intensity of scattering is inversely proportional to the fourth power of the wavelength. Thus shorter, blue wavelengths are scattered more than longer, red wavelengths. Carbon dioxide in the atmosphere absorbs wavelengths longer than 2300 nm, and water vapor attenuates those between 720 and 2300 nm. At the short end of the spectrum, the mid-UV (290–320 nm) is absorbed by ozone, with the result that wavelengths reaching the earth's surface on a clear day range from the near UV, through PAR, to the near infrared. The total irradiance reaching the earth's surface is also affected by the sun's angle. As it shifts from the zenith position (90° solar angle) toward the horizon, the total irradiance decreases; moreover, the wavelength of maximum energy shifts from about 470 nm at 90° to 650 nm at 11.3°.

Table 2.2. *Reflection of sunlight from the sea at various solar angles*

Altitude of the sun (°)	10	20	30	40	50	60–90
Percentage reflected	28	14	8	6	5	4

Source: Holmes (1957), in *Treatise on Marine Ecology and Paleoecology*, vol. 1, pp. 109–128, with permission of the Geological Society of America.

Light hitting the sea surface is reduced by two processes: reflection and absorption. The percentage reflected depends on the angle of the sun to the water, and hence also on the state, or roughness, of the water. Reflection from a smooth sea with the sun near its zenith is only about 4% of the total light (sun plus sky), whereas with a sun altitude of 10° reflection is 28% (Table 2.2). With an overcast sky reflection is about 10% regardless of the altitude of the sun (Holmes 1957). Waves can decrease reflectivity at low solar altitude by increasing the angle of the water surface to the sun, but if the sea is rough enough to produce white caps, reflectivity increases to about 31% (Holmes 1957, Jerlov 1968). Under sunny skies waves can cause considerable heterogeneity in the subsurface light field by temporarily focusing the sun's rays to certain spots and away from others, especially in the top few meters of water (Fig. 2.5).

As solar energy penetrates the oceans it is altered in both quality and quantity. The attenuation results from absorption and scattering, as in the atmosphere. Water itself absorbs maximally in the infrared and far red, above 700 nm. Irradiance at wavelengths greater than 1300 nm is totally absorbed in the top 10 mm of water (Jerlov 1976), while far red (ca. 750 nm) has been measured down to 5 to 6 m depth (Smith & Tyler 1976). Absorption of radiant energy by particles in the water, such as phytoplankton, will depend on the pigments they contain. Scattering by particles larger than 2 μm contributes to attenuation by increasing the length of the optical path of quanta once they are in the water, thereby increasing the opportunities for absorption. Smaller particles and sea salts do not contribute appreciably to attenuation in the visible region (Jerlov 1976).

Attenuation of light by various processes and particles results in a number of classes of seawater, related to transmittance of radiant energy (Jerlov 1976). The oceans can be divided into two broad categories, green

Figure 2.5. Heterogeneity of near-surface underwater light. (a) Photograph looking up through a *Macrocystis* canopy; (b) recordings of irradiance at the surface, at 1 m (in the region of greatest heterogeneity), and at the seabed, 7 m below the surface. (a, courtesy of and © 1983 John Pearse; b, courtesy of and © 1983 Valrie Gerard)

(a)

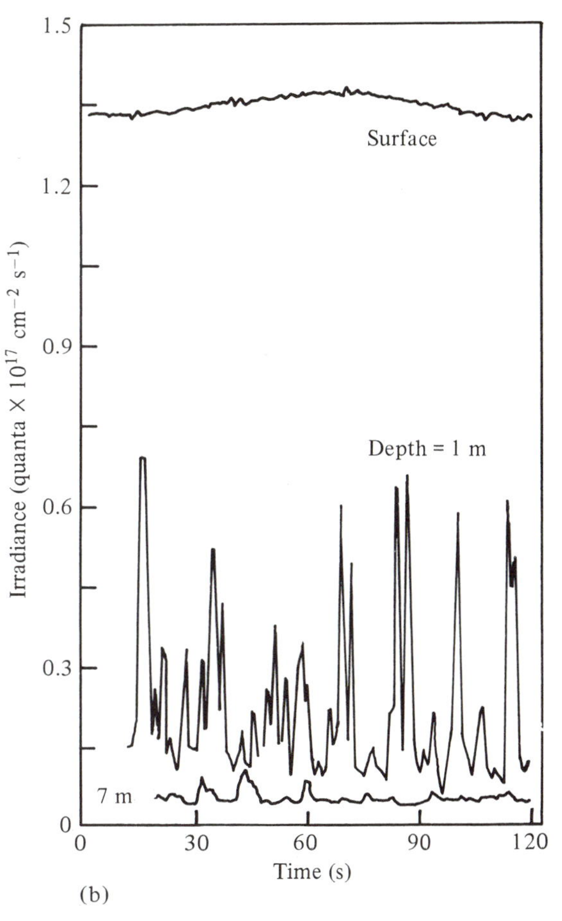

(b)

coastal waters and blue oceanic waters, with many variations in each of these categories (Morel & Smith 1974). Coastal waters have a characteristic green color because of absorption at shorter wavelengths by plant pigments and by dissolved organic substances ("Gelbstoff") that absorb strongly in the blue wavebands. Gelbstoff comes from terrestrial humic material brought to the seas by rivers, and is also produced in the sea by algae. Jerlov (1976) distinguished five ocean water types and nine coastal water types, as exemplified in Figure 2.6. Here it can be seen that in the clearest water (oceanic type I), the maximum transmittance (about 98.2% of surface irradiance) is at approximately 475 nm. The water type that transmits the least solar energy is coastal type 9. In such water the maximum transmittance occurs at about 575 nm (green) and is only 56% of the total irradiance at the surface. Coastal waters have the greatest concentrations of organic matter, including chlorophyll. Smith & Baker (1978) have incorporated the contribution of chlorophyll-like pigments into their optical classification of water types. Transmission of wavelengths around 675 nm, where chlorophyll absorbs in vivo, reaches 66% of surface energy in oceanic waters. In coastal waters (type 9), transmission at 675 nm is only 40% (Lüning 1981a). The effect of wavelength dependence of transmittance is seen in the changes in spectra at various depths (Fig. 2.7).

2.3.2 Artificial light sources

There are two principal kinds of lamps used in laboratory work: incandescent and fluorescent. In the former, light is produced by a filament, usually of tungsten, and most of the radiant energy produced is in the infrared, above 760 nm (Fig. 2.8a). Thus the light from these lamps is quite different from sunlight. Fluorescent lamps have a phosphor coating inside the tubes that converts invisible 253.7 nm radiation from a mercury arc into longer wavelengths by fluorescence. They are widely used in growth chambers because they produce less heat than incandescent lamps and, more important for biological experiments, some have more natural spectral qualities (compare Figs. 2.2 and 2.8b–d). An important

Figure 2.6. Percentage of transmittance downward of irradiance of various wavelengths in Jerlov's different optical water types. The types range from clear oceanic (type I) to turbid coastal (type 9). (Jerlov 1976, reprinted with permission from *Marine Optics* © 1976 Elsevier Science Publishers)

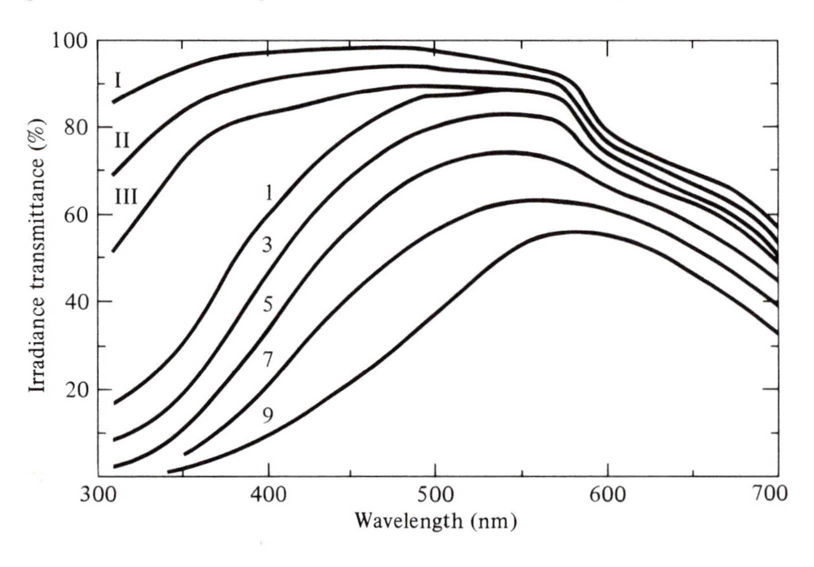

Figure 2.7. Energy spectra of natural light at various depths in the northern Baltic Sea. Compare with solar energy at the surface and with plant absorption spectra (Fig. 2.3). (Jerlov 1970, reprinted with permission from *Marine Optics* © 1976 Elsevier Science Publishers)

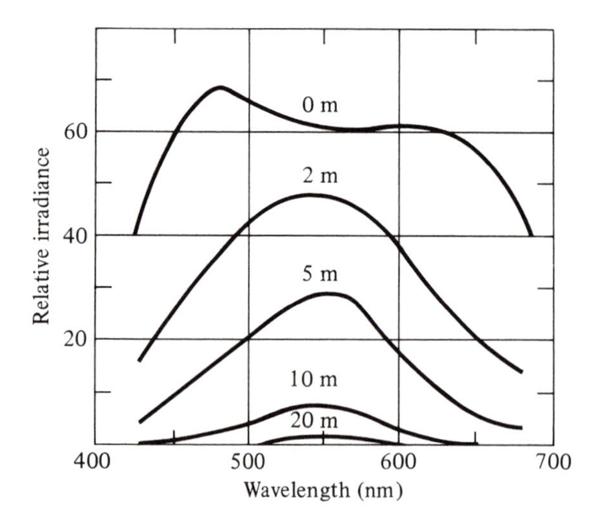

point in long-term laboratory experiments is that the quality and especially the quantity of light from most fluorescent bulbs changes with time.

2.4 The photosynthetic apparatus and pigments

2.4.1 Chloroplasts

Seaweed groups differ not only in their accessory pigments but in the very structure of the chloroplasts. Algae in general have a great variation in size, shape, and number of chloroplasts, particularly in freshwater species (Brawley & Wetherbee 1981). Most seaweed chloroplasts are disk-shaped (Fig. 2.9 a–c), similar to higher plants. However, the more primitive Bangiophycidae have lobed chloroplasts which differ in several respects from chloroplasts in the Florideophycidae. The austral brown alga *Splachnidium* is unusual in having a stellate plastid (Fig. 2.9d) (Asensi et al. 1977). The functional significance of the different shapes is not understood. All chloroplasts are bounded by an outer envelope, a 12–20 nm wide double membrane. In the Phaeophyceae the whole chloroplast is further enclosed within a double membrane sac of endoplasmic reticulum (ER) (Fig. 2.9b). Ribosomes may be attached to the outer surface of the ER membrane. In some plants the chloroplast ER is continuous with the nuclear envelope. In the Rhodophyceae and most of the Chlorophyceae, the chloroplast is bound only by the double chloroplast membrane. Some tropical coenocytic green algae have a distinctive integument around the chloroplast that apparently prevents their disruption. Isolated chloroplasts from *Codium* and *Caulerpa* do not swell or burst in distilled water. Treatment with detergent will remove chlorophyll and internal membranes but leave the integument intact. This integument may prevent the chloroplasts of these species from being digested when they are eaten by sacoglossan mollusks, thus allowing the chloroplasts to continue photosynthesis in a symbiotic relationship with the animal (Greene 1970).

The chloroplast envelope encloses the matrix or stroma, in which are embedded flattened membrane-bound sacs called thylakoids (or photosynthetic lamellae) in which the pigments and electron transport machinery are arranged (Fig. 2.9a, b). The fact that the thylakoids are

Figure 2.8. Emission spectra of (a) an incandescent bulb; (b–d) various fluorescent bulbs: (b) cool white, (c) warm white, (d) daylight bulbs. (After IES Lighting Handbook 1956)

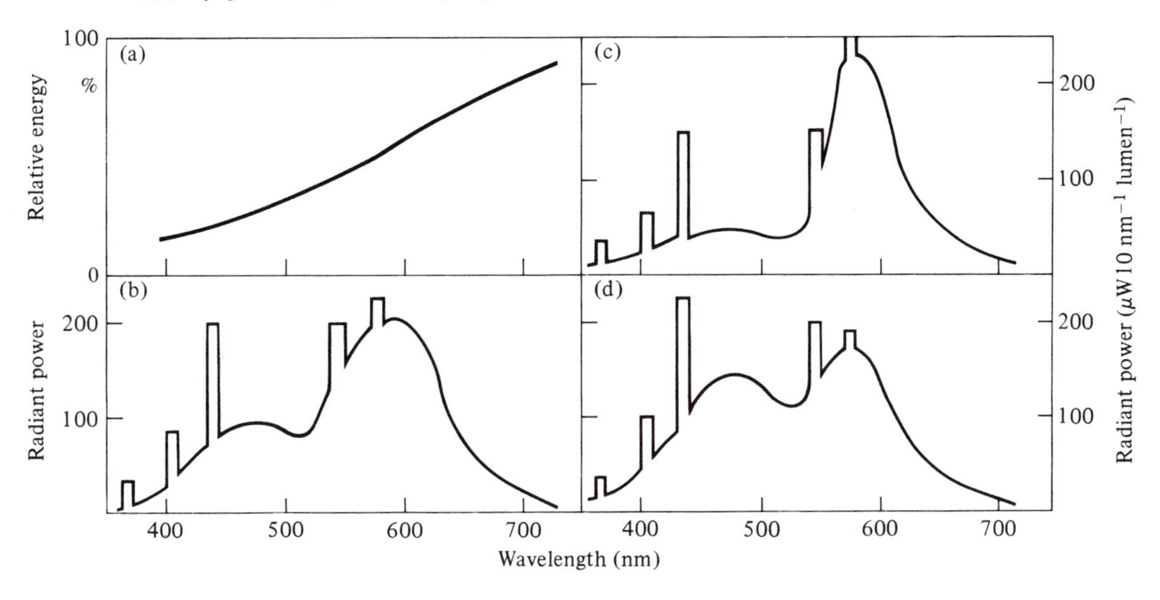

completely enclosed sacs is important in the conversion of light energy to ATP (see Sec. 2.5.3). The arrangement of the thylakoids distinguishes the chloroplasts of the three seaweed groups. In the red algae thylakoids lie more or less lengthwise and individually, about 20 nm apart and parallel in the stroma. In species of the more advanced Florideophycidae there is usually an outermost thylakoid that passes around the chloroplast and encloses the internal set (Fig. 2.9a). On the outside of red algal thylakoids (and also on the photosynthetic lamellae of Cyanophyceae) are the phycobiliprotein-containing phycobilisomes (see Sec. 2.4.3). The brown algae have thylakoids associated in groups of three, with gaps of 2 to 4 nm between adjacent thylakoids within each band (Fig. 2.9b). Green seaweeds have bands consisting of 2 to 6 or more thylakoids. Many species of green algae have thylakoids stacked into grana, similar to higher plants, with some thylakoids extending through the stroma from one granum to another. This type can be seen in *Acetabularia* and *Caulerpa*. The simplest type among the green algae is found in *Spongomorpha*, where the thylakoids are predominantly in pairs.

Besides thylakoids, seaweed chloroplasts also contain ribosomes, genophores or DNA, and (in some species) pyrenoids. The genophore is defined as a gene-containing body, similar to that found in viruses and bacteria (Bisalputra 1974). In the brown algae it forms a ring just inside the encircling band of three thylakoids (Fig. 2.9b). In other algae the genophores are scattered about. Besides directing the division and movement of chloroplasts, the DNA, along with the ribosomes, is involved in the synthesis of proteins, including the larger subunit of the main photosynthetic enzyme, ribulose-1,5-bisphosphate (RuBP) carboxylase. Pyrenoids contain RuBP carboxylase and in the Chlorophyceae perhaps also starch synthesizing enzymes. The majority of Chlorophyceae have pyrenoids, usually one within each chloroplast. Starch may be deposited in grains forming a shell around the pyrenoids, or (in some Caulerpales) may be stored in modified chloroplasts called amyloplasts. Often one or two thylakoids pass through the pyrenoid. Species of green algae that lack pyrenoids tend to deposit starch grains at random, as do higher plants. In red algae, floridean starch is deposited in grains outside the chloroplast, but these grains usually congregate near the pyrenoid if there is one. Brown algal pyrenoids project from the chloroplasts. Laminaran, the brown algal starch, is stored in the cytoplasm but not in visible grains.

2.4.2 Chlorophylls, carotenoids, and phycobiliproteins

Three groups of pigments – chlorophylls, phycobiliproteins, and carotenoids – are directly involved in photosynthesis. Chlorophyll *a* is essential; no plant without it is known to trap light energy for photosynthesis. Most of the other pigments, which make up the bulk, are in the light-harvesting antenna along with much of the chlorophyll *a*. These absorb photons and funnel the resulting excitation energy to chlorophyll *a* in the reaction centers.

Five chlorophylls are known to occur in seaweeds: chls *a*, *b*, c_1, c_2, and *d*. Besides chl *a*, which is common to all seaweeds, chl *b* is found in Chlorophyceae, chls c_1 and c_2 in Phaeophyceae, and chl *d* has been reported in Rhodophyceae (Ragan 1981). The structures of the chlorophylls are shown in Figure 2.10; all are tetrapyrrole rings with Mg^{2+} chelated in the middle. Chlorophylls *a*, *b*, and *d* share the possession of a long fatty-

Figure 2.9. Plastids of seaweeds. (a) Plastid of the red alga *Laurencia spectabilis*, showing parallel single thylakoids and one thylakoid (arrow) surrounding the others, just inside the plastid membrane. (b) Plastid of the brown alga *Fucus* sp. showing characteristic triple thylakoids, the genome (G), and endoplasmic reticulum (ER) surrounding the plastid.

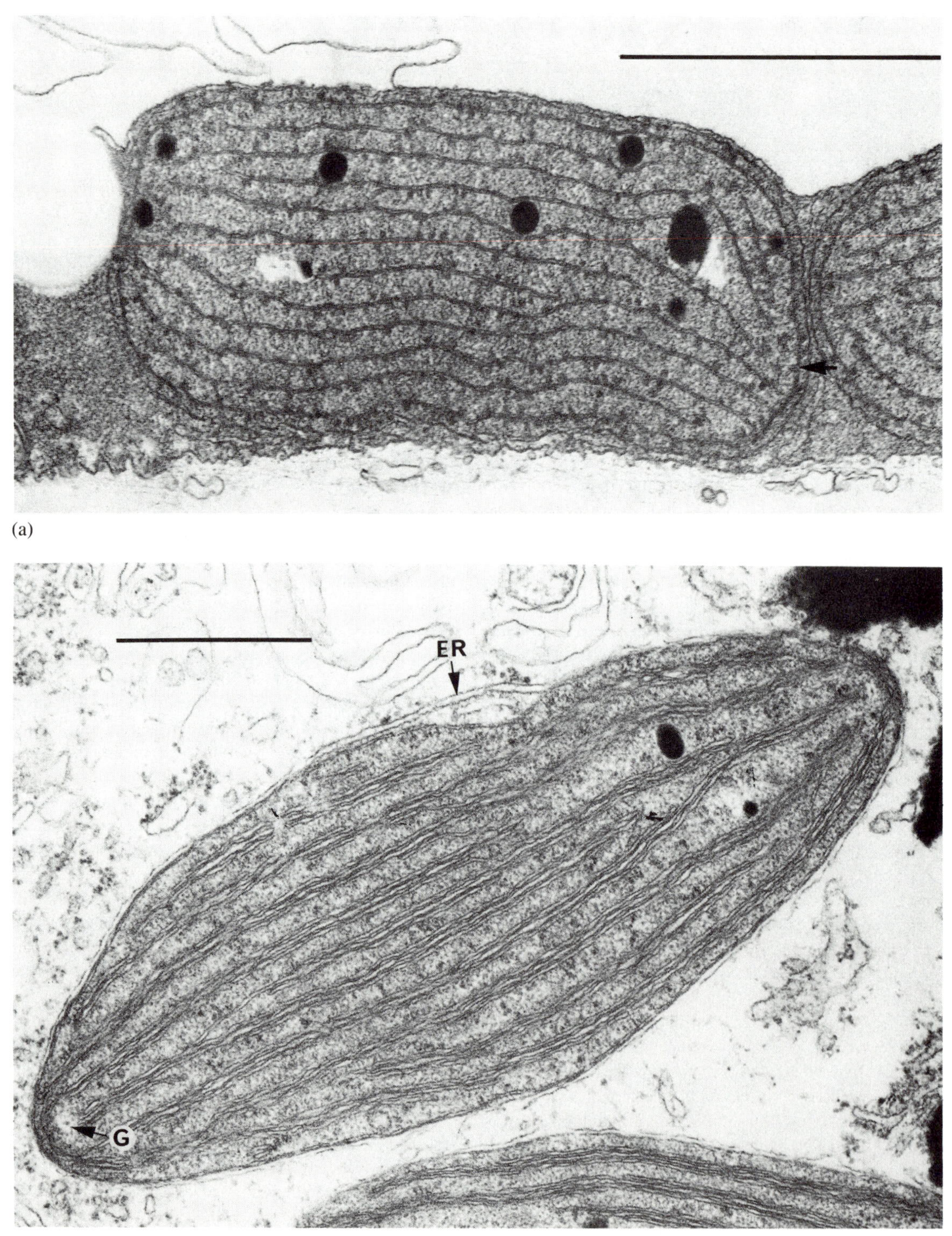

Figure 2.9 (*cont.*) (c) Freeze-etched plastid of *Laminaria*. (d) Apical and subapical cells of the brown alga *Splachnidium* showing stellate plastid in the subapical (lower) cell, physodes (P) and nucleus (N) in the apical (upper) cell, and plasmodesmata between the cells. Scale: 1 μm. (a, c, and d courtesy of T. Bisalputra, b courtesy of L.J. Veto)

(c)

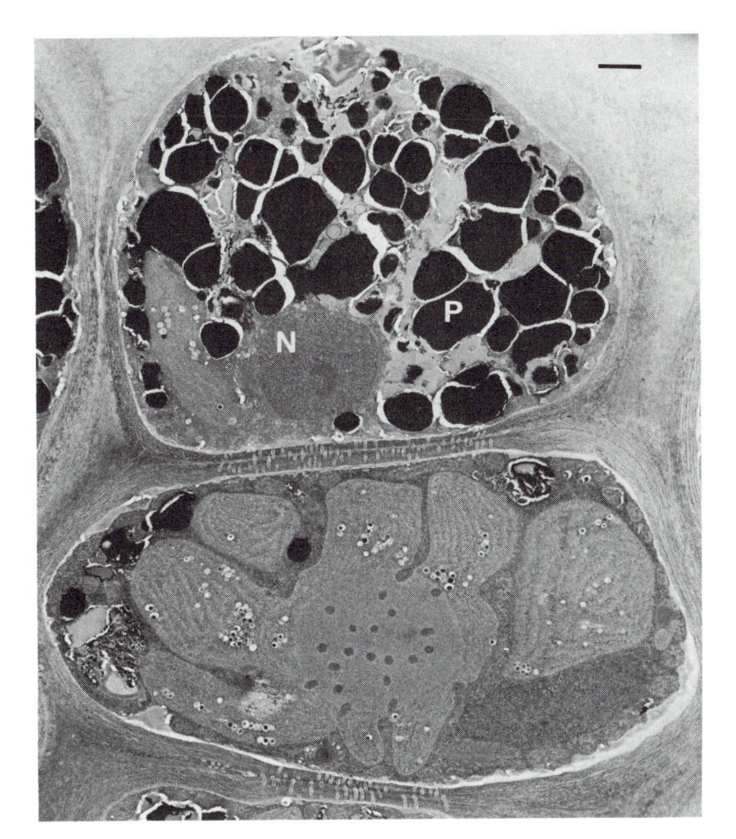

(d)

acid phytol tail ($C_{20}H_{39}COO$-). The phytol tail is presumed to be involved in attaching the pigments to the thylakoid membrane. Chlorophylls c_1 and c_2 lack the tail and hence are about two-thirds the size of chl *a* or *b* (610 vs. about 900 daltons) (Prézelin 1981). The other differences in the structures of the chlorophylls consist of different radicals attached at the positions marked R_1, R_2, and R_3 in Figure 2.10.

Seaweeds have a wide variety of carotenoids (Table 2.3). They are C_{40} molecules, known as tetraterpenes, consisting of eight branched C_5 isoprenoid units, joined in such a way that the linking of the units is reversed in the center (Fig. 2.11a–d). Two types of plant carotenoids occur: the carotenes, which are hydrocarbons, and the xanthophylls, which contain one or more oxygen molecules. β-Carotene is the most commonly occurring carotene in seaweeds, and the red and green algae also have ϵ-carotene (Weber & Wettern 1979). Until recently the major carotenoids of red algae were known to be α-carotene, β-carotene, lutein, and zeaxanthin. Later, several species of *Gracilaria* were found to have a different major carotenoid, antheraxanthin (Brown & McLachlan 1982), ranging from 2 to 50% of the total. The prevailing view had been that carotenoid epoxides such as antheraxanthin were lacking in the red algae (Liaaen-Jensen 1978). β-Carotene is present in low concentrations in all

Figure 2.10. Structures of the chlorophylls active in photosynthesis, also showing the conjugated double bond system (stippled). Chlorophylls *a*, *b*, and *d* (left-hand diagram) all have the phytol tail but differ in the side chains at R_1 and R_2 as listed in the figure. Chlorophylls c_1 and c_2 lack the phytol tail (right-hand diagram) and differ from each other in the group at R_3.

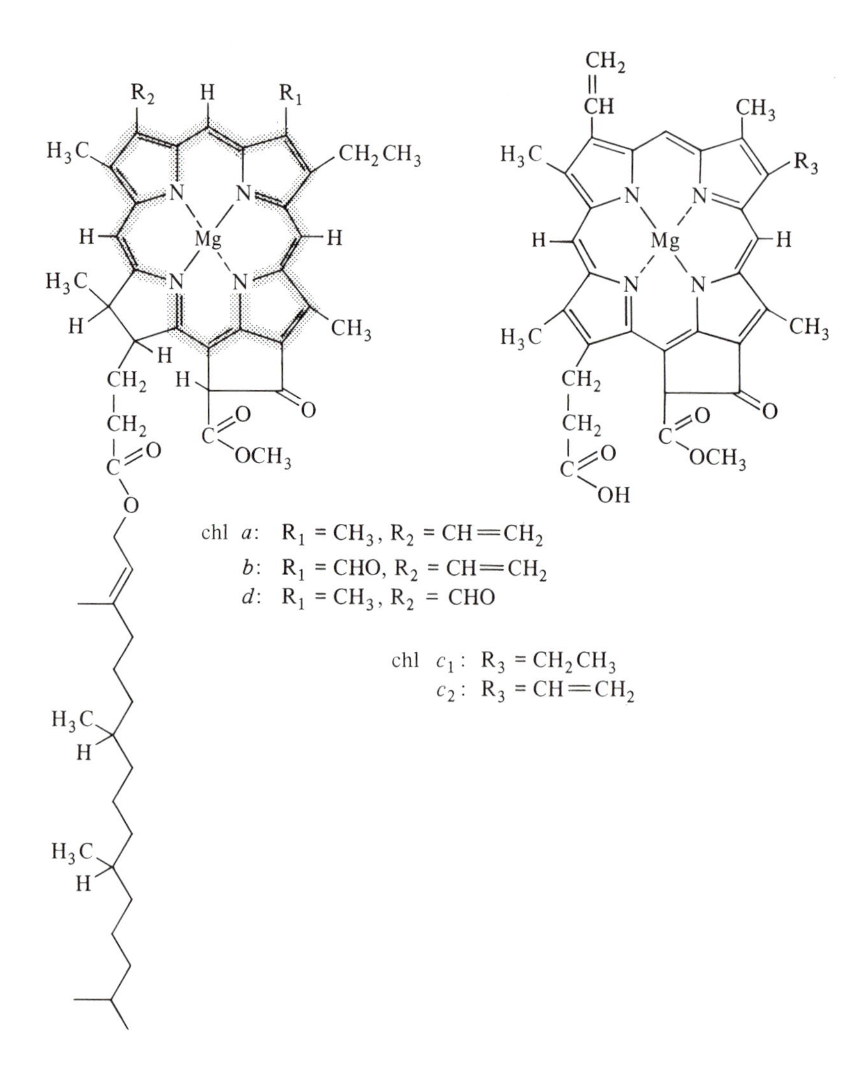

Table 2.3. *Distribution of pigments among seaweed classes*

Pigment	Rhodophyceae	Chlorophyceae	Phaeophyceae
Chlorophylls			
a	+	+	+
b	−	+	−
c	−	−	+
d	+[a]	−	−
Biliproteins	+	−	−
Carotenes			
α	±	±	±
β	+	+	+
Major xanthophylls			
Violaxanthin	−	+	+
Lutein	+[a]	+	−
Neoxanthin	−	+	−
Siphonaxanthin	−	+[b]	−
Fucoxanthin	−	−	+

Symbols: + = present; − = absent; ± = present in some species.
[a]Present in at least a few species.
[b]Present in siphonous orders.

the Phaeophyceae, but their major carotenoid is the xanthophyll, fucoxanthin, which is active in photosynthesis. Fucoxanthin absorbs in vivo well into the green region of the spectrum, which accounts for the brown color of these algae. The only other xanthophylls known to function in photosynthesis in seaweeds are siphonaxanthin (Fig. 2.11d) and siphonein, which occur in the siphonous members of the Chlorophyceae (Yokohama 1981). Peridinin, a xanthophyll that occurs in dinoflagellates, also plays a role in photosynthesis. While certain carotenoids participate in photosynthesis, another role of prime importance is the protection of chlorophyll against photooxidation in bright light. It has been shown that chlorophyll is destroyed in bright light in algal and higher plant mutants that lack carotenoids. In seaweeds such a function would be particularly important for plants living high in the littoral zone and exposed to direct sunlight (see also Sec. 9.3.1).

Biliproteins consist of linear tetrapyrrole chromophores (bilins) covalently bound to protein chains. The phycobiliproteins are found only in the Rhodophyceae, Cyanophyceae, and in some species of Cryptophyceae. There are two bilins, phycocyanobilin (PCB) and phycoerythrobilin (PEB) (Fig. 2.11e, f). In contrast to the tetrapyrrole ring compounds, they contain no metal ion. When first discovered these pigments were named according to the organisms in which they were found (reviewed by O'hEocha 1965). For example, R-phycocyanin (R-PC) and R-phycoerythrin (R-PE) were first found in the Rhodophyceae. C-PE and C-PC were found in the Cyanophyceae. More recent work has shown that the phycobiliproteins are composed of two main protein subunits, to which varying numbers of PCB and PEB chromophores are attached. In most phycobiliproteins there are light and heavy peptide chains (α and β subunits), usually in a molar ratio of 1 : 1; two phycobiliproteins have a third type of subunit (Ragan 1981).

The relationship between pigments and the energy they absorb can be seen in their absorption spectra (Fig. 2.3), a recording of the peaks of absorption in vitro across the PAR region. The absorption spectra of whole plant tissues give valuable information on the pattern of energy absorption in nature, but they do not give specific information about the quantities of the different pigments or their specific absorptive properties. The physical and chemical characteristics of individual pigments must be studied after the pigments have been extracted from the plant. Chlorophylls and carotenoids are usually extracted in acetone or methanol, although other solvents have also been used (e.g. Jensen 1978), whereas the phycobiliproteins are water soluble. Once a pigment has been purified it may be crystallized and the absorption coefficients can be determined for a given amount in a specified solvent. Some seaweeds have proven particularly difficult to extract completely, notably the thick kelps that have large quantities of the cell wall polysaccharide alginic acid. The extraction can be facilitated by starting with dimethyl sulfoxide (DMSO), which readily penetrates membranes (Seely et al. 1972, Duncan & Harrison 1982). There are marked differences in absorption peaks among solvents and also differences when compared to the living plant. In acetone, the most commonly used solvent, chlorophyll *a* has a red maximum at 665 nm and a second peak in the blue at 420 nm, the Soret band. In vivo, the maximum absorption peaks are always shifted to longer wavelengths. Moreover, chlorophyll *a* is known

Figure 2.11. Structures of the major photosynthetic carotenoids (a–d) and the chromophores of phycobilins (e, f) found in seaweeds.

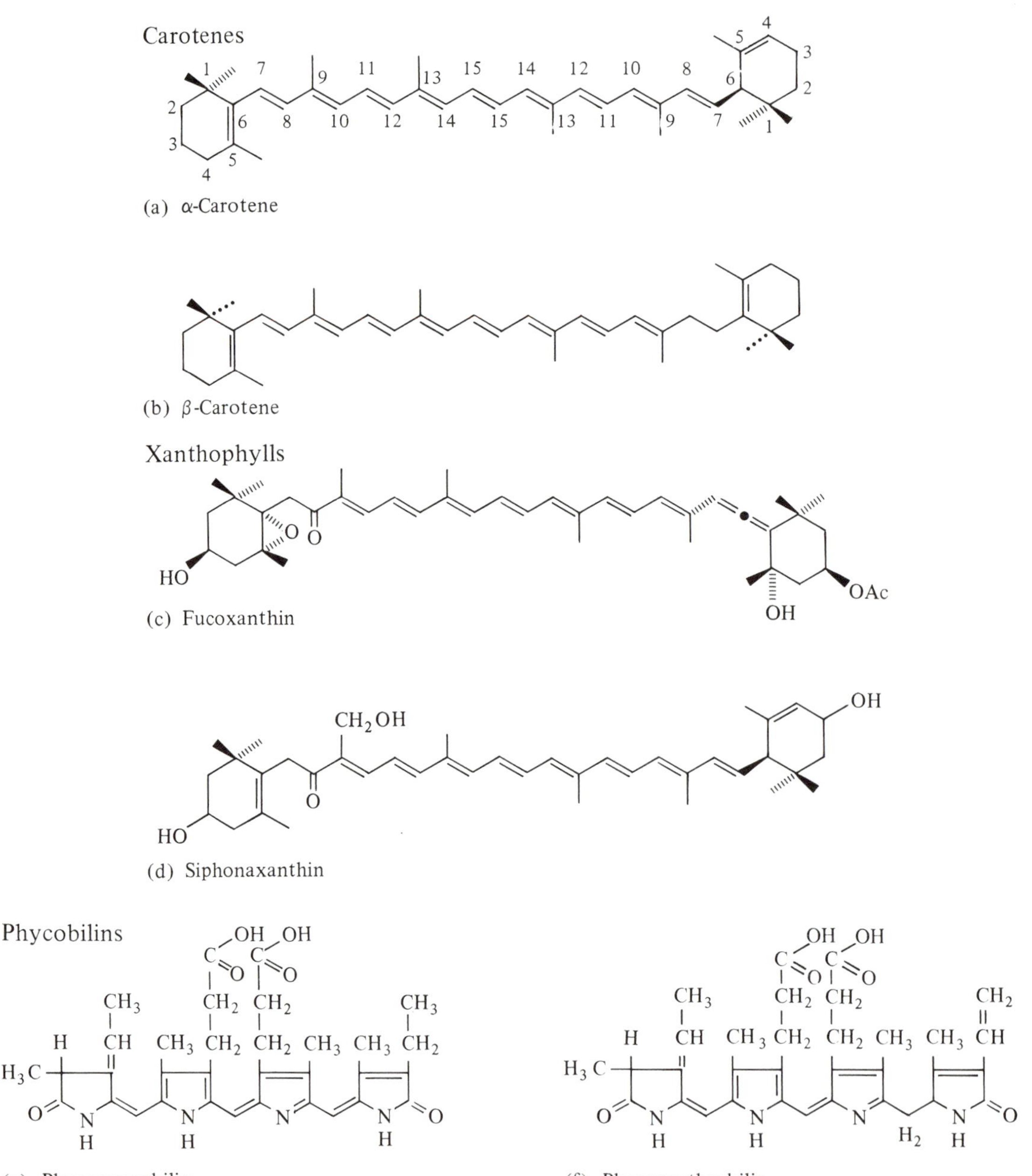

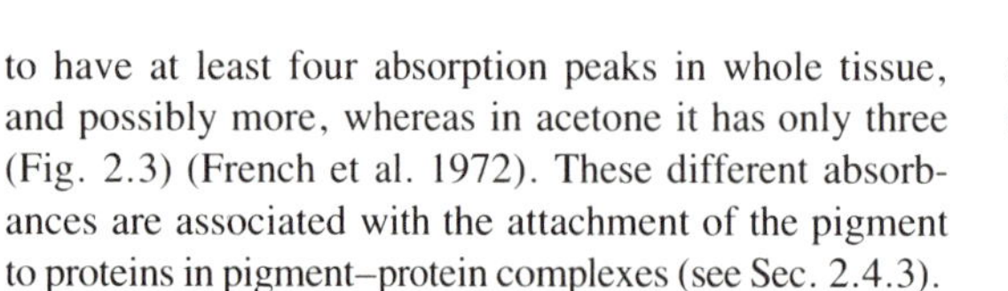
to have at least four absorption peaks in whole tissue, and possibly more, whereas in acetone it has only three (Fig. 2.3) (French et al. 1972). These different absorbances are associated with the attachment of the pigment to proteins in pigment–protein complexes (see Sec. 2.4.3).

The absorption spectra of the individual pigments clearly demonstrate how plants are able to use solar energy from about 400 nm to 700 nm. To better understand how pigments work as energy trappers, one must study how they are organized in the chloroplast and in functional complexes in and on the thylakoid membranes.

2.4.3 Thylakoid membranes and pigment–protein complexes

The arrangements of thylakoid membranes differ among chloroplasts of red, green, and brown seaweeds and are different from higher plants. Yet the thylakoid membrane itself is essentially similar in all plants. It is a flattened, saclike structure consisting of a bilayer membrane surrounding an aqueous space. All the pigments and enzymes that function in trapping light energy and converting it into chemical energy are located in the thylakoid membranes, with the exception of the biliproteins, which are in phycobilisomes. The pigments and compounds involved in electron transport are assumed to be held in close and specific proximity to each other. The pigments are in two groups: a few chlorophyll *a* molecules (approximately 1%) in reaction centers are involved in electron transfer; the bulk of the chlorophyll *a* and all the remaining photosynthetic pigments are in light-harvesting antennae. Together, the reaction centers and antennae constitute photosystems; these plus the electron transport apparatus make up hypothetical photosynthetic units (PSUs) (Fig. 2.12).

The evidence that much of the chlorophyll *a* is in the antennae came from work of Emerson & Arnold (1932), who were the first to establish that about 8 quanta are needed for the release of 1 molecule of O_2. Then, using light flashes strong enough to excite every chlorophyll molecule, they showed a maximum yield of 1 O_2 per 2400 chlorophyll molecules, rather than the expected 1 O_2 per 8 chlorophyll molecules. The PSU was originally conceived as having 2400/8 = 300 chlorophyll *a* molecules plus 1 chlorophyll *a* in the reaction center. However, many workers now consider that groups of antennae are associated with several reaction centers in the ratio of 300:1 (Clayton 1980).

There are two reaction centers. The pigments in them have slightly different red absorption maxima. That in photosystem I (PSI) has a peak at 700 nm, whereas that in PSII has a peak at 680 nm. Each reaction center is thought to be associated with a different array of antenna pigments; a model of how these might be arranged is shown in Figure 2.12. In each case, excitation energy is passed to pigments with absorption peaks at longer wavelengths (the reason will be explained in Sec. 2.5.1). The evidence for two photosystems came from the demonstration, in the early 1960s, that cytochrome f, one of the molecules known to be involved in photosynthetic electron transport, became reduced (accepted electrons) in red light, but became *oxidized* in far-red light. The conclusion was that cytochrome f had to be between two photosystems, one of which (PSII) had a lower absorption peak than the other (PSI) (Avron 1981).

The relationships between pigments and proteins, and their positions in the thylakoids, have been studied via two kinds of experimental approaches. One approach involves fractionating the membranes and isolating the pigment–protein complexes by centrifugation of the fragments on a differential sucrose gradient; this is followed by treatment with detergents, chromatography, and electrophoresis. As the procedures were refined, a major antenna pigment–protein complex was discovered, which contained roughly equimolar amounts of chlorophylls *a* and *b*. This complex, named LHa/b (light-harvesting a/b), had no photochemical activity. It served rather to transfer excitation energy to both photosystems, chiefly PSII. All the chlorophyll *b* present in chloroplasts is found in LHa/b, along with about 60% of the total thylakoid chlorophyll and about half the thylakoid protein. Brown seaweeds have chl *a*/chl *c*/fucoxanthin–protein complexes in place of LHa/b. Anderson & Barrett (1979) found two main light-harvesting complexes in several brown algae: an orange fucoxanthin–chl a/c_2–protein complex, and a green complex with chlorophylls *a*, c_1 and c_2, plus the xanthophyll violaxanthin. Complexes with PSI and PSII activity have been isolated from higher plants, and Anderson & Barrett (1979) isolated a P700-chlorophyll *a*-protein complex from their Phaeophyceae.

Figure 2.12. Working hypothesis of the arrangement of pigments in the light-harvesting antennae for PSII (left) and PSI. (After Govinjee & Braun 1974)

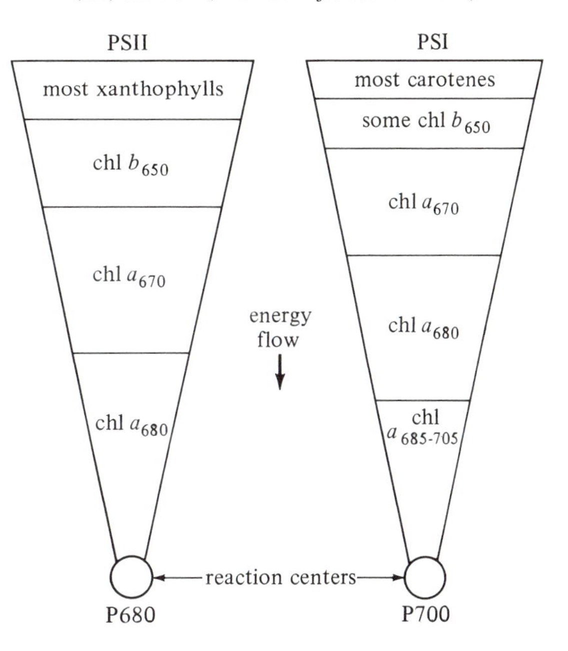

The other approach involves freeze fracture and freeze etching of chloroplasts. When frozen plastids are struck, they tend to fracture along membrane surfaces (Fig. 2.9c). Each surface has particles of characteristic sizes, which are associated with different parts of the photosynthetic apparatus (Hiller & Goodchild 1981). A model showing the function and spatial relations of the particles is shown in Figure 2.13. In addition to the PSI, PSII, and LHa/b particles, the membrane also contains the coupling factor by which ATP is phosphorylated (Sec. 2.5.3); the coupling factor has a base complex (CF_0) in the membrane and surface particle (CF_1).

In the red algae and Cyanophyceae the accessory pigments, the phycobiliproteins, are associated into visible globules on the outer surfaces of the thylakoids (Gantt 1975). Phycobilisomes are probably the functional equivalent of the LHa/b complex and are aligned on the membrane surface in such a way as to transmit energy effectively to the reaction centers. The chlorophyll *a*–containing reaction centers are within the thylakoid membranes in these algae as in other plants. Phycobilisomes contain no chlorophyll, but comprise phycocyanins, phycoerythrins, and allophycocyanin in stacked rods. Besides the polypeptides of the phycobiliproteins, which account for about 85% of the protein in phycobilisomes, there are colorless polypeptides associated with the rods (Fig. 2.14) (Lipschultz & Gantt 1981). These may serve to bind or stabilize the phycobiliproteins in the rods (Gantt 1981). The shapes and sizes of phycobilisomes depend on the organism. The number per unit area is also variable, ranging from 400 μm^{-2} in the red alga *Porphyridium cruentum* to 1200 μm^{-2} in the blue-green *Oscillatoria brevis*. Some phycobilisomes are disk-shaped, whereas others are almost as broad as high (Gantt 1981). The arrangement of the phycobiliproteins in the phycobilisomes of a red alga, *Rhodella violacea*, and a blue-green, *Pseudanabaena* sp., is shown in Figure 2.14. Six stacked rods of phycoerythrin and phycocyanin fan out from a distinct core of allophycocyanin. Allophycocyanin is assumed to be at the base because it is released last during progressive disassociation of phycobilisomes and because for energetic reasons the transmission of energy would go from phycoerythrin to phycocyanin to allophycocyanin to the reaction center. Another assumption is that the core units in the hemidiscoidal phycobilisomes of red algae are close to the photosynthetic reaction centers.

Figure 2.13. A model of the organization of the components of the photosynthetic apparatus in the grana and stroma regions of green plant thylakoid membranes. Photosystem I and II complexes (PSI and PSII), light-harvesting complex of PSII (LHCP). Coupling factor (CF_1) and coupling-factor base complex (CF_0) are involved in photophosphorylation. (With permission from Hiller & Goodchild 1981, Thylakoid membranes and pigment organization, in *The Biochemistry of Plants*, vol. 8, pp. 1–49. Copyright 1981 Academic Press Inc. [London] Ltd.)

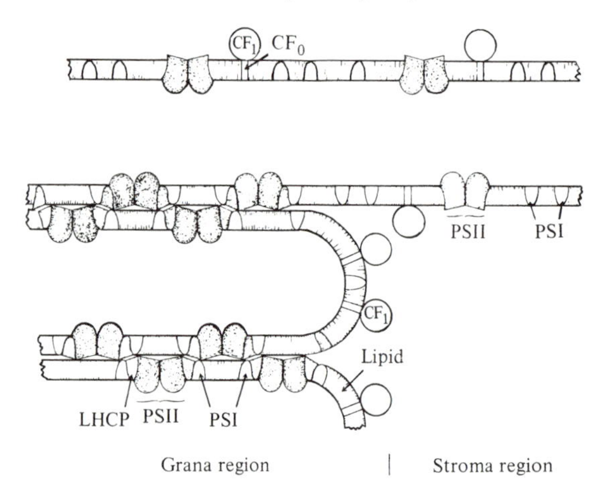

2.5 Photosynthetic electron transport and photophosphorylation

Photosynthesis consists of two major groups of reactions. The first is the capture of light energy and its conversion to chemical potential as ATP and NADPH; this group, the "light reactions," is considered in this section. The second group, the "dark reactions," is the sequence of reactions by which this chemical potential is used to fix inorganic carbon; this group will be dealt with much later, in Chapter 7. The light reactions fall into three processes: energy absorption, energy trapping, and generation of ATP and NADPH. The reactions have been studied chiefly in angiosperms and green algae, and the assumption that they are the same in all algae may need to be modified. Recently, Popovic et al. (1983) found differences in plastids from several brown algae.

2.5.1 Energy absorption

Photosynthesis begins with the absorption of a photon by an antenna pigment and the formation of electronic excitation within the pigment molecule. The next step is the transfer of excitation energy among the pigments until it is trapped at the reaction center.

Before a pigment receives a photon, it is in the ground state. Upon absorption of a photon, the electrons

Figure 2.14. Structure of phycobilisomes. Structure on left represents thin hemidiscoidal phycobilisomes found in *Rhodella violacea* and *Pseudanabaena*; that on right represents *Porphyridium cruentum*. The black dot represents a protein common to the phycobilisome and the membrane that is presumed to anchor the *Porphyridium*-type phycobilisome. (From Gantt 1981, reproduced with permission from the *Annual Review of Plant Physiology*, vol. 32, © 1981 by Annual Reviews, Inc.)

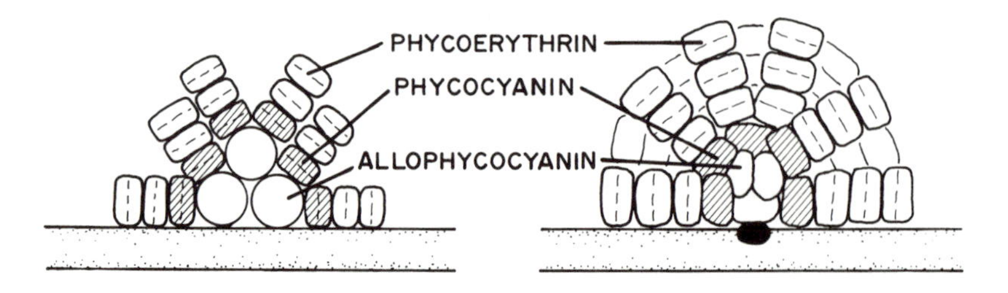

are rearranged within the nuclear framework and the molecule is in a more energetic, unstable excited state. Only a few of the most labile electrons in the molecule are moved: those in the conjugated double bond system (Fig. 2.10). Incident radiation must have the exact amount of energy to move the electrons into the excited state; other radiation is transmitted rather than absorbed. The greater the energy in the photon, the higher the excited state, as illustrated in Figure 2.15a–d (Sauer 1975). However, no photochemistry is done until the molecule has "thermally relaxed" to a lower excited state. For this reason, the high energy of blue light is of no more use in photosynthesis than the less energetic longer wavelengths.

The ground and excited states, shown in Figure 2.15e, comprise many sublevels or substates, indicated

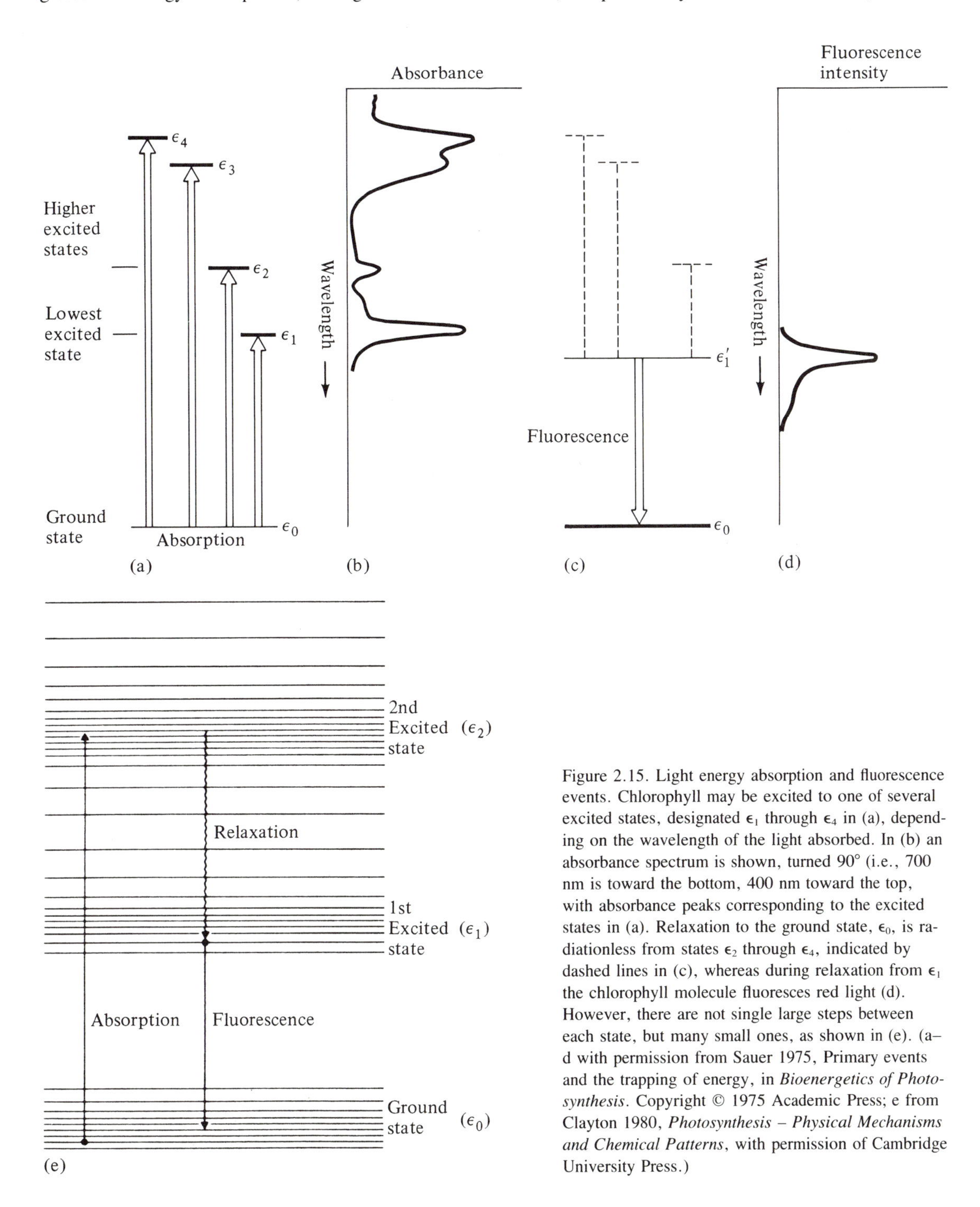

Figure 2.15. Light energy absorption and fluorescence events. Chlorophyll may be excited to one of several excited states, designated ϵ_1 through ϵ_4 in (a), depending on the wavelength of the light absorbed. In (b) an absorbance spectrum is shown, turned 90° (i.e., 700 nm is toward the bottom, 400 nm toward the top, with absorbance peaks corresponding to the excited states in (a). Relaxation to the ground state, ϵ_0, is radiationless from states ϵ_2 through ϵ_4, indicated by dashed lines in (c), whereas during relaxation from ϵ_1 the chlorophyll molecule fluoresces red light (d). However, there are not single large steps between each state, but many small ones, as shown in (e). (a–d with permission from Sauer 1975, Primary events and the trapping of energy, in *Bioenergetics of Photosynthesis*. Copyright © 1975 Academic Press; e from Clayton 1980, *Photosynthesis – Physical Mechanisms and Chemical Patterns*, with permission of Cambridge University Press.)

by parallel horizontal lines. Once a molecule is excited, several competing processes can occur. The molecule can undergo thermal relaxation to achieve thermal equilibrium with its surroundings (Sauer 1975). The molecule goes through many transitions between substates, losing small amounts of energy as heat at each step. This is a rapid process, usually taking about 10^{-12} s (Clayton 1980). Alternatively, the molecule may fluoresce, whereby it undergoes spontaneous decay directly to the ground state, emitting radiation of slightly longer wavelength than was absorbed (because some energy is lost as heat). Red fluorescence of chlorophyll can be seen by illuminating it in solution. The most important means by which photosynthetic pigment molecules lose the energy absorbed is the transfer of that energy from molecule to molecule of chlorophyll and eventually to the reaction center. This does not occur by emission of a photon by fluorescence of one molecule and absorption by the next, but by a more intimate process called resonance transfer. In the process of de-excitation, the oscillation of one molecule is coupled to a sympathetic oscillation in a neighboring molecule, causing the latter to become excited. If the molecules are identical, a cluster of chlorophyll molecules, for example, they can be considered a "supermolecule" in which the excited state is a property of the aggregate, not of any single molecule (Clayton 1980).

2.5.2 Electron transport

When the excitation energy reaches the chlorophyll *a* in the reaction center, it causes an electron to be not merely raised to a higher energy level but to be transferred out of the chlorophyll to an acceptor molecule, which is thereby reduced. The flow of electrons may be cyclic, involving only PSI, or noncyclic from water to NADPH via both photosystems (Fig. 2.16). In noncyclic electron transport (the so-called Z-scheme), the first photosystem in the sequence, PSII, generates a strong oxidant, which oxidizes water, and a weaker reductant. Electrons from the oxidation of water flow to the reaction center, rereducing it. PSI generates a strong reductant that reduces NADP to NADPH, plus a weaker oxidant. The ability of a compound to accept or donate electrons is measured by its redox potential: a stronger reductant (electron donor) has a more negative redox potential, whereas an avid electron acceptor has a pos-

Figure 2.16. Paths of electron transport in green plant photosynthesis. The ordinate shows the approximate redox midpoint potential of each component. Electrons flow from water, via compounds Z and perhaps S, to photosystem II (P680). The electron acceptor I is reoxidized by a quinone, Q_a, whence electrons flow past a plastoquinone (PQ) pool, cytochrome f, plastocyanin (PCy), and other compounds to photosystem I (P700). From ferredoxin (FD), cyclic flow takes electrons back to the PQ pool, whereas noncyclic flow ends in reduction of $NADP^+$. (From Clayton 1980, *Photosynthesis – Physical Mechanisms and Chemical Patterns*, with permission of Cambridge University Press)

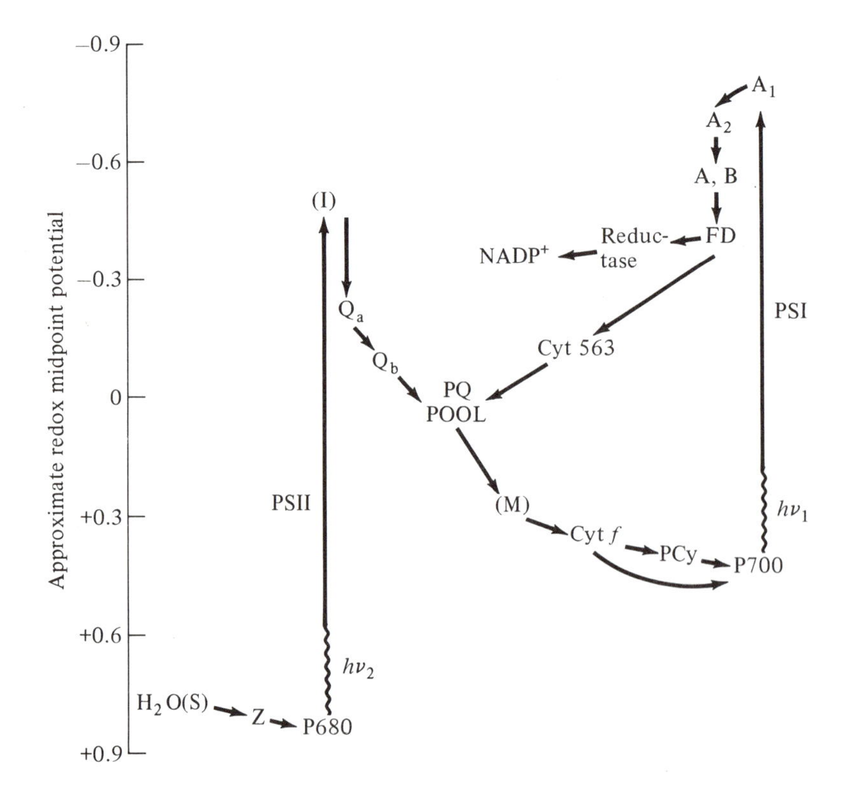

Figure 2.17. Redox changes in metal and coenzyme electron carriers, showing oxidized and reduced forms. The coenzyme illustrated is a flavin (as in FMN, for example).

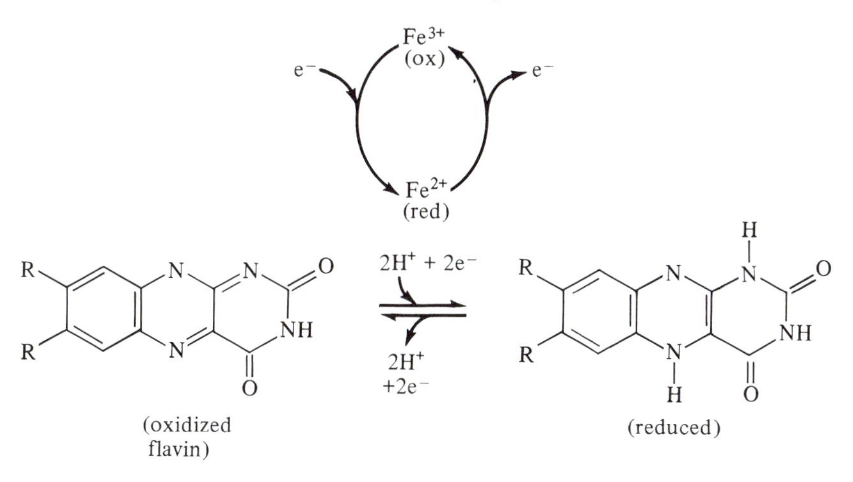

itive potential. As seen in Figure 2.16, the reaction centers of PSI and PSII are electron acceptors; electrons can be removed from them only by the energy of light. In contrast, the first compounds that accept electrons from the reaction centers (designated I and A_1 in Fig. 2.16) have a great tendency to pass on electrons, and thus to reduce other compounds. The flow of electrons down the redox potential gradient from I to PSI is coupled to the phosphorylation of ATP. The flow from A_1 may go to NADP or back to the reaction center of PSI. There is evidence that I is a modified, magnesiumless chlorophyll, phaeophytin *a*, and that A_1 may be a chlorophyll monomer (Clayton 1980). However, with new methods of rapid kinetic spectroscopy and low-temperature electron spin resonance, it appears that several electron acceptors function in a complex as the primary electron acceptor of PSI (Malkin 1982). The sequence of electron transport compounds from I to PSI begins with two quinones, Q_a and Q_b, which are blocked by the herbicide DCMU [3−(3′, 4′-dichlorophenyl)−1,1−dimethylurea]. Then there is a plastoquinone pool, which carries both protons (H^+) and electrons and is thought to be crucial to phosphorylation. Electrons in cyclic phosphorylation also enter this pool. Other compounds in the chain have iron or copper, which change valence (e.g., from Fe^{3+} to Fe^{2+}) when they accept an electron (Fig. 2.17). Such compounds include cytochrome f and the copper-containing plastocyanin. In the chains from A_1 are other iron-sulfur–containing compounds, designated A_2 and A,B, and iron-containing ferredoxin.

The most familiar part of photosynthesis is oxygen evolution, yet this part remains the least understood. P680, the reaction center of PSII, donates its electron to I and is rereduced in approximately 30 ns by an unidentified compound Z. The mechanism by which Z obtains electrons from water is also unknown, except that manganese ions, apparently six per PSII, are essential.

Figure 2.18. The cycle of S-states of the water-splitting PSII complex. (See text for details.) (From Clayton 1980, *Photosynthesis – Physical Mechanisms and Chemical Patterns*, with permission of Cambridge University Press)

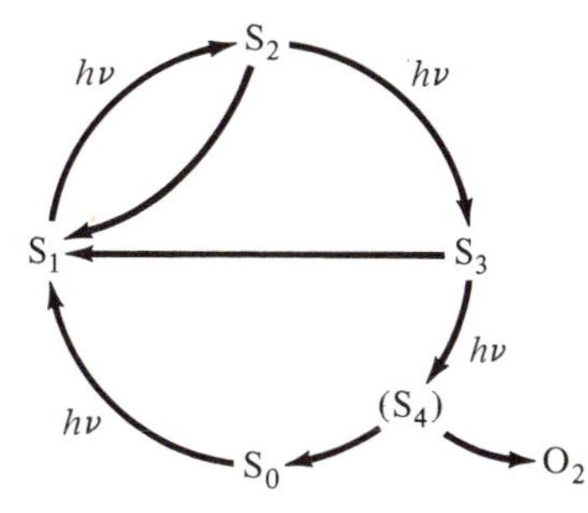

Whereas the reaction centers pass on electrons one at a time, the release of oxygen yields four electrons at once:

$$2H_2O \rightarrow O_2 + 4\,H^+ + 4\,e^-$$

Kok et al. (1970), on the basis of experiments with short light flashes, formulated a scheme of S-states to explain this difficulty. They postulated that each PSII enzyme complex becomes progressively more oxidized with each photochemical event until four electrons have been donated. S_0 to S_4 represent the successive oxidation states (Fig. 2.18). At S_4, water is split and O_2 released, and the enzyme returns to the reduced S_0 state. S_2 and S_3 may decay back to S_1. This explanation is clearly far from complete and the enzyme(s) have yet to be isolated and identified.

Although electron transport between water and NADP is usually depicted as a chain consisting of PSII, PSI, and a complement of intersystem transfer components, Velthuys (1980) suggests that the two individual photosystems may have a different physical reality. Arguments in favor of this view include: (1) the two photosystems have separate pigment systems; (2) they are physically separable, for example, by the mild detergent digitonin; (3) they do not occur in a strict 1:1 ratio. Velthuys proposed that the Z-scheme be modified so that plastoquinone and plastocyanin are represented as a movable link between the two photosystems, without being associated with a specific photosystem. This new scheme is illustrated in Figure 2.19. The new component (III) is not involved in a light reaction.

2.5.3 Photophosphorylation

Electrons can be used to reduce compounds, such as NADP, but not to phosphorylate. The reaction

$$\text{ADP} + \text{inorganic phosphate} \rightarrow \text{ATP} + H_2O$$

is coupled to the flow of electrons through the simultaneous transport of protons. It depends also on the integrity of the thylakoid sac and it can be induced by raising the pH outside the thylakoid. As explained by the chemiosmotic theory of Mitchell (1974), protons are transported via the plastoquinone pool from the outside to the inside of the thylakoid (Fig. 2.20). Since the thylakoid membrane is impervious to protons except at special places, H^+ accumulates, resulting in an electrochemical gradient across the membrane. Probably one H^+ is transported for every electron, though this is not certain. The special places in the membrane through which the electrochemical imbalance can be relieved are the coupling factor particles (Fig. 2.13), which include ATPase. The mechanism by which proton flow is coupled to ATP phosphorylation is not well understood. Eight quanta, four absorbed by each photosystem, can transport eight protons to the inside of the thylakoid sac. ATPase is believed to phosphorylate one ADP for every three H^+. The energy efficiency of photosynthesis in plants can be computed thus:

Input: 8 quanta near 680 nm; energy = 8 × 1.82 eV = 14.6 eV.

Output: 4 electrons from H_2O to NADPH, spanning 1.15 eV redox potential; energy = 4 × 1.15 = 4.6 eV; plus 8 H^+ giving 8/3 ATP at 0.34 eV each; energy = 0.9 eV. Total output = 5.5 eV.

Efficiency = 5.5/14.6 = 38%

This efficiency is remarkable considering the complexity of the system (Bidwell 1979). The efficiency of cyclic phosphorylation alone is 25%. Energy losses occur in the conversion of light energy to chemical energy and in the stabilization of the excited states of the pigments and intermediates, that is, trapping the energy of the migrated electron so that it cannot fall back into its previous position.

Figure 2.19. Modified photosynthetic electron transport scheme showing the third component (III), which is not light absorbing, in the photosynthetic membrane. (From Velthuys 1980, reproduced with permission from the *Annual Review of Plant Physiology*, vol. 31, © 1980 by Annual Reviews, Inc.)

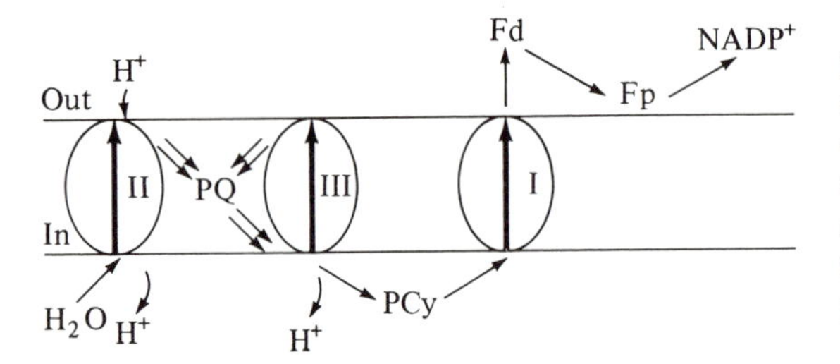

Figure 2.20. Patterns of electron and H^+ transport in relation to the thylakoid membrane. PQ pool corresponds roughly to system III of Figure 2.19, and ATPase is located in the coupling factor (CF_0-CF_1) shown in Figure 2.13. Four quanta absorbed by each photosystem can deposit eight H^+ inside the thylakoid, and ATPase is believed to convert one ADP to ATP for every 3 H^+ that move back through the coupling factor. (CH_2O) denotes carbohydrate fixed in the Calvin cycle. (From Clayton 1980, *Photosynthesis – Physical Mechanisms and Chemical Patterns*, with permission of Cambridge University Press)

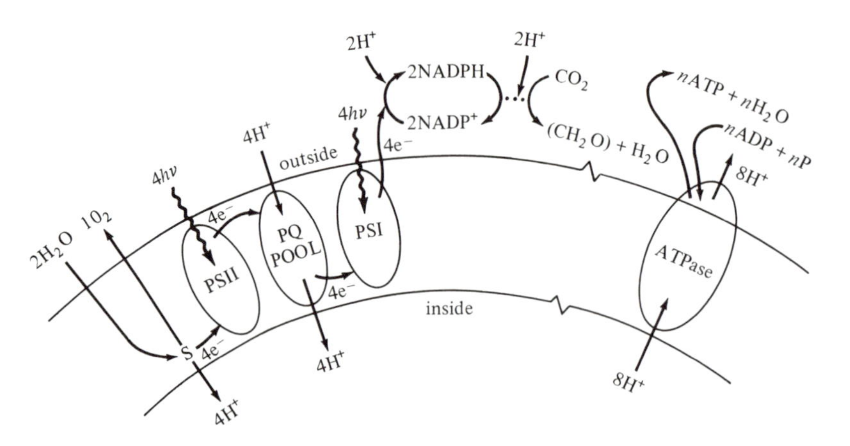

2.6 Measuring photosynthesis and respiration

2.6.1 Photosynthesis

The external manifestations of photosynthesis and respiration are CO_2 and O_2 exchange between the plant and the water (or air). Determination of the photosynthetic rate usually includes estimation of respiratory rate, since respiration causes gas exchanges in directions opposite from photosynthesis. The rates of the metabolic process can be estimated by measuring rate of exchange of O_2 in water or CO_2 in air, or by mixing radioactive tracer carbon, ^{14}C, with the CO_2 or HCO_3^- supply in water or air. Each method has its particular advantages and limitations, which are discussed in technical manuals, including Strickland (1960), Vollenweider (1969), Hellebust & Craigie (1978), and certain research papers including Littler (1979), Carpenter & Lively (1980), and Peterson (1980).

Most studies of gas exchange in water employ some version of the light- and dark-bottle technique, in which the alga(e) or portions of them are enclosed in bottles filled with seawater. One bottle is completely blackened, the other is transparent; both bottles are then incubated under the same conditions in the laboratory or in situ. If oxygen is being determined by the chemical Winkler method, an initial bottle is also required. The oxygen electrode technique allows continuous measurement, whereas the Winkler method measures at the end of a fixed time. The incubation must be long enough for measurable change in the O_2 content of the water to occur, or for the alga to accumulate measurable ^{14}C. Durations of 1 to 8 h are common. Some peculiar conditions produced by enclosing a small volume of water preclude very long incubations. (These problems include bacterial growth and activity, oxygen saturation, and water stagnation.)

Routinely, 300 mL bottles are used, but very much larger containers have been employed for whole kelps – Hatcher's (1977) chambers held 76 L. In small bottles, the O_2 concentration changes more rapidly, but bacterial growth on the walls is a proportionately larger problem. In ^{14}C experiments, the size of the container is less important, but does influence the quantity of tracer that must be used and hence the cost. In ^{14}C experiments on macrophytes, the labeled photosynthetic products must be extracted from the tissue; long incubation times allow more ^{14}C to be deposited in insoluble compounds and require the tissue to be dissolved or burned. The total inorganic carbon content of the water must be measured to give the proportion that is radioactive. Measurements of CO_2 exchange in air can be made with an infrared gas analyzer system, which may also include a radiation detector for simultaneous determination of $^{14}CO_2$ (as a means of measuring photorespiration). Like the oxygen electrode, this method allows continuous measurements, and the same sample can be used for measurements in both light and darkness. Further technical considerations are beyond the scope of this textbook, but some of the biological aspects require discussion.

Measurements of the rate of reactions as complex as photosynthesis and respiration, which are affected by a great many variables, present a major problem in comparability. Each measurement must be interpreted in terms of these variables, including irradiance, age of the tissue, nutrient levels, temperature, and pH. The result is that absolute comparisons are never possible except within carefully controlled groups of experiments. Moreover, there is the awkward choice of denominator: dry weight, wet weight, per unit chlorophyll *a*, area, etc. (Lobban 1981). Šesták et al. (1971) pointed out that "the rate of a process yields maximum information about the process itself if it is expressed on the basis of a plant characteristic which limits or at least strongly influences the process." In accordance, Ramus (1981) recommends expressing photosynthesis as carbon flux per unit of chlorophyll, even while recognizing that the amount of chlorophyll and the rate of photosynthesis do not bear a constant relationship even at saturating irradiances. Respiration is probably best expressed on the basis of total protein, since it is a process taking place in the cytoplasm portion of the cells, rather than on the basis of dry weight, which includes the cell walls.

In the simplest theory, the ratio of oxygen liberated to carbon fixed, called the photosynthetic quotient, PQ, should be unity because 6 moles of CO_2 plus 6 moles of water fixed into 1 mole of hexose leaves 6 moles of oxygen:

$$6CO_2 + 6H_2O \rightarrow C_6H_{12}O_6 + 6O_2$$

However, this equation conceals a great complexity of photosynthetic and metabolic pathways. Moreover, there is not a direct connection between the O_2 released and the CO_2 (or HCO_3^-) fixed: O_2 is derived from the splitting of water at the beginning of photosynthetic electron transport, whereas CO_2 is fixed in the Calvin cycle. Measurements of PQ have been frequently made for phytoplankton and have shown its value to be often approximately 1.2, rather than 1.0 (Strickland & Parsons 1972). However, there can be considerable variation, depending in part on whether the fixed C is going mainly into carbohydrates (as in the equation above), fats, or proteins and whether photosynthetic energy is being directed to uses such as NO_3^- reduction. Few measurements of PQ have been made for seaweeds, but Hatcher et al. (1977) reported a range in *Laminaria longicruris* from 0.67 to 1.50.

One difficulty in relating O_2 production to CO_2 fixation – and in measuring photosynthesis by oxygen methods – is the Mehler reaction, or pseudocyclic photophosphorylation, in which ATP is generated but O_2 is *consumed* as well as released. The process is obviously difficult to detect because of all the O_2 liberated by the Hill reaction of photosynthesis, but it can be demonstrated in the absence of CO_2 in the laboratory and is thought to occur also in nature, although the extent is uncertain (Raven & Beardall 1981). The electrons from water (which result in O_2 release) flow through PSII (and perhaps PSI) but their ultimate acceptor is oxygen, in the reaction

$$O_2 + 2e^- + 2H^+ \rightarrow H_2O_2$$

Taking account of the $1/2O_2$ liberated in the Hill reaction, the net equation is

$$H_2O + \tfrac{1}{2}O_2 \rightarrow H_2O_2$$

No NADPH is formed in the Mehler reaction.

2.6.2 Respiration

Whereas photosynthesis consumes CO_2 and produces O_2, respiration uses O_2 and releases CO_2. Since both processes occur when light is available, the gas exchange rate in light, which measures net or apparent photosynthesis, must be corrected for respiration *in the light*, to give gross photosynthesis. The correction for respiration has usually been taken from the dark rate. There has been considerable disagreement over the rates of respiratory gas exchange in the light and in darkness, but the weight of data now indicates that the rate is much lower in the light, although the metabolic pathways operate continuously (Raven & Beardall 1981, Noggle & Fritz 1983). In higher plants, most of the CO_2 released in the light comes from photorespiration, a quite different process (Sec. 2.6.3). These findings create difficulties for productivity studies, particularly when photosynthesis is low, because the measurement of respiratory gas exchange in light is difficult. When tissue portions are used rather than the whole thallus, a complication arises from respiration of wounded tissue, which is considerably higher than normal for many hours after the cut is made (Hatcher 1977, Littler 1979).

2.6.3 Photorespiration

Another complication in gas exchange measurements is the uptake of O_2 and release of CO_2 due to photorespiration. This process has been fairly well studied in angiosperms, where it appears to be quite important except in species with C_4 acid metabolism (see Bidwell 1983). (In C_4 photosynthesis CO_2 is initially incorporated into a four-carbon acid, and later released to be fixed by RuBP carboxylase.) The extent to which photorespiration occurs in algae, and particularly seaweeds, is a matter of controversy. Photorespiration arises because RuBP carboxylase can also act as an oxygenase, binding O_2 rather than CO_2 to RuBP, and the process includes the metabolic pathway followed by the C_2 portion of RuBP. The carboxylase activity is competitively inhibited by O_2, so that photorespiration is more pronounced at high O_2 or low CO_2 concentrations. The literature on glycolate metabolism in freshwater and marine plankton is voluminous, but the ecological significance of photorespiration in these algae remains in doubt (Harris 1980). Freshwater algae, with metabolic pathways similar to C_3 plants, show gas exchange characteristics of C_4 plants: little external manifestation of photorespiration. This may be due to a biophysical mechanism for concentrating CO_2 at the chloroplast (Morris & Darley 1982, Beardall et al. 1982). A potential indicator of photorespiration is the CO_2 compensation point, the CO_2 concentration at which the plant shows no net CO_2 uptake in the light. In the absence of photorespiration this concentration should be zero. However, some angiosperms do not manifest photorespiration because they can refix escaping CO_2. Few measurements of photorespiration indicators have been made on seaweeds (e.g., Tolbert & Osmond 1976, Burris 1977), but evidence so far suggests that it does occur, particularly under high O_2 concentration. The best evidence for the presence of photorespiration is Akazawa & Osmond's (1976) demonstration that RuBP carboxylase in crude extracts of two green seaweeds can act as an oxygenase. The activity of the oxygenase was only 1% that of the carboxylase. However, because marine plants are usually exposed to high CO_2/HCO_3^- (see Sec. 7.1.1) and low O_2, photorespiration is probably of little importance in productivity (Kremer 1981a).

2.7 **Light and photosynthetic rate**

2.7.1 P versus I curves

The relationship between photosynthetic rate and irradiance is shown in a P versus I curve (Figs. 2.21, 2.26). There are two important points in such curves: the saturation and compensation points. In darkness the rate of oxygen evolution or carbon fixation will be negative, owing to respiration. As irradiance is increased, a point is reached at which photosynthetic gas exchange just balances respiration. This is the compensation point, which is usually defined on a 24-h basis. The compensation point can also be presented as a depth at which irradiance produces no net photosynthesis for a given plant, but such a value can only be based on average irradiance in situ, and plants are often found below their calculated compensation depth. As irradiance is increased further, the rate of photosynthesis increases linearly (Fig. 2.21a). The initial slope, α, is a useful indicator of quantum yield. However, eventually the curve levels off, as photosynthesis becomes saturated, reaching a maximum, P_{max}.

The level of irradiance needed to saturate a species shows some correlation with habitat: intertidal species require 400 to 600 $\mu E\ m^{-2}\ s^{-1}$ (full sun is approximately 2000 $\mu E\ m^{-2}\ s^{-1}$), upper- and mid-sublittoral species saturate with 150 to 250 $\mu E\ m^{-2}\ s^{-1}$, and deep-sublittoral species require less than 100 $\mu E\ m^{-2}\ s^{-1}$ (Lüning 1981a). However, the upper side of a thick thallus will become saturated long before the lower side, and P vs. I curves for such species reach the asymptote very gradually.

Saturating irradiance can be defined by the value I_k, the point at which the extrapolated initial slope crosses P_{max}. Under certain conditions, very high irradiance causes photoinhibition (Fig. 2.21a). The causes of photoinhibition include photooxidation of components of the photosynthetic apparatus; carotenoids may function to mitigate this problem, allowing some seaweeds to photosynthesize in full sunlight. At least in part because of photoinhibition, the photosynthetic rate does not always follow the incident irradiation (I_0) curve. Prediction based on instantaneous measures (P vs. I curves) would be for photosynthesis to increase from dawn until I_0 exceeded I_k, to remain level until I_0 again fell below I_k in the afternoon, and then to decrease toward dusk. However, there is often a noontime depression of photosynthesis rate, with only partial recovery in the later afternoon. Morning and afternoon P vs. I curves do not always correspond (Ramus & Rosenberg 1980, Ramus 1981). As much as 70% of daily photosynthesis may take place before noon when peak I_0 is much greater than I_k. There is, in many seaweeds, an effect of preirradiation on rate of photosynthesis. In some species (e.g., *Dictyota dichotoma*), preirradiation also affects whether the plastids are on the faces or sides of the cells, but the position of the plastids is not the cause of the effect on photosynthesis. In other seaweeds (e.g., *Alaria esculenta*), there is an effect on photosynthesis and no movement of plastids (Nultsch et al. 1981).

The P vs. I curve is a model of photosynthetic rate. The rate can be standardized to a unit of biomass or a unit of chlorophyll. The model can be used to predict the change in rate if one of the parameters, such as pigment concentration, changes. For instance, an alga may acclimate to low irradiance by increasing the number of PSUs (or reaction centers), or by increasing the size of each PSU. The model predicts that if the *number* of PSUs is increased (Fig. 2.21b), the maximum rate of photosynthesis (P_{max}) will be increased and more light will be needed to saturate photosynthesis (hence a higher I_k). On the other hand, if *size* of the PSU is increased without changing the number of traps (Fig. 2.21c), the maximum rate will remain unchanged and less light will be required to saturate photosynthesis; the PSU in this case will be more efficient. Few seaweeds have been studied to determine whether they adjust number or size of PSU (or both). Among those that have been studied, *Ulva lactuca* changes PSU number whereas *Porphyra umbilcalis* changes both (Ramus 1981). Correlations have also been shown between irradiation and plastid size or thylakoid density, but the effects of such structural changes on photosynthesis have not been established.

Figure 2.21. Model light saturation curves for photosynthesis (P) versus incident irradiance (I_o). (a) General model, defining P_{max} = maximum photosynthesis; P_g = gross photosynthesis; P_n = net photosynthesis; R = respiration; I_k = saturating irradiance level. (b) Model of adjustment of the photosynthetic unit to extreme low and high I_o by changing the number of PSUs and not their size. (c) Model of adjustment to extremes of irradiance by changing the size of a fixed number of PSUs. (From Ramus 1981, with permission of Blackwell Scientific Publications)

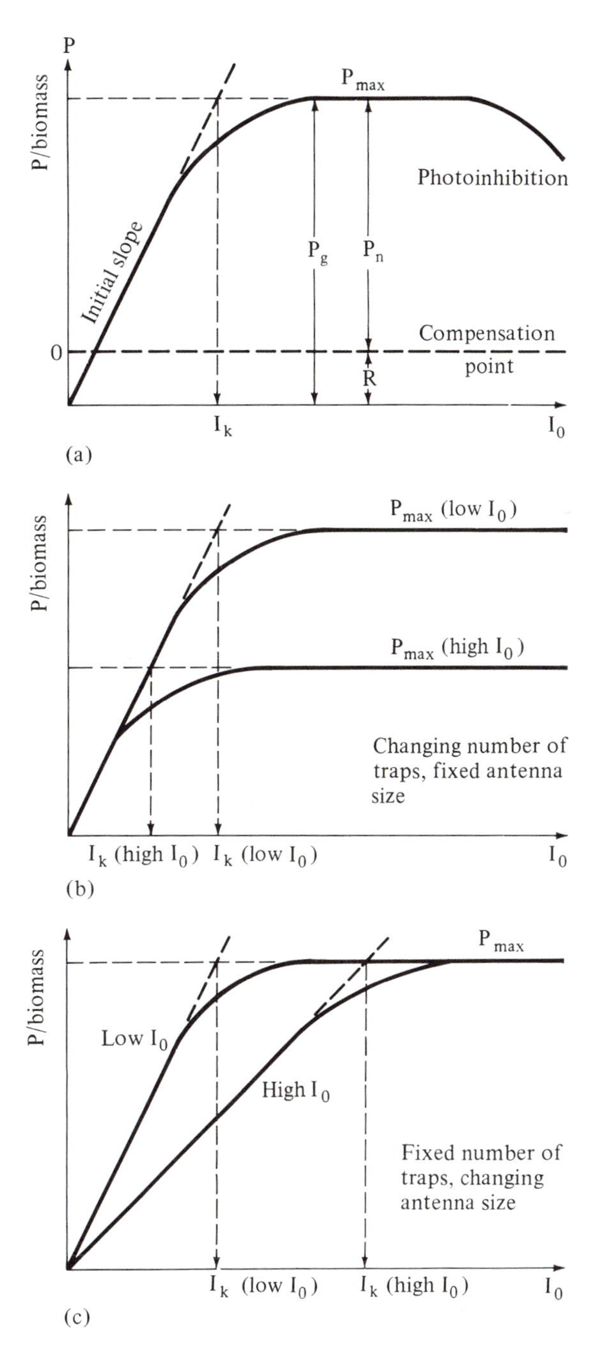

2.7.2 Circadian rhythms

Circadian (diel) rhythms affect several aspects of cellular activity in a number of species. Among the activities affected are photosynthetic gas exchange, some

Figure 2.22. Photosynthetic rhythms. (a) Light-saturated photosynthetic rate for *Ulva lactuca* kept in continuous dim light (2 mW cm^{-2}). Along the abscissa Eastern Standard Time and the hours of darkness in the collection environment. (b) Diurnal changes in photosynthetic capacity of *Porphyra yezoensis*. The solid line shows plants transferred to continuous light after the first day (light–dark cycles shown on lower abscissa). The broken line shows plants transferred to a reversed light–dark cycle (upper abscissa) and shows the reversal of the endogenous rhythm. (a from Mishkind et al. 1979, with permission of the American Society of Plant Physiologists; b from Oohusa 1980, with permission of Walter de Gruyter & Co.)

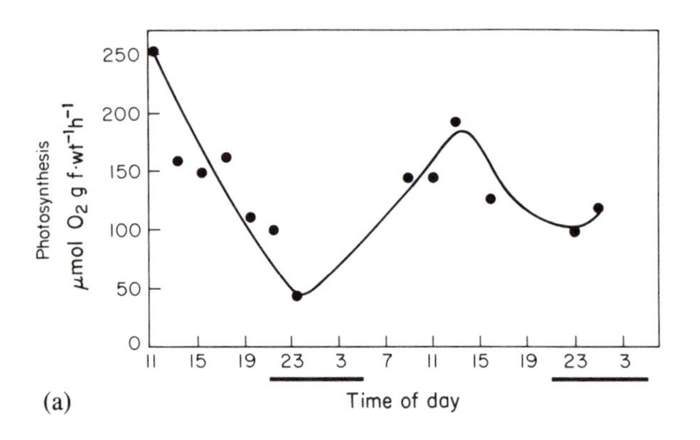

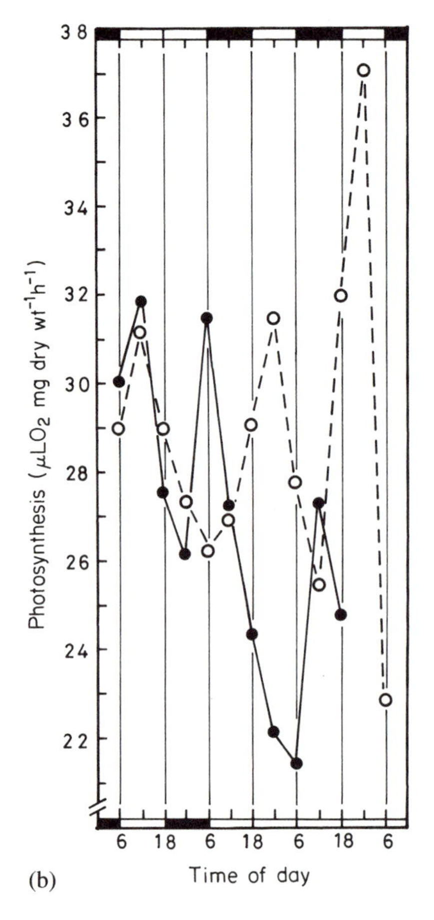

enzymes, and cell division. These rhythms have periods of 21 to 27 h (hence circa-dian) and are entrained to 24 h by light stimuli such as the time of dawn or dusk (Hillman 1976, Sweeney & Prézelin 1978). The diurnal changes are due to an endogenous rhythm and not the result of environmental variables. This can be shown by transferring plants to continuous (usually dim) light and taking samples for measurement of light-saturated photosynthetic rates at various times parallel to the original light–dark cycle. Characteristically, the rate is highest during a time corresponding to the middle of the light period and lowest in what would have been the middle of the dark period (Fig. 2.22). Over several days of continuous illumination the period of the rhythm gradually changes because there is no entraining stimulus (Mishkind et al. 1979, Oohusa 1980). If the light-dark cycle is reversed, the phase of the endogenous rhythm also shifts, as shown in Figure 2.22b. In *Ulva* there is also a diel migration of chloroplasts between the sides and faces of the cells (Britz & Briggs 1976), but this does not regulate the diel rhythm of photosynthesis (Nultsch et al. 1981). The rate-limiting step in electron transport, and hence a probable site of circadian control, is in one of the steps between plastoquinone and PSI (see Fig. 2.16) (Mishkind et al. 1979). Further points of control may lie in certain enzymes. Yamada et al. (1979), using the brown alga *Spatoglossum pacificum*, showed rhythms in several enzymes of the Calvin cycle, including RuBP carboxylase, fructose-1,6-bisphosphate phosphohydrolase, mannitol-1-phosphate phosphohydrolase, and ribose-5-phosphate isomerase (see Fig. 8.5). The relationship of these enzyme activities to the circadian clock is likely to prove complex. For example, RuBP carboxylase (in higher plants) is activated by light-driven Mg^{2+} fluxes (Jensen & Bahr 1977).

The current model of the mechanism of circadian rhythms is Sweeney's (1974) membrane model, based on studies of *Acetabularia*, which holds that the rhythms are generated by a feedback loop consisting of active transport across organelle membranes and the distribution of molecule(s) between organelles and cytoplasm. The organelle membrane is predicted to be slowly permeable to some molecule or molecules, *X*. Thus diffusion tends to equalize the concentrations of *X* in the organelle and cytoplasm. When equilibrium is reached, active import of *X* by the organelle is initiated and continues until a certain critical internal concentration of *X* causes a change in membrane properties so that active transport stops. In the absence of active uptake, diffusion gradually reestablishes equilibrium. Light is known to affect membrane properties and could thus reset the rhythm, for instance, by initiating active transport of *X*. Each rhythm may depend on different key substances in the appropriate organelle; for example, Mg^{2+} might affect RuBP carboxylase activity. K^+ has also been suggested as a possibility for *X* (Sweeney & Prézelin 1978). Among the features of *Acetabularia* behavior that led Sweeney to her model is that the rhythm of photosynthesis continues in enucleated fragments, yet if a nucleus and cytoplasm from two plants with diametrically opposed rhythms are combined, the rhythm gradually shifts from that of the cytoplasm to that of the nucleus. According to the model, this can be explained on the basis of both nucleus and chloroplasts transporting *X*; the nucleus of *Acetabularia* has a relatively large volume, and whether it is leaking or importing *X* it will have a major effect on the concentration of *X* in the cytoplasm. Hence the flux of *X* between cytoplasm and chloroplast and the timing of the photosynthetic rhythm gradually change.

2.7.3 Action spectra

A standard measure of seaweed photosynthetic efficiency is the action spectrum: photosynthesis as a function of energy absorbed at discreet wavelengths (Figs 2.23, 2.24). If all the pigments were involved with equal efficiency in photosynthesis, the curve of photosynthesis versus wavelength should closely parallel the curve of absorbance versus wavelength. The first such measurements were made a century ago in elegant experiments by Engelmann (1883, 1884), who irradiated red, brown, green, and blue-green algal cells with narrow wavebands obtained through a prism. He measured evolved oxygen by the position of aerotactic bacteria, which moved to the areas of greatest O_2 evolution. Much later, Haxo & Blinks (1950) published their now-famous absorption and action spectra of red, green, and brown seaweeds obtained with light of only 10-nm wavebands from a monochromator.

One notable feature of action spectra is that, despite continued absorbance of light beyond 700 nm, photosynthesis decreases rapidly beyond about 675 nm, the so-called red drop (Fig. 2.23). Photosynthesis can be increased in this region as well as in the blue region by the addition of small amounts of shorter wavelength light. The magnitude of the enhancement is much greater than would be caused by the extra quanta. Enhancement can be shown by measuring photosynthesis (P) at two wavelengths, λ_1 and λ_2, singly and together and calculating

$$\text{enhancement} = \frac{P(\lambda_1 + \lambda_2)}{P\lambda_1 + P\lambda_2}$$

Values greater than 1 indicate enhancement (Emerson et al. 1957; Ramus 1981). Enhancement is an important phenomenon in the field, where light is always broadband. Notice also that the supplementary light at 546 nm used in Fork's (1963) experiment (Fig. 2.23) is of a wavelength near the peak of the deep water spectrum. Thus, action spectra such as those produced by Haxo & Blinks have limited ecological value (Ramus 1981).

Figure 2.23. Absorbance and action spectra for *Porphyra perforata*, with and without supplementary background light of 546 nm. (From Ramus 1981, after Fork 1963, used with permission of D.C. Fork)

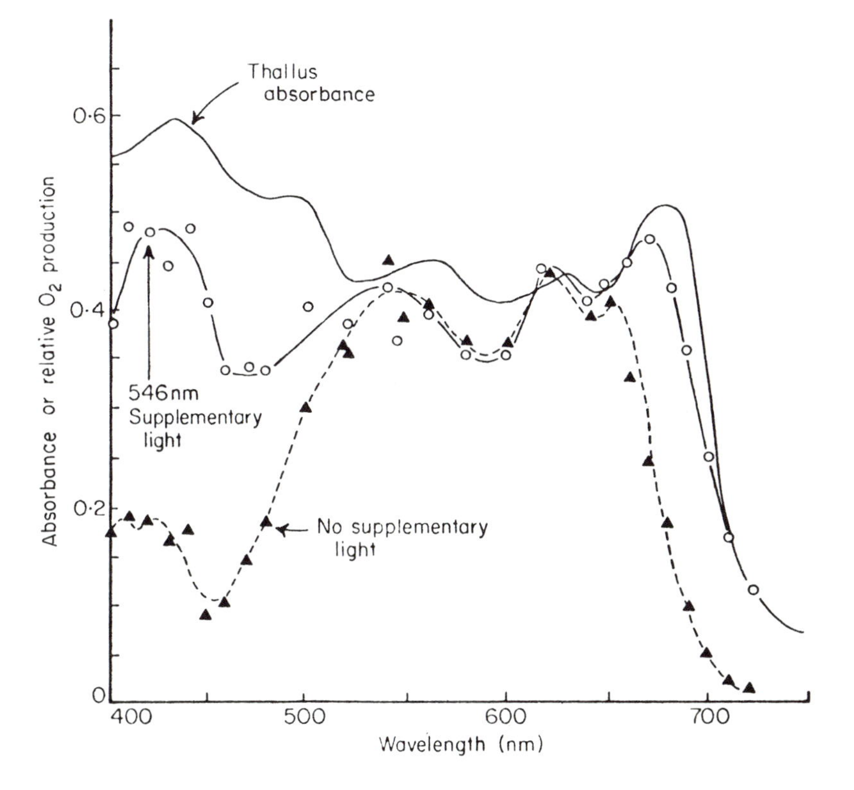

Figure 2.24. Comparison of the (a) absorption and (b) fluorescence spectra of *Ulva japonica*, which has siphonaxanthin, and *U. pertusa*, which lacks it. (From Kageyama et al. 1977, with permission of Y. Yokohama)

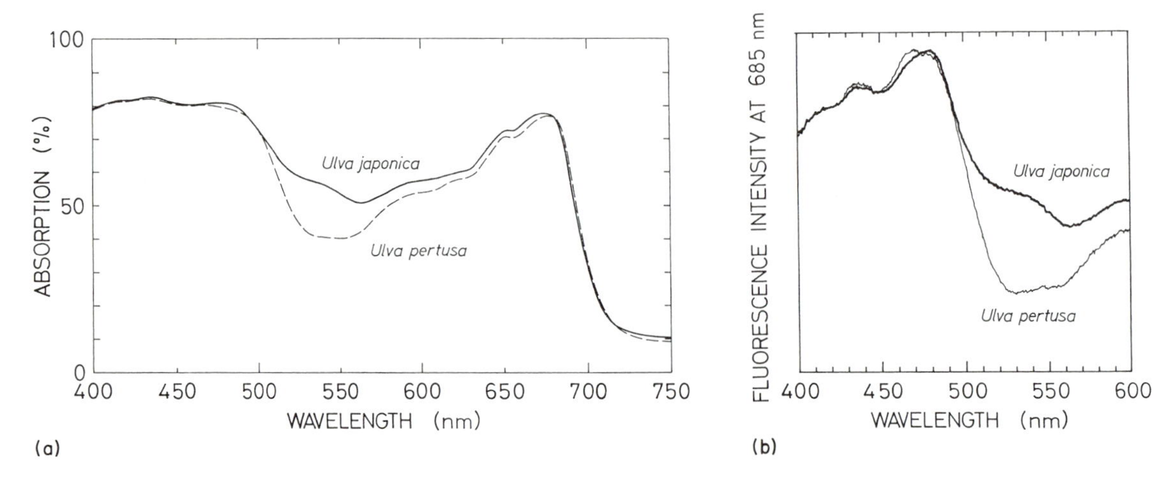

In their study of *Ulva japonica* and *U. pertusa*, Kageyama et al. (1977) measured action spectra by fluorescence. *U. japonica*, a deep-growing species, has siphonaxanthin, whereas the shallower water *U. pertusa* does not. The absorption spectra of the two species (Fig. 2.24) show a difference at about 540 nm, attributed to absorption by siphonaxanthin. In the action spectra of chlorophyll *a* fluorescence there is a shoulder at 540nm in *U. japonica*, corresponding to siphonaxanthin absorption. Such studies help determine the role of accessory pigments in photosynthesis and their activity in underwater irradiance.

2.7.4 Chromatic adaptation and vertical distribution of seaweeds

The irradiance conditions under which seaweeds grow range from full sunlight in the exposed intertidal zone, to spectrally limited, dim light in deep water. The depths to which seaweeds grow are determined for the most part by the amount of available light. In clear oceanic waters (type I of Jerlov), coralline and green algae grow as deep as 175 m, where irradiance is about 0.05% of that at the surface. In the Mediterranean Sea corallines are found at 120 m, where they receive about 0.1% of surface light, calculated to be about 12.6 E m^{-2} yr^{-1} (Lüning 1981a).

Algae growing in caves or at depths where irradiance is 0.9 to 21 μE m^{-2} s^{-1} have several means of surviving in low light. Most importantly, they have efficient light capture and conversion to chemical energy, plus a restricted dark respiration rate. Large seaweeds, such as kelps, need considerably more light energy than delicate forms. Lüning (1981a) suggested that they need about 70 E m^{-2} yr^{-1}. In North Sea waters (type 7), off the island of Helgoland, kelps grow to a depth of 8 m, but in the clearer type III water off the northwest coast of France, *Laminaria* grows as deep as 25 m, and in the very clear (type IA) waters off Corsica as deep as 95 m. Many investigations have been done since the late 1800s to learn how the pigment composition and photosynthetic capacity of seaweeds relates to the great variety of light conditions under which they grow.

Engelmann's work did not stop with his excellent action spectra. In 1902 he and Gaidukov showed that a blue-green alga, *Oscillatoria sancta*, changed color when grown in certain wavebands. In green light, it synthesized more phycoerythrin, which absorbs at green wavelengths, and it appeared red. When grown under orange light, the *Oscillatoria* produced more phycocyanin. These findings led to the theory of complementary chromatic adaptation, which holds that an organism growing in narrow wavebands will synthesize pigments complementary to those wavelengths, given that it has the genetic ability to do so. The theory was carried further to suggest that pigment composition of marine plants, particularly seaweeds, determined the depths to which they could grow. Red algae, with phycobiliproteins, would be expected to have the greatest range because they could make use of green light in deep water, whereas the green algae, with only chlorophylls, would be found primarily near the surface. In many locations this is the *general* pattern, but it is by no means the rule.

There were many opponents to the theory from the beginning. Oltmanns (1905) particularly insisted that total irradiance had a greater role in establishing vertical distribution of attached seaweeds than did the spectral quality. In 1928 two Russian scientists, Lubimenko & Tichovskaya, investigating seaweeds in the Black Sea, concluded that total quantity of pigment, rather than specific pigments, determined an alga's capability to photosynthesize in the low light at 50 m depth. Much evidence collected over the years refuted the theory of chromatic adaptation, both in laboratory experiments (Yokum & Blinks 1958; Waaland et al. 1974) and in field studies (Rhee & Briggs 1977; Ramus et al. 1977). In most cases, pigment concentrations were found to increase or decrease in response to low or high irradiance, rather than changing with respect to wavelength. Often the accessory pigments would show greater response to light changes than did chlorophyll *a*. However, complementary chromatic adaptation does occur in blue-green algae (Bogorad 1975), though even there it is restricted to certain species.

Ecological studies have also given support to the "anti-chromatic adaptation theory." First, scuba divers showed that red algae do not necessarily or exclusively occupy the greatest depths to which seaweeds can be found. Green algae were found growing at the limit of seaweed distribution off Malta and in the Bahamas (Larkum et al. 1967; Lang 1974). Transplant experiments showed that green algae could photosynthesize as well as red algae in very low irradiance. Seaweed physiologists recently provided theoretical as well as empirical evidence that seaweeds do not synthesize pigments complementary to the color of the incident irradiance (Dring 1981; Ramus 1983). Although green algae absorb relatively poorly in the green region of the spectrum, they do absorb there, as can be seen in Figure 2.24. As Ramus (1978, 1981) has pointed out, if a seaweed has enough pigment or is thick enough to be opaque, it is optically black and the particular pigment complement is unimportant. Thick thalli, or those within which there is much scattering and reflection of light, absorb more completely because the optical path is longer and the chances of a photon striking a plastid are greater. Thus there seems to be little evidence that complementary chromatic adaptation is important in determining the depth distributions of seaweeds.

2.8 **Light, growth, and development**

A prerequisite for growth is that the total energy and carbon fixed must exceed that used in respiration (maintenance metabolism). For most seaweeds the balance can be calculated on a 24-h basis, but for species such as some kelps that store carbohydrates during periods of nitrogen shortage, the balance may have to be calculated over several months. Growth is a complex phenomenon and subject to many input variables besides light, as will be discussed below. First, the irradiance requirements of growth will be considered. (Photoperiod effects are separate and are reserved until Sec. 10.4.1.)

Of particular interest in growth versus irradiance relationships is the compensation point, since this should be related to the depth to which the plant can grow. Brown algal gametophytes have been favorite objects for studies of minimum irradiance requirements for growth and reproduction because of their extreme shade environment and ease of culture. Chapman & Burrows (1970) showed that development of *Desmarestia aculeata* gametophytes depends on the mean daily irradiance, $I_0 \times$ photoperiod/24. (They actually measured illuminance, in lux, but the conclusions are the same.) At the two lowest irradiances gametophytes did not mature, though they survived and were able to develop later when I_0 was increased (Fig. 2.25). More detailed studies were done on kelp gametophytes by Lüning & Neushul (1978). Again, a threshold level of irradiance was required for gametogenesis. (Kelp gametophytes can become fertile when only one or a few cells in size; they apparently do not have to achieve a minimum size before they can reproduce. Indeed, large gametophytes indicate poor conditions for reproduction.) Vegetative growth was saturated in very weak light (4 W m^{-2} or about 20 μE m^{-2} s^{-1} or

Figure 2.25. Development of *Desmarestia* under various mean daily illuminances. Stages: 1 = settled zoospore; 2 = zoospore with germ tube or vegetative cells; 3 = fertile gametophyte; 4 = uniseriate sporophyte; 5 = uniseriate sporophyte with a meristem; 6 = corticated sporophyte. Approximate mean daily irradiances in μE m^{-2} s^{-1} (converted from illuminances): 30; 12; 4; 0.3; and 0.1. (From Chapman & Burrows 1970, with permission of Blackwell Scientific Publications)

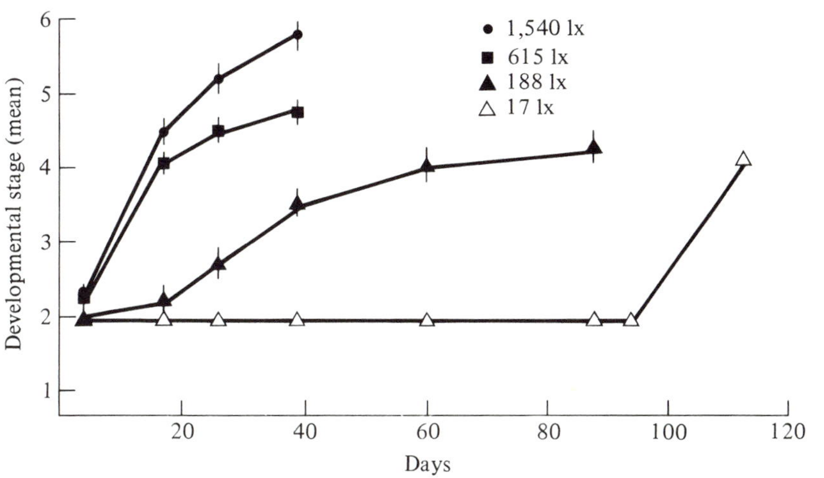

1000 lux), but for reproduction two or three times this irradiance was required. Blue light (400–500 nm), alone or as part of white light, is required for reproduction; in red light gametophytes grow only vegetatively. The ability of these plants to grow vegetatively in extremely weak light and to reproduce only when irradiance increases markedly provides an advantage to populations in retaining space after the canopy plants (parent sporophytes) are lost.

Blue light seems to be generally required for normal development of seaweeds, so far as is known. In work on *Fucus* spp., McLachlan & Bidwell (1983) even found that light in the range 575–625 nm (provided by orange and yellow-green filters) is deleterious to excised apices and to embryos after about 2 weeks of growth. This was not a photosynthetic problem, since tissues survived darkness without harm. Both longer and shorter wavelengths allowed continued growth and prevented necrosis. The basis for the importance of blue light is not understood. In phytoplankton and other plants, blue light generally promotes synthesis of protein, RNA and DNA, whereas red light promotes carbohydrate synthesis (Voskresenskaya 1972, Raven 1974). Normal growth obviously depends on a balance between these two extremes. Blue light is known to stimulate activities of certain enzymes in freshwater algae (but not yet in seaweeds) (Senger 1980).

2.9 Other factors affecting photosynthesis and respiration

Of the two major processes of photosynthesis, light trapping is a photochemical process, the rate of which is influenced by irradiance but not by temperature. Carbon fixation is an enzymatic process, strongly affected by temperature, pH, and salinity. It is also affected by light, inasmuch as many of the enzymes are light activated. Respiration, like carbon fixation, is an enzymatic process. Temperature shows little effect on photosynthetic rate as long as irradiance is below saturation, since the photochemical processes limit the overall rate in this portion of the P vs. I curve (see Fig. 2.26). Respiration and light-saturated photosynthesis respond to temperature by approximately doubling with every rise of 10 C (up to inhibitory or lethal temperatures). A significant question, particularly in regard to seaweeds living at high latitudes (low temperature, low light), concerns their winter respiration rate. Is this lower than one would predict from summertime measurements of Q_{10}? The few attempts to answer this question have indicated that Q_{10} tends to be greater in winter than in summer. The effect of temperature is greater, thus respiration at a given low temperature is lower, in winter specimens than in summer specimens (Yokohama 1973, Kremer 1981a). This change appears to be no more pronounced in polar than temperate species (Drew 1977). Moreover, the relation between pH and temperature (see Sec. 3.3.1) needs to be taken into account in such studies.

Salinity has a broad influence on metabolism and its effects can thus be seen in photosynthesis (see Fig. 4.6) and respiration rates. Some of these effects stem from reduced inorganic carbon or Ca^{2+} availabilities, others probably from the energy costs in osmotic adjustment, others from as yet unidentified causes. External pH affects photosynthesis through its effect on the ratio of CO_2 to HCO_3^- (see Fig. 7.2). Internal pH affects rates of reactions through its effects on enzyme activities.

Figure 2.26. Effect of irradiance on apparent photosynthesis (oxygen evolution measured manometrically) of *Chondrus crispus* (winter specimens) at 5 C and at 15 C. (From Mathieson & Norall 1975, with permission of Springer-Verlag)

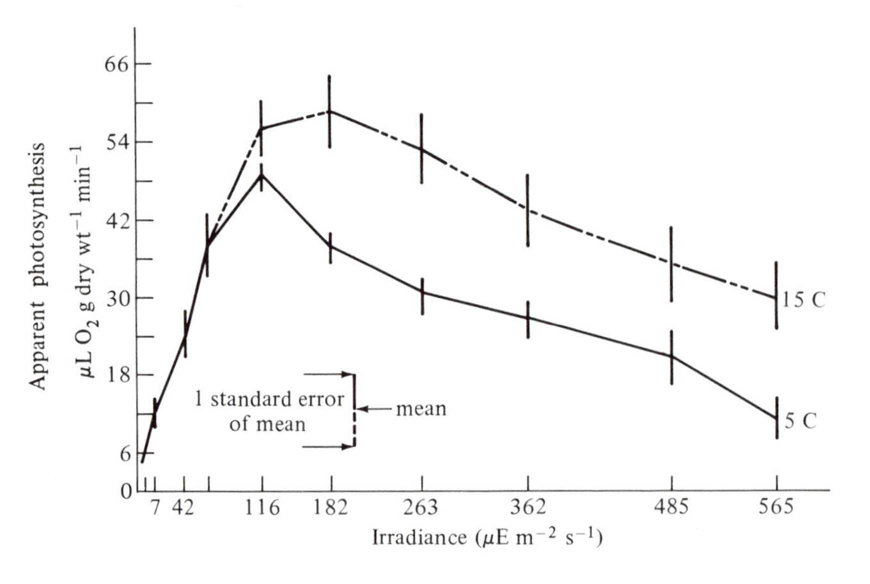

Gas exchange is influenced by the concentrations of CO_2/HCO_3^- and O_2; this is part of the reason for the low photosynthetic rate in air of many seaweed species (Sec. 9.3.1). Clearly, the rate of carbon fixation depends on the availability of carbon, that is, on the carbonate equilibrium, the dissolution of CO_2, and the effects of temperature and pH on these processes (Sec. 7.1.1). Oxygen is required for respiration, as the ultimate electron acceptor in oxidative phosphorylation; its concentration is also a factor in photorespiration as previously discussed. The effect of oxygen concentration on overall photosynthetic and respiratory oxygen exchange is illustrated in Figure 2.27. As oxygen concentration rises, photosynthesis decreases (Dromgoole 1978, Littler 1979). This phenomenon in nature is probably of importance only in tidepools where the water on a sunny day can become supersaturated due to algal photosynthesis (Kalle 1972). It is an important problem in laboratory measurements of photosynthesis. The response of respiration to oxygen concentration is a hyperbolic curve (Fig. 2.27b), which is not surprising since O_2 is a substrate in an enzyme (cytochrome oxidase) reaction.

Nutrient metabolism also has the potential to alter photosynthetic and respiration rates, since nutrient uptake is an active (i.e., energy-requiring) process. In the light, NADPH and ATP from photosynthesis may be used directly. In phytoplankton, when nutrient-limited cells are given a pulse of the limiting nutrient, photosynthetic gas exchange is shut down during the period of rapid nutrient uptake (Lean & Pick 1981). Furthermore, some nutrients, notably NO_3^- and SO_4^{2-}, require reduction before they can be incorporated into proteins or nucleic acids. Photosynthesis provides the carbon skeletons and reducing power for incorporation of these nutrients, so feedback controls can be expected to maintain a balance between energy production and energy use. Nutrient, trace metal, or vitamin limitation may slow down photosynthesis. Nonnutrient elements and other pollutants may reduce photosynthesis and alter respiratory rates (Chap. 8).

In natural communities there are many species of algae of various morphological forms. Littler and coworkers have recently begun to assess the relationship of photosynthesis to morphology (Littler & Littler 1980, Littler & Arnold 1982, Littler et al. 1983). Macrophyte morphologies can be arranged into six groups: (1) thin tubular and sheetlike; (2) delicately branched; (3) coarsely branched; (4) thick blades and branches; (5) articulated corallines; and (6) encrusting algae (Fig. 2.28) (see also Fig. 9.16a). Each class of seaweeds has representatives in most of these groups, which suggests that the forms are adaptive. Fitness may in part be conferred through reduction of grazing damage or through increased competitive ability. A significant part of testing such hypotheses is to assess the productivity potential of each a priori functional group. Littler & Arnold's (1982) results

Figure 2.27. Photosynthesis and respiration of *Carpophyllum* spp. as functions of oxygen content of seawater. (a) Apparent photosynthesis of four samples of *C. maschalocarpum* at 20 C and irradiance of 39–43 W m^{-2}, expressed as a percentage of the rate in 20% saturated water (1.03 mL O_2 L^{-1}). (b) Respiration (R) of primary axes of four species of *Carpophyllum* at 20 C. (From Dromgoole 1978, reprinted with permission from *Aquatic Botany*, vol. 4, © 1978 Elsevier Scientific Publishing Co.)

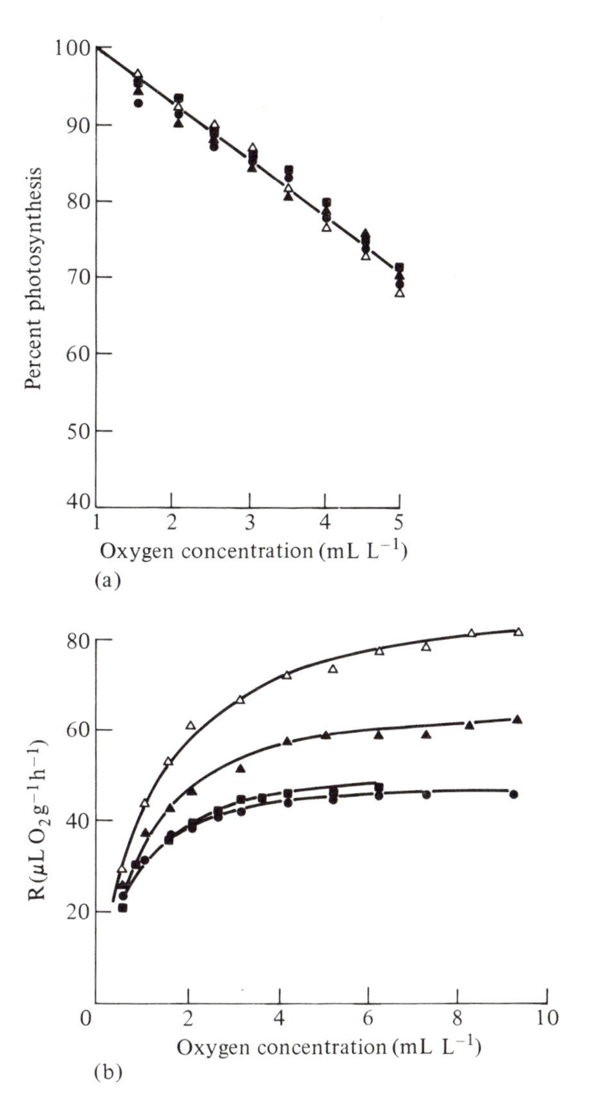

(Fig. 2.28) demonstrate that the sheet group has the highest mean net photosynthesis. This is not surprising since in general all the cells are photosynthetic, have direct access to carbon supplies in the water, and cause little self-shading. There was a twofold decrease in mean rate between each of the first five groups, while the encrusting species had extremely low rates. There was, of course, considerable overlap from group to group: for

Figure 2.28. Photosynthetic performances of algae in six functional-form groups. Experiments on plants from more than one of six collecting sites, from southern and Baja California, are shown separately. Mean rate in milligrams of C fixed per gram dry weight per hour is shown for each group. C = Chlorophyta; P = Phaeophyta; R = Rhodophyta. (From Littler & Arnold 1982, with permission of *Journal of Phycology*)

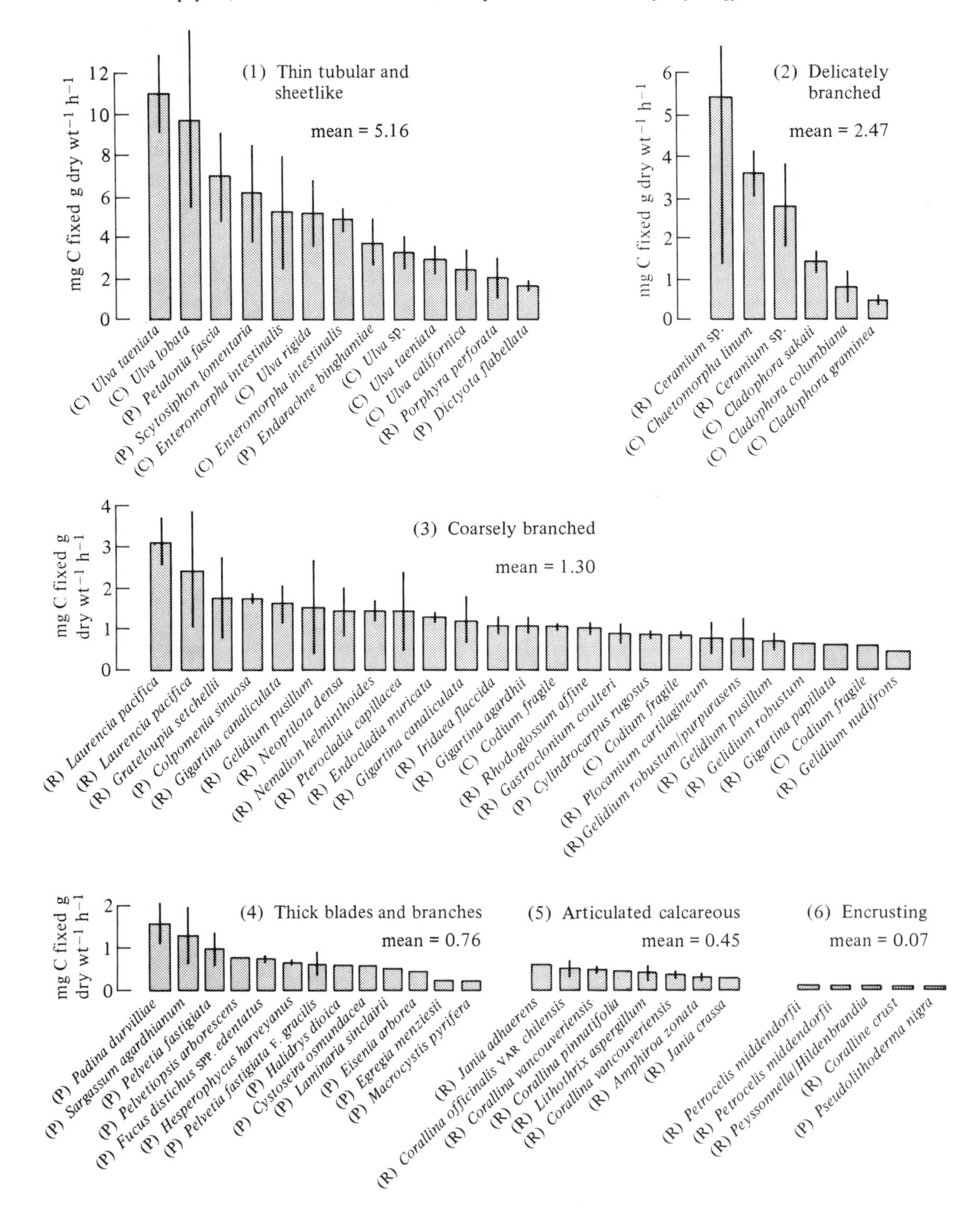

instance, the rate for *Laurencia pacifica* (group 3) is higher than that for *Porphyra perforata* (group 1).

2.10 **Primary production**

Primary production is the rate of net incorporation of carbon into organic compounds. It includes carbon retained in the plants, and organic carbon released as exudates or as pieces of tissue (or of entire plants if population productivity is being considered). It does not include carbon returned to the environment as CO_2 via respiration or photorespiration. In those rare, mostly laboratory, situations where exudation is very low and virtually no tissue is lost through breakage or grazing (i.e., the rates of these losses are not more than a small percentage of net photosynthesis), one can determine net production by measuring biomass at the beginning and end of a suitable time interval. One can arrive at an instantaneous value for primary production by multiplying net photosynthetic rate per unit of plant material by the standing stock (biomass) or number of units of plant material. Unfortunately, instantaneous values are of limited use and several practical difficulties make the estimation of monthly or annual production complicated and inaccurate. Brinkhuis (1977) compared three methods of estimating annual production and illustrated some of the problems and potential. Representative hourly productivity rates are seen in Figure 2.28. Published estimates of annual production indicate that seaweeds are roughly on a par with terrestrial agricultural production; comparative figures for seaweeds, seagrasses, saltmarshes, and terrestrial ecosystems are reviewed by Ferguson et al. (1980).

In addition to the problems inherent in measuring the hourly rate of photosynthesis, this rate in nature is not constant over a day, nor from day to day, nor season to season, because of changes in irradiance. Some plants also show diurnal changes in photosynthetic rate even under uniform laboratory conditions (Kageyama et al. 1979, Mishkind et al. 1979, Ramus 1981). Furthermore, the number of hours of daylight per day is not constant (except at the equator). Since a very small percentage of the season or year or lifetime of the organism is experimentally measured, errors in extrapolation from typical 4- or 8-h measurements of photosynthesis become very large. Several features of the plants further complicate the extrapolation of sample photosynthesis rates to plant or population productivity. The proportion of nonphotosynthetic tissue can change as a plant ages, thus affecting photosynthetic rate on any weight basis, and can vary between individuals (e.g., wiry versus fleshy plants of *Gigartina canaliculata* recorded by Littler & Arnold 1980). While there is little change in photosynthetic rate of gametes, zygotes, and young sporophytes of *Laminaria* (Kremer & Markham 1979), the rate does change along an adult frond, where tissue ontogenetic age increases from the meristem (Küppers & Kremer 1978). Results from similar studies on *Fucus* (McLachlan & Bidwell 1978, Khailov et al. 1978, Küppers & Kremer 1978) show much the same trend, although there is an apparent increase in rate for zygotes over their first few days. The reproductive condition of the tissue is a further complicating factor (Littler & Arnold 1980). Finally, the determination of either exudation or tissue loss is a study in itself.

The rates of exudation by various seaweeds under most natural conditions remain a matter largely of conjecture. Considerable controversy has been generated by various studies that have attempted to assess exudation under experimental conditions. Some experiments have shown that exudation is 30 to 40% of net assimilation in seaweeds (Khailov & Burlakova 1969, Sieburth 1969). Other experiments show it to be much less: a few percent down to less than 1%, except under stress (Moebus & Johnson 1974, Moebus et al. 1974, Harlin & Craigie 1975, Brylinski 1977). In all experiments there is a risk that exudation (or its lack) is an artifact of the method. The roles of the exuded materials are not known.

Tissue loss is more easily documented and understood than exudation. Tissue loss stems from two groups of factors: (1) extrinsic, including direct or indirect grazing damage, physical abrasion, and microbial degradation; (2) intrinsic, including shedding of old fruiting branches (e.g., by fucoids) and erosion of old tissue at the tips of *Laminaria*. Loss of tissue from populations also includes loss of entire plants. Surprising quantities of tissue are abscised by *Fucus* and *Ascophyllum* following reproduction. In the former, not only the receptacles but also the internodes beneath are shed (Knight & Parke 1950). *Ascophyllum* receptacles can account for half the biomass of the plant (Josselyn & Mathieson 1978). *Laminaria* blades continuously lose old tissue from the tips; the entire blade can be turned over one to five times per year (Mann 1973). Infection by an ascomycete, *Phycomelaina laminariae*, has been found to accelerate breakdown of old *Laminaria* tissue (Schatz 1980). *Laminaria hyperborea* provides a somewhat special case in that a distinct new blade is produced in late winter at the base of the old blade. Carbon, and perhaps nitrogen, is salvaged from the old blade for the new growth before the old blade is shed. Lüning (1969) demonstrated that the new blade could form in darkness with the old blade present, but if the old blade was cut off there was virtually no new growth even in light. Subsequently, Lüning et al. (1973) demonstrated translocation of newly fixed carbon from old to new blades, but no one has yet shown with tracers that *old* carbon or amino acids are salvaged. Analogy with senescence of angiosperm leaves suggests that salvage might occur in kelps, since there is translocation.

Exuded material and pieces of tissue enter the food chain where they feed microorganisms, herbivores, detritivores, and filter feeders. Exudates may remain as dissolved organic carbon or may flocculate and contribute to particulate organic carbon in the sea (Conover

1978). The rate of breakdown of seaweed detritus and the significance of this material in the higher trophic levels is a subject beyond the scope of this text. The interested student is referred to recent pertinent papers, including Albright et al. (1980), Linley et al. (1981) and Newell & Field (1983). The significance of recycling of minerals from beached or buried seaweeds has been studied by Koop et al. (1982) and Birch et al. (1983), among others.

2.11 **Synopsis**

Light is part of the electromagnetic spectrum and travels both as a wave and as packets of energy called quanta or photons. The part of the spectrum that is photosynthetically active radiation is from 350 or 400 nm to 700 nm. The energy of photons is inversely proportional to their wavelength, so that blue (ca. 450 nm) are more energetic than red (ca. 650 nm). The quantity of light arriving, or flux, is called irradiance and is measured in $\mu E\ m^{-2}\ s^{-1}$ or $\mu W\ m^{-2}$. Instruments that measure illuminance – light as perceived by the human eye – are inappropriate for work with plants, because the sensitivity of plant pigments to light is quite different. In the sea, light is attenuated by scattering and absorption, with the red end of the spectrum being more strongly reduced. Thus the light in deep water tends to be blue or green. Several ocean water types have been defined on the basis of light quality in them.

Photosynthetic pigments in plants always include chlorophyll *a*. In addition, each group of algae has a characteristic array of accessory pigments, including other chlorophylls, carotenes, xanthophylls, and, in red algae among the seaweeds, phycobilins. A few of these, including carotene, do not contribute to light harvesting for photosynthesis. Each pigment has a characteristic absorption spectrum, which is a graph of the probability of its absorbing a photon of a given wavelength. Chlorophylls and phycobilins occur in pigment–protein complexes. Phycobilins are clustered in phycobilisomes on the thylakoid membrane surface; other accessory pigments and the rest of the photosynthetic apparatus are integral parts of the membrane. The arrangement of thylakoids within the chloroplasts differs among phyla. Other features of the chloroplast include a small genome made of chloroplast DNA and the carbon-fixing enzyme ribulose bisphosphate carboxylase, which in some groups is in pyrenoids.

Although there are such differences in chloroplast structure and pigment complement among seaweed groups, thylakoid membranes and the process of photosynthesis are essentially the same in all plants. The photosynthetic apparatus consists of two different reaction centers with slightly different red absorption peaks. Photosystem I has peak absorption at 700 nm, PSII at 680 nm. The reaction centers consist of one or a pair of chlorophyll *a* molecules with a specialized protein linkage. Light is funneled into the reaction centers by the accessory pigments and most of the chlorophyll *a*, which are arranged in antennae with those absorbing at longer wavelengths thought to be closer to the reaction center. In the reaction center, the energy causes an electron to be donated to an electron transport compound and the chlorophyll to be oxidized. Rereduction of the chlorophyll in PSII takes place by electrons derived from water, which results in release of oxygen. PSI is rereduced by electrons from PSII in noncyclic electron transport, or by an electron returning in cyclic transport. In each case a series of electron transport compounds capable of reversible oxidation-reduction reactions is involved, and the flow of electrons down the redox potential gradient is coupled to phosphorylation of ADP to yield ATP. The mechanism of photophosphorylation is thought to be the transfer of H^+ inside the thylakoid sac and the discharge of this electrochemical gradient through coupling-factor particles in the membrane.

The rate of photosynthesis is strongly dependent on irradiance level. At the compensation point, gross photosynthesis equals respiration and therefore net photosynthesis equals zero. At the saturation point, photosynthesis is maximum. At very high irradiance, photosynthetic rate may decline again due to photoinhibition.

Action spectra show the rate of photosynthesis at different wavelengths, and usually follow the curve of the absorption spectrum. However, photosynthesis in a narrow waveband is often enhanced by addition of small amounts of short wavelength light. In nature, light is always broadband.

Seaweeds can acclimate to differences in light quality and quantity. Quantity of pigment or density of photosynthetic units can be increased, and sometimes the ratio of accessory pigments to chlorophyll *a* changes. However, the old idea that seaweeds change color to complement the color of the light in their habitat has been discredited.

The rate of photosynthesis depends on numerous other factors, including inorganic carbon supply, temperature, pH, circadian rhythms, and age of the tissue. The rate can be measured by following CO_2 uptake or O_2 release by illuminated tissue, but account has to be taken of respiratory use of O_2 and release of CO_2 in the light. Respiration is usually measured in darkness but occurs, to a much smaller extent, in the light. Another cause of CO_2 release and O_2 uptake is photorespiration, the result of competition between O_2 and CO_2 for binding to RuBP carboxylase; this process may not be important in marine algae. O_2 may also be consumed in pseudocyclic phosphorylation in the light.

The estimation of primary production requires measurements not only of photosynthesis, respiration, and photorespiration (if any), but also of tissue loss and organic carbon exudation rates. Seaweeds can be divided into functional-form groups such as crustose forms and thin sheets, which have characteristic levels of productivity.

3 Temperature

3.1 Introduction

Whereas light may be considered the most important physical factor in the environment of photosynthetic plants, temperature is undoubtedly the most fundamental factor for all organisms because of its effects on molecular activities and properties, and hence on virtually all aspects of metabolism. Living organisms are rarely at thermal equilibrium with their environment (Campbell 1977) and the important temperature is that inside the cell, not that of the surroundings. However, the internal temperatures of seaweeds and other poikilothermic organisms are usually near the temperature of their surfaces or of the surrounding air or water. This approximation must be assumed because most temperature probes, even so-called microclimate instrumentation (Forstner & Rützler 1970), are much too large for use within the average cell. A probe must be relatively small compared to the subject or it may cause a significant change in temperature.

3.2 Natural ranges of temperature

3.2.1 Ocean temperatures

Ocean surface temperatures change in two broad ways. First, they decrease toward higher latitudes, from about 28 C in the tropics to 0 C toward the poles, although this trend is markedly affected by ocean currents. Because of the California Current, for example, fairly uniform cool temperatures prevail in the seawater along much of the west coast of North America, even though land temperatures change considerably. Second, the seasonal changes in ocean temperatures are larger at mid-latitudes. Shallow bays of Prince Edward Island, in the Gulf of St. Lawrence, freeze over in winter but warm up to 22 C or more in summer. In the tropics and at the poles the annual temperature range is often less than 2 C (Kinne 1970).

A further complication in ocean water temperature regimes is the presence of thermal stratification. Rather sharp temperature boundaries, called thermoclines, may develop, especially in sheltered waters, between layers of water (Fig. 3.1). They are often accompanied by a rapid change in salinity (halocline), the two combining to give a change in density (pycnocline). Typically, but not always, the surface water is warmer and less salty than the deeper water on which it floats. Vertical mixing is surprisingly slow unless wave action is vigorous. In places with moderate to large tidal amplitude, such pycnoclines (which are often only a meter or two below the surface) will sweep portions of the shallow subtidal and lower intertidal zones, causing rapid temperature and salinity changes with each ebb and flood of the tide.

3.2.2 Intertidal temperature regimes

The principal environmental feature of the intertidal zone is the regular exposure to atmospheric conditions. Temperature regimes are thus much more complex than subtidal regimes. Myriad microenvironments result from the many factors that affect the local and organisms' temperatures. Some factors, such as shading, affect the influx of heat to an organism, whereas others, such as evaporation, affect heat efflux. The major source of heat in the intertidal zone during ebb tide is direct solar radiation. Irradiance may be reduced through shading by clouds, water, other algae and shore topography (including overhangs, crevices, and the direction of slope). Small-scale topographic features also give shelter from breezes, hence from evaporative cooling. Two examples of actual algal temperatures recorded during emersion on hot days are given in Figure 3.2. *Endocladia muricata* (Fig. 3.2a) is a stiff tufty plant: the temperature of the interior of the clump, which is shaded yet open to air flow, remains considerably cooler than the air or open rock surface (Glynn 1965). *Porphyra fucicola* (Fig. 3.2b), on the other hand, is flattened against the rock surface like a little solar panel and, on a calm day such as that illustrated, becomes much hotter than the air (Biebl 1970). These graphs also show the sharp drop in temperature

Figure 3.1. Thermoclines and haloclines as seen in temperature and salinity profiles at a station between Copenhagen and Elsinore, Denmark. Rapid changes in temperature (■) and salinity (●) are seen at 4 m and 14 m below the surface, at the boundaries between three overlying water masses. (Modified from Friedrich 1969, with permission of Gebrüder Borntraeger)

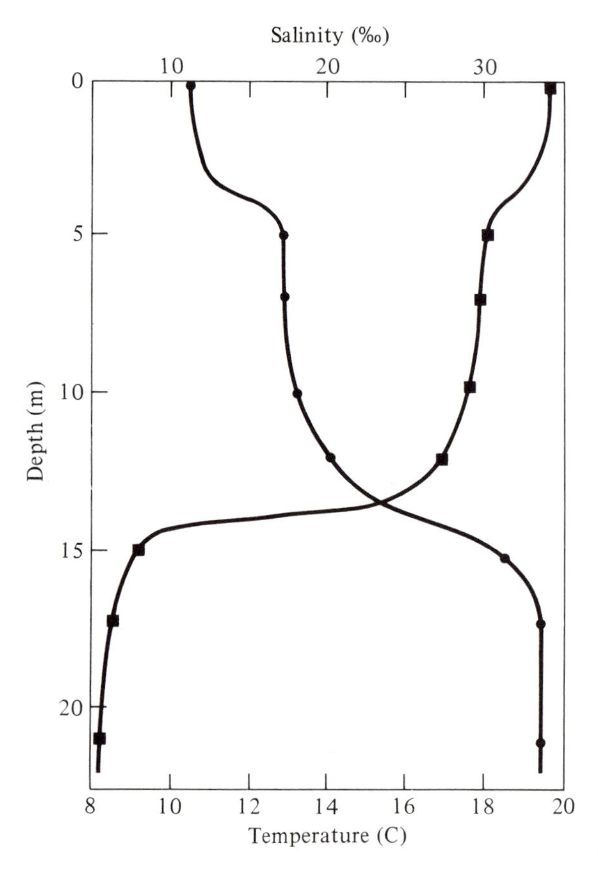

as the tide covers the plants. Notice that the *Porphyra* thallus surface temperature falls from 33 C to 13 C in a matter of minutes as the water reaches it. The sharp temperature gradient across the cell membrane at such times may provide substantial energy for ion movement, but no work has been done on this possibility except for a theoretical framework and methodology provided by Thorhaug & Katchalsky (1972).

Other variables affecting the temperature of intertidal plants are the time of day at which low water occurs and the extent of heating or cooling due to waves. At dawn or dusk there will be little heating by the sun, whereas if the low tide occurs in the middle of the day heating (and also desiccation) can be extreme. In summer the water may be cooler than exposed rock and algae, as in Figure 3.2b. In winter, seawater is frequently warmer than the air, so it can thaw algae that have been frozen during exposure to subzero air. However, if the rocks are cold enough the waves will freeze onto them, embedding the seaweeds in ice. Wave splash can also affect algal temperature by maintaining a supply of water for evaporation.

3.2.3 Littoral pools

Water left in pools by the ebbing tide initially has the same conditions as the mass of seawater. Conditions change during exposure because of the influence of the atmosphere; the longer the exposure and the higher the pool's surface to volume ratio, the greater the changes that will have accumulated before the tide refloods the pool. Most of the factors that affect the temperature of exposed rock pools also affect their salinity. Changes in tide pool temperatures are more drastic during the day and in summer (Fig. 3.3) because the chief source of heat is solar energy. However, air temperature may warm or cool the pool (even freeze it). Water added to the pool as runoff, rain, or snow may heat or cool the pool. Tide pools have little or no mixing and thus may easily become stratified (Fig. 3.4), especially if the pool is being warmed. (If the pool is being cooled some mixing may be brought about, since seawater becomes denser as it cools and sinks through warmer water.) Freshwater floats on the pool surface because of the difference in salinity; if runoff is warmer than seawater it will float on that account too. Temperature stratification formed in the day, in the absence of salinity stratification, usually breaks down at night.

3.3 **Biochemical and physiological effects of temperature**

3.3.1 Chemical reaction rates

Heat is the energy of molecular motion; at higher temperatures molecules have more energy and reactions proceed faster. The effect is summarized in a value called the temperature coefficient, Q_{10}: the ratio of the reaction rates at (t + 10) C and at t C. Values of Q_{10} for enzyme-catalyzed reactions range from about 1.1 to 5.3, but are usually around 2 (West et al. 1966). In other words, reaction rates approximately double for a 10 C rise in temperature. However, if the rate of an enzyme-catalyzed reaction is measured over a broad range of temperatures, a peak is found. This optimum temperature and the sharpness of the peak also depend on pH and the purity of the enzyme. The reason the reaction rate does not keep doubling is the increasing rate of thermal denaturation of the enzyme above a critical temperature (Lehninger 1975; Fitter & Hay 1981). On cooling, the enzyme may regain its active conformation or it may be permanently damaged.

The effect of temperature on enzyme activities and the bearing this has on seaweed physiology is illustrated in a study by Küppers & Weidner (1980) of six *Laminaria hyperborea* enzymes from diverse metabolic pathways. Ribulose bisphosphate carboxylase is given as an example in Figure 3.5. Enzymes were extracted from the kelp and their activities measured under standard conditions, which included a temperature of 25 C. Sea-

Figure 3.2. (a) Temperature observations of three microhabitats in the high intertidal *Endocladia-Balanus* association at Monterey, California, as related to low water exposure. Also shown is the air temperature at a nearby weather station during the observation period. The horizontal bar and line at the top of the graph show, for the level observed, the approximate duration of submerged (cross-hatching), awash (clear), and exposed (line) periods. (b) *Porphyra fucicola* thallus temperature during ebb tide on a calm, sunny day. (a from Glynn 1965, with permission of the Zoological Museum, Amsterdam; b from Biebl 1970, with permission of Springer-Verlag)

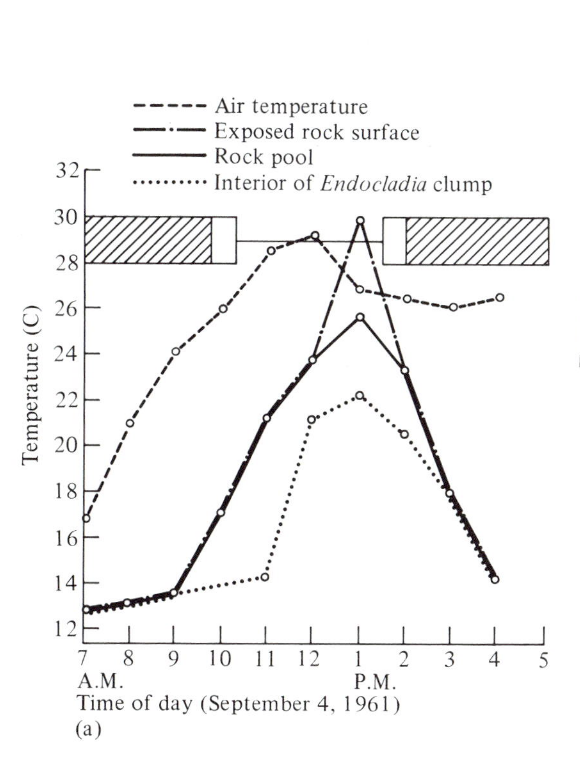

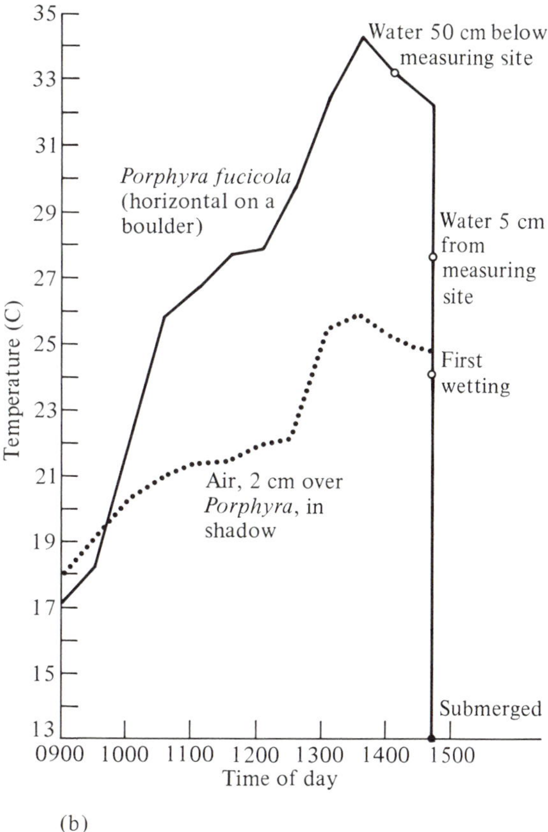

sonal changes were found in all six enzymes, with peaks generally in February to April (Fig. 3.5a). In nature, of course, the temperature is not constant. By determining the effect of temperature on in vitro activities of each enzyme and by recording seawater temperatures, Küppers & Weidner were able to calculate what the enzyme activities would have been in the living kelp on the shore. There was again a seasonal change in each enzyme, but the peaks were now in August (Fig. 3.5b). The interpretation of these results was that in summer the kelp could have high enzyme activities with smaller numbers of enzyme molecules because of favorable temperatures; in spring, during a period of high growth rate, high metabolic rate could be achieved by an increase in the total amount of enzyme present. This is in contrast to conclusions drawn for phytoplankton by Morris & Glover (1974). Moreover, a second factor affecting the activities of these enzymes in *Laminaria* is nitrogen availability in the seawater (Wheeler & Weidner 1983).

A change in enzyme quantity is only useful as a response to temperature if the enzyme is working near its maximum rate, V_{max}. Since intracellular metabolite concentrations are generally well below those that will saturate the enzyme, a more effective means of increasing enzyme rate is a change in $K_{0.5}$. ($K_{0.5}$ is the substrate concentration giving half V_{max} of a regulatory enzyme.) Studies in the 1960s on enzymes from fish suggested that temperature might act as an allosteric modulator of regulatory enzymes. Several problems arose with this theory, the most important being the role of pH in activity changes. Many enzymes are highly sensitive to pH (although others are not). Thus, experimenters have generally taken care to keep pH constant with a buffer, including during experiments on temperature effects. However, the pH of a buffer changes with temperature, and the neutrality pH of water (i.e., where $[H^+] = [OH^-]$) also changes with temperature. Thus intracellular pH is likely to change with temperature. Recent reviews (Somero

Figure 3.3. Temperature changes in three pools at different heights in the intertidal zone near Halifax, Nova Scotia, May 8–9, 1970. Times of high tide indicated by arrows, darkness by the cross-hatched bar. Pools 1 and 2, between neap and spring higher high waters, were generally flushed twice daily, except for pool 1 during periods of calm or neap tides. Pool 3, 0.4 m above extreme high water, was washed only during severe storms and contained no perennial macroscopic algae. (From Edelstein & McLachlan 1975, with permission of Springer-Verlag)

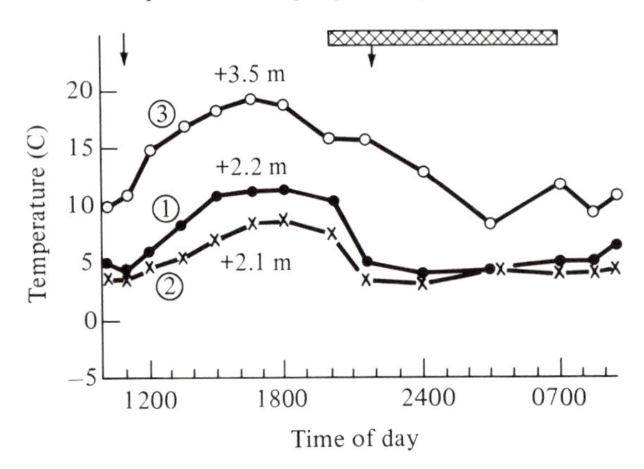

Figure 3.4. Stratification in a tide pool. The pool, on San Juan Island, Washington State, was exposed at 0730 h; this temperature profile was made at 1515 h, the time of maximum stratification. Sunny day; air temperature 20 C; seawater temperature 11 C. (From Carefoot 1977, with permission of the author)

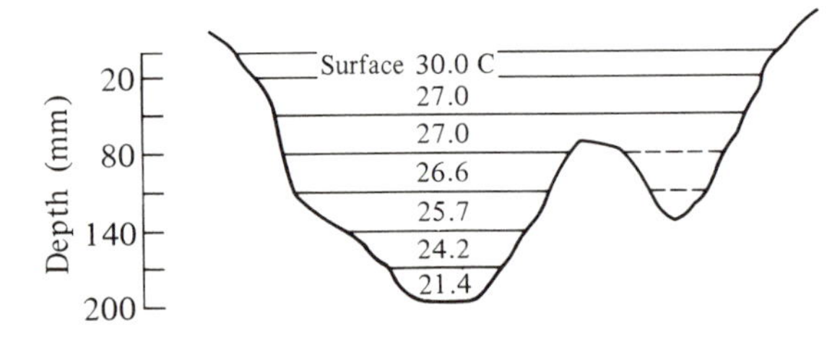

1981, White & Somero 1982) suggest that this pH change is the real cause of temperature effects on enzymes; these studies on animal enzymes should be extended with research on seaweed enzymes.

The interaction between various environmental factors and enzyme control and growth forms a complex web, which is explored in the following sections, working through progressively higher organizational levels.

3.3.2 Metabolic rate: temperature effects at the cellular level

When rate versus temperature measurements of complex reactions, such as photosynthesis and respiration, are made, the overall rate is a composite of all the individual reaction rates. If there is a rate-limiting reaction, it is not necessarily the same one at all temperatures. The effect of a given temperature change is not the same on all metabolic processes because of differing temperature sensitivity of enzymes and the influences of other factors, including light, pH, and nutrients. For example, increased respiration at high temperatures is due partly to breakdown in cell compartmentalization, which allows stored materials and degradative enzymes to mix (Fitter & Hay 1981).

Numerous studies have investigated the effect of temperature on photosynthesis, respiration, and growth under otherwise uniform conditions. Not surprisingly, maximum rates have often been found to correlate with the temperature regime in the alga's habitat. However, there have been some reported instances in which the optimum temperature (i.e., that giving maximum rate) was not near the natural conditions. For instance, Fries (1966) found optimum temperatures for growth of three red algae in axenic culture to be 20 to 25 C, whereas the water temperatures in the vicinity, even in summer, rarely rose above 15 C. She speculated that a reason for the discrepancy may have been the absence of bacteria from the cultures: seaweeds may use growth substances produced by their associated microflora, and marine bacteria grow best at lower temperatures. In other words, there is a physiological optimum for the alga alone and

Figure 3.5. Activities of ribulose-1,5-bisphosphate carboxylase extracted from the proximal part of *Laminaria hyperborea* blades. (a) Seasonal change in standard enzyme activity at 25 C. (b) Seasonal change in temperature-adjusted enzyme activity. (From Küppers & Weidner 1980, with permission of Springer-Verlag)

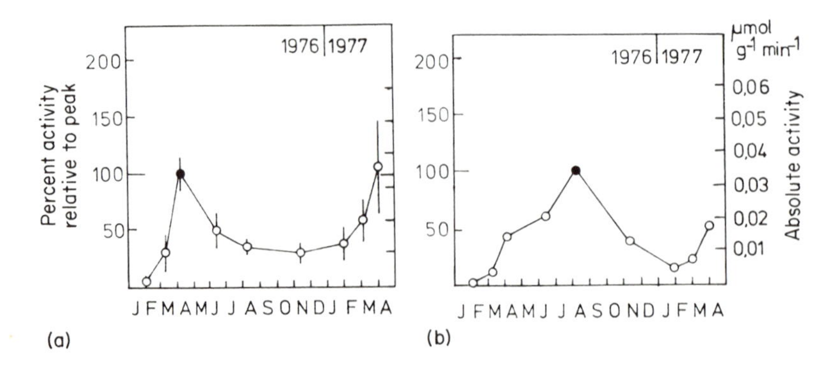

an ecological optimum in nature where it is interacting with bacteria and fungi. Lehnberg's (1978) data, shown in Figure 9.25, suggest that ecological optima may also have much to do with levels of interacting environmental variables.

Poikilotherms, including plants and lower animals, cannot regulate their internal temperature. Nevertheless, their metabolic relationship to temperature may change over the short term and long term. Küppers & Weidner's work on kelp enzymes, previously discussed, is a case in point. Through these responses seaweeds become acclimated to temperature changes. (Acclimation is a passive process. Organisms do not wilfully alter or "optimize" their metabolism.) The rate of change of temperature is thus as important a factor as the absolute temperature: if the rate of change is slow the organism may become acclimated. Seasonal changes in photosynthesis and respiration have been shown in several seaweeds by comparing performances of summer and winter plants under identical conditions. Newell & Pye (1968) found evidence of a low Q_{10} (less than 1.2) in that part of the temperature range to which several intertidal seaweeds were acclimated. In summer, the relatively flat part of the respiration versus temperature curves was in the range of 10 to 20 C; in winter it was shifted to lower temperatures. The effect of these changes, both diurnally and seasonally, was to minimize the effects of temperature fluctuations on respiration. Newell & Pye were careful to note that their results did not demonstrate that temperature itself was responsible for inducing the seasonal changes. Similar data were obtained by Mathieson & Norall (1975) in a detailed study of the physiology of *Chondrus crispus*. They showed that at a given irradiance, apparent photosynthesis (APS) is maximum at a lower temperature in winter specimens than in summer specimens (Fig. 3.6). More important, APS in cold water is higher in winter plants than in summer specimens; and summer plants maintain near-peak APS through warmer temperatures than can winter plants.

Other data on effects of temperature on photosynthesis and respiration of seaweeds have been reported by, among others, Adey (1970), Yokohama (1972), and Durako & Dawes (1980). Different species, and even different populations of the same species, show diverse responses to temperature. Thus, the data just presented are merely examples to illustrate the principles involved; they should not be taken as "typical," nor as representing all seaweeds.

Other effects of temperature at the cellular level include changes in rates of nutrient uptake (see Chap. 6) and, at least in some phytoplankters, changes in chemical composition of cells (Aaronson 1973, Morris & Glover 1974, Goldman & Mann 1980).

3.3.3 Growth rates and temperature: the organism level

At progressively higher levels of organization the effects of temperature become more difficult to interpret and more easily confounded by effects of other environmental factors and by the variety of temperature effects themselves. Nevertheless, studies of the effects of temperature on the whole organism, in all its complexity, are useful since in nature the whole organism lives or dies, grows to a greater or lesser extent, and succeeds in reproduction to a greater or lesser degree, depending on the environmental influences it, as a whole, receives.

Studies of algal growth fall into two broad categories. First are studies in which correlations are sought between growth and temperatures (usually only of the water), both measured in the field. In fortuitous cases, it may be possible to separate, at least partially, the effects of temperature from other major environmental variables such as light, nutrients, salinity, and water motion. Second, the rate of growth is recorded at various temperatures under otherwise uniform conditions in the laboratory. Growth rate is certainly affected by temperature and, as for individual enzyme reactions, there is generally a peak or high plateau above and below which

Figure 3.6. Apparent photosynthesis of *Chondrus crispus*: oxygen output measured manometrically. Effect of temperature on winter and summer specimens from −12 m at an irradiance of 116 $\mu E\ m^{-2}\ s^{-1}$. (From Mathieson & Norall 1975, with permission of Springer-Verlag)

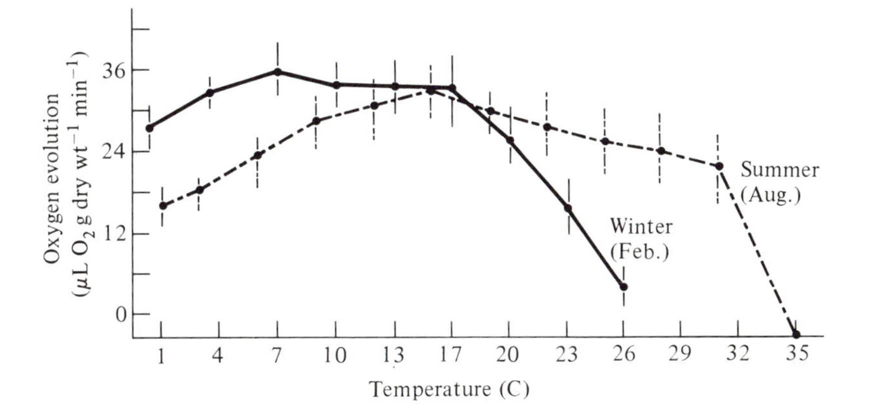

growth rate falls off (see, for example, Fortes & Lüning 1980). However, in situations where light or nutrients are limiting, temperature may have little effect on growth (see Fig. 2.26).

Within one species there may be considerable genotypic variation in temperature tolerance and optimum temperatures for growth (as for other responses). Such variation may be great enough for geographically diverse populations to appear as distinct strains or races. *Phenotypic* variation within each population is also likely. However, in *Ectocarpus siliculosus*, the populations are not distinct from one another as races (or "ecotypes"); Bolton (1983) described the gradual change as an ecocline. This species is very widely distributed in eastern North America, from Texas to the high Arctic, but the range of genetic variability shifts gradually from population to population along a temperature gradient. Evidence for the genotypic basis of variation was that the various geographic isolates had been maintained in culture at a uniform temperature of 20 C for several years.

Temperature optima vary among species and among strains, and between heteromorphic life history stages (Sec. 3.4). Changes with age of the thallus have also been noted. For example, the optimum for cultivated *Porphyra yezoensis* drops from 20 C at the time of conchospore germination to 14 to 18 C for thalli 10 to 20 mm high, and still lower for larger thalli (Tseng 1981). Moreover, the temperature optimum of an alga under laboratory conditions, where temperature is kept constant, may appear narrower than it is in nature (even if other conditions are equal). A possible example of this is a strain of freshwater green alga, *Scenedesmus*, which in the laboratory will not grow above 34 C, yet in nature grows without inhibition at peak temperatures of 45 C (Soeder & Stengel 1974). Many flowering plants grow better under differing day and night temperatures, a phenomenon known as thermoperiodicity (Noggle & Fritz 1983). While marine environments do not experience regular diurnal temperature fluctuations, the effect of fluctuating temperatures on seaweed growth and temperature tolerance needs to be studied.

An interesting laboratory study of temperature effects, which also illustrates the importance of heterogeneity in the environment, is Strömgren's (1977) study of short-term effects of temperature changes on growth of Fucales. Using a laser-beam technique that permitted him to measure growth over a matter of only hours, he transferred specimens of five fucoids from preincubation conditions, of 6 to 8 C, to higher temperatures, up to 35 C. The graphs in Figure 3.7a show growth rate of *Ascophyllum nodosum* as a function of temperature at various times after transfer, compared to the controls kept at 6 to 8 C. In the first hour the temperature had a tremendous effect, with growth increases, at 35 C, 20 to 30 times the control. The magnitude of the effect decreased with time, however, and at the highest temperature growth ceased. In the 8 to 24-h interval the effect

Figure 3.7. Growth of apices of *Ascophyllum nodosum* following transfer from 6 to 8 C to various higher temperatures, compared with growth rate of controls left at 6 to 8 C. (Irradiance approximately 65 $\mu E\ m^{-2}\ s^{-1}$.) (a) Percentage of change in growth rate versus temperature at five times after transfer. (b) Change with time of growth rate after transfer to 20 C. (a from Strömgren 1977, with permission of Elsevier Biomedical Press; b redrawn from data in a)

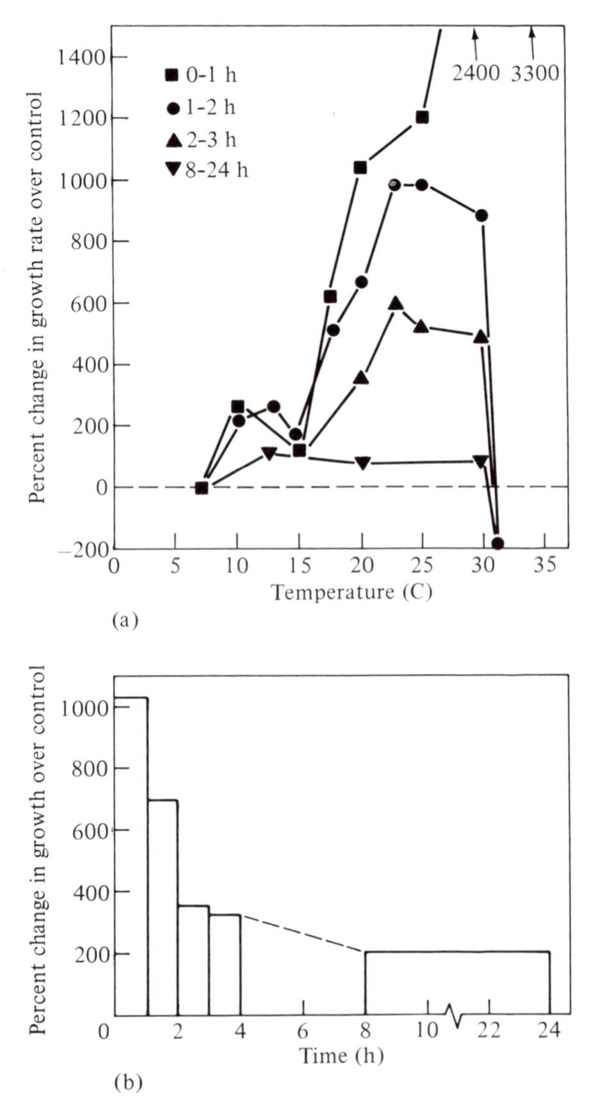

of temperature was quite small, compared with the initial response, but was still about twice the control rate. The change at one temperature, 20 C, with time after transfer from 6 to 8 C, is shown in Figure 3.7b. Irradiance in these experiments was well below saturating levels; nevertheless, temperature had an effect. The very rapid elongation in the first hour or two was attributed by Strömgren to increased use of stored metabolites rather than to increased photosynthetic rate. The relatively stable growth rate versus temperature after 8 to 24 h showed

Figure 3.8. Growth and stipe elongation of juvenile kelp sporophytes after 6, 12, and 18 days in culture at 5 C, 10 C, or 17 C. *Lh, Laminaria hyperborea*; *Ld, L. digitata*; *Ls, L. saccharina*; *Sp, Saccorhiza polyschides*. (From Kain 1969, The biology of *Laminaria hyperborea, J. Mar. Biol. Ass. U.K.*, vol. 49, pp. 455–473, with permission of Cambridge University Press)

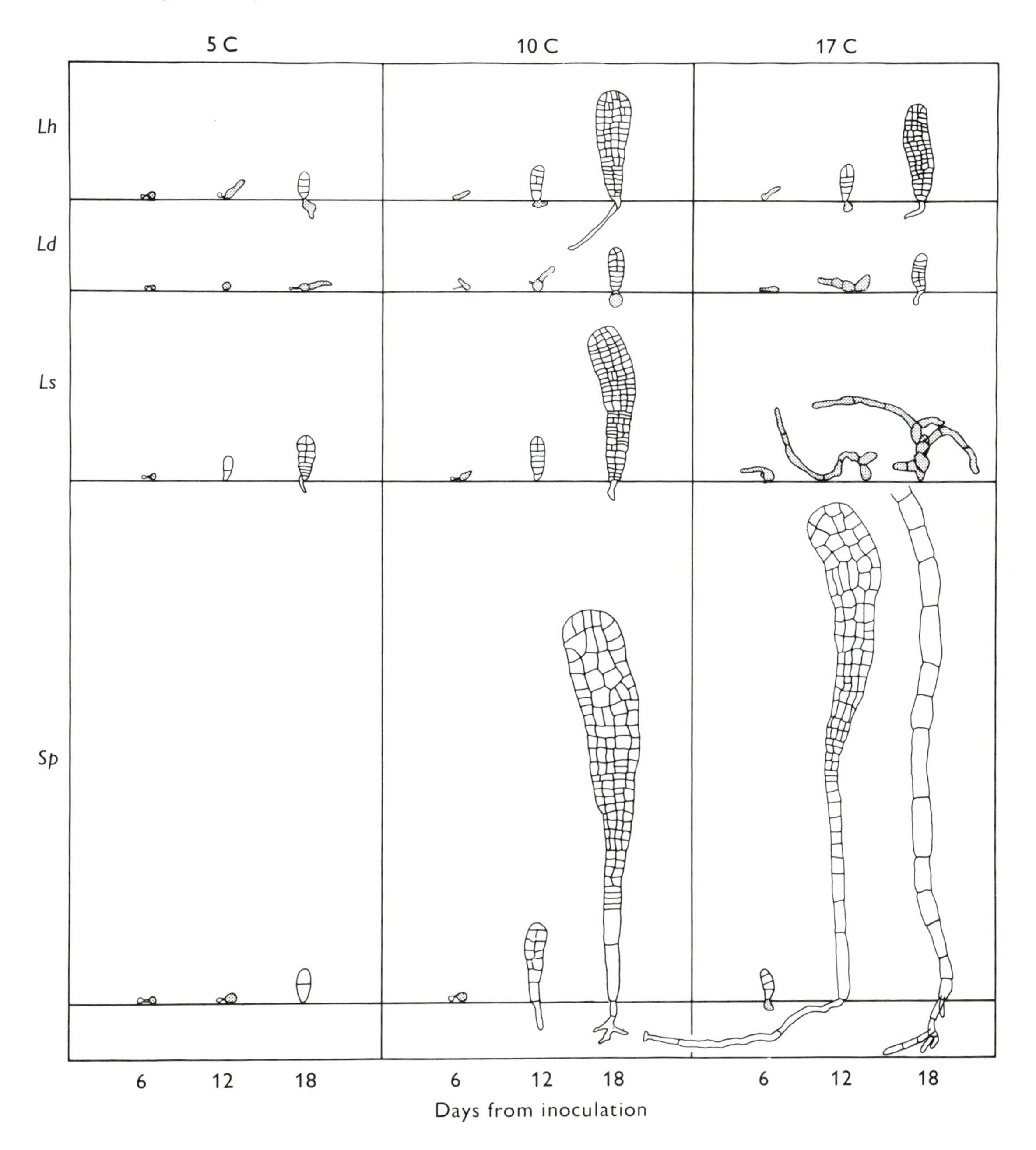

that storage use had largely ceased by this time and growth was due to light-limited photosynthesis. Such short-term changes in growth rate are important in species that may be subjected to sudden, brief temperature increases during daily emersion at ebb tide.

Norton (1977) has suggested that the tremendous growth rate of *Sargassum muticum* at high temperatures is what makes it so invasive in relatively warm waters. A clear example of such an advantage has been recorded by Kain (1969) between *Saccorhiza polyschides* and *Laminaria* spp. Gametophytes and young sporophytes of *S. polyschides* grow faster at 10 C and 17 C than those of *Laminaria hyperborea, L. digitata,* or *L. saccharina* (Fig. 3.8), whereas growth of *Saccorhiza* at 5 C is slower than that of *L. hyperborea* and *L. saccharina*. Not only does *Saccorhiza* grow faster in warmer water, its cells are markedly larger (at all temperatures), and those in the stipe especially elongate greatly, giving the young sporophyte better access to light. At 5 C *L. saccharina* is furthest ahead after 18 days; *L. hyperborea* does equally well at 10 C and 17 C; while *L. digitata* grows most slowly at all temperatures. *Saccorhiza* has a more southerly distribution than *Laminaria* species in Europe, which correlates with its poor growth in cold water. Correlations of the relative growth rates of the *Laminaria* species with their distributions were not so clear. Kain rationalized this as partly due to taxonomic difficulties with the species, making distribution records doubtful, but there may also be diverse ecological strains within species.

Apart from its effect on growth rate, temperature can affect plant morphology, although examples of this are not well documented. Gessner (1970) commented: "The information available on temperature effects on size and external and internal structures of marine plants is very limited and rather vague. General trends . . . have not yet become apparent aside from the fact that many algae tend to attain a larger final size in the colder parts of their . . . distribution." Temperature-induced changes in morphology may occasionally confound taxonomic determinations. While such a problem has not been reported yet, Garbary et al. (1978) found that temperature affected axis width and cell size in *Ceramium rubrum*. These are not taxonomic criteria in this genus but they are in some other genera.

Organisms have ranges of temperature, or any environmental variable, that they can withstand. Within the range there is an optimum peak or plateau, as we have seen. Stress can be defined physiologically as suboptimal or supraoptimal levels of any environmental variable, particularly when the levels are far from optimal. Strain is the response of an organism (or material in general) to stress. In a mechanical analogy, stress is the force applied to stretch a rubber band, strain is the stretching of the rubber. Levitt (1972) defines two kinds of strain: elastic strain, which is completely reversible when the stress is removed; and plastic strain, which produces some permanent change in the organism. Injury can thus be defined as the result of plastic strain. Since the point at which injury begins is difficult to determine, tolerance studies usually seek the stress level at which 50% of a group of organisms (or cells) is killed. Yet, even the point of death is difficult to recognize. Although dead matter is physiologically inactive, cell processes tend to come slowly to a halt unless subjected to extreme conditions, and organisms show remarkable ability to recover from strain or injury if returned to favorable conditions.

Various physiological criteria have been used as measures of the extent of the stress on metabolism. Photosynthesis and respiration have been most commonly used, owing to the ease with which they can be measured, although a low metabolic rate does not necessarily indicate injury (plastic strain) as opposed to elastic strain. Hayden et al. (1972) developed an electrical impedance technique for studying chilling, desiccation, and other injury in plants, and MacDonald et al. (1974) used the method to assess injury to *Ascophyllum nodosum* and *Fucus vesiculosus*. Injury was indicated by a sudden increase in the slope of the impedance-cooling curve and by lower impedances after thawing than before freezing. *Ascophyllum* and *F. vesiculosus* tolerated chilling down to −20 C without any indication of injury, and *Ascophyllum* showed no injury by this criterion with up to 70% desiccation.

Most seaweeds are killed if they become frozen. However, the presence of solutes in water lowers its freezing point, and the high concentration of salts in cytoplasm provides some protection against freezing of intracellular water. Tissue water is not completely frozen until −35 C to −40 C. During progressive cooling, ice crystals form first outside the cells. This tends to draw water out of the protoplasts, causing dehydration, unless cooling is very rapid, when the protoplasts may freeze. Damage is also caused by mechanical disruption of cell components by ice crystal formation (Bidwell 1979). Damage to the tonoplast will be especially injurious, since toxic materials stored in the vacuole could thus be released and poison the cell.

Intertidal algae in cold temperate and polar regions must be resistant to subzero temperatures and are able to withstand a certain amount of freezing. In the Arctic, *Fucus vesiculosus* may survive for several months at −40 C (Gessner 1970). The ability of leafy *Porphyra* thalli to withstand −20 C has proven of great use to Japanese mariculturists, who store nets covered with young plants at this temperature, as insurance against loss of the crop on nets in the sea (Miura 1975; Tseng 1981). Water content of the thalli has a pronounced effect on survival, however. Half-dried thalli are much more tolerant of freezing than fully hydrated thalli (Fig. 3.9), perhaps because there is less mechanical damage by intracellular ice. Freezing resistance of higher plant parts, such as seeds, is also greater if the tissue has low water content. *Porphyra pseudolinearis* males have the most extreme known low temperature tolerance: 50% survived −70 C for 24 h (Terumoto 1964).

While the effects of freezing are easy to explain, neither damaging effects of chilling temperatures nor chilling resistance are well understood in any plants. Damage may result from low-temperature sensitivity of proteins in susceptible species (Gessner 1970; Graham & Patterson 1982), or it may result from a phase change in membrane lipids leading to the inactivation of membrane-bound enzymes. Lyons (1973) suggested that higher proportions of unsaturated fatty acids in the membrane lipids of temperate species of higher plants lead to more stable membranes and less risk of chilling injury (see also Lyons et al. 1979).

Heat and cold hardiness (survival following a 12

Figure 3.9. Percentage cell survival in thalli of *Porphyra tenera* frozen at different temperatures. (A) Freezing in seawater after cooling at a rate of about 50 C min^{-1}. (B) Freezing in seawater after cooling at about 10 C min^{-1}. (C) Freezing half-dried thalli (water content = 30% thallus weight) after cooling at about 10 C min^{-1}. (From Gessner 1970, after Migata 1966, in *Marine Ecology*, Vol. 1, pt. 3. Copyright 1966 John Wiley & Sons, Ltd. reprinted by permission)

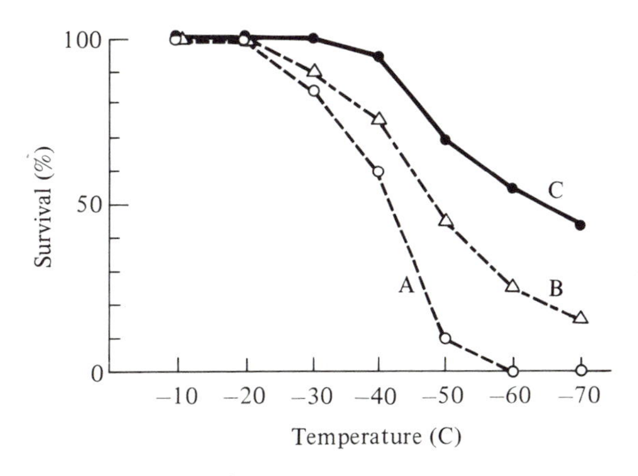

h test in water) of littoral and subtidal algae were investigated extensively by Biebl (1970). Heat and cold hardiness both increased from the littoral fringe to subtidal. Hardiness of littoral algae was related to air temperature and irradiance, whereas hardiness of subtidal algae was related to water temperature. Heat hardiness of subtidal algae increased from the pole toward the equator but – according to Biebl's experiments – even in cold seas was never less than 22 C; the maximum was 32 C. (Yet temperate seaweeds in culture tend to die when growth chambers break down and warm up to room temperature.) Maximum for littoral algae, in water, was 40 C for some tropical species, but for most species was 30 to 32 C. Cold hardiness of subtidal algae did not go below 0 C even in the Aleutians (Biebl did not study in the Arctic), and toward warmer waters decreased rapidly. Minimum tolerable temperature for some algae in Puerto Rico was as high as 16 C.

Heat damage appears to begin by the failure of one or a few thermolabile enzymes. As the temperature increases further, more and more enzymes become denatured. (Because the tertiary structure of enzymes is vital to their function, denaturation need not proceed to the extreme of coagulation for the enzyme to be inactivated.) Heat stability of enzymes seems to be partly related to the temperature at which they were formed. Since there is constant turnover of protein molecules, the enzymic machinery may thus become gradually acclimated to changing temperature. Moreover, recent evidence indicates that diverse organisms respond to thermal shock by initiating synthesis of special heat-shock proteins that seem to protect against cellular damage (Schlesinger et al. 1982).

Temperature-induced damage to thalli has been implicated in the restriction of algae both vertically on the shore and horizontally along coastlines. Schonbeck & Norton (1978, 1980) described temperature damage to the high intertidal fucoids *Pelvetia canaliculata* and *Fucus spiralis* as consisting of reddish spots of decaying tissue developing some 10 days after thermal stress, along with narrowed apical growth and reduced rates of elongation and weight gain. High temperature damage in both species was less severe at lower humidities (i.e., when the plants were drier). Plants recovered from this damage unless it was extreme. Adverse effects of unusually high temperatures have been noted in populations of *Macrocystis pyrifera* in California, where warm water promotes black rot disease (Andrews 1976), and of cultivated *Porphyra* in Japan (Tseng 1981). The effects of thermal pollution on individuals and communities are elaborated in Section 8.2.

3.3.4 Temperature and life histories

Temperature has quantitative effects on reproduction in a number of seaweeds (Dring 1974, Lüning 1980) and qualitative effects on the life history. Many seaweeds, especially in waters with a large annual range of temperature, have very dissimilar sporophytes and gametophytes, or they produce a crustose or filamentous microthallus and an erect leafy or bladelike macrothallus (Bold & Wynne 1978). Frequently, one form is present at one season, the other at the opposite season. For instance, the leafy stage of *Porphyra* species is generally a winter form, whereas the microscopic conchocelis stage is the summer form. Such seasonal shifts in phase or form may be cued by abiotic environmental variables with seasonal extremes: light and temperature are the obvious candidates. It may be difficult in nature to determine which of these two factors is more important or whether both play roles, but laboratory studies have elucidated the relative contributions in a few species. However, heteromorphy may also be a response to predation. Lubchenco & Cubit (1980) and Slocum (1980) suggested that grazing controls the formation of erect phases versus the presence of crusts or boring phases of a variety of intertidal seaweeds. In other instances there may be interacting cues, as Dethier (1981), in a similar study on *Scytosiphon/Petalonia-Ralfsia*, has found. In this case, the crust (*Ralfsia*) was both grazer resistant and grazer dependent. Limpet grazing, which removed potential competitors of *Ralfsia*, was essential to persistence of the crust. The crust reproduced throughout the year but the erect phases appeared only during winter. Both crust and erect phases were most abundant in winter, when grazing was minimum.

In some species, different steps of reproduction have different temperature optima. In the conchocelis stage of *Porphyra tenera* in Japan, the temperature optimum for monosporangium formation is 21 to 27 C, whereas for monospore release it is 18 to 21 C (Kurogi & Hirano 1956; see Dring 1974). Chen et al. (1970) found that conchosporangia of *P. miniata* from Nova Scotia were formed at higher temperatures (13 to 15 C in this case) but conchospores were released only in low temperatures (3–7 C) and short days. Dring (1974) commented that studies of "spore production" at different temperatures are liable to reveal only a compromise between maxima for several processes.

In *Ectocarpus siliculosus*, which has a very flexible life history, Müller (1963) found that plants produced unilocular sporangia at lower temperatures (< 15 C) and plurilocular sporangia at higher temperatures (> 15 C) (Fig. 3.10). (All plants grew from spores from plurilocular sporangia and are assumed to have been all the same ploidy level.) Light intensity and daylength also affected the proportion. Since unilocular sporangia are frequently apomeiotic in this species and in other brown algae, this shift from unilocular to plurilocular sporangia does not necessarily imply a shift from sexual to asexual reproduction. Indeed, unilocular and plurilocular sporangia can occur on the same individual (see Fig. 10.11). Furthermore, from a population viewpoint, the formation of sporophytes and gametophytes, if it does occur, would perhaps be of small significance, since the generations are – morphologically, at least – identical. Another plant with essentially isomorphic generations is *Sphacelaria furcigera*, studied by Colijn & van den Hoek (1971) (summarized by Bold & Wynne 1978). The sporophyte is slightly more robust than the gametophyte. In this case there is a definite shift from sexual to asexual reproduction: formation of propagules by both generations takes place at higher temperatures, while gametangia are formed in cooler water (also dependent on daylength).

Several species in the Dictyosiphonales have been reported to show temperature control of their morphological form (Wynne & Loiseaux 1976). The most thoroughly studied of these is *Desmotrichum undulatum*, a member of the Punctariaceae and, according to Rietema & van den Hoek (1981), synonymous with *Punctaria latifolia*. Rhodes (1970) found that pluriseriate, straplike macrothalli, present in nature during winter, produced plurilocular sporangia in culture at 21 C (i.e., under summer conditions); the zoospores germinated into creeping filamentous microthalli (Fig. 3.11). At 21 C these microthalli reproduced themselves via zoospores from plurilocular sporangia; however, at 6 C they formed wide, erect filaments that subsequently developed into pluriseriate blades with uniseriate tips. No sporangia were seen in cultures at 6 C. Rietema & van den Hoek (1981) elucidated some of the complexities of the development. Zoospores from the straplike thallus formed microthalli under a wide range of temperatures: 4 to 30 C. These microthalli produced erect (macro-) thalli as follows (Fig. 3.11): at high temperature, 20 to 30 C, growth of the

Figure 3.10. Percentages of unilocular and plurilocular sporangia on *Ectocarpus siliculosus* after 26 days culture at various temperatures. (From Müller 1963, courtesy of the Zoological Station of Naples)

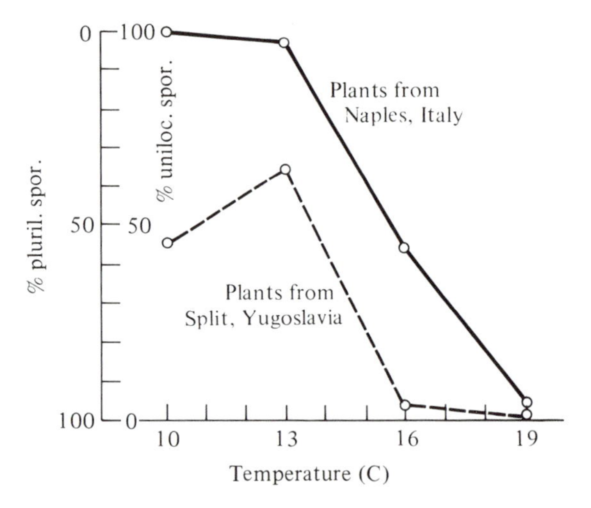

macrothallus was rapid and brief, only a uniseriate filament being produced before zoosporogenesis; at 16 to 20 C the basal part of the erect filaments became pluriseriate; at still lower temperatures, and especially in long days, zoosporogenesis was long delayed and the full straplike macrothallus was able to form.

Among the green algae, *Ulothrix, Urospora,* and *Monostroma* have been found to alter phase in response to temperature (Lüning 1980, Tanner 1981).

Probably many of the seaweeds that have environmentally cued phase shifts respond to both temperature and light (Dring 1974, Lüning 1980, 1981a). This was evident in some of the examples just discussed, and a further example is the initiation of growth of macrothalli from microthalli of *Dumontia contorta* (Rietema 1982). Initiation is strictly daylength-controlled but the initials do not grow out unless the temperature is less than 16 C.

3.4 Geographic distribution of seaweeds

The flora of seaweeds changes from region to region of the world, just as terrestrial floras do, with each species having its own particular limits. In any given region the flora may comprise several distributional groups that extend various distances from polar or equatorial centers (Fig. 3.12) (Humm 1969, Druehl 1981, van den Hoek 1982). The study of the distribution of plants is called phytogeography. Among the physical factors potentially limiting seaweed distributions are salinity and substratum, which have important but localized influences, light, and temperature. Latitudinal changes in irradiance are considered to be unimportant in seaweed distribution. Biological limiting factors are considered in Section 9.4.3. The principal environmental variable with which phytogeographers have been con-

Figure 3.11. Life history of *Desmotrichum undulatum*, showing effects of temperature adduced by Rhodes (1970) (broken arrows) and by Rietema & van den Hoek (1981) (solid arrows). (Redrawn from Rietema & van den Hoek 1981, from *Marine Ecology Progress Series*, with permission of Inter-Research)

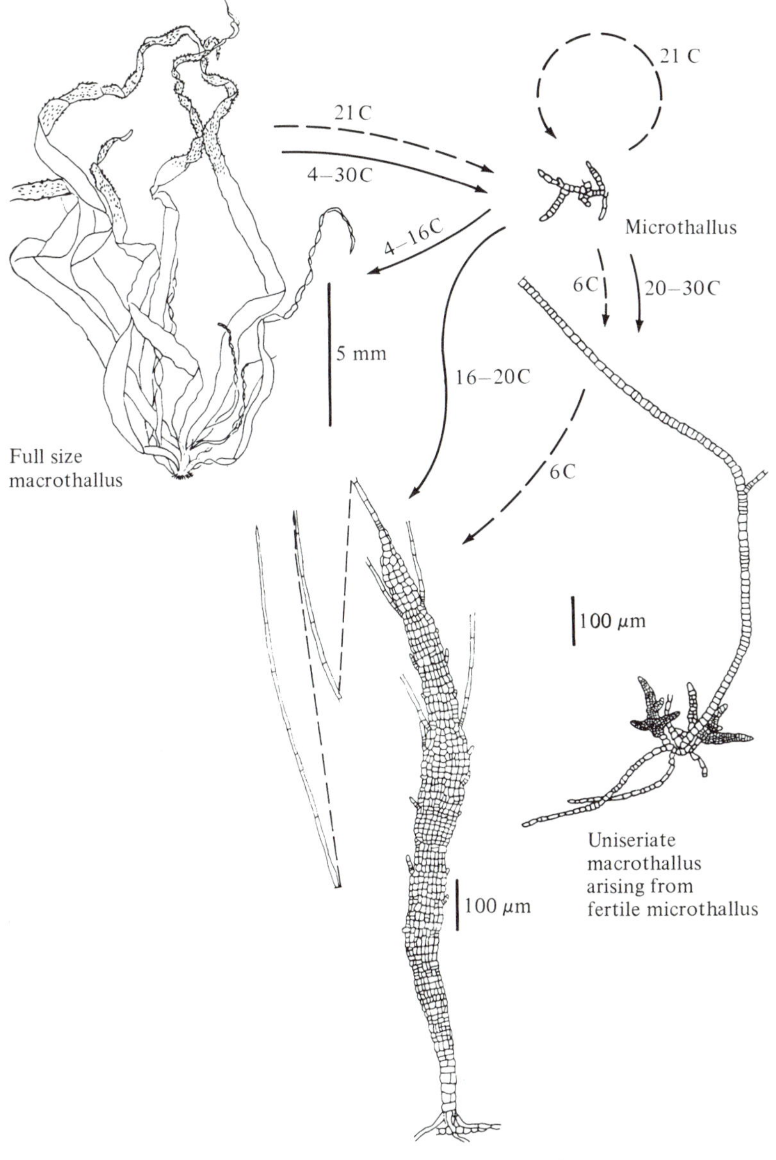

cerned is temperature. However, much argument has centered on whether mean or extreme temperatures are the more important in limiting species distribution.

Work on temperature-defined zones of seaweed distribution stemmed from work on terrestrial floras carried out in the latter part of the nineteenth century. Setchell (1915) divided oceans into nine zones, which were, from north to south: Upper Boreal, Lower Boreal, North Temperate, North Subtropical, Tropical, South Subtropical, South Temperate, Lower Austral, and Upper Austral (similar zones have been defined by Michanek 1979). These zones were defined by 5-C ranges of surface water temperatures of the warmest month (drawn as isotheres on maps), except for the two polar zones, which covered 10-C ranges. Setchell later introduced the use of mean coldest month temperatures (isocrymes) to define provinces within the broader zones. He observed that in open ocean waters not influenced by currents there is an annual temperature range of some 5 C at a given spot, making the annual temperature range over one zone 10 C. He

Figure 3.12. Distributional groups of marine algae along the east coast of North America, showing three break points where there are distinct changes in the flora. Four groups constitute the cold water flora, three groups the warm water flora. (From Humm 1969, with permission of Blackwell Scientific Publications)

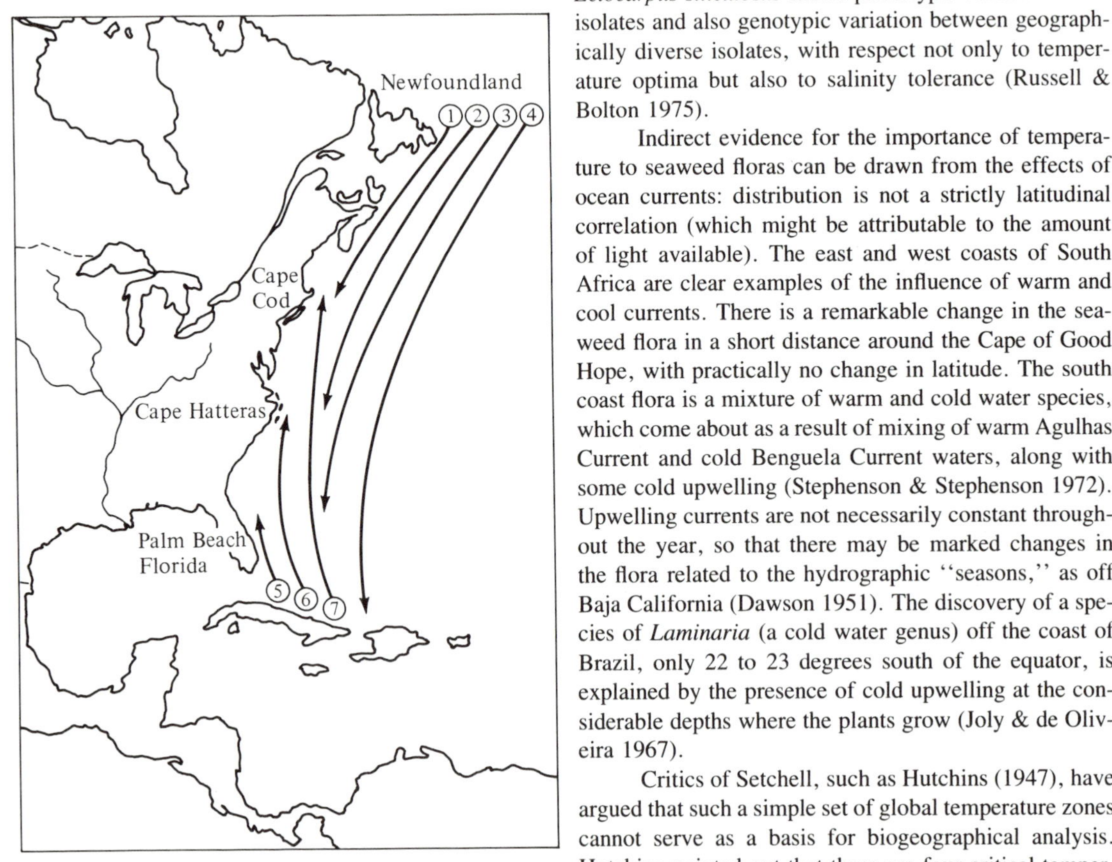

maintained that 10 C was the normal range for activity of some species of seaweed, calling such seaweeds stenothermal. Species with tolerance greater than 10 C he called eurythermal (Setchell 1920). Setchell suggested that some eurythermal species survived in more than one zone because temperatures suitable for reproduction (the most critical stage(s) in the life history) were to be found at some time. For example, *Ascophyllum nodosum*, which spans three zones (Upper Boreal to North Temperate), reproduces in late winter in Long Island Sound but in summer in Greenland. Much more recently, West (1972) concluded that strains of *Rhodochorton purpureum* are genetically selected for temperature dependency of sporulation related to latitude. (At different latitudes, the daylength requirement for algae whose reproduction is controlled by photoperiod can change, as found by Lüning [1980] in *Scytosiphon* [see Fig. 10.10]. Since both temperature and daylength change with latitude, either or both potentially affect the time of year at which reproduction takes place.) Bolton's (1983) study, described in Section 3.3.3, showed that stenothermy and eurythermy cannot be distinguished solely on the basis of geographic distribution. Setchell's terms imply phenotypic variability rather than the existence of discrete genotypic races. The very broadly distributed brown alga *Ectocarpus siliculosus* shows phenotypic variation within isolates and also genotypic variation between geographically diverse isolates, with respect not only to temperature optima but also to salinity tolerance (Russell & Bolton 1975).

Indirect evidence for the importance of temperature to seaweed floras can be drawn from the effects of ocean currents: distribution is not a strictly latitudinal correlation (which might be attributable to the amount of light available). The east and west coasts of South Africa are clear examples of the influence of warm and cool currents. There is a remarkable change in the seaweed flora in a short distance around the Cape of Good Hope, with practically no change in latitude. The south coast flora is a mixture of warm and cold water species, which come about as a result of mixing of warm Agulhas Current and cold Benguela Current waters, along with some cold upwelling (Stephenson & Stephenson 1972). Upwelling currents are not necessarily constant throughout the year, so that there may be marked changes in the flora related to the hydrographic "seasons," as off Baja California (Dawson 1951). The discovery of a species of *Laminaria* (a cold water genus) off the coast of Brazil, only 22 to 23 degrees south of the equator, is explained by the presence of cold upwelling at the considerable depths where the plants grow (Joly & de Oliveira 1967).

Critics of Setchell, such as Hutchins (1947), have argued that such a simple set of global temperature zones cannot serve as a basis for biogeographical analysis. Hutchins pointed out that there are four critical temperatures: (1) the minimum for survival, which might set the winter poleward boundary of the species; (2) minimum for reproduction, controlling the summer poleward boundary; (3) maximum for reproduction, controlling the winter equatorward boundary; and (4) maximum temperature for survival, determining summer equatorward boundary. Van den Hoek (1982) added two more potential boundaries, those limiting growth poleward and equatorward. Thus a species may be restricted by survival temperatures north and south, reproduction temperatures north and south, or by one survival and one reproduction limit. A further complication is the effect of short-term extremes of temperature. Infrequent unusually cold or hot periods may wipe out seaweed populations at the edges of their ranges. Gessner (1970) argued that such brief extremes, not average warmest or coldest month temperatures, are the decisive circumstances controlling algal distributions.

One of the principal drawbacks of the method of relating flora to temperature zones is that species distributions are tabulated with the remotest recorded finding of each species. As Michanek (1979) pointed out; "What such findings actually express is first how intensely an area has been investigated, secondly how rich it is in different biotopes and finally to what extent it contains enclaves with local conditions reminiscent of those prevailing in some other more or less distant region ... Such examples ... represent exceptions, they are the products of local conditions, presumably occurring only in favored years, and observed quite accidentally."

Direct evidence for temperature control of seaweed distributions comes from studies on tolerances of seaweeds to extreme temperatures, and from studies of temperature effects on algal life histories. This evidence has been recently drawn together by van den Hoek (1982), and it indicates strongly that temperature effects on survival, growth, and reproduction are a primary cause of species distributions. By way of example, the geographic distribution of *Desmotrichum undulatum* (see Fig. 3.11) can be interpreted as follows (Rietema & van den Hoek 1981): The northern limit is determined by summer temperatures warm enough to permit development and fruiting of macrothalli; southern limits are set by winter temperatures too high for development of macrothalli (in warmer water the microthallus may exist without the macrothallus but would not be noticed).

Fralick & Mathieson (1975) found that *Polysiphonia* species in an estuary in New Hampshire fell into two categories based on their distributions and temperature optima: (1) cold water species, restricted to the outer coast and the mouth of the estuary, had peak photosynthesis at 21 to 24 C but active photosynthesis also at 5 C; these plants exhibited thermal injury (expressed as a sudden rise in respiration) above 25 C; and (2) warmer water species, which also had a wider tolerance of salinity, penetrated further into the estuary; these plants had peak photosynthesis at 27 to 30 C and little photosynthesis below 10 C. Thermal injury in the second group began at 30 C.

Among the other factors that affect species distributions are substratum and biological interactions. An example of substratum influence is seen on the east coast of the United States, where a long region of largely sandy shores causes distinct discontinuities from the adjacent rocky shore floras (van den Hoek 1975). On the Pacific coast of North America there are north–south gradients in the abundance of vegetatively propagating red algae, in grazing intensity, and in the frequency of space-clearing disturbances (floating log action); this complex of factors might account for the latitudinal variation in intertidal community structure along that coast (Sousa et al. 1981).

Although certain aspects of seaweed distribution have been at least partly explained through the theories of temperature zonation just described and through observations of more local effects of other factors, there remain some enigmatic phytogeographical observations. Among these, the ratio of algal classes to one another has attracted attention but so far little explanation. The ratio of the number of species of Rhodophyceae to Phaeophyceae increases toward the equator, although this trend does not hold in the Antarctic, where there is a "disproportionately" large number of red algae. Cheney (1977) revised this R/P ratio to include the green seaweeds: he suggested that R + C/P is a useful indicator of the nature of a regional flora, with tropical floras having a ratio of 6 or more, cold water floras 3 or less. The biological significance of this and the reason for it are unknown. Since some areas, such as the polar regions, have been little studied, the ratios may change as more information becomes available on smaller, ephemeral, deep water, or rare species.

3.5 Synopsis

Temperature has many profound effects on seaweeds, owing ultimately to its effects on molecular structure and activity. Seawater surface temperatures vary with latitude and ocean currents. The annual range in the open ocean is often only some 5 C, but in shallow waters of estuaries and bays it may be much greater. Moreover, intertidal seaweeds are exposed to atmospheric heating and cooling during low tide.

Biochemical reaction rates approximately double for every 10 C rise in temperature, but enzyme reactions show peak activities at optimum temperatures above which changes in tertiary or quaternary structure inactivate and ultimately denature the enzymes. Photosynthesis, respiration, and growth, being sequences of enzyme reactions, also have optimum temperatures, but the effects of temperature are not uniform across all processes. These optima vary between and within species. At these more complex levels other environmental variables have larger effects and may overshadow the effects of temperature. Metabolic rates may become acclimated to gradually changing temperatures. Freezing kills many algae, especially if ice crystals form in the cells. However, many intertidal algae can withstand temperatures well below zero, especially if their cells are partially desiccated. On the other hand, tropical algae are killed by low temperatures above 0 C.

In regions of extreme seasonal temperature change some seaweeds have life history events cued by temperature (and also by photoperiod). Through the effects on life history and the temperature-range tolerances of seaweeds, temperature affects the geographic distribution of seaweeds and is probably the principal large-scale regulatory factor; salinity, wave action, and substratum play important but local roles in phytogeography.

4 Salinity

4.1 **Salinity and the properties of seawater**

Salinity can be defined simply as grams of salts per kilogram of solution (parts per thousand; ‰). Yet this simple definition belies the physical, chemical, and biological complexity of this "factor." From the physical point of view, the complexity lies in the relationships that seawater density, light refraction, and electrical conductivity bear to salinity (and also to temperature) (Kalle 1971). The aspects of salinity of biological significance are ion concentrations, density of seawater, and, especially, osmotic pressure.

Chemically, seawater is not simply a sodium chloride solution, although Cl^- makes up 55% by weight of the dissolved salts and Na^+ 30.6%. There are many other elements in seawater; 74 of the 89 naturally occurring elements had been detected up to 1970 (Kalle 1971) and the others are expected to be there in concentrations below present detection limits. The balance of positive and negative ions is not even: there are more cations than strong anions, making the sea slightly alkaline; its pH is 8.0 to 8.3 on the average. The elements present are categorized as major elements if their concentrations are more than about 1 mg L^{-1}, trace elements if present in smaller quantities. With the exception of a few elements, such as nitrogen and phosphorus, which are in great biological demand, the proportions of elements in seawater remain remarkably constant. This constancy enables seawater to be characterized on the basis of a relatively simple measure such as salinity.

The various physical and chemical relationships provide several means of determining salinity. Although the definition appears to give a simple method of measurement, it is not possible in practice to get an accurate weight for the total salts simply by evaporating a quantity of seawater. Ammonium ions escape as ammonia and chloride ions tend to escape as hydrogen chloride, whereas some salts, such as calcium sulfate, are very hygroscopic and difficult to dry completely. The international standard method is based on the chemical constancy of seawater. In theory, any element could be measured but chlorine is the most abundant, so chlorinity is determined (see Strickland & Parsons 1972) and salinity is calculated from the relationship: S‰ = 1.8065 Cl‰ (Sharp & Culberson 1982). This definition assumes, among other things, that all organic matter is oxidized and carbonates are converted to oxides. Less time-consuming methods rely on secondary properties of seawater related to salinity: oceanographers routinely use a salinometer to measure electrical conductance (Parsons 1982; Gieskes 1982). Other means include the refractometer, which measures light refraction, and the hydrometer, which measures density. Although from the chemical point of view the titration method is the most accurate, it requires a considerable volume of water. For measurement of small quantities of water, such as the surface film on a seaweed, the refractometer may be preferable, since it requires only a few drops of water. Still, some water, such as that in intercellular spaces, is inaccessible to measurement by any means.

Methods for measuring salinity do not give precise values for particular elements. Although the chemical composition of seawater is relatively constant, biologists are frequently interested in those elements that, because of their utilization by organisms, fluctuate the most. Each element must be determined separately, and sometimes several methods may be needed to quantify various forms of the element (see Chap. 6).

Although the biological effects of salinity are complex, owing to the various effects they have on the physical properties of water, the most important effects are the consequences of movement of water molecules along water potential gradients and flow of ions along electrochemical gradients. These processes take place simultaneously and both are regulated in part by the semipermeable membranes that surround cells, chloroplasts, mitochondria, and vacuoles. The details of this exchange are presented in Section 4.3.1, following a brief description of the salinity environments in which seaweeds occur.

4.2 Natural ranges of salinity

4.2.1 Oceans

The salinity of open ocean surface water is generally 34 to 37‰, lower off areas with great rainfall (e.g., the northwest coast of North America) and higher in subtropical areas of high evaporation and low rainfall (Groen 1980). Certain seas have markedly higher or lower salinities: the Mediterranean, because there is high evaporation and little freshwater influx, has salinities of 38.4 to 39.0‰; the Baltic, essentially a gigantic estuary, is notably brackish, particularly at the surface, ranging from 10‰ near its mouth to 3‰ or less at the northern extreme.

In coastal waters, especially those that are partially cut off from the ocean or subject to heavy runoff, salinity is characteristically 28 to 30‰ or lower, even far along a coast from a major river mouth. Slight latitudinal trends in salinity are overshadowed in coastal areas by freshwater influx, hence geographic distribution of seaweeds as related to salinity is a local, not a global phenomenon. In areas with marked seasonal differences in rainfall, or with winter snow periods followed by spring melt, salinity, especially of surface water, may change dramatically over the year. Lower salinity water, being less dense, floats on the heavier, saltier water. Because of the slow mixing of layers, even a small freshwater influx may have a pronounced, if very local, effect. It can affect individuals as strongly as a major influx but it will affect fewer individuals.

4.2.2 Estuaries

Regions where freshwaters mix with seawaters are described as estuarine. Such regions include not only river mouths but also embayments of various kinds (e.g., fjords, the Bay of Fundy, the Waddenzee) (Anderson & Green 1980). Indeed, by this definition the seas along the shores of British Columbia and Norway and off giant river mouths such as the Mississippi and Amazon are estuarine. From the viewpoint of seaweed ecology, however, there is nothing to be gained by considering these latter regions to be estuarine. Pearse & Gunter (1957) suggested that the dividing line between marine and brackish waters be at the point where the curves for freezing temperature and temperature of maximum density versus salinity intersect: at 24.7‰, −1.3 C. Kinne (1971), in contrast, defines brackish water as having salinities of 0.5 to 30‰, seawater as 30 to 40‰.

Salinities in river mouths depend on the proportions of river water and seawater: these proportions depend on the state of the tide and the state of the river, and change both daily (Fig. 4.1) and seasonally. For example, the annual range of salinity that *Cladophora* aff. *albida* tolerates in Peel Inlet in Western Australia is 2 to 50‰ (Gordon et al. 1980). In addition to the gradient along the estuary, salinity lower on the shore is often more variable than higher on the shore, because the higher shore is often covered only by the surface, fresher water, whereas the lower shore is alternately under river water and saltwater (Anderson & Green 1980).

Figure 4.1. Salinity changes during 24-h periods in spring at two sites in Burrard Inlet, British Columbia. Values are averages of three days in separate years. (From Hsiao 1972, with permission of Simon Fraser University)

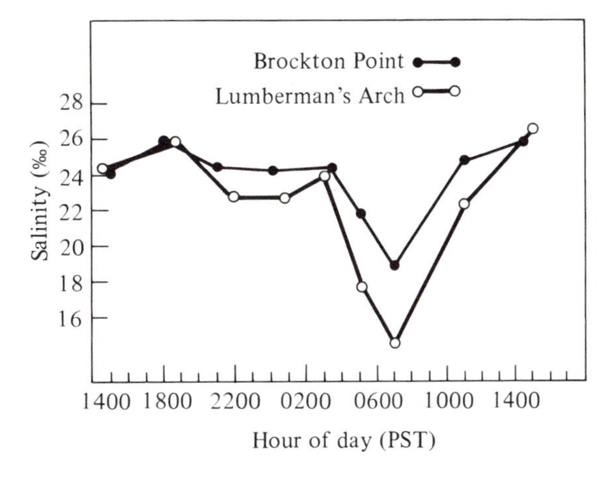

4.2.3 Intertidal ("open coast") habitats

During exposure to the atmosphere, seaweeds on open rock surfaces and in tide pools may be subjected to frequent salinity fluctuations. Evaporation causes an increase in salinity of water in the surface film of seaweeds and, more slowly, in tidal pools. In contrast, rain, snow, and freshwater streams cause a reduction in salinity. Since freshwater floats on saltwater and since a long period of evaporation is generally necessary to effect any significant change in salinity of pool water, salinity changes little in mid and low intertidal pools except during extremely hot days or torrential downpours. High intertidal pools, which are inundated infrequently or receive seawater only from wave splash, may become very brackish in rainy weather or strongly hypersaline in hot, dry weather. Sharp increases in tide pool salinities come about as a result of freezing, since salts are initially excluded from the freezing layer, and concentrated in the remaining liquid (see Edelstein & McLachlan 1975). Seaweeds are found in pools with salinities from about 0.3 to 2.2 times normal (i.e., 35‰) seawater, that is, about 10 to 77‰ (Gessner & Schramm 1971). Algae out of water are also subjected to salinity changes due to evaporation of the surface film and intercellular water, but these effects are difficult to separate from the effects of dehydration.

4.3 Biochemical and physiological effects of salinity

4.3.1 Water potential and ion movement

To understand the physiological effects of salinity, one must understand the basic principles of water potential. (More detailed accounts can be found in plant physiology textbooks, e.g., Bidwell 1979, Noggle & Fritz 1983.) Movement of molecules requires free energy. Molecules may have free energy as a result of temperature, concentration, pressure, gravity, and other forces. The free energy a substance has is called its chemical potential, and the chemical potential of water is called water potential, denoted by the Greek letter ψ. With reference to a cell surrounded by a solution there are several components to ψ. A minor component (under most circumstances) is the matric potential (ψ_m), a measure of the forces that bind water molecules to colloidal material (including proteins and cell walls). Osmotic potential, ψ_π, is the potential of water to diffuse toward a solution. The osmotic potential of pure water is zero. Anything dissolved in water lowers the osmotic potential. The more particles there are in solution, the more negative the osmotic potential. Water flows down the potential gradient, that is, toward the more negative ψ_π. Effectively, water movement results in a dilution of more concentrated solutions. The decrease in ψ_π is proportional to the number of particles dissolved, regardless of their size. Each dissociated ion of a salt counts as one particle, so that, ideally, a molar solution of sodium choride has twice the osmotic potential of a molar solution of sucrose. (In practice, a small correction factor, the activity coefficient, must be included in the calculation.) The concentration of solutions to be used for osmotic measurements is not given in molarity (moles per liter of solvent at 20 C) but in *molality* (moles per kilogram of solvent), since addition of solute molecules dilutes the solvent molecules. By using molality, we refer always to the same number of solvent molecules.

As water flows into a plant cell, it pushes against the wall and creates a pressure. The tendency of water to move as a result of pressure is called the pressure potential, ψ_p. The pressure potential of water outside the cell is defined as zero at atmospheric pressure but would be positive for plants underwater, owing to hydrostatic pressure. At equilibrium, when net water flow is zero,

$$\psi_\pi(\text{outside}) + \psi_p(\text{outside}) = \psi_\pi(\text{inside}) + \psi_p(\text{inside})$$

(Bidwell 1979). Pressure potential is a property of the water, but as water presses against the cell walls, the cell walls react with an equal and opposite pressure, which is called turgor pressure. Note that turgor pressure is a property of the cells, not of the water. If external pressure potential is negligible, turgor pressure at equilibrium equals the difference between the osmotic potentials inside and outside the cell. Finally, the term osmotic pressure, Π, refers to the concentration (chemical potential) of solutes. The trend in osmotic *pressure* is thus the exact opposite of the trend in osmotic *potential* (ψ_π): a more concentrated solution has a greater osmotic pressure but a lower osmotic potential.

In systems containing charged particles there is not only a chemical potential of the solutes but also an electrical potential. There is a tendency for the numbers of positive and negative charges to come to equilibrium. Gradients in these two potentials are not always in the same direction. The net passive movement of ions across cell membranes depends on the combined electrochemical gradient and is also complicated by the fact that many molecules cannot freely cross the membranes. Moreover, cells actively import and export ions, across both the plasmalemma and the tonoplast. Many cells actively exclude Na^+, whereas Cl^- is not actively pumped or else is imported (Gutnecht & Dainty 1968, MacRobbie 1974). The resulting electrical imbalance is partly satisfied by uptake of nutritionally useful cations, such as Mg^{2+} and certain trace metals. Seawater has a large osmotic pressure because of all the salts dissolved in it, but cells maintain even higher concentrations of particles and the resulting turgor pressure is important for cell growth (see Cosgrove 1981).

What happens when a cell is placed in a solution with which it is not in equilibrium? (A solution of lower solute concentration [higher ψ_π] is called hypotonic, of higher concentration, hypertonic.) The answer to this question is important in considering the effects of salinity changes on cells. If the cell is placed in hypertonic solution (ψ_π[outside] less than ψ_π [inside]), water will flow out of the cell (Fig. 4.2). At first, with reduction of turgor pressure, the cell will become flaccid. Then, as the cytoplasm and vacuole shrink further, the plasmalemma will tear away from the cell wall. The damage to the plasmalemma caused by this process, plasmolysis, is usually irreparable. There are some seaweeds, however, that can survive plasmolysis (Biebl 1962). If the cell is placed in hypotonic solution, water will enter the cell (and ions will leave), causing it to swell and, if the difference in osmotic potentials is great enough, to burst. (Seaweed cells lack contractile vacuoles with which to expell water.) Again, this rupture is fatal. During the first few minutes of submersion in distilled water, seaweed thalli rapidly lose ions from their "free space" (intercellular spaces and cell walls) as the solution in the free space comes to equilibrium with the medium (Gessner & Hammer 1968). Since the osmotic potential of seaweed cells is more negative than that of seawater, sometimes much more negative, salinity must be greatly increased before seawater becomes hypertonic (ψ_π [outside] greater than ψ_π [inside]) (Fig. 4.2). The ability of seaweeds to tolerate high salinity (i.e., their ability to avoid plasmolysis) depends on the difference between internal and external osmotic potentials and on the elasticity of the cell wall. (As long as the wall can collapse, the plasmalemma will not be torn away from it.) In reduced salinity the turgor pressure will increase (Fig.

Figure 4.2. Diagram to show the general trend in relationships between internal and external osmotic potential (ψ_π) in various salinities. (Based on a diagram in Bidwell 1979, *Plant Physiology*, 2nd ed., reused with permission of Macmillan Publishing Co.)

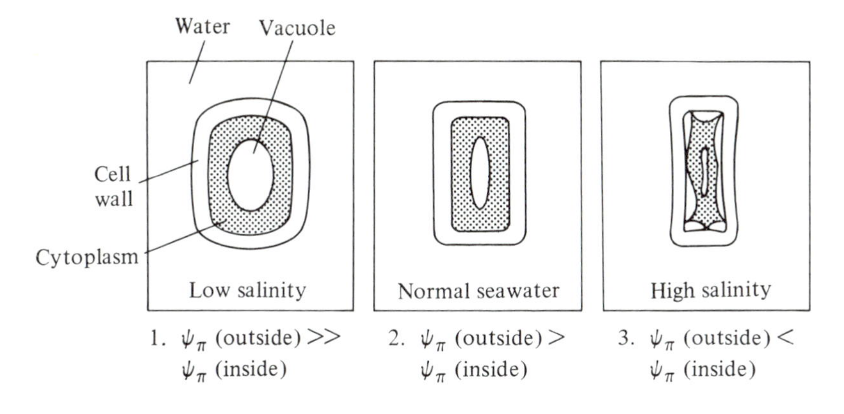

4.2). Cells will expand as long as their walls are elastic but, since normal seawater is already hypotonic with respect to the inside of the cell, strain will increase with any reduction in salinity. The strength of the cell walls and the ability of the cells to make their internal osmotic potential less negative will determine their resistance to low salinity.

4.3.2 Cell volume and osmotic control

Cell volume control is clearly vital to wall-less cells (some phytoplankton; Hellebust 1976), but even some seaweeds have been reported to alter their internal osmotic pressure in response to salinity changes. These observations have led to the hypothesis that such seaweeds are able to regulate their cell volume. By increasing or decreasing ψ_π (inside) in response to ψ_π (outside) the cells can control their volume and turgor pressure. Some studies support this hypothesis, whereas others indicate that, at least in some species, such changes in ion or metabolite concentrations do not affect cell volume.

Algal cells may alter their internal osmotic pressure (Π_i) by pumping ions in or out or by the interconversion of monomeric and polymeric metabolites (Hellebust 1976). Energetically, a change in ion concentration is likely to be cheaper, but if cytoplasmic enzymes and ribosomes cannot tolerate wide fluctuations in ionic composition of the cytoplasm, acclimation may have to be on the basis of innocuous molecules such as photosynthates (Bisson & Kirst 1979). Some plankters use both means. *Platymonas subcordiformis*, which has a nonrigid wall, compensates by altering ionic composition when a change in external osmotic pressure (Π_o) is small, by changing mannitol concentration when the change in Π_o is large (Kirst 1977; see also Bisson & Kirst 1979). The relative importance of inorganic ions and mannitol for osmoregulation is greater in estuarine populations of *Pilayella littoralis* than marine populations (Reed & Barron 1983). In algae whose cells are largely filled by a vacuole, such as the siphonous green *Valonia*, changes in turgor pressure of the vacuole dominate the overall cell turgor changes. Nevertheless, water content of the cytoplasm is also important. Of course, any change in ψ_π of the vacuole will affect water movement between the vacuole and the cytoplasm, just as ψ_π of seawater affects the cytoplasm from the other side. In the seaweed *Griffithsia monilis* turgor pressure is regulated in the vacuole by changes in ionic composition (K^+, Na^+, and Cl^-), whereas pressure in the cytoplasm is regulated in addition by concentration changes in digeneaside (the main photosynthetic product) (Bisson & Kirst 1979). Kremer (1979b) argued that red algae do not use low molecular weight photosynthates for osmoregulation but his conclusions have been challenged by Reed et al. (1980b). As well as inorganic ions, a tertiary sulfonium compound, β-dimethylsulfoniopropionate (DMSP), has been shown to be involved in the salinity responses of *Ulva lactuca* and to be present in osmotically significant amounts in about a quarter of other species tested (Reed 1983). In higher plants and marine invertebrates, analogous quaternary nitrogen compounds are involved in osmoregulation (Dickson et al. 1980). Seventeen seaweeds, including reds, greens, and browns, studied by Kirst & Bisson (1979) all maintained fairly constant turgor pressure over a wide range of external ψ_π by changing internal concentrations of K^+, Na^+ and Cl^-, especially in the vacuole. Activity of an inwardly directed Cl^- pump regulated by Π_o seemed to be the principal means of turgor pressure control in these species. In the studies cited, the seaweeds were subjected to constant changed salinities, whereas in nature salinities fluctuate either abruptly or on a more or less sinusoidal curve. Dickson et al. (1982) studied the responses of *U. lactuca* to both kinds of fluctuation and, although there were some puzzles in their data, they were able to conclude the following. Changes in osmolality (Π_i) closely followed salinity fluctuations, reducing turgor pressure changes (Fig. 4.3a–d). Cellular K^+, Na^+, Cl^-, SO_4^{2-}, and DMSP concentrations also closely followed salinity fluctuations

Figure 4.3. Responses of *Ulva lactuca* to fluctuating salinity regimes. The background stippling traces the changes in salinity; two regimes were used: sinusoidal (a, c, e) and abrupt changes (b, d, f). Changes in tissue osmolality (Π_i) (a, b), apparent turgor pressure (c, d), and K^+ concentration on a tissue water basis (e, f) are given for experiments run in light (○ – ○) or in darkness (● – ●). (From Dickson et al. 1982, with permission of Springer–Verlag)

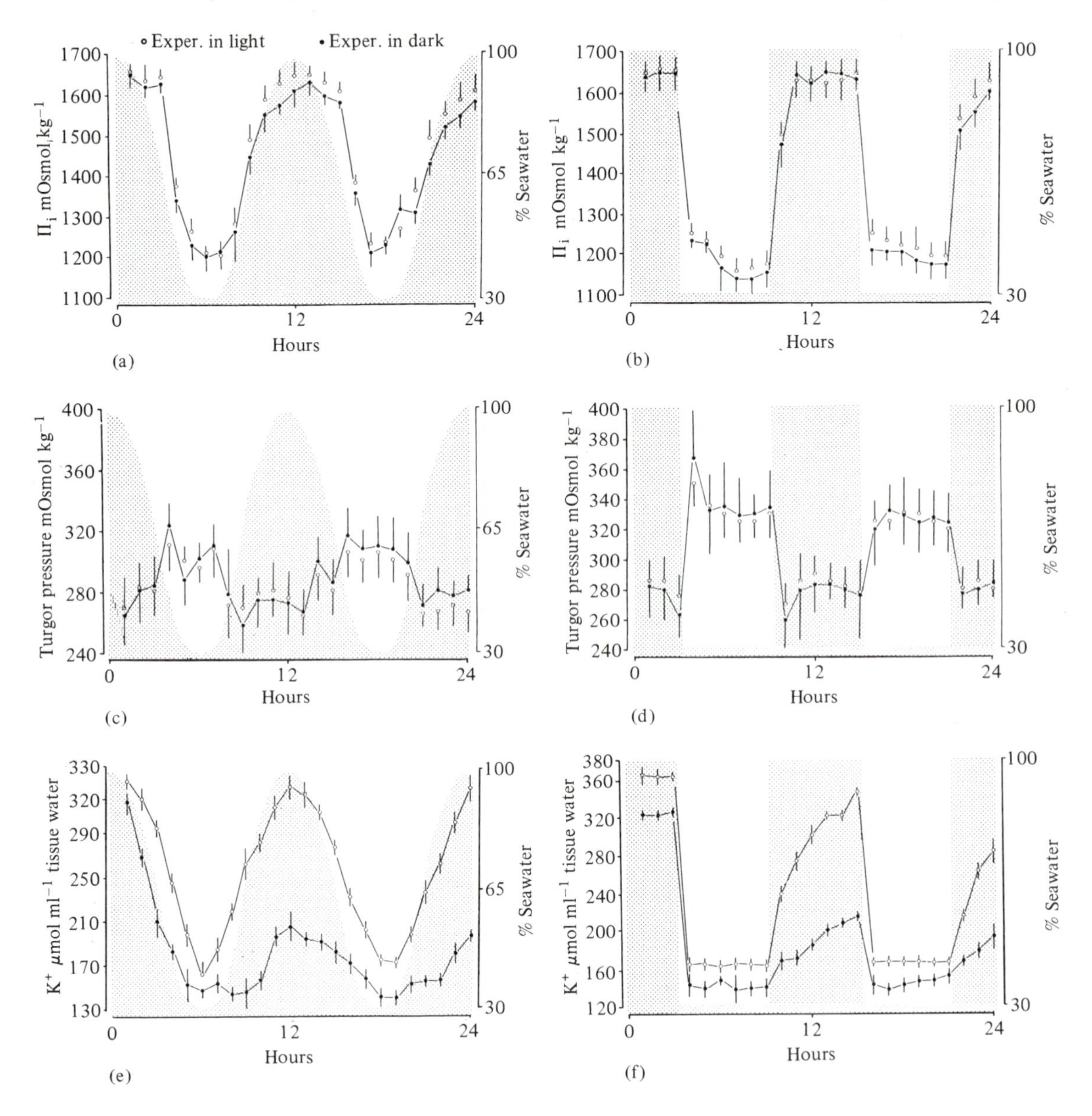

except that K^+ reentry into cells was slow in darkness (Fig. 4.3e, f). Mg^{2+} was not lost from cells in decreasing salinity. *Porphyra* under fluctuating salinities has been studied by Reed et al. (1980c).

The biochemical basis of changes in metabolite concentration has been partly worked out by Kauss (1973) and Kauss et al. (1978) in the wall-less freshwater flagellate *Poterioochromonas malhamensis*, which responds to osmotic pressure changes by adjusting its content of isofloridoside relative to its polymeric storage polysaccharide, chrysolaminaran (Fig. 4.4). The enzyme isofloridoside phosphate synthase exists as an inactive proenzyme as long as the cell is in a stable osmotic condition, but when osmotic pressure fluctuates, another enzyme cleaves off part of the proenzyme, enabling it to bond galactose (in the form of UDP-galactose, from polymeric glucan) to glycerol phosphate (also from glucan), giving isofloridoside phosphate. When external osmotic pressure rises, the synthase drives the reaction toward isofloridoside; when Π_o falls, the reaction favors formation of the polymer. The change in external osmotic pressure is apparently sensed not by the enzyme(s)

Figure 4.4. Sequence of events in osmoregulation by the wall-less flagellate *Poterioochromonas malhamensis*. (Redrawn from diagrams in Kauss 1978 and Kauss & Thomson 1982, with permission of Elsevier Biomedical Press and Pergamon Press Ltd.)

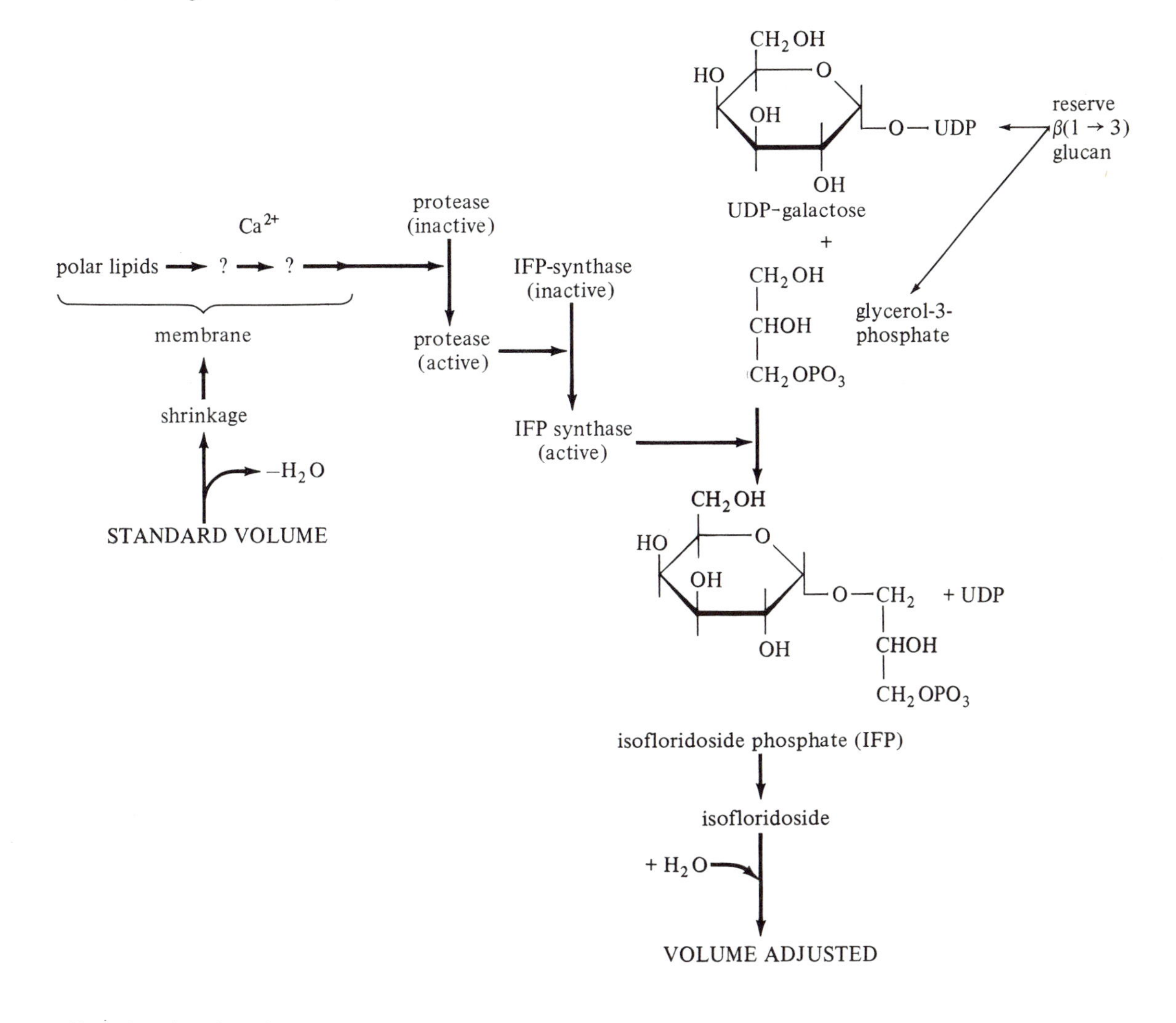

affected; rather there is some other sensor, possibly a membrane component, that responds to a pressure stimulus by producing some "controller" substance(s), possibly involving Ca^{2+} and the calcium-binding protein calmodulin, and these in turn regulate the enzymes (Fig. 4.4) (Kauss & Thomson 1982).

Porphyra purpurea, in contrast to many seaweeds, has a nonrigid cell wall composed chiefly of mannan and xylan rather than of cellulose. The cell wall polymers are arranged as granules rather than as ordered microfibrils. This species does not regulate its cell volume or turgor pressure, so that salinity fluctuations cause water to flow in or out of the cells. There is only a slight swelling and shrinking of the cell walls in response to salinity. In concentrated seawater there is a straight line relationship between the reciprocal of the pressure and the cell volume (Fig. 4.5) (Reed et al. 1980a). In diluted seawater the relationship becomes a curve because of the constraint of the cell wall. In 2‰ seawater the cells are swollen and pressed together, becoming polygonal rather than rounded; they also develop prominent vacuoles. In 105‰ seawater the cells are shrunken and dark (owing to concentration of the pigments). Even though *P. purpurea* does not regulate turgor or volume, it and other species of *Porphyra* do show changes in ionic composition (especially K^+ and Cl^-) and in metabolite concentrations in response to salinity changes (Reed et al. 1980b, Wiencke & Läuchli 1981). How these responses contribute to osmotic acclimation is not known.

4.3.3 Other effects of salinity on cells

Although osmoregulation primarily involves adjustment of internal osmotic pressure (Π_i) and hence water potential (ψ_π), salinity and osmotic strength of seawater also have other effects. These other effects may be directly or indirectly related to adjustment of Π_i (e.g.,

Figure 4.5. Effect of salinity on cell volume of *Porphyra purpurea*. (Salinities expressed as the reciprocal of the osmotic pressure: high salinity to the left.) (From Reed et al. 1980a, with permission of Oxford University Press)

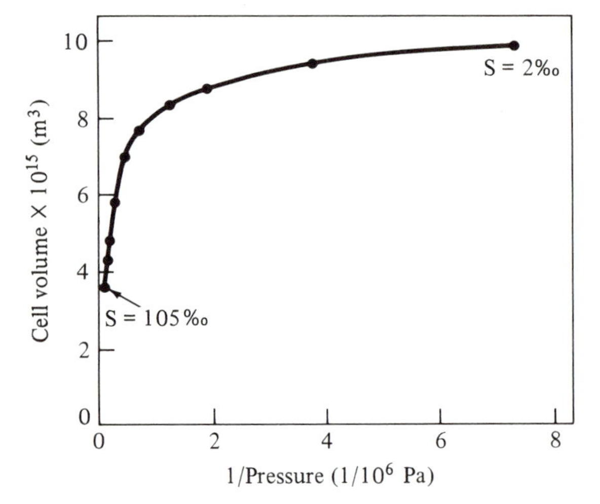

Figure 4.6. Apparent photosynthesis of estuarine *Polysiphonia lanosa* and *P. elongata* as a function of salinity at 5 C and 15 C, showing optima near full seawater salinity. (From Fralick & Mathieson 1975, with permission of Springer-Verlag)

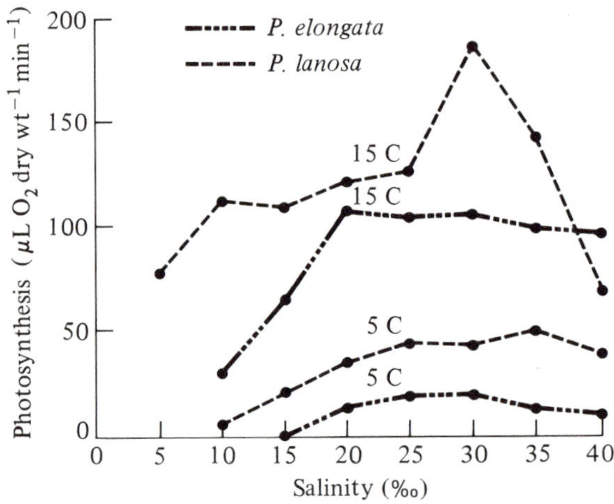

consequences of internal ion concentrations), or related to ionic composition of the seawater. A few salinity effects on cells will be described in this section.

Photosynthesis, respiration, and growth all tend to have optimum salinities (Fig. 4.6), just as they have optimum temperatures. Numerous examples of salinity optima are given by Gessner & Schramm (1971). In addition, a few studies have been made of the interactions among salinity, temperature, and light (e.g., Lehnberg 1978). In Lehnberg's study, of *Delesseria sanguinea* from the Baltic, the data on photosynthesis and respiration were presented as two- and three-dimensional maps (Fig. 4.7, Fig. 9.25), showing regions where one factor or another was limiting metabolic rate. In Figure 4.7 different responses of juvenile and adult thalli are evident: among other things, photosynthesis of adult thalli at low salinities is restricted to a very narrow temperature zone. Although these plants came from seawater of 15‰, their peak photosynthesis was at full seawater salinity.

Lowered salinities often stunt growth of seaweeds and have variable effects on branching (Norton et al. 1981). At the cellular level, Reed et al. (1980a) have noted that cell division of *Porphyra purpurea* is inhibited in concentrated seawater. Lowered salinity also promotes changes in the chemical composition of seaweeds, such as fucoids, which have been studied in the field (Munda 1967) and in the laboratory (Munda & Kremer 1977). Mannitol, ash, and chloride contents, and dry weight as a percentage of fresh weight, declined, whereas protein increased (Table 4.1). The decline in mannitol was attributed to decreased photosynthesis in reduced salinities.

Several authors have attempted to determine why diluted seawater causes a decline in photosynthesis. There is a sharp drop in photosynthetic rate of several marine plants, including *Ulva lactuca*, transferred from seawater to tap water, and a corresponding sharp return to normal when transferred back to seawater. This has been explained as an effect of carbon supply (CO_2 and HCO_3^-) (Hammer 1968, Gessner & Schramm 1971). Recently, Dawes & McIntosh (1981) undertook to explain why photosynthesis of the red alga *Bostrychia binderi* is temporarily greater in water of certain Florida estuaries than in either full seawater or seawater diluted with distilled water (Fig. 4.8). They found that the estuaries are all fed by springwater, the significant components of which are Ca^{2+} and HCO_3^-. Although *Bostrychia* dies if left too long in very low salinity water, the improved photosynthesis in springwater-diluted seawater enables the plants to survive short periods (a few days) of very low salinities better than in estuaries not fed by springwater. Calcium makes the plasmalemma less permeable to other ions, thus reducing the loss of ions that takes place (in addition to water influx) when cells are placed into dilute seawater (Gessner & Schramm 1971). Eppley & Cyrus (1960) found that lack of Ca^{2+} in freshwater resulted in rapid loss of K^+ from *Porphyra perforata*. Yarish et al. (1980) found that Ca^{2+} and K^+ were limiting factors for photosynthesis of estuarine red algae. While Ca^{2+} is unlikely to be absent from brackish water, it may be too low in rainwater, which affects emersed intertidal algae.

Gessner (1971) recorded the photosynthesis rates in seawater of two marine algae after pretreatment in distilled water or 1 M mannitol solution for various times and found that in *Halymenia floresia* mannitol effectively protected the photosynthetic apparatus from low salinity damage (Fig. 4.9a), whereas in *Dictyopteris membranacea* both osmotic strength and ionic composition changes affected photosynthesis (Fig. 4.9b). Guillard & Myk-

Figure 4.7. Temperature-salinity maps of photosynthesis rate of *Delesseria sanguinea*. (a) young thalli; (b) adult plants. The curves show equal photosynthesis rates (in the same way that contours plot equal heights on a map); the rates plotted, in mg O_2 g dry wt^{-1} h^{-1}, are given as numbers on the curves. The maps are divided by dashed lines into regions showing where temperature (T), salinity (S), or both (T/S) are limiting. (From Lehnberg 1978, with permission of Walter de Gruyter & Co.)

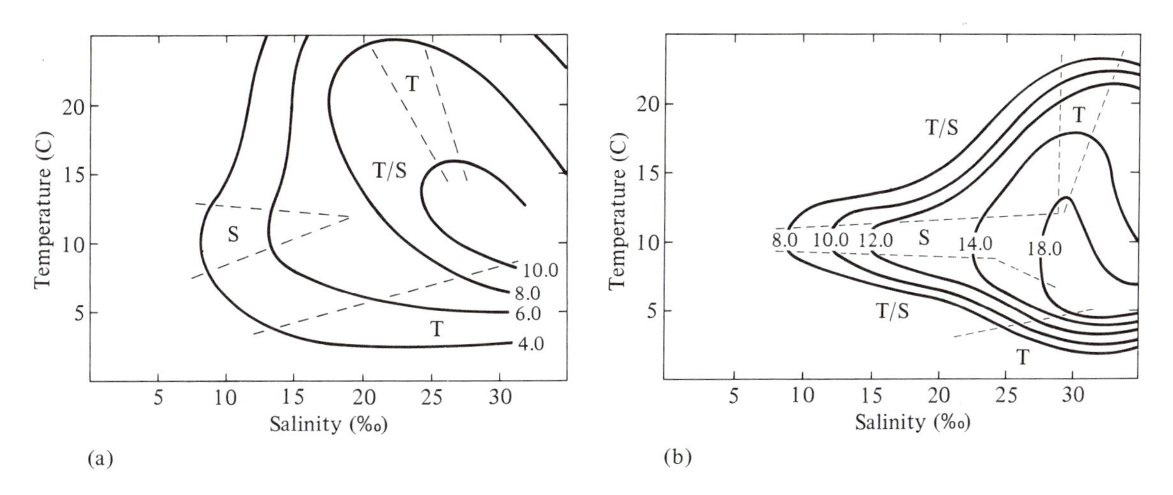

Table 4.1. *Changes in chemical composition of* Fucus vesiculosus *(summer material) after 19 days' incubation in seawater of different salinities (control = 31.03‰)*

Salinity (‰)	Dry weight in % of fresh weight	Percentage dry weight of Ash	Mannitol	Protein
31.03	17.45	22.60	14.57	15.67
20.94	15.70	22.50	6.61	15.31
15.53	13.21	18.20	5.81	15.54
10.67	12.50	16.20	5.38	16.31
5.15	10.80	15.90	3.18	17.74

Source: Munda & Kremer (1977), with permission of Springer-Verlag.

lestad (1970) separated the effects of osmotic pressure and ionic strength on growth rate of a marine diatom by comparing growth in diluted seawater with growth in solutions where the osmotic strength was restored by adding NaCl or sucrose (Fig. 4.10). The results showed that the reduction in growth rate was a consequence of reduced osmotic strength of the seawater. Guillard & Myklestad did not offer an explanation of how this effect might operate. The effect of hypotonic medium on isolated chloroplasts of *Caulerpa* is disruption and the loss of stromal proteins, such as glucose-6-phosphate dehydrogenase and glutamate dehydrogenase, which are not attached to membranes. Ribulose-1,5-bisphosphate carboxylase, which is held in the pyrenoid, was not released by this treatment (Wright & Grant 1978).

4.3.4 The organism level: tolerance and acclimation

If frequencies of occurrence of various intensities of temperature and irradiance are plotted for the world, unimodal curves are obtained, whereas for salinities a bimodal curve is obtained with one peak for freshwater habitats and one for marine; brackish water habitats are relatively less common (Gessner & Schramm 1971). When organisms become acclimated to a new range of conditions, they generally lose the ability to perform as well under the previous conditions. This phenomenon has, in the course of evolution, resulted in two separate groups of organisms: freshwater and marine. Very few species or even genera are able to cross the so-called salinity barrier. One might suppose that brackish habitats, being mixtures of freshwater and seawater, might be populated by a mixture of freshwater and marine algae, but this is not the case. Brackish waters, even down to 10‰ or less, are populated by marine algae. Invasion of salt marshes by vascular plants takes place from the land, whereas the algae have invaded from the sea.

Intertidal seaweeds are generally able to tolerate seawater from 10 to 100‰; subtidal algae are less tol-

Figure 4.8. Photosynthetic rates of *Bostrychia binderi* at 28 C after 3 days in seawater diluted with various amounts of distilled water or spring water. (From Dawes & McIntosh 1981, with permission of Springer-Verlag)

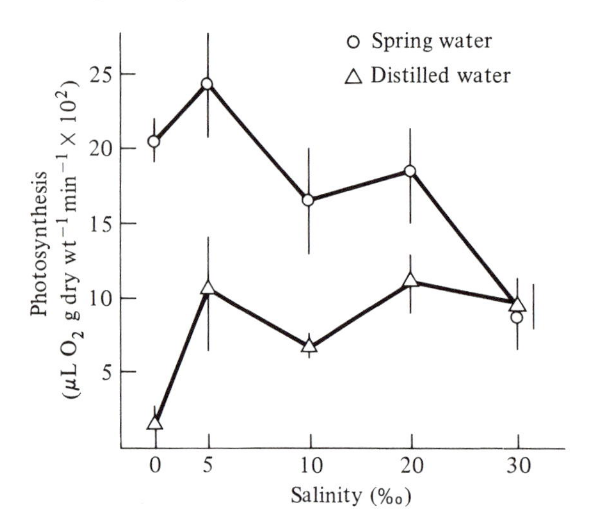

erant, especially to increased salinities, withstanding generally 18 to 52‰ (Biebl 1962, Gessner & Schramm 1971).

Acclimation of a species to higher or lower than normal salinities may result from the development of genetically diverse populations (ecotypes) or by phenotypic change (without genetic change) either in individuals or through successive mitotic spore generations; or both processes may take place within one species (Yarish et al. 1979). Genetic variation can be inferred if (a) populations of plants from different salinity regimes show different tolerances to salinity ranges, and if (b) progeny of laboratory cultured plants show responses to salinity similar to that of the original isolates. Reed & Russell (1979), using regeneration of pieces of *Enteromorpha intestinalis*, tested the tolerance of populations from maritime pools 100 m from the top of the intertidal zone (salinity virtually zero), from high intertidal pools, and from open intertidal rock. The range of experimental salinities was 0 to 136‰. Intertidal zone plants showed the smallest salinity tolerance, with a peak at 34‰ (Fig. 4.11a). High intertidal pool plants had broad salinity tolerance, with a plateau from 0 to 51‰ or more (Fig. 4.11b). Maritime pool plants had broad tolerance (Fig. 4.11c). Tolerance to duration of exposure to various salinities was also broader in high intertidal pool populations than in plants from lower down the shore. The critical test to show that these variations were genetic and not phenotypic was to culture swarmers from each population and then compare their respective salinity responses. The progeny showed responses similar to the parents, demonstrating that the variation was genetically maintained. (Contrast Geesink & den Hartog's experiments, below.) Bolton (1979) reached a similar conclusion for *Pilayella littoralis*. Various populations from the head of an estuary to the sea showed differing tolerances, especially to very low salinities, even after culture for two months or more in full seawater. Yarish et al. (1979) concluded that their two estuarine red algae showed ecotypic variation and that some of the ecotypes also had some capacity for phenotypic variation.

Figure 4.9. Photosynthesis, as a percentage of the maximum control O_2 release, of two algae in seawater, after pretreatment for various times in distilled water or 1 M mannitol. (a) The red alga *Halymenia floresia*; (b) the brown alga *Dictyopteris membranacea*. (Modified from Gessner 1971, with permission of Walter de Gruyter & Co.)

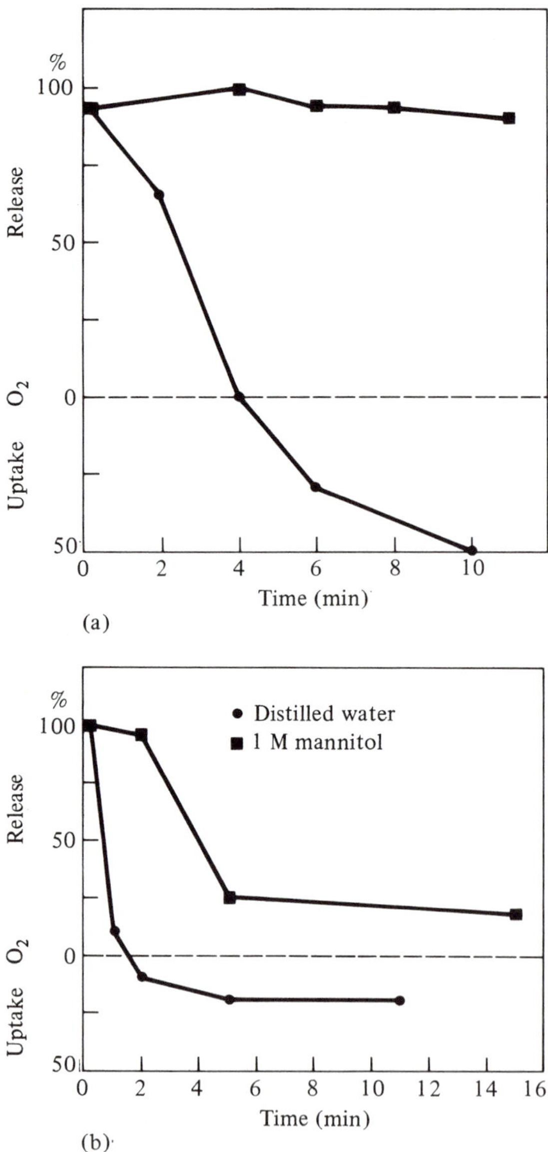

Figure 4.10. Growth rates of the diatom *Thalassiosira oceanica* at various salinities, when seawater was simply diluted with distilled water or when, following dilution, the osmotic strength of the medium was restored to 30 g L^{-1} with NaCl or sucrose. (From Guillard & Myklestad 1970, with permission of *Helgoländer Meeresuntersuchungen*)

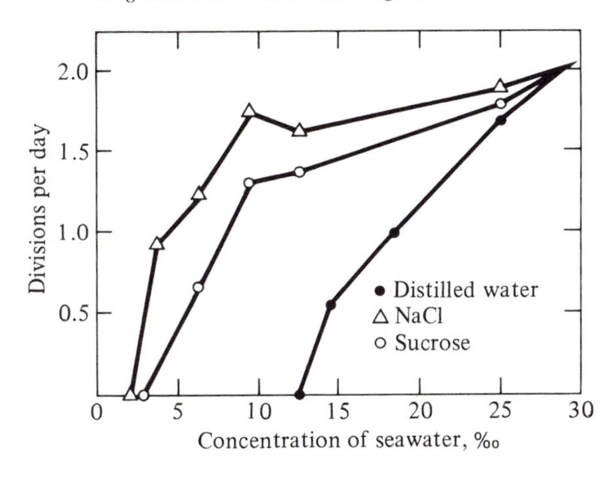

Figure 4.11. Rhizoid production by three populations of *Enteromorpha intestinalis* in response to salinity. (a) Eulittoral zone population; (b) littoral fringe plants; (c) plants from maritime pools influenced only by sea spray. (From Reed & Russell 1979; reprinted with permission from *Estuarine & Coastal Marine Science*, vol. 8, pp. 251–258, © Academic Press Inc. [London] Ltd.)

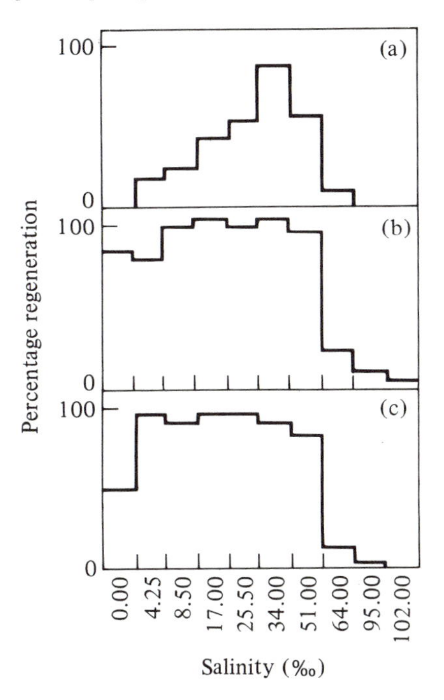

Phenotypic acclimation has been shown in *Bangia* by Geesink and den Hartog (den Hartog 1972b, Geesink 1973). *B. fuscopurpurea* is a marine form, *B. atropurpurea* a freshwater form. The optimum salinity for growth of sporelings was that in which the spores had formed. By transferring plants shortly before spore formation into water of 10 to 20% lower salinity, den Hartog and Geesink were able to acclimate *B. fuscopurpurea*, over a number of generations, to complete fresh water. *B. atropurpurea*, could be acclimated to full seawater in a similar way. (The authors concluded that these are the same species.) Although *Bangia* grows in both freshwater and seawater, it is rare in estuaries, occupying sites only at the mouth and the head. Den Hartog surmised that *Bangia* cannot tolerate the widely fluctuating salinities occurring in mid-estuary. Geesink and den Hartog's experiments might suggest that *Bangia* could have invaded freshwater habitats by progressively moving up estuaries, but the field observations suggest that it could not. Den Hartog (1972b) suggested that probably freshwater populations came about when patches of marine habitats were cut off from the sea (e.g., by dikes or by land uplift) and gradually became freshwater habitats. The Zuyder Zee in the Netherlands was closed from the sea in 1932, becoming the Ysselmeer. After a decade or more, *Bangia* was flourishing, whereas other marine plants, lacking the ability to adapt to freshwater, had died out.

4.4 **Salinity and geographic distribution**

There is a strong gradient in salinity along an estuary but there is often a temperature gradient also. Biotic factors or available substratum may also restrict distributions of seaweeds. Many descriptions of algal distribution in estuaries, bay systems, and fjords have been carried out: for example, by Munda (1978) in Iceland, by Widdowson (1965) in British Columbia, by Silva (1979) in California, and by Mathieson et al. (1981) in New England. These studies tend to show that red and brown algae do not penetrate as far as green seaweeds into estuaries (see also Gessner & Schramm 1971, Druehl 1981). While in many such studies distribution patterns have been attributed to salinity gradients, physiological evidence is scarce. The distribution of *Phymatolithon calcareum* into the mouth of the Baltic Sea is restricted by low salinity because low Ca^{2+} concentration restricts calcification (King & Schramm 1982). On the other hand, the distribution of *Polysiphonia* species along the Great Bay Estuary System in New Hampshire is regulated by temperature limits of the species, not by their salinity tolerances (Fralick & Mathieson 1975).

Temperature–salinity diagrams are a useful way of describing the water climate of a place and may give useful insight into causes of distribution patterns. The distribution of *Macrocystis integrifolia* in British Columbia is controlled by salinity and temperature (plus wave action), and in high salinity the plants can withstand higher temperatures (Druehl 1978). Although the mean

Figure 4.12. Temperature–salinity diagrams for two locations in British Columbia, one of which (Nootka) supports growth of *Macrocystis integrifolia*, the other (Entrance I.) does not. The crosses show mean and standard deviations for annual temperatures and salinities. The outlines trace monthly temperature–salinity coordinates, with solid lines around winter conditions, broken lines around summer conditions. The contrast in the two water climates is belied by their annual means. (From Druehl 1981, with permission of Blackwell Scientific Publications)

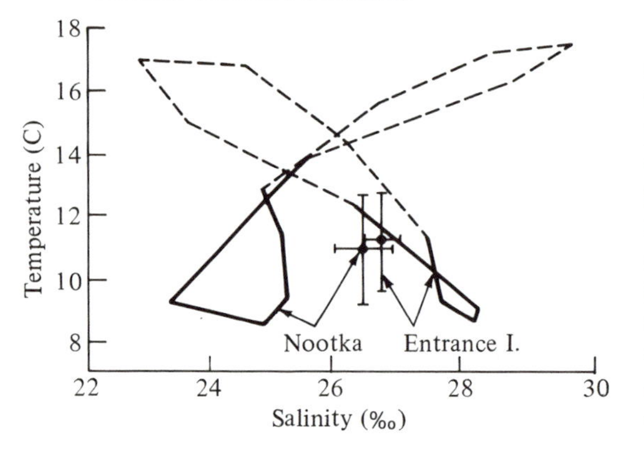

annual temperatures and salinities for Nootka and Entrance Island are nearly identical, only Nootka supports growth of *Macrocystis*. The T–S diagram (Fig. 4.12) shows that at Nootka, when temperature is high, salinity is also high, whereas at Entrance Island, when temperature is high, salinity is low. This is thought to explain the distribution of *Macrocystis*, although physiological tests have not been carried out.

4.5 **Synopsis**

Natural salinities in marine and brackish waters range from about 10 to 70‰, although 25 to 35‰ are most common. The components of salinity that are important to seaweed physiology are the total concentration of dissolved salts and the corresponding water potential, plus the availability of some specific ions, notably calcium and bicarbonate. The internal cell pressure of many seaweeds with rigid cell walls is regulated through active movement of ions across membranes or by interconversion of monomeric and polymeric compounds. Seaweed distribution in estuarine areas can be regulated by salinity, but temperature may have an overriding control or may have a synergistic interaction with salinity.

5 Water motion

The waters of the seas are in constant motion. There are the great ocean currents, tidal currents, waves, and other forces down to the small-scale circulation patterns caused by density changes. The force of water motion is a direct environmental factor, but water motion also affects other factors, including nutrient availability, light penetration, temperature, and salinity changes. Since an understanding of the effects of water motion on seaweeds requires an understanding of the physical nature of waves and currents, the chapter begins with a brief review of this aspect of oceanography, based largely on Sverdrup et al.'s (1942) classic textbook and on Tricker's (1980) essay. Good introductory accounts have been published by Vogel (1981) and Koehl (1982); the student is also referred to current oceanography texts (e.g., Thurman 1978, Gross 1982) for more details. Following this introduction, the effects of water motion are examined at the cellular level on gas and mineral exchange and on spore settlement, then at the larger scales of whole plants and of populations.

5.1 Waves and currents

5.1.1 Waves

Ocean waves are generated by wind blowing over the sea surface. Certain characteristics of waves, such as height, period, and wavelength, depend on the speed and duration of the wind and the distance of open water over which the wind has blown. The last of these is referred to as "fetch." As waves approach the shore, their speed decreases. When water depth becomes small compared to wavelength, which is the case for swells near shore, the velocity is proportional to the water depth; the energy lost from velocity goes into increasing wave height. The movement of water as a wave passes is circular in the open ocean; the radius of the circle decreases with depth until there is no effect of the wave at all. Close to the seabed, vertical motion of water is restricted, so that where the seabed is less than some 30 m from the water surface, ocean swells "feel" the bottom. The circular motion of water is progressively flattened until at the seabed it is simply horizontal, back and forth motion – surge (Fig. 5.1a). The movement of water over the seabed causes surface shear stress, τ_0 (Fig. 5.1b), a force tending to rip objects, including seaweeds, from the seabed. This wave force is constantly varying (Fig. 5.1b).

Waves tend to approach a beach nearly at right angles, whatever their direction was at sea. If their approach is oblique, one end of the wave drags first on shallower seabed and is slowed down; as a result the wave swings around parallel to the shore. The energy of waves tends to be concentrated on promontories and diminished in bays; hence headlands are more exposed to wave action.

The extent of wave action on a shoreline depends not only on the size of the waves but also on the slope of the shore. The most extreme conditions are likely to occur on a moderate, unbroken slope. There, as on a sandy beach, the waves (especially swells) break and surge over a quarter or a third of the shore, even in calm weather. Forces as high as 100 tons m^{-2} have been recorded in storms (Riedl 1971). On open coasts, vertical faces are usually subject to less force than are moderate slopes, except in stormy weather, for the waves may simply slop up and down without breaking (Lewis 1964).

The measurement of wave action presents practical difficulties, owing to the destructive ability of waves. Phycologists have been principally interested in measuring surge, in connection with subtidal seaweeds (Charters et al. 1969), wave force (Jones & Demetropoulos 1968), and frequency of wetting (Druehl & Green 1970) in connection with intertidal plants. The instruments made by Charters et al. (1969) and Jones & Demetropoulos (1968) were purely mechanical: water pulled on a drogue and a spring balance measured the force applied, just as if a weight had been applied. Druehl &

Figure 5.1. Waves and surge. (a) Change with depth in path of water motion from elliptical to horizontal. The values in this example were obtained at a site off Santa Catalina Island, California. A 5-m horizontal excursion of water was measured at a depth of 6 m when a 2-m wave moved past at a velocity of 8 m s^{-1}, a wavelength of 120 m, and a wave period of 15 s. Horizontal and vertical velocities in the upper elliptical path are indicated. (b) Calculated surface shear stress (τ_0), in dynes per square centimeter, at the center of a smooth flat rock, 2 m long, in 6 m depth, due to the surge produced by a wave of 1-m height and 15-s period. + and − refer to direction of shear. (a from Neushul 1972, with permission of Japanese Society of Phycology, b from Charters et al. 1973, reprinted with permission from *Limnology & Oceanography*, vol. 18, pp. 884–896, © 1973 American Society of Limnology and Oceanography)

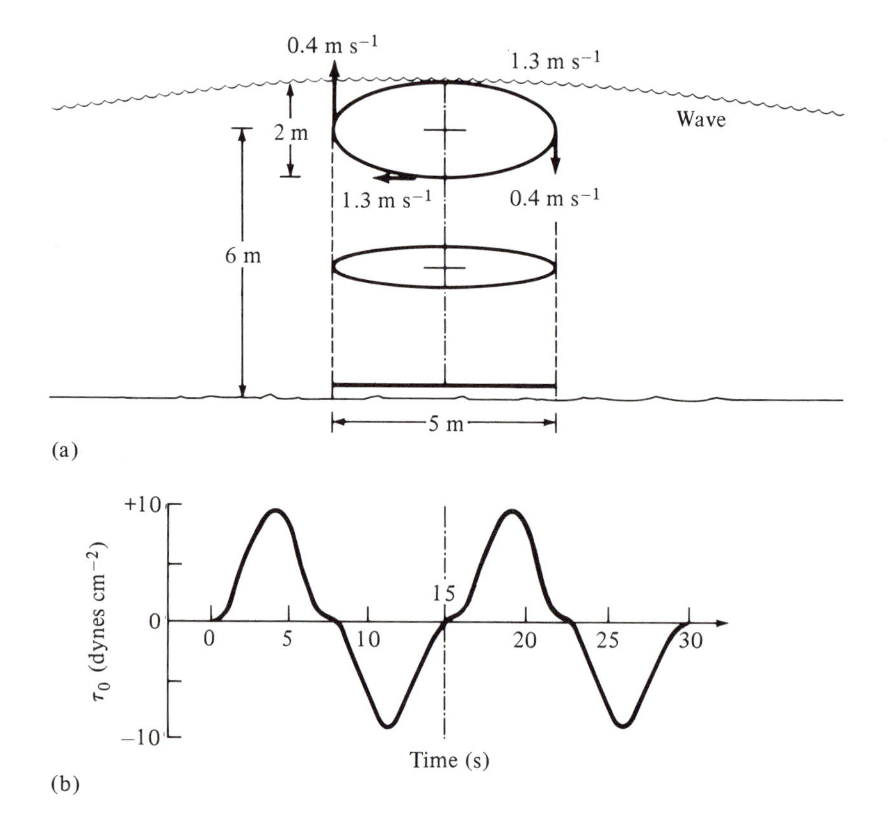

Green (1970) were able to connect their "surf-sensor" to a strip chart recorder, to get a record of the number of times a spot was inundated during low tide. The instrument worked on the basis of the electrical conductivity of seawater: when a wave covered two metal rods it completed an electrical circuit and set off the recorder.

Owing to the expense and the technical difficulties in direct measurement of wave action, less direct ways are frequently used. Relative wave action (or water motion in general) has been measured by erosion of plaster of Paris from "clod cards" (Doty 1971), plaster balls (Mathieson et al. 1977, Gerard & Mann 1979), and tethered concrete blocks (Craik 1980). Another commonly used method is the construction of exposure indices based on the number of compass degrees of water unobstructed at some arbitrary distance from the site, as determined from a map (e.g., Baardseth 1970). This method relies on the relation of wave action to fetch, but does not take account of the other factors involved in generating or modifying wave action. Indirect methods, and particularly exposure indices, are ultimately limited because of the circular argument that underlies them. The calibration or rationalization of the method is dependent on the distribution of organisms, which the method should help to explain. Yet another approach is the biological exposure scale, which has the advantage of integrating all the complex components of wave exposure, including wave impact pressures, wetting effects of wind-driven spray, presence of sedimentation in the intertidal zone, mobility of loose stones, and nutrient availability (Lewis 1964, Dalby 1980). Unfortunately, biological scales also have the shortcoming of circularity. There is really no substitute for direct measurements. The kind of measurements needed depends on the nature of the problem to be solved; at the population level the integrated effects of all components may be important but at the physiological level more specific data are needed.

5.1.2 Currents

Currents range in magnitude from the great ocean currents, chiefly of interest in temperature effects on phytogeography (Chap. 3), to small-scale flow around and over surfaces. This section considers principally tidal currents; the nature of flow around objects is taken up in the next section.

Physical oceanographers have devised instruments for measuring currents in large bodies of water and mathematical principles for calculation of what cannot be measured. Neither the instruments nor the equations are practical in the seaweed zone, where currents change direction frequently owing to surge and topography. Oceanographic instruments are of some use in tidal rapids for obtaining a general impression of water velocity, but more appropriate instruments for measuring in this zone are a small digital flow meter such as used by Mathieson et al. (1977) or a thermister-type of instrument (Forstner & Rützler 1970). Forstner & Rützler's meter operates on the principle that heat loss of a small, heated thermister, relative to the ambient water temperature, yields the current speed. The probe is small enough that it does not significantly interfere with water motion on this scale. Extreme current speeds are attained only in narrow channels; an example is Seymour Narrows, at the mouth of Burrard Inlet, Vancouver, British Columbia, where maximum predicted speed is 2.5 to 3.5 m s^{-1} (5–7 knots) on a spring tide. Currents of 0.5 m s^{-1} (1 knot) are generally considered strong (see Fig. 5.2). The flow of a current along a shore is complicated by topography, as is shown in Figure 5.2; eddies develop in every indentation. Moreover, the velocity of a current decreases rapidly close to the seabed, owing to friction and to eddy viscosity (the latter a property of turbulence). Thus, at the seabed, water motion will be due as much to turbulence as to the general flow of water.

Figure 5.2. Water currents along an irregular shore at four stages during a tidal cycle, showing the complexity of water flow. The sizes of the arrows indicate relative current strengths. The seven transects marked by dashed lines were categorized into five current regimes (a–e): (a) sheltered back eddy areas with no measurable currents; (b) moderate ebb (0.20–0.25 m s^{-1}) and very reduced flood; (c) moderate ebb (0.35 m s^{-1}) and flood (0.20 m s^{-1}); (d) strong ebb (0.40–0.60 m s^{-1}) and very reduced flood; (e) strong ebb (0.80 m s^{-1}) and moderate flood (0.20–0.25 m s^{-1}). The most exposed transects (d and e) exhibited erratic pulsations of water motion during ebb tide. (From Mathieson et al. 1977, with permission of Walter de Gruyter & Co.)

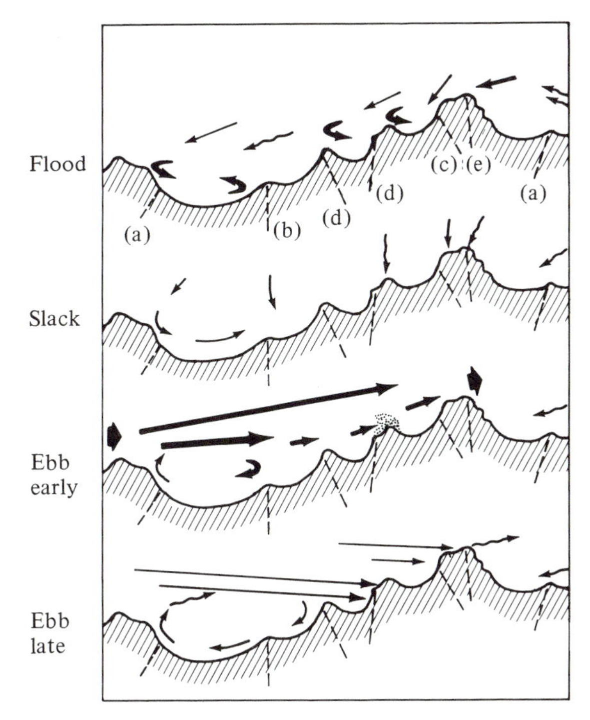

5.2 Water flow over surfaces

5.2.1 Laminar and turbulent flow

To understand the exchange of gases and uptake of minerals by seaweeds, we need to know something about water flow very close to the seaweed surface, since it is from the water layer immediately adjacent to it that a seaweed takes up nutrients. When a current flows over irregular topography of seaweed vegetation, eddies form. Large eddies are unstable and tend to break down into smaller and smaller eddies, as expressed in this rhyme by L.F. Richardson (quoted by Campbell 1977):

Great whirls have little whirls
That feed on their velocity;
And little whirls have lesser whirls,
And so on to viscosity.

The following brief general account is based on Streeter (1980), with reference also to Neushul (1972), Charters et al. (1973), and Wheeler & Neushul (1981).

There is a velocity gradient from the unimpeded current to the surface of the object, where it is zero. Since the surface is effectively slowing down the water, there is a stress set up against the surface, called fluid (or surface) shear stress, τ_0 (see Fig. 5.1b). (Shear refers to a force parallel to a surface, as opposed to compressive forces, which are perpendicular.) You can readily feel shear stress by holding your hand parallel to the ground out the window of a moving car. The layer of water near the surface where the current velocity has been reduced by the presence of the surface is called the velocity boundary layer (i.e., with reference to the boundary of the fluid).

Water flow along a surface may be laminar or turbulent if the surface is smooth, but it is nearly always turbulent if the surface is rough. Laminar flow means flow of water in layers (laminae) with velocities decreasing toward the surface. In turbulent flow there is vertical mixing as well as the horizontal flow. In both cases there is a thin layer of motionless water right against the surface. When a current flows over an obstacle, eddies are set up at the edges, as shown in Figure 5.3. If the obstacle is a smooth, flat plate, the water flow over the middle of the plate will be laminar at first. Shear slows down the water flow next to the surface and these flow layers drag against the current, causing the thickness of the laminar layer to increase (Fig. 5.4a); this happens regardless of whether the surface is smooth or rough. Some actual data are given in Table 5.1. Sooner or later, depending on surface roughness, the laminar layer becomes unstable and passes through a transition stage to turbulent flow. If the surface of the obstacle is rough, turbulence develops more quickly (Fig. 5.4b). The layer of motionless water that remains against the outer surface of an algal cell wall may be as thin as 5 μm for unicells or up to 150 μm for large algae, even in rapidly stirred water, according to estimates reviewed by Smith & Walker (1980).

Figure 5.3. Water flow over surfaces. Diagrams showing large-scale flow over flat blocks and 45° prisms (triangular cross-section). E = eddies, R = regions of reduced water velocity. When the current direction reverses, the eddy pattern is also reversed. (From Foster 1975, with permission of Springer-Verlag)

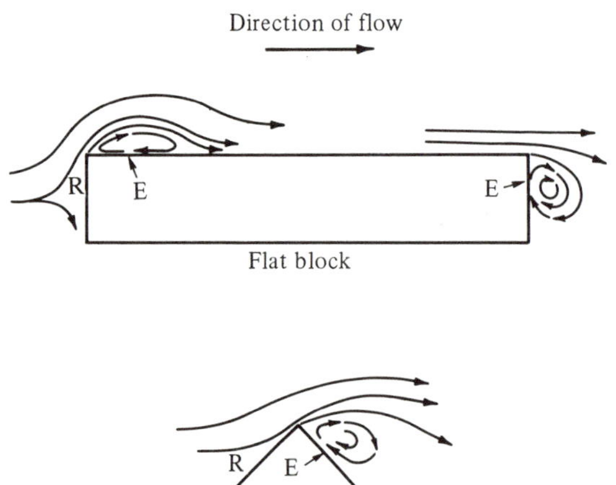

In addition to the velocity gradient next to any surface in a current, there are, next to plant surfaces, gradients of nutrient concentrations. The thickness of this *diffusion* boundary layer depends on the diffusion rate of the particular nutrient in water, and is not the same as the thickness of the velocity boundary layer. The thickness of the two types of layer are related by the Schmidt number, *Sc* (Table 5.2: equations 1 and 2). For nutrients such as nitrate and phosphate in seawater the diffusion boundary layers are about one-eighth the velocity boundary layer (Charters in Neushul 1972).

A model of water flow and nutrient exchange over the surface of *Macrocystis pyrifera* blades has been developed by Wheeler (1980). In his model the thickness of the diffusion boundary layer is a function of the dimensions of the blade (length and area presented to the flow) and the velocity of the flow (Table 5.2: equations 3 and 4). Calculated thicknesses are up to 1 mm if flow is laminar but much thinner if flow is turbulent (Fig. 5.5a). The thickness of the layer varies with the current velocity (Fig. 5.5b; Table 5.2: equation 5) and along the length of the lamina (Fig. 5.5a). Turbulence develops over a *Macrocystis* blade at velocities as low as 10 mm s^{-1}. In nature the current itself is very likely to be turbulent, enhancing mixing at the seaweed surface. Little is known at present about how the edge of a seaweed thallus deflects, fans out, or focuses water flow passing over it, nor about turbulence set up by adjacent upcurrent plants (Norton et al. 1982).

Thalli composed of cylindrical branches, as are many of the smaller seaweeds, present quite a different set of hydrodynamic features: here flow is over a series

Figure 5.4. Laminar and turbulent flow revealed by dye streams. (A) Laminar flow over a flat plate; (B) turbulent flow over a *Macrocystis* blade. Dye was injected upstream (to the left) of the object into a current of 20 mm s^{-1} velocity. (From Wheeler 1980, with permission of Springer-Verlag)

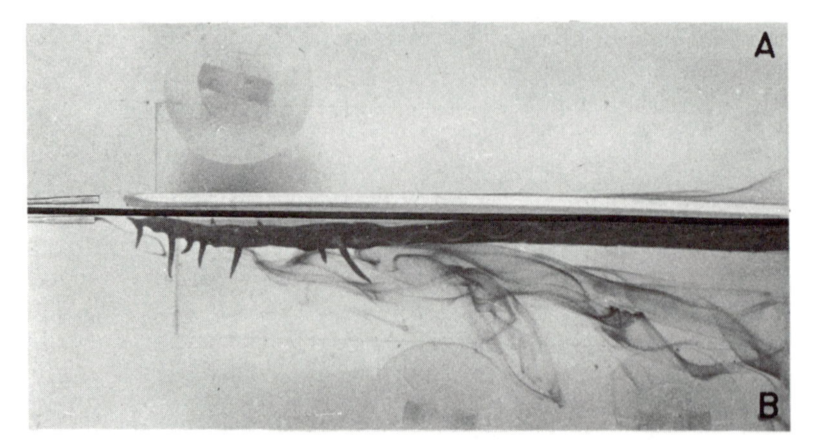

Table 5.1. *Thickness of the boundary layer and magnitude of the shear force on a surface (the platform of a waterbroom) at various distances from the point at which a jet of water strikes it, showing increase in boundary layer thickness and decrease in shear force*

Distance (mm)	5	20	60	100	140	180
Boundary layer thickness (μm)	889	1,397	2,667	4,191	4,445	5,080
Shear force on surface (τ_0, dynes 100 mm^{-2})	15.86	11.94	6.08	4.04	3.36	2.57

Source: Norton & Fetter (1981), The settlement of *Sargassum muticum* propagules in stationary and flowing water. *J. Mar. Biol. Ass. U.K.*, vol. 61, pp. 929–940, with permission of Cambridge University Press.

Table 5.2. *Equations defining various aspects of water flow, gas exchange, and spore sinking rate*

1. $\frac{\text{Thickness of velocity boundary layer}}{\text{Thickness of diffusion boundary layer}} = Sc^{1/3}$
2. $Sc = \mu/\rho_w D$
3. $T_l = 3\, Sc^{-1/3}\, Re_x^{-1/2}\, X$
4. $T_t = 2.49\, Sc^{-1/4}\, Re_x^{-7/8}\, X$
5. $Re_x = Ul/\mu$
6. $J = D\,(C_a - C_s)/\,T$
7. $J = (C_a + K_s + rV_{max}) - [(C_a + K_s + rV_{max})^2 - 4\, rC_aV_{max}]^{1/2}\, /\, 2r$
8. $R = (C_a - C_s)/r$
9. $V_t = 2g\,(\rho_s - \rho_w)\, r^2/\, 9\mu$

Symbols: C_a = ambient concentration; C_s = concentration at the cell surface (hence $C_a - C_s$ is the concentration gradient across the boundary layer); D = diffusion coefficient of a particle; g = acceleration due to gravity; J = flux of molecules across diffusion boundary layer; K_s = half-saturation constant of an enzyme; l = characteristic length of object; r = radius of spore; r = resistance of boundary layer; R = respiration; Re_x = Reynolds number (defined in equation 5); Sc = Schmidt number (defined in equation 2; taken as constant by Wheeler 1980); T, T_l, T_t = thicknesses of diffusion boundary layers in general or for laminar or turbulent flow; V_{max} = maximum velocity of enzyme reaction; V_t = terminal sinking velocity; U = current velocity; X = half the mean length of object; μ = seawater viscosity; ρ_s = spore density; ρ_w = seawater density.
Source: Based on equations in Coon et al. (1972), Dromgoole (1978), Wheeler (1980), and Streeter (1980).

Figure 5.5. Theoretical curves for boundary layer thicknesses (T) and diffusion resistances of a *Macrocystis pyrifera* lamina, based in part on equations 3 and 4 in Table 5.2. (a) Diffusion boundary layers over a kelp blade with water flow of 50 mm s^{-1} if there is laminar flow (T_l) or turbulent flow (T_t); below the graph is a drawing of the lamina; (b) laminar (T_l) and turbulent (T_t) diffusion boundary layer thicknesses at a point 250 mm along the blade, plotted for different water velocities; boundary layer resistance to the diffusion of HCO_3^- is calculated on the right; dashed line is estimated transition from laminar to turbulent boundary layer. (From Wheeler 1980, with permission of Springer-Verlag)

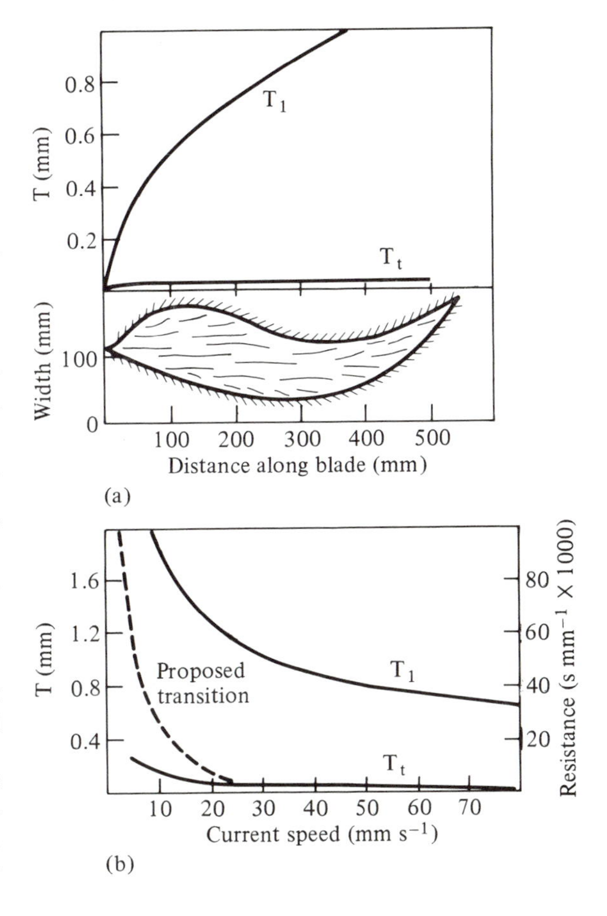

of narrow, closely spaced, rodlike surfaces. The effect of such a thallus on water flow through it has been studied with *Gelidium nudifrons* (Anderson & Charters 1982). At all velocities the thallus as a whole damps large-scale turbulence in the water. At low velocities the water leaving the thallus has smooth flow. Above a critical velocity, 60 to 120 mm s^{-1} (depending on the diameter and spacing of the branches), the branches create microturbulence. Waves and currents may produce no effective turbulence for plants of this form, so that the microturbulence the plants create is probably very important to them for gas and nutrient exchange. If this is so, and since flow at low water velocities is smooth, one would expect *G. nudifrons* to do poorly in areas where water velocity was consistently below 120 mm s^{-1}. *G. nudifrons* was a relatively easy thallus to study since its fronds are stiff and smooth; plants that are lax (e.g., *Ectocarpus*) or clothed with short lateral branches (e.g., *Platythamnion*) would present still other hydrodynamic characteristics.

5.2.2 Gas exchange and nutrient uptake

How does water flow affect gas exchange? Photosynthesis, respiration, and growth of seaweeds is greatly restricted in stagnant water, as the simple experiment shown in Figure 5.6 illustrates. (Several other examples are presented in Schwenke's 1971 review.) If the reduced water flow in the boundary layer forms resistance to transport of molecules, then the thinner this layer, the smaller the resistance, and the easier it is for fresh nutrient molecules to get to the seaweed surface. The concept of resistance comes from electrical theory, and in biological terms it is the concentration gradient of a substance from the fluid stream to the seaweed surface, divided by the flux, or rate at which these molecules cross the boundary layer. Resistance is also related to the thickness of the boundary layer. The flux, J, of molecules across the diffusion boundary layer can be calculated from equation 6 (Table 5.2). The diffusion coefficients of the various dissolved molecules increase with the mass of the molecule and the temperature. At a given temperature those for CO_2, HCO_3^-, and O_2 are not greatly different (Broecker 1974). Using equation 6, Wheeler predicted possible rates of photosynthesis at various current velocities and inorganic carbon concentrations (Fig. 5.7). He recognized that the rates could not keep increasing linearly because the uptake and photosynthetic enzymes would become saturated. This phenomenon could be incorporated into his model with the Michaelis-Menten equation (see Dromgoole's study, below).

The current velocity at which current ceases to restrict the overall reaction rate, U_p, is an important value. Theoretical calculations, plus observations on actual *Macrocystis* blades, indicated that this velocity is 36 to 60 mm s^{-1}. The value depends on V_{max} since for higher values of V_{max} the flux of molecules must be greater. The effect of V_{max} on U_p shows that as water velocity decreases the influence of the boundary layer on V increases. Within this region, turbulence can be important, since the boundary layer is correspondingly much thinner, and the flux that much greater when flow is turbulent (Figs 5.5, 5.7). Two features of the *Macrocystis* blade, its rugosity (wrinkled surface) and its marginal spines, affect water flow; rugosity also increases strength. However, rugosity and number of spines of *M. integrifolia* both increase toward more exposed areas, where water velocities are high and often greater than U_p. Thus in rough water these anatomical features may serve to reduce drag rather than to increase turbulence (Wheeler 1980, Norton et al. 1982), but in areas of low to medium current they may enhance nutrient uptake.

The effects of stirring or not stirring water in laboratory measurements of photosynthesis and respiration have come under scrutiny in two recent studies: Dromgoole (1978) and Littler (1979). Dromgoole predicted the effects of various thicknesses of boundary layers on respiratory oxygen uptake at various concentrations of O_2, taking into account the enzymic nature of uptake and metabolism. His prediction was that when the boundary layer was thin, in vigorously stirred water, the curve of respiration versus the ambient O_2 concentration would be hyperbolic, as are curves of enzyme rate versus substrate concentration (Fig. 5.8a). This means that when the boundary layer is thin the cells are readily saturated

Figure 5.6. Effect of water motion on metabolic rate of *Fucus serratus*, measured as O_2 consumption in a Warburg apparatus. The apparatus was alternately shaken and static. Three replicate experiments are shown. (From Schwenke 1971, in *Marine Ecology*, vol. 1, pt. 2, reprinted by permission of John Wiley & Sons, Ltd.; modified from Nath, *Bot. Mar.*, Vol. 10, p. 198, used with permission of Walter de Gruyter & Co.)

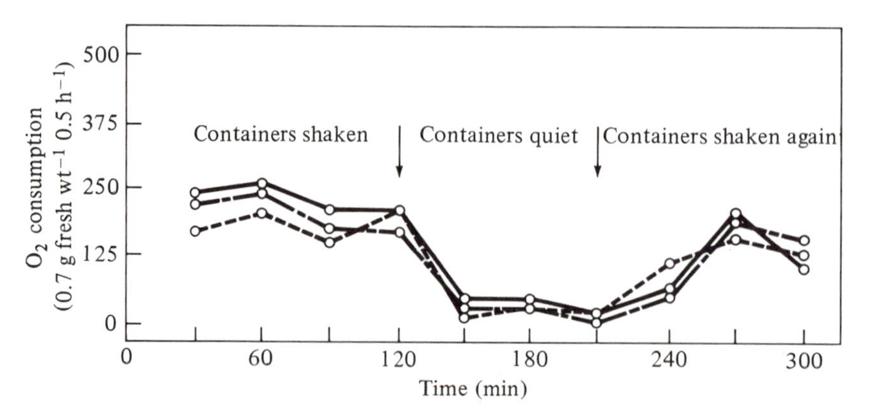

with O_2. When the layer is thick, the shape of the curve flattens, since great O_2 concentrations are needed to overcome boundary layer resistance and achieve saturating concentrations at the cell surface. Experiments with *Carpophyllum maschalocarpum* (Sargassaceae) bore out the predictions (Fig. 5.8b). The concentration of O_2 at the cell surface, C_s, can be worked out if respiration rate, R, resistance of the boundary layer, r, and the ambient O_2 concentration, C_a, are all known (Table 5.2: equation 8). In practice, both C_s and r will often be unknown, but the predictions are useful in understanding observed responses of seaweeds.

Figure 5.7. Rates of transport of O_2 (evolved in photosynthesis), across laminar (a) and turbulent (b) boundary layers, calculated from equation 6 in Table 5.2; current velocities in mm s^{-1}. (From Wheeler 1980, with permission of Springer-Verlag)

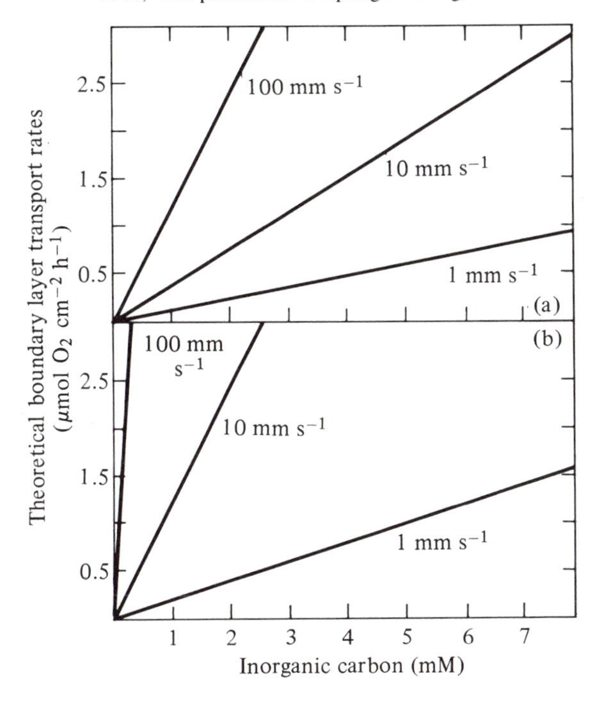

The importance of water motion to nori growth and its interaction with nutrient availability was documented by Matsumoto (1959). Normal growth of the nori depends on a current of 0.2 m s^{-1} in normal seawater. If the water is nutrient-rich, a current of 0.1 m s^{-1} suffices; if the water is nutrient-poor, 0.3 m s^{-1} is necessary. Water velocities of 0.4 to 0.5 m s^{-1} were generally detrimental (Fig. 5.9). Matsumoto also found that increasing frond densities quickly reduced the growth rate, a problem that was partly corrected by moderately increased water flow (to 0.3 m s^{-1}). The combined influence of nutrient (nitrogen) concentration and water flow on internal reserves of nitrate in *Laminaria saccharina* was described by Chapman et al. (1978). Nutrient uptake is explored further in Chapter 6.

5.3 Surge and wave action: water motion and the organism

5.3.1 Spore settling

For a new seaweed generation to begin, the reproductive cells of the parent must get to a surface and stick to it. Some reproductive cells, such as zoids of green and brown algae, have a limited ability to swim. Other cells, such as red algal spores and green or brown algal aplanospores, are nonmotile. (For the sake of simplicity here motile and nonmotile cells are referred to as "spores" without regard to whether the cells are actually spores, eggs, or zygotes.) The discussion in the previous section showed that if a spore can settle onto a surface

Figure 5.8. Calculated and measured effects of ambient oxygen concentration (C_a) on respiration (R). (a) Effects calculated from equation 7 in Table 5.2, for a range of resistances (r) from 0 to 100 min 10 mm^{-1}. The highest value for r represents an increase in the effective boundary layer thickness of 1200 μm. Other values used in the equation were V_{max} = 0.1; K_s = 0.50 μL O_2 cm^{-3}. (b) Measured effects of oxygen concentration on respiration of primary axes of *Carpophyllum maschalocarpum* under stagnant or vigorously stirred conditions. (From Dromgoole 1978, *Aquatic Botany*, vol. 4, © 1978 Elsevier Scientific Publishing Co.)

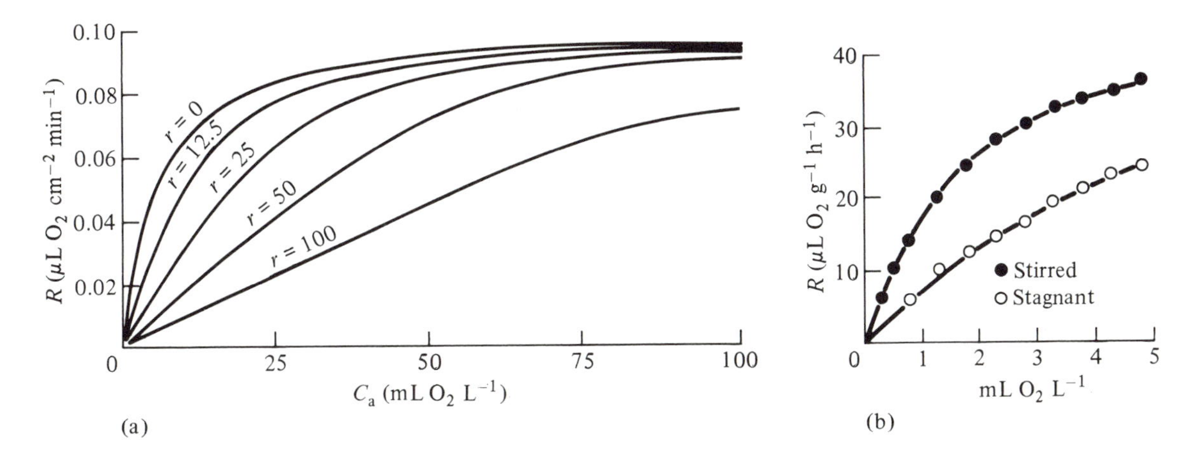

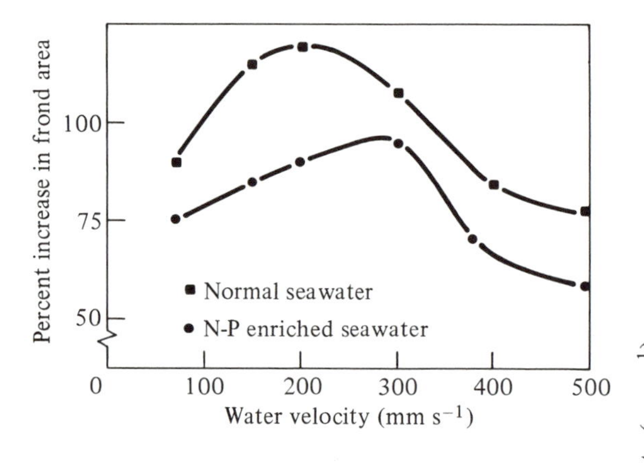

Figure 5.9. Effect of water velocity on growth of *Porphyra tenera* over a 2-week period under two different nutrient conditions: normal seawater and N-P enriched seawater. (Drawn from a table in Matsumoto 1959)

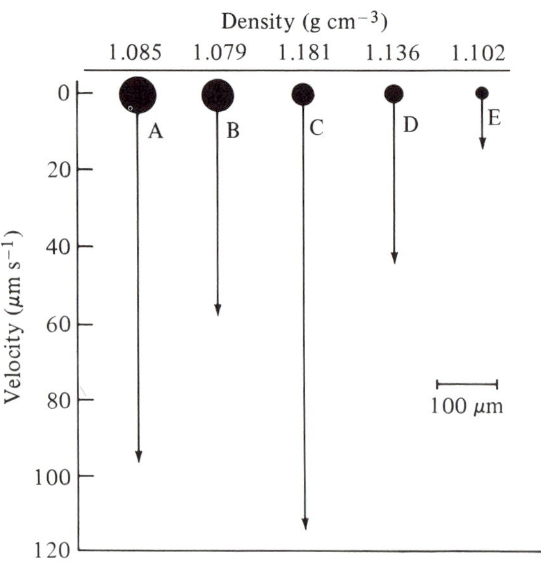

Figure 5.10. Interaction of size, density, and sinking velocity in carpospores of (A) *Cryptopleura violacea*, (B) *Myriogramme spectabilis*, (C) *Sarcodiotheca gaudichaudii* (reported as *Agardhiella tenera*), (D) *Gelidium robustum*, and (E) *Callophyllis flabellulata*. Density of seawater = 1.03 g cm^{-3}. (From Coon et al. 1972, with permission of the University of Tokyo Press)

it will be in a layer of still water or very reduced flow. Estimates of the thickness of the nonmoving layer (only part of the total boundary layer) are 5 to 150 μm for various surfaces. Velocity increases with distance from the surface. If these estimates are correct, the thickness is in the range of spore size. Measurements of red algal spore sizes and sizes quoted from the literature by Coon et al. (1972) range from 15 to 120 μm. According to Neushul (1972), algal spores are small enough to occupy the slow-moving and nonmoving layers of water and thus have time to attach. To get into this "safe zone," however, spores must travel through moving water. More important than horizontal flow are the upward components of turbulence.

Nonmotile spores have various forces acting on them (Coon et al. 1972). The force of gravity tends to pull objects downward at ever-increasing speeds, but drag also increases with speed, so that a maximum (terminal) velocity is reached. Equation 9 (Table 5.2) relates terminal velocity, V_t, to the densities (specific gravities) of the spore and water, the viscosity of water, the radius of the spore (not counting any surrounding mucilage layer), and the acceleration due to gravity. Coon et al. (1972) were able to record V_t by taking time-exposure photomicrographs and to measure spore radius, and thus to calculate spore density. Their results are shown in Figure 5.10. The calculated densities are the averages for the spore plus its invisible mucilage sheath. Boney (1975), who demonstrated the sheath (Fig. 5.11), suggested that while the sheath probably increases the drag on the sinking spore, it may also aid the spore in lodging in crevices once it does reach the seabed. The fastest-sinking spores are not the largest: fastest was *Sarcodiotheca gaudichaudii* (reported as *Agardhiella tenera*), which sank at 116 μm s^{-1}; this is about 10 times faster than phytoplankton cells sink but much less than typical

Figure 5.11. Diagram of a sinking *Polysiphonia lanosa* spore, with the mucilage sheath revealed by India ink. (From Boney 1975, Mucilage sheaths of spores of red algae, *J. Mar. Biol. Ass. U.K.*, vol. 55, pp. 511–518, with permission of Cambridge University Press)

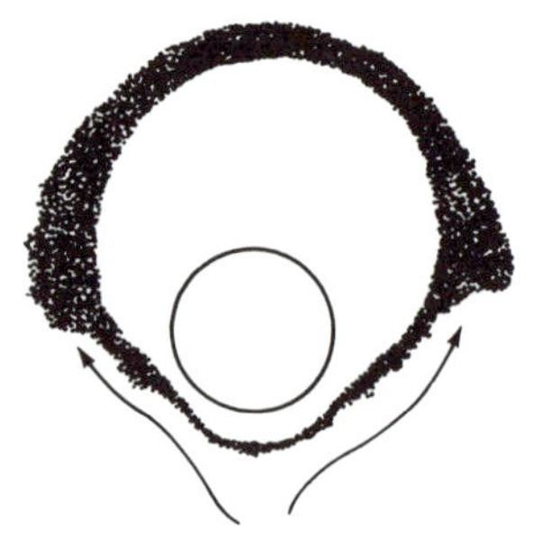

current velocities. Neushul (1972) estimated that it would take a *Cryptopleura* carpospore 10 min to fall through perfectly still water from the cystocarp on the adult plant to the seabed. Spores are likely to remain in the water column for some time, which aids dispersal.

Motile spores are better able to reach the seabed, but swimming speeds are slow compared to water currents. North (1972) recorded *Macrocystis* zoospore velocities of approximately 5 mm s^{-1} and Suto (1950) reported speeds for various zoids of only 125 to 300 μm s^{-1}, but clearly swimming is an advantage over merely sinking, in terms of attachment efficiency. Many zoospores, including those of *Macrocystis*, swim randomly, changing direction frequently. Zoospores with eyespots can orient with respect to light and may be negatively phototactic, thus swimming toward the seabed. Nevertheless, even very slow currents can drastically reduce spore settlement, especially on smooth surfaces such as glass slides.

Natural surfaces, in contrast to glass slides, are typically rough. Evidence from a number of experiments shows that surface roughness, even though it increases turbulence, is an important factor in spore settling. A detailed study has been made of the effect of surface roughness on settling of *Sargassum muticum* propagules in moving water (Norton & Fetter 1981). These scientists used a special apparatus called a waterbroom, in which a jet of water from a fixed nozzle flowed onto and along a plate into which microscope slides could be recessed. Water flow at various distances from the nozzle was measured (Table 5.1). Sand grains sorted to particular sizes were used to create rough surfaces. Norton & Fetter found that settlement of *Sargassum* propagules was best on a surface with a mean depression depth of 800 μm (Table 5.3), no matter what the water speed (range from 0.22 to 0.55 m s^{-1}). The propagules attach not because of sinking but because of turbulent deposition. The reason suggested for the drop in attachment in the largest depression size was that this size would be big enough to be swept clean by water flow, rather than creating depositional eddies. Small algae already growing on the seabed create an algal turf, which provides a place for spores to lodge. Of course, where spores can settle, so can sediment. The reduced water velocity in eddies also provides a more favorable settling environment, while at the same time providing turbulence and good nutrient availability to growing plants.

Experiments in the field do not distinguish roughness from other effects. For example, Harlin & Lindbergh (1977), who put out plates with three grades of particles and a smooth control quadrant in the intertidal zone, found little permanent algal growth on the control quadrants (although ephemeral species had evidently settled). The observed distributions demonstrated the effects of surface texture on development of seaweed populations but the experiments did not distinguish between an effect on settlement and the effects of differential drying rates or grazer action on germlings (Norton & Fetter 1981).

Several seaweeds have evolved interesting means of improving the chances of spore settlement. *Nereocystis* blades float far above the seabed, but the entire sorus, which sinks readily, is shed before the spores are released (Bold & Wynne 1978). *Postelsia*, which grows in very high-energy intertidal habitats, releases its spores when the first waves of the incoming tide splash over the plants; water and spores flow down channels in the drooping blades and drip onto the rock and parent plant holdfast. Spores settle in about 30 min, before the tide completely covers them (Druehl & Green, unpublished). *Fucus* releases its eggs still held together in the oogonium: the mass of eight eggs sinks faster than would a single egg. *Sargassum muticum*, which has become a weed in several parts of the world, has a very effective settling mechanism. Eggs released from the conceptacles cluster on the outside of the receptacle, where they are fertilized and develop into small germlings, usually without rhizoids, before they drop to the seabed. As a result of their relatively large size (mean 156 μm), these propagules sink at an average rate of 530 μm s^{-1} in still water (Deysher & Norton 1982), some 5 to 10 times faster than the red algal spores shown in Figure 5.10. Once rhizoids start to grow out, they increase drag and slow the sinking rate (Norton & Fetter 1981).

After spores initially contact and stick to a surface they begin to improve their adhesion by secreting and hardening polysaccharide and protein mucilages and by developing rhizoids. These processes take time; thus, experiments designed to dislodge settled spores show that with a certain shear stress (given in Fig. 5.12 as water pressure) the number of spores that are washed away decreases the longer they have been allowed to settle. Other experiments show that the degree of shear stress that a settled spore can withstand increases with time. The rapidity with which spores settle and with

Table 5.3. *The effect of substratum roughness on the percentage of propagules of* Sargassum muticum *settling on the platform of a waterbroom (see Table 5.1)*

Mean depth of depressions (μm)	0.01	67	141	275	422	589	801	1,190
Expt. 1	1.3	4.5	8.2	9.4	12.0	20.2	34.4	10.0
Expt. 2	2.9	5.1	10.8	11.3	12.3	14.5	31.1	12.0

Two independent experiments were run, with sample sizes of 1620 and 767 respectively; within each experiment several water velocities were used and the results pooled.
Source: Norton & Fetter (1981), The settlement of *Sargassum muticum* propagules in stationary and flowing water. *J. Mar. Biol. Ass. U.K.*, vol. 61, pp. 929–940, with permission of Cambridge University Press.

Figure 5.12. Effect of shear stress (measured as hosing pressure in millimeters of mercury) on adhesion of *Enteromorpha intestinalis* zoospores from two different parents, allowed to settle in still water for various lengths of time (in darkness at 20 C). The percentage of spore detachment was calculated from numbers of settled cells counted after an initial rinse and after hosing for 10 s at various pressures. (From Christie et al. 1970, with permission of the Annals of Botany Company)

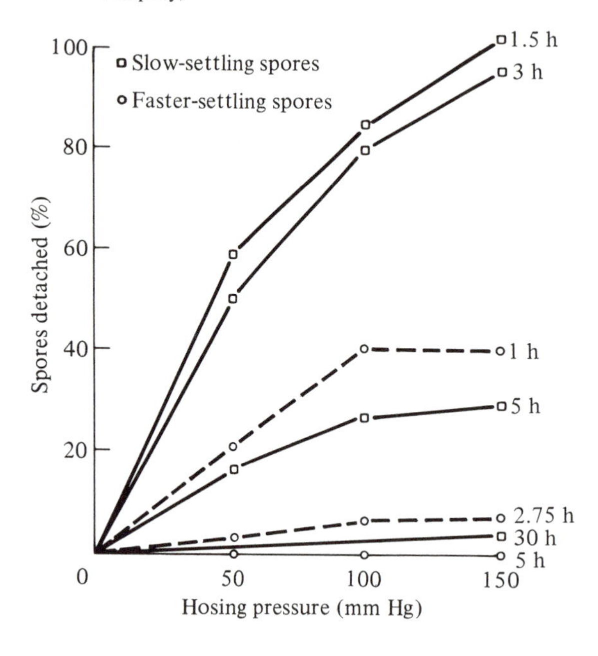

which they anchor firmly is affected by environmental variables such as temperature and salinity (Christie & Shaw 1968) and also varies considerably both within a species (Fig. 5.12) and between species. Whereas spores of *Enteromorpha intestinalis* and *Laminaria hyperborea*, among other species, swim for many hours, permitting wide dispersal (Jones & Babb 1968, Christie & Shaw 1968, Kain 1964), *Postelsia palmaeformis* spores settle very quickly, permitting clumping. The normal course of events is for spores to settle, attach, and then germinate. However, Kain (1964) noted that *Laminaria hyperborea* zoospores sometimes lose their motility and start to germinate while still in the water column; this would not preclude their subsequent attachment. North (1976) used the fact that very young sporophytes of *Macrocystis pyrifera* are sticky to transfer germlings from cloth culture substrata to the seabed.

5.3.2 Effects of sediment and sand

Wherever water motion conditions are most suitable for algal spore settling, they are also likely to be favorable for sediment settling. Spores that settle on sediment particles are apt to be washed away before long, especially as they grow up into faster-moving water layers. Some algae, especially inhabitants of tropical lagoons such as *Penicillus* and *Halimeda*, are adapted to gain holdfasts in sand and silt, but the majority of seaweeds require a firm substratum. Moreover, if sediment settles on top of spores, the spores will be shaded and may be smothered. Sediment can also scour settled spores.

The interaction of sediment and water motion on *Macrocystis pyrifera* spore settlement and gametophyte survival was studied by Devinny & Volse (1978). They found that even small amounts of sediment introduced before or along with the spores greatly reduced the percentage of spores able to settle and grow on glass slides. The interaction between spores and sediment is effectively one of competition for space. When cultures were shaken, either from the time when spores and sediment were added, or starting one day later, survival was significantly reduced. Shaking also reduced survival in the absence of sediment.

One might expect that scouring and burial of habitats by sand would prevent seaweed growth. Certainly, isolated rocks on a sandy beach have very few species of algae (Daly & Mathieson 1977) and these tend to be robust, psammophilic (sandloving) perennials or else ephemerals such as *Enteromorpha, Ectocarpus*, and colonial diatoms. The ephemerals are able to settle when scouring is at a minimum and the rocks bare, and to reproduce and disappear before scouring begins again. Sand movement on beaches is typically seasonal. Sand builds up in spring and is washed into the subtidal in autumn (Fig. 5.13). Several seaweeds have adapted to resist scouring and even months of burial. The species include *Gymnogongrus linearis, Laminaria sinclairii, Phaeostrophion irregulare,* and *Ahnfeltia* spp. from the west coast of North America, and a *Polyides–Ahnfeltia* association and *Sphacelaria radicans* on the east coast of North America (Markham & Newroth 1972, Markham 1973, Sears & Wilce 1975, Daly & Mathieson 1977). Characteristics of these algae include tough, usually cylindrical thalli with thick cell walls; great ability to regenerate, or an asexual reproductive cycle functionally equivalent to regeneration (Norton et al. 1982); reproduction timed to occur when plants are uncovered; and physiological adaptations to withstand darkness, nutrient deprivation, anaerobic conditions, and H_2S. The nature of the physiological adaptations is at present unknown.

Another effect of sediment and sand in the water is turbidity, the principal consequence of which is a reduction in light transmittance (Hagmeier 1971, Wilber 1971); this has been illustrated in Figure 2.6.

5.3.3 Form and function in relation to wave action

The interaction between adult plants and water motion has two major aspects: tolerance to wave action and morphological adaptations to reduce the effect of wave force.

To withstand wave force, a seaweed must be tough enough to avoid being shredded by direct water motion

Figure 5.13. Habitat of *Laminaria sinclairii* on a sandy beach in Oregon (a) in April with little sand present; (b) in July with rocks almost buried in sand. Arrows point to the same rock. (From Markham 1973, with permission of *Journal of Phycology*)

(a)

(b)

and to avoid being abraded by thrashing against rock; it must be supple enough to bend with the wave force; and it must be securely anchored to the substratum. The strength of the holdfast has been briefly studied by Barnes & Topinka (1969) and Norton et al. (1982), although such studies are difficult to carry out quantitatively owing to the problem of gripping the thallus without weakening it. Barnes & Topinka found that considerably less force was required to pull *Fucus vesiculosus* from barnacles than from rock, particularly when the plants were large. Plants growing in calm water are often poorly attached; Norton et al. found that *F. vesiculosus* and *Ascophyllum* growing on the seaward side of boulders were harder to pull off than those on the leeward, more sheltered side. In general, the strength of the holdfast seems to be related to the wave force experienced. Water motion may stimulate holdfast development and unusually severe storms rip out weaker plants.

Although seaweeds have low tensile strengths (the maximum load per unit area they can support while being stretched), they have various means of reducing their drag and thereby the load on the holdfast (Norton et al. 1982). Such means are remarkably effective, especially in stipes of *Nereocystis luetkeana*, which become very narrow and elastic at the base. The stipe bases seem incongruously small compared to the holdfast, yet they can withstand the strain imposed by most waves until they are damaged by grazing or abrasion (Koehl & Wainright 1977). Several genera, including *Macrocystis, Nereocystis,* and *Chorda* occasionally show stipe coiling (Fig. 5.14), but although these coils look like shock-absorbers, few stipes have them, and the factors that cause them have not yet been investigated. The slippery polysaccharide films over seaweed surfaces may also help to reduce drag, as well as providing protection against abrasion. Many larger seaweeds either have narrower blades in more exposed habitats (Druehl 1978, Gerard & Mann 1979) or have the blade split into narrow strips (as in *Laminaria digitata* and *Lessonia*). *Fucus vesiculosus* becomes progressively smaller and less vesiculate in more exposed habitats (Fig. 5.15); however, the extent to which such changes are due to wave exposure has been brought into question by Russell (1978). He pointed out that if the plants do respond to wave action by becoming shorter, narrower, and less vesiculate, there should be strong correlations between the morphological features as well as between each and a scale of wave exposure. Russell's data demonstrated that only some of the expected correlations exist. For example, vesicle number does not correlate with frond width or plant fresh weight. The relationship between vesiculation and wave exposure also needs investigation in consideration of Hurka's (1974) hypothesis that gas vesicles serve to reduce entanglement of long fronds by restoring thalli to the vertical as the shear force reaches zero between forward and backward surge.

The ability to bend is also important in reducing drag. Even apparently stiff plants such as *Corallina* are flexible because of their noncalcified genicula. Good examples of such adaptation are adult *Eisenia arborea* (Charters et al. 1969, Neushul 1972) and *Lessonia nigrescens* (Koehl 1982). The stiff stipes of these plants branch and each branch has a bundle of blades arrayed at the tip (Fig. 5.16). When the water is calm, this morphology holds the blades spread out for maximum light harvest. When there is a surge, the blades in each clump layer together and the stipe bends so that the whole thallus becomes streamlined. Water tends to flow around rather than through the bundles of blades and this, too, reduces drag. Neushul (1972) also concluded that the changes in morphology through which kelps pass during development correlate with their growth through various water motion zones. An alternative means of resisting hydrodynamic forces is to be able to stretch and recoil without breaking; *Durvillaea* and the stipes of *Nereocystis* fall into this category (Koehl 1982).

Figure 5.14. Coiling at the base of a *Macrocystis integrifolia* stipe; the blade shown is a frond initial. Coin is 19 mm diameter.

Figure 5.15. Diagrammatic representation of thallus variation in *Fucus vesiculosus*, illustrating features associated with different degrees of wave action. (From Russell 1978, with permission of The Systematics Association)

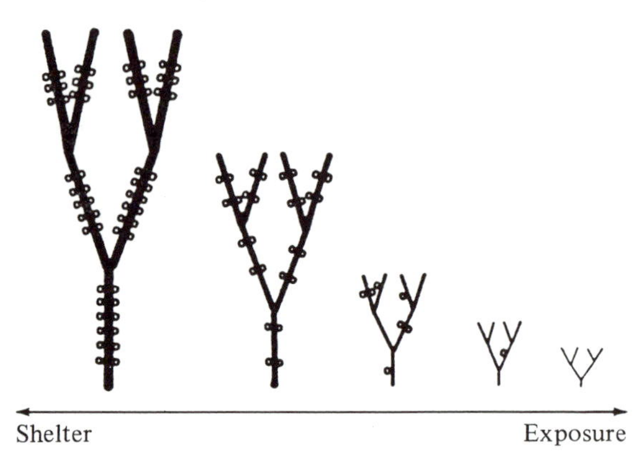

Specimens growing in more exposed habitats are generally more robust than plants in more sheltered water (Price 1978, Gerard & Mann 1979, Norton et al. 1982). *Callithamnion* species respond to increasing wave action by becoming bushier and more heavily corticated (Price 1978). Such morphological changes also create taxonomic problems.

Among the green algae, some of the calcified genera show adaptations to water motion. Thalli of *Penicillus dumetosus* and *P. pyriformis*, species characteristic of moderately exposed areas, are rounded when growing in deeper water where wave action is slight, in shallow water if the direction of the waves changes frequently, and in culture. But in shallow areas in which waves come from one predominant direction the stipe and bushy capitulum are flattened and oriented with the flat side perpendicular to the direction of the waves. Such orientation is also seen in some other genera, with blades that are always flat (e.g., *Udotea*; Friedmann & Roth 1977). Other species of *Penicillus*, characteristic of sheltered waters, do not show flattening or thallus orientation. A rather different response to water motion has been found in another siphonous green, *Codium fragile*, by Ramus (1972). Plants grown in culture first form free filaments. These filaments will unite only if the culture is shaken. They first bind into knots, which become axis primordia, which then grow into the familiar multiaxial thallus. If shaking is discontinued the filaments start to grow apart again. The mechanisms of these effects in *Penicillus* and *Codium* are unknown.

The interaction between wave exposure and plant morphology has been studied at the genetic level in *Laminaria* by Chapman (1973b, 1974). According to prevailing taxonomic criteria, there were two nondigitate species of *Laminaria* in Nova Scotia: *L. saccharina* with short, solid stipes, and *L. longicruris* with long, hollow stipes. Chapman (1974) reported evidence against such a taxonomic distinction: in his studies he found no reproductive isolation between the species, since they are fertile at the same time and can interbreed. In addition, Chapman (1973b) found that there was a correlation between stipe characters and wave exposure, with stipes becoming shorter and solid in rougher habitats. However, he has not formally changed the taxonomy, because it depends also on knowing the relationships to the Nova Scotia plants of *L. saccharina* in Europe and on the west coast of Canada. The question then was how much of the variation in morphological characters, or phenotype, was due to the environment and how much to the genetic make-up, or genotype, of the plants. Transplanting adult plants from one habitat to another would be of limited value, since plants taken from sheltered to exposed hab-

Figure 5.16. Positions of *Eisenia arborea* stipe and blades in calm water (left) and with a surge current (right). (Drawn by Lynn Yip)

itats are usually destroyed by water motion (Chapman 1974, Druehl 1978, and Gerard & Mann 1979, among others, have noted this). Characters, especially of permanent organs like the stipe, may be fixed early in development and may not respond later to the environment.

In an attempt to answer the question, Chapman (1974) cultured gametophytes from the two extreme populations and then made crosses between them ("hybrids" on the basis of current taxonomy). He split the resulting families of young F_1 sporophytes between the two habitats and compared stipe characteristics within and between families. The basis for the analysis is this: stipe characters are not inherited in simple Mendelian fashion but are assumed to be determined by multiple genes acting together without dominance. If variation is caused entirely by the environment, the appearance of the offspring will bear no relation to the appearance of the parents: in this case the offspring outplanted in the exposed habitat should all have short, solid stipes, while their siblings, outplanted in sheltered water, should all have long, hollow stipes. If, on the other hand, genotype were all-important, all the progeny would look alike, no matter where they were planted. (The F_1 generation from hybrids would in this case be intermediate between the parental types.) At the one extreme heritability is zero, at the other it equals one. Variation within and between families in two environments can be compared by making a number of crosses and placing half of each resulting family into each of the two environments. Most populations show aspects of genetic and environmental control of phenotype, and the relative importance of each control can be determined from the ratio of genotypic variance (that between families) to phenotypic (within-family) variance. The maximum information about each component of variance between two populations can be gained by determining heritability within each exposure population in order to evaluate heritability in the hybrids. Chapman (1974) found heritability of his F_1 hybrids to be relatively high compared with published data on other organisms. He concluded that genotype plays an important role in determining the phenotype of the plants between exposed and sheltered populations. His results do not demonstrate that the environment plays no role, only that it is not large enough to completely mask the additive genetic component. Indeed, he found between-family variation to be smaller in F_1's planted in the exposed site than at the sheltered site, indicating that wave exposure restricts the extent of phenotypic variation. This may constitute a selection pressure that could reduce genetic variability.

Other unrelated characters of *Laminaria* stipe and blade have also been examined for heritability. Mucilage canals have very low heritability, hence are essentially environmentally controlled (except in a Nova Scotian population that breeds true!) (Chapman 1975). Bullations of the blade of *Laminaria* affect water flow over the surface, so one might expect water motion to have an important role in their heritability. It turns out, however, that their inheritance is Mendelian, with bullate dominant over smooth: purebred offspring of smooth populations (from the Isle of Man, U.K.) produced only smooth plants, even in the habitat of the bullate plants (Helgoland, Germany), while hybrids were always bullate (Lüning 1975, Lüning et al. 1978).

An important lesson here is that, because plants are so different from humans and the animals with which we are most familiar, every assumption made about plants needs to be checked by observation and experiment. In the words of Evans (1972), "only too often the response of a plant . . . is anything but what a human being would expect." Similar caution has been advised by Norton et al. (1982): "Attempts to interpret the functional significance of a particular morphology or to elucidate the factors that brought it about often founder in a sea of guesswork, rationalizations and teleology. Often an explanation that sounds quite reasonable at first hearing is based on little or no factual evidence."

5.4 **Water motion and populations**

5.4.1 Tidal rapids populations

Tidal rapids provide conditions intermediate between sheltered and exposed coast conditions, yet in some ways they provide better conditions than either. Swift water currents maintain an ample supply of nutrients and sweep away silt, yet there is not the buffeting associated with wave action. Morphological adaptations to water motion such as described in the previous section are found in tidal rapids plants. For example, *Macrocystis integrifolia* blades from tidal rapids are intermediate in size and shape between sheltered and exposed plants (Druehl 1978). Plants in tidal rapids frequently grow to immense sizes if the current velocities are moderate, perhaps because of the nutrient supply, but in very rapid currents water motion causes large plants to flap about, and plants in such conditions are often as stunted as in extreme wave exposure (Norton et al. 1982).

The qualitative effects of currents on seaweed vegetation can be illustrated by an example from one of the most thoroughly studied tidal rapids in the world, those at the entrance to Lough Ine, County Cork, Ireland (Kitching & Ebling 1967). Although these rapids are somewhat atypical in having a large population of *Saccorhiza polyschides* (Lewis 1964), other features are typical of loch and fjord channels of western Ireland and Scotland. As water velocity increases from inside the loch toward the channel, calm water plants such as *Halidrys siliquosa* and *Laminaria saccharina* gradually give way to plants characteristic of moderately exposed shores, such as *Himanthalia elongata*. In the fastest currents, such as over the sill in the middle of Lough Ine Rapids, where current velocity reaches 2.6 m s^{-1}, the water becomes turbulent and exposed coast plants such as *Laminaria digitata* and *L. hyperborea* generally appear. In some rapids *Halidrys* may persist into rapid currents, flourishing side by side with *L. digitata*, an unusual combination of sheltered and exposed coast plants (Lewis 1964).

5.4.2 Wave action and populations

The floras of exposed and sheltered shores frequently differ, sometimes very markedly, as illustrated in Figure 5.17 (Shepherd & Womersley 1981). In this

Figure 5.17. Relations between marine plant communities, water motion, and depth in a southern Australian bay. (a) Map of the distribution of the seaweed and seagrass communities; (b) mosaic chart of algal distributions on rocky bottoms in relation to water movement and depth. (Reprinted from Shepherd & Womersley 1981, *Aquatic Botany*, vol. 11, p. 305, © 1981 Elsevier Scientific Publishing Co.)

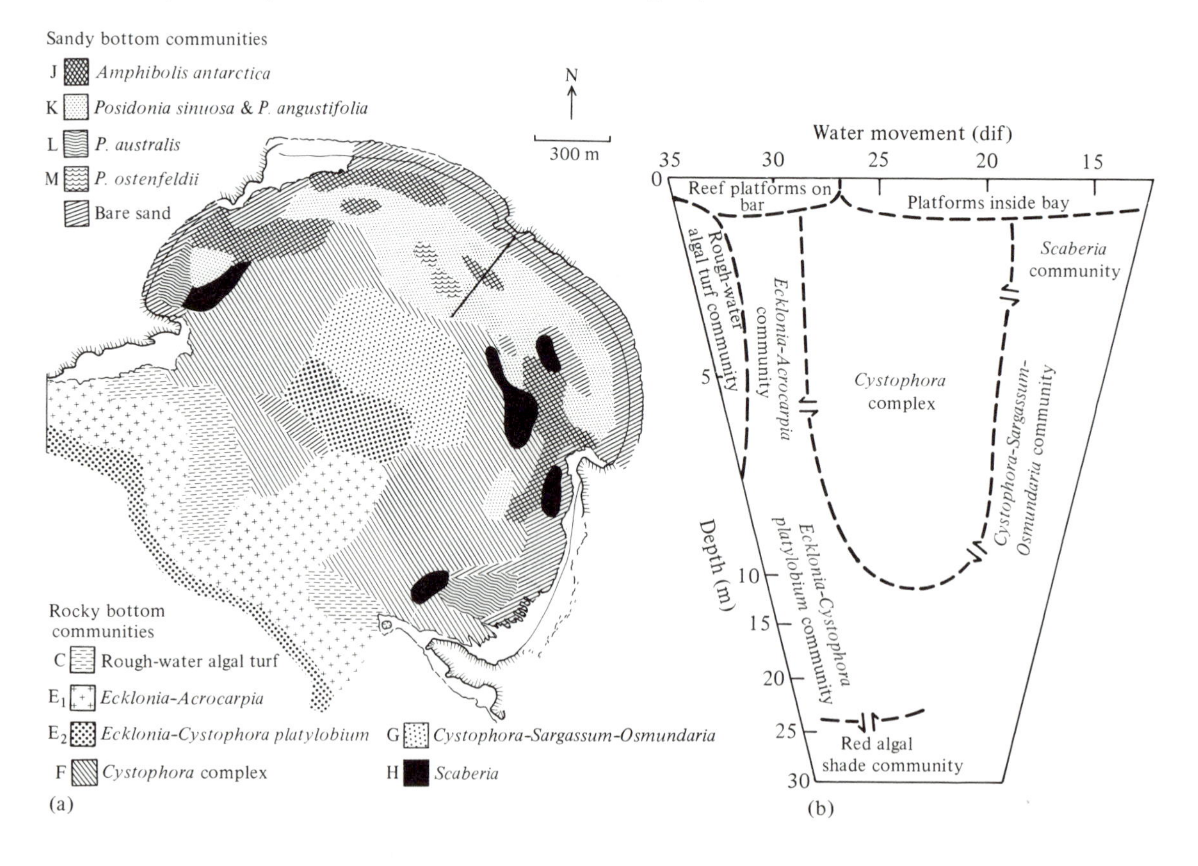

example, from subtidal Australian habitats, the vegetation of a small bay was mapped (Fig. 5.17a) and the water movement at numerous spots assessed with the clod card method so that, among other analyses, community structure could be seen in relation to depth and water motion (Fig. 5.17b). Such correlations do not demonstrate that wave severity per se determines the distribution of seaweeds, any more than the distribution of seaweeds along an estuary is necessarily determined by salinity.

Wave-exposed sites are characterized by large water forces, little or no sediment settling (though sometimes sand scouring), and thorough mixing, causing small variations in temperature, salinity, and nutrients. Sheltered habitats, in contrast, are characterized by small hydrodynamic stresses, siltation, and water stratification, causing marked daily or seasonal changes in temperature, salinity, and nutrient concentrations. Once water flow is adequate for nutrient uptake, water movement becomes a question of force to seaweeds. As illustrated in Section 5.3.3, water motion can induce morphological changes in seaweed thalli.

The abundance of *Laminaria* populations around Vancouver Island, British Columbia, is partially correlated with wave exposure. Salinity and temperature are important in restricting the growth of *L. groenlandica* to outer coast habitats. Within its range, two morphological forms are found. Long stipe plants are restricted to areas of heavy surf, short stipe plants grow in moderately exposed habitats (contrast this with stipe lengths of *L. saccharina–longicruris* discussed in Sec. 5.3.3). *L. saccharina* around Vancouver Island tolerates a broad range of temperature and salinity but does not grow in the intertidal zone when there is heavy surf. It does grow subtidally at some very exposed sites (Druehl 1967).

One very interesting adaptation to wave force is seen in *Postelsia palmaeformis*, which lives in the mid-intertidal zone of only the most wave-exposed headlands of the northwest coast of North America. The habitat is dominated by mussels. The persistence of *Postelsia* depends on periodic disasters and catastrophes to clear space on the rock for their spores to settle. The disasters are the periodic removal by waves of clumps of *Postelsia* and the mussels to which they have attached. The catastrophes are the sporadic denudation of patches of shoreline by logs, which are hurled against the rocks by waves and crush the organisms (Dayton 1973, Paine 1979). (''Disaster'' and ''catastrophe'' are used here in Harper's [1977] sense: a catastrophe is an infrequent extreme event that destroys large portions of a population; a disaster is a more localized but more frequent event.) Such disturbances keep the ecosystem unstable and are thought to permit *Postelsia* to maintain populations in spite of the competitive dominance of the mussel. According to Dayton's hypothesis, sporophytes growing on barnacles attached to mussels and on algae create drag and also smother the barnacles, so that the surf carries them all off, clearing a patch of rock on which more *Postelsia* spores (from nearby plants) can settle and grow to maturity. Druehl & Green (unpublished, but see Carefoot 1977) have challenged this interpretation. As they see the situation, spores released in fall overwinter as gametophytes or very small sporophytes among the smaller algae and under the mussels. In spring, those new sporophytes that are among small algae are able to grow up, whereas those under mussels are inhibited by low light until (or unless) the mussels are removed. Paine (1979) predicted that *Postelsia* patches are unable to persist on cleared spaces because the plants are annuals, because mussels encroach on the patches from the periphery, and because chiton grazing restricts development of young plants.

Disturbance is also important to the maintenance of *Ecklonia radiata* kelp beds in Australia, but in this case the kelp is the dominant organism in a subtidal habitat (Kirkman 1981). *Ecklonia* maintains its dominance as long as disturbances are small (disasters rather than catastrophes), affecting only one or two plants. Shade-tolerant juveniles are present in the understory; when small clearings are created these quickly grow up and replace the plants that were lost. If the disturbance is catastrophic, removing a large part of the *Ecklonia* canopy, spores of shade-intolerant species such as *Sargassum* are able to outcompete *Ecklonia*. New sporophytes of the kelp that settle in the areas remain as juveniles until the canopy is removed by disaster.

In spite of adaptations to withstand wave force, seaweeds are regularly pulled off the substratum and cast ashore as ''drift.'' In some places, such as Scotland and Prince Edward Island, Canada, some species are cast ashore in commercially harvestable quantities. In many seaweed populations water motion is probably as important as grazing damage in mortality of adult plants. Mortality is increased in *Macrocystis* beds in California by drifting plants that entangle and pull off other plants (Gerard 1976).

5.5 Synopsis

A certain amount of water motion is necessary to maintain an effective supply of nutrients; beyond that, the force component of water motion becomes critical. In the intertidal zone a component of wave action is its wetting effect. Wave action is difficult to quantify, particularly if integrated effects of all components are to be measured vis-à-vis distribution of intertidal communities.

Water moves over plant surfaces as a current, with decreasing velocity toward the surface because of the drag created by the surface. Water flow may be laminar or turbulent, and the slow-moving velocity boundary layer is much thinner if flow is turbulent. Thus gas and mineral exchange is greater when flow is turbulent. Because of nutrient uptake by plants there are also diffusion boundary layers, which are considerably thinner than the velocity boundary layer.

Spore (and zygote) settlement depends a great deal on water motion. Although the boundary layer may be as thick as a spore, spores cannot settle into that layer except under perfectly still conditions, and appear rather to be deposited by eddies if the size of the irregularities in the surface is appropriate. Following settlement and attachment, spores must produce a firm holdfast. At first they are susceptible to being removed from the surface by water currents.

Water motion may involve sediment movement. While sediment is generally deleterious to algae, some species are able to tolerate long periods of sand burial.

Some seaweeds show morphological changes when moved from rough to calm water, and the interaction of genotype and wave environment has been elucidated in a *Laminaria* species. Finally, wave action plays a role in local geographic distributions of populations, both via the abilities of plants to withstand shear stress and via disturbance of communities.

6 Nutrients

Seaweeds require inorganic carbon, water, light, and various mineral ions for photosynthesis and growth. In this chapter, mechanisms of uptake, nutrient requirements, and metabolic roles of essential nutrients (excluding C, H, and O) will be examined. The importance of nutrient uptake and growth kinetics will be discussed in terms of their effect on chemical composition, growth, development, and distribution of macroalgae. Particular emphasis will be placed on nitrogen because it is the element most frequently limiting to seaweed growth. Even though seaweeds are larger (multicellular) than phytoplankton and usually attached to a substratum, their nutritional requirements are similar; therefore, some discussion of phytoplankton nutrition is also included.

6.1 Nutrient requirements

6.1.1 Essential elements

The development of defined culture media for growing algae axenically has allowed the testing of a variety of elements to determine which are essential. The criteria for an absolute requirement of an element were established by Arnon & Stout (1939):

1. a deficiency of the element makes it impossible for the alga to grow or complete its vegetative or reproductive cycle;
2. it cannot be replaced by another element;
3. the effect is direct and not due to interaction with (e.g., detoxification of) other, nonessential elements, stimulation of epiflora, or the like (Levitt 1969).

C, H, O, N, P, Mg, Fe, Cu, Mn, Zn, and Mo are considered to be required by all algae (O'Kelley 1974, DeBoer 1981). S, K, and Ca are required by all algae but can be replaced by other elements. Na, Co, V, Si, Cl, B, and I are required only by some algae. All the major constituents of seawater, except for Sr and F, are required by macroalgae (DeBoer 1981). There is a tendency to consider the requirements of all algae to be similar, but the heterogeneity of macroalgae makes generalizations about their nutritional requirements difficult.

Up to 21 elements are required for the main metabolic processes in plants (see Table 6.6), but more than double this number have been reported to be present in seaweeds. Therefore the mere presence of an element in the seaweed is not proof that the element is essential, nor is the amount present indicative of the relative importance of the element. Generally, essential (and nonessential) elements are accumulated in the tissues of algae above their concentration in seawater, giving rise to concentration factors of up to a thousandfold (Table 6.1). Some elements are absorbed in excess of requirements, whereas others are taken up but not used. The elemental composition of the ash of macroalgae is similar to that of phytoplankton. The effects of nutrient supply on chemical composition are discussed in Section 6.8.1, and the metabolic roles of elements in Section 6.5.

6.1.2 Vitamins

Some seaweeds require trace amounts of one or two organic carbon compounds for normal growth; these compounds do not act as a carbon source for algal growth. This type of nutrition is referred to as auxotrophy, and the organic compounds are vitamins. Phytoplankton also require the same vitamins as seaweeds, but most higher plants synthesize their own vitamins and do not depend on environmental sources. Representative concentrations of vitamins in seawater are shown in Table 6.2.

The three vitamins that are routinely added to culture media are B_{12} (cyanocobalamin), thiamine, and biotin (Fig. 6.1). Of these, B_{12} is the most widely required by seaweeds (Table 6.3), which may be related to the fact that B_{12} is present in seawater in lesser amounts than thiamine and biotin. Thiamine is known to be required only by *Acetabularia acetabulum* (*A. mediterranea*) (the only macroscopic marine green alga studied so far); no requirement has yet been found for biotin among the

Table 6.1. *Concentrations of some essential elements in seawater and in seaweeds*

Element	Mean concentration in seawater (μg g^{-1})	Concentration in dry matter: Mean (μg g^{-1})	Concentration in dry matter: Range (μg g^{-1})	Ratio of supply in seawater to concentration in tissue
Macronutrients				
H	105,000	49,500	22,000–72,000	2.1×10^{0}
Mg	1,290	7,300	1,900–66,000	1.8×10^{-1}
S	905	19,400	4,500–82,000	4.7×10^{-2}
K	406	41,100	30,000–82,000	1.0×10^{-2}
Ca	412	14,300	2,000–360,000	2.9×10^{-2}
C	27.3[a,b]	274,000	140,000–460,000	1.0×10^{-4}
N	0.488[a,c]	23,000	500–65,000	2.1×10^{-5}
P	0.068	2,800	300–12,000	2.4×10^{-5}
Micronutrients				
B	4.39	184	15–910	2.4×10^{-2}
Zn	0.004[a]	90	2–680	4.4×10^{-5}
Fe	0.003[a]	300	90–1,500	1.0×10^{-5}
Cu	0.002[a]	15	0.6–80	1.7×10^{-4}
Mn	0.001[a]	50	4–240	2.0×10^{-5}

[a]Considerable variation occurs in seawater. [b]Dissolved inorganic carbon.
[c]Combined nitrogen (dissolved organic and inorganic).
Source: DeBoer (1981), with permission of Blackwell Scientific Publications.

Table 6.2. *Average vitamin concentrations (ng L^{-1}) in Pacific Ocean waters*

Water	Vitamin B_{12}	Thiamine	Biotin
Scripps Institution pier	2.9	15	3.8
Coastal	1.6	9	2.6
Central Pacific	0.1	8	1.3

Source: Provasoli & Carlucci (1974), with permission of Blackwell Scientific Publications.

seaweeds. One macroscopic freshwater green alga (*Draparnaldiopsis salishensis*) has recently been shown to require B_{12} (Johnstone 1977). Because only a few seaweeds have been studied, one must not draw generalizations from the limited data in Table 6.3.

Thiamine and B_{12} are both complex molecules (Fig. 6.1), and some phytoplankton require only part of the molecule. In the case of thiamine, the requirement may be specific for the thiazole moiety or for the pyrimidine moiety, or the alga may require both moieties, the whole molecule, or homologues of thiazole or pyrimidine (Provasoli & Carlucci 1974). In the case of B_{12}, requirements may be satisifed with the corrin group alone, by the corrin group plus any nucleotide side chain, or by the whole B_{12} molecule. Algae requiring only the corrin group would have the greatest advantage in the sea since many bacteria produce cobalamins of some kind, but only 15 to 30% produce the entire B_{12} molecule.

6.1.3 Limiting nutrients

Over 100 years ago an agriculturalist named Liebig stated that "growth of a plant is dependent on the minimum amount of foodstuff presented." This statement has come to be known as "Liebig's law of the minimum." The nutrient available in the smallest quantity with respect to the requirements of the plant will limit its rate of growth, if all other factors are optimal. A comparison of the relative number of atoms indicates that N, P, and Fe occur in seaweeds in higher concentrations than in seawater (Table 6.1). Further evidence that N, P, Fe, as well as Cu, Zn, Mn, and C might limit algal growth is that concentrations of these elements in seawater vary considerably due to biological activity. Recent experiments have shown that out of these possible limiting elements, nitrogen is the element most frequently limiting the growth of seaweeds (Topinka & Robbins 1976, DeBoer & Ryther 1977) and phytoplankton (Ryther & Dunstan 1971).

The concentration of the limiting nutrient gives some indication of whether the nutrient is limiting, but the nutrient supply rate or the nutrient turnover time is more important in determining the magnitude or degree of limitation. For example, if the concentration of the limiting nutrient is low but the supply rate is almost equal to the uptake rate of the algae, the algae will be only slightly nutrient limited.

Table 6.3. *Vitamin requirements of seaweeds*

	No. species tested	B_{12}	Thiamine	Biotin
Chlorophyceae	1	R	R	0
Phaeophyceae	9[a]	0	0	0
	1	R	0	0
Rhodophyceae	10	R	0	0

R = required; 0 = not required.
[a]Four of the nine species are stimulated by vitamin additions.
Source: DeBoer (1981), with permission of Blackwell Scientific Publications.

Figure 6.1. Chemical structures of the three vitamins required for growth of macroalgae. (Reprinted with permission from Bauernfeind & DeRitter 1970, *Handbook of Biochemistry*, 2nd ed. © CRC Press, Inc., Boca Raton, Fla.)

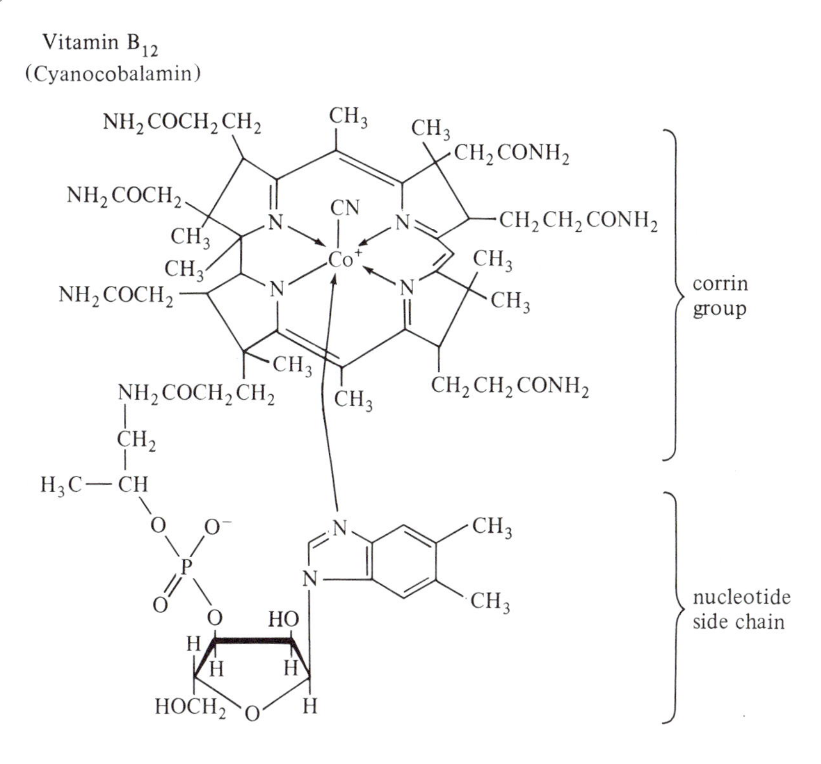

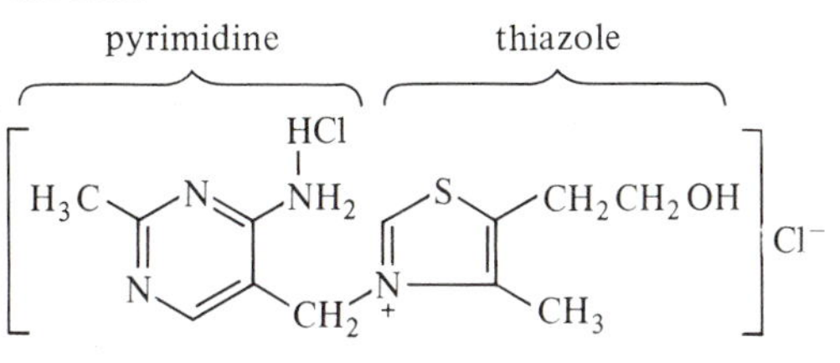

Biotin

Table 6.4. *Typical nitrate, nitrite, ammonium, and phosphate concentrations at each of two depths in the three major oceans*

		Nutrient concentration (μM)			
Ocean	Depth (m)	NO_3^-	NO_2^-	NH_4^+	PO_4^{3-}
Atlantic	5	0–20	0–0.5	0–3	0.1–1
	4,000	c. 20	<0.5	c. 4	c. 2
Indian	5	0–20	0–0.5	0–3	0.1–1
	4,000	c. 35	<0.5	c. 4	c. 3
Pacific	5	0–20	0–0.5	0–3	0.1–1
	4,000	c. 40	<0.5	c. 4	c. 3

Source: Parsons & Harrison (1983), with permission of Springer-Verlag.

The possibility of the limitation of growth by two nutrients simultaneously, or dual nutrient limitation, has been explored. The original concept proposed by Droop (1973) was for a multiplicative effect by the two nutrients, but experimentation has revealed that it is an either–or effect with only one nutrient limiting growth at one time (Droop 1974, Rhee 1978). However, the ratio of two nutrients (e.g., N/P) required by one algal species may be quite different from the ratio required by another species. Consequently, one species may be nitrogen limited while another species may be phosphorus limited, illustrating limitation by a single nutrient with competition for different nutrient resources (Tilman et al. 1982).

6.2 Nutrient availability in seawater

Concentrations of various elements in seawater differ considerably (Tables 6.1, 6.2, 6.4). Those in the nanomolar (nM) range, except nitrogen and phosphorus, are considered micronutrients or trace elements for nutritional purposes. Elements occurring in larger quantities are frequently referred to as macronutrients. The units that are used to express concentrations vary. Generally marine scientists use micromolar (μM), although microgram-atom per liter (μg-at L^{-1}) is still used. Freshwater nutrient concentrations are generally expressed as μg L^{-1} or parts per billion (ppb). To convert from μg L^{-1} to μM, one must multiply μg L^{-1} by the atomic weight of the element. For example, the atomic weight of nitrogen is 14, so 14×1 μg L^{-1} = 1 μM nitrogen. For combined forms, 1 μM nitrogen = 1 μg-at L^{-1} NO_3^- or NH_4^+, but 2 μg-at L^{-1} urea, because urea has two nitrogen atoms per molecule.

Aspects of nutrient cycles have been discussed elsewhere (Redfield et al. 1963, Riley & Skirrow 1965, Riley & Chester 1971, Parsons & Harrison 1983), and only a few basic principles will be reviewed here. The important features of the nitrogen cycle are summarized in Figure 6.2.

Nitrogen is the element that most frequently limits algal growth in the sea, and the important ions for use by seaweeds are nitrate and ammonium. Even though the most abundant form of nitrogen in the sea is dissolved nitrogen gas (N_2), seaweeds are unable to use N_2. However, nitrogen fixation is associated with *Codium decorticatum* (Rosenberg & Paerl 1981). The nitrogenase activity has been attributed both to nitrogen-fixing bacteria (e.g., *Azotobacter* sp.) and heterocystous blue-green algae (e.g., *Calothrix* sp.), which occur as epibionts, and not to the seaweed. The measured rate of nitrogen fixation was equivalent to about 7% of the daily nitrogen requirement of the seaweed, but there is no evidence that the host plants obtain significant amounts of this fixed nitrogen. *Codium isthmocladum* does not have this associated nitrogen-fixing flora.

Processes that bring nitrogen into the euphotic zone where it can be used directly in autotrophic production by plants include (1) physical advection primarily in the form of nitrate from below the nutricline; (2) atmospheric input of ammonia either in rain or through N_2 fixation by bacteria and blue-green algae; (3) ammonium from bacterial decomposition of sediments, the magnitude of which may be enhanced by burrowing animals or a salinity intrusion into the interstitial waters of the sediments; and (4) in coastal areas, inputs of nitrogen from land drainage, sewage, and agricultural fertilizers. Regeneration of nitrogen in the water column occurs as a result of two largely separate processes; one involves bacteria and the other results from excretion by marine fauna, particularly ammonium by the zooplankton community (Fig. 6.2). In small, shallow estuaries, decomposition of extensive mats of *Enteromorpha* may dominate the nitrogen cycle for short periods in summer (Owens & Stewart 1983).

Chemical and physical methods of determining so-called phosphate have indicated that total dissolved phosphorus in seawater consists of a rather heterogeneous group of inorganic phosphate species and phos-

Figure 6.2. Schematic representation of the nitrogen cycle of the sea. PON = particulate organic nitrogen; DON = dissolved organic nitrogen; DIN = dissolved inorganic nitrogen. (From Turpin 1980, with permission of the author)

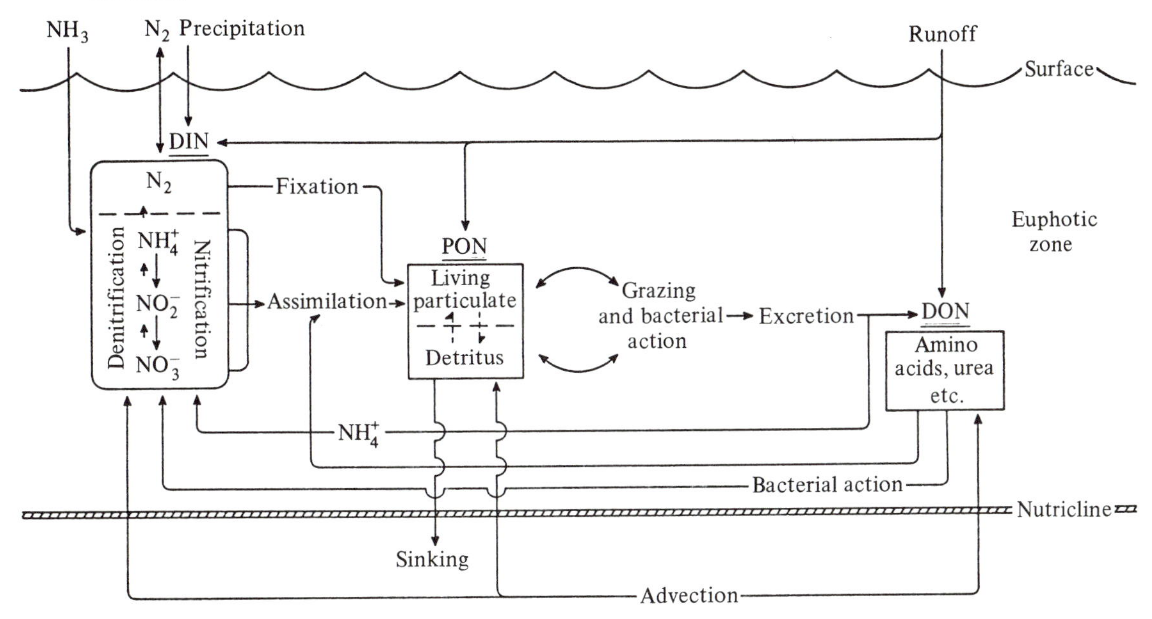

phorus-containing organic compounds. At the pH of seawater, phosphorus exists in primarily three free ionic species in equilibrium. At pH 8, 20 C, HPO_4^{2-} accounts for 87% of the free ions, PO_4^{3-} 12%, and $H_2PO_4^-$ 1% (Riley & Chester 1971). Orthophosphate ions (PO_4^{3-}) form metallophosphate complexes (e.g., with Ca^{2+} and Mg^{2+}) or they can combine with organic compounds; therefore free orthophosphate represents less than one-third of the total inorganic phosphate in seawater. Several techniques have been used to distinguish among forms of organic phosphorus in seawater, but there are problems with the techniques and the separation is not always clear (Strickland & Parsons 1972).

Iron, the fourth most abundant element in the earth's crust, is one of the least soluble metals in oxygenated waters. For this reason iron concentration may limit growth of seaweeds, but the very complicated aquatic chemistry of iron is a major impediment to our understanding of iron uptake by algae. The speciation of iron in seawater is complex (Byrne & Kester 1976). At the pH of seawater (ca. 8.2) the ferric ion combines with hydroxyl ions, forming ferric hydroxide, which is relatively insoluble ($K_{sp} \approx 10^{-38}$M). Therefore much of the iron is maintained in solution only by the formation of complexes with natural chelators or ligands such as humic acid. Recent measurements in the western North Pacific indicate that dissolved iron concentrations are 2.2 to 3.8 $\mu g\ L^{-1}$, of which 80 to 90% is organically bound iron (Sugimura et al. 1978). In many cases, past experimental results are now known to be inconclusive because of the following problems:

1. Adsorption of iron onto the surfaces of the experimental containers (especially glass) could lead to the erroneous conclusion that the seawater being tested was iron deficient (Lewin & Chen 1971, Paasche 1977).
2. Addition of a chelator (e.g., EDTA) may stimulate algal growth, again suggesting that iron limitation has been overcome because the chelator made the iron more available. However, the chelator may have only served to eliminate or reduce the adsorption of iron onto the container walls, or to detoxify other metals such as copper and zinc by forming a metal complex.
3. An increase in particulate iron cannot be ascribed to uptake by the algae because it may be caused by abiotic processes in the medium or passive adsorption onto the cell surface.
4. Freshly added iron may form a precipitate in the medium and consequently coprecipitate and adsorb other metals, reducing their availability to algae (Huntsman & Sunda 1980).
5. The ratio of chelator : trace metal is important. It is now clear that the activity of the free metal ion generally determines availability or toxicity. Chelators, by complexing with free metal ions, reduce but never totally eliminate their activity. The chelator acts as a metal buffer, releasing more ions into the medium as other ions are taken up by the algae, but under optimal conditions never allowing toxic levels to be reached. However, if too much chelator is present in relation to the trace

metal concentration, the activity of the metal may be reduced below that required for optimum growth.

Natural copper concentrations range are 1 to 4 μM in coastal waters (Schmidt 1978, Boyle 1979, Lewis & Cave 1982). However, it is the concentration of the free ion that is important, not the total copper concentration. There is no suitable method of determining the concentration of free ion and therefore bioassays are used to determine "biologically available" copper which is equated with the free ion concentration (Sunda & Guillard 1976, Whitfield & Lewis 1976, Lewis & Cave 1982).

Early reports indicated that the addition of a synthetic chelator EDTA alleviated poor growth of phytoplankton in newly upwelled seawater (Barber & Ryther 1969). Subsequently, additions of iron or manganese were also shown to stimulate growth (Barber et al. 1971). Recent experiments under defined conditions have verified that the toxicity of copper to phytoplankton is determined by cupric ion activity and that chelated copper is not directly toxic (Sunda & Guillard 1976). The cupric ion activity in some waters that are low in organic chelators has been shown to be toxic to some phytoplankton species (Cross & Sunda 1977). Recent measurements of the copper complexing ability of exudates of several macroalgae suggest that the exudates may aid in complexing and in some cases possibly detoxifying copper when it is present in high concentrations (Sueur et al. 1982). Similarly, Ragan et al. (1979) found some detoxification of zinc by brown algal polyphenols.

When these recent observations are applied to the problem of determining the cause of poor algal growth in recently upwelled water, earlier hypotheses must be revised. Unlike other trace metals such as Cd, Cu, Ni, and Zn, whose concentrations increase with depth in the oceans, manganese concentrations usually decrease (Bender et al. 1977). Therefore newly upwelled water may have relatively high cupric ion and low manganous ion concentrations, producing an unfavorable Mn^{2+}/Cu^{2+} ion ratio for phytoplankton growth. As the water "ages" the cupric ion concentration decreases and the manganous ion concentration increases, so that the toxic effect of the water is overcome, without the intervention of chelators (Sunda et al. 1981).

The lack of a rapid analytical technique to determine vitamin concentrations has resulted in relatively few measurements (Table 6.4). A bioassay is generally used to measure vitamins, but it is laborious and suitable bioassay organisms are not known for some moieties (Provasoli & Carlucci 1974). Further difficulties arise in the analysis of vitamin B_{12} because some algae produce a substance that binds the vitamin and makes it unavailable for uptake (Swift 1980). The exogenous vitamins required by some seaweeds may be obtained from the seawater, or they may come directly from microorganisms living on seaweed surfaces. In addition a few *phytoplankters* are known to produce vitamins during part of their growth cycle (Provasoli & Carlucci 1974).

6.3 Pathways and barriers to ion entry

6.3.1 Adsorption

Ions enter cells by moving across the boundary layer of water surrounding the cell (Section 5.2.2). The route of ion entry into cells is to the cell surface, then passing through the cell wall and plasmalemma into the cytoplasm. The thickness of the boundary layer may affect the uptake rate of an ion because if turbulence around the thallus is low, the boundary layer is thick and uptake may be limited by the rate of diffusion across this layer. The cell wall, unlike the plasmalemma, does not generally present a barrier to ion entry. When a macroalga is placed in nutrient medium there may be an initial rapid uptake that does not require energy. This observation is generally attributed to diffusion into the space that is exterior to the plasmalemma; this space is generally termed the diffusion free space. Ions can readily be removed from the diffusion free space by washing the alga or keeping it in nutrient-depleted medium for some time.

Some ions, especially cations, may not reach the plasmalemma because they become adsorbed to certain components of the cell wall. Polysaccharides and proteins have sulfate, carboxyl, and phosphate groups from which protons can dissociate, leaving a net negative charge on these compounds in the cell wall. In effect, these macromolecules act as cation exchangers; consequently, large amounts of cations can be adsorbed from the environment. Haug & Smidsrød (1967) suggested that concentrations of Ca, Sr, and Mg in brown algae are largely the result of ion exchange between seawater and the acid polysaccharide, alginate, in the cell walls; the amount of these ions in the cytoplasm and vacuole is a relatively small portion of the total in the whole thallus. The affinity of alginic acid for different metals is as follows (Haug 1961):

Pb>Cu>Cd>Ba>Sr>Ca>Co>Ni, Zn, Mn>Mg

This cation adsorption by seaweeds is referred to as a Donnan exchange system, and the volume in which these cations are confined is the Donnan free space. The diffusion free space plus the Donnan free space is called the apparent free space. It is the volume of the cell wall into which ions can diffuse or be absorbed (Epstein 1972, Lüttge & Higinbotham 1979). The apparent free space can be measured quantitatively in physiological experiments and is generally used in a physiological context, whereas the term "apoplast" is used in discussing translocation and is based partially on morphology. In higher plants, the apparent free space extends to the endodermis of the roots; in seaweeds it includes the cell walls and intercellular spaces. Only a few measurements of apparent free space have been made in seaweeds, but these

Figure 6.3. Three modes of movement of ions or molecules across a membrane. The relative concentrations are indicated by the relative number of dots on either side of the membrane. (From Avers, *Cell Biology*, 2nd ed.,

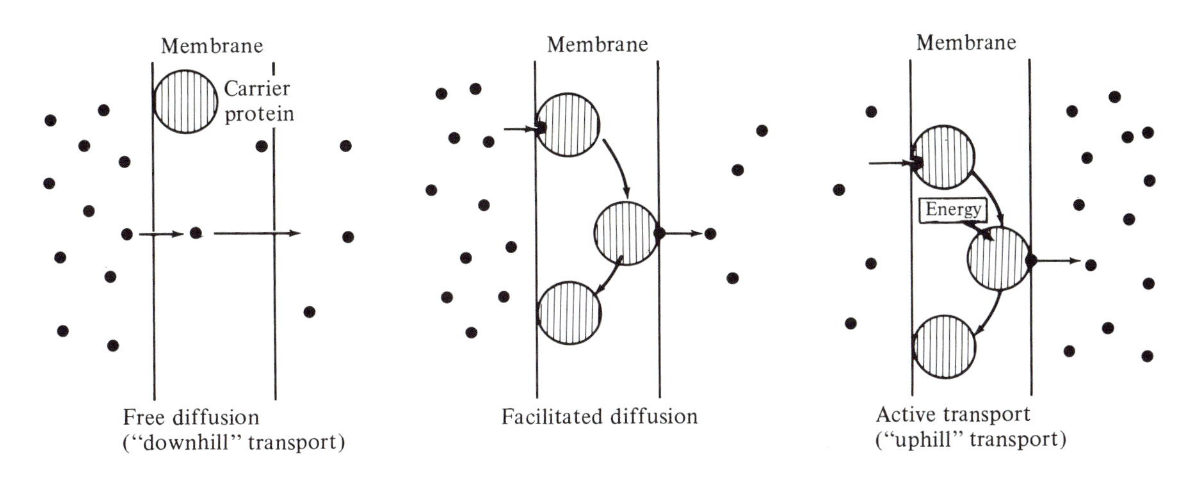

measurements indicate that the space can be significant, up to 20 to 30% of the cell volume for the red alga *Porphyra perforata* (Eppley & Blinks 1957, Gessner & Hammer 1968, Hammer 1969).

6.3.2 Passive transport

Algal plasma membranes consist of polar lipid bilayers interspersed with proteins. Nonelectrolytes (uncharged particles) diffuse through membranes at a rate proportional to their solubility in lipid and inversely proportional to their molecular size. Dissolved gases move more freely than most solutes. As a result, many important gases (e.g., CO_2, NH_3, O_2, and N_2) cross lipid bilayers by dissolving in the lipid portion of the membrane, diffusing to the other lipid–water interface, and dissolving in the aqueous phase on the other side of the membrane. Uncharged molecules such as water and urea are also highly mobile. However, molecules such as NH_3 may be trapped inside cells when they are converted to ions. The pK_a for NH_3/NH_4^+ is 9.2, and in seawater only 5 to 10 % of the total ammonium is present as NH_3. At a higher pH, which may be found in dense cultures or restricted tidal pools, the percentage of NH_3 can increase to 50% (at pH 9.2) or more, allowing rapid diffusion into the cell. Since the pH of cytoplasm is only 7 to 7.5, most of the NH_3 that enters is protonated to NH_4^+ and cannot diffuse back across the membrane. This mechanism can account for approximately 10% of net uptake (Walker et al. 1979). Theoretically, passive accumulation of ammonia by diffusion and acid-trapping in the cytoplasm/vacuole could result in a more than thousandfold accumulation in marine diatoms (Wheeler & Hellebust 1981). Presumably the same mechanism operates in macroalgae.

Molecules diffuse down a free energy or chemical potential gradient, hence the term "downhill" transport (Fig. 6.3). The rate of diffusion varies with the chemical potential gradient or the difference in activity (approximately equivalent to the concentration) across the plasmalemma (Table 6.5: equation 1). The permeability coefficient of the membrane (P) is proportional to the diffusion coefficient of the molecule and inversely proportional to the thickness of the membrane (Table 6.5: equation 2).

Ions (charged particles or electrolytes) usually have a much lower permeability than uncharged molecules. The charge on the ion makes it difficult to penetrate a membrane with active or charged groups that either repel or attract (immobilize) the ions. In addition, ions are usually strongly lipophobic and hydrophilic and their particle size is frequently increased by a substantial layer of water of hydration. Both of these properties tend to decrease the rate of diffusion. Finally, since ions are charged, the driving force for ion movement involves an electrical term as well as a chemical (concentration gradient) term; the combined effect is referred to as an electrochemical potential gradient. The electrochemical driving force, E, across a membrane can be calculated for any ion with the Nernst equation (Table 6.5: equation 3). The net transmembrane electrochemical potential gradient at equilibrium is the sum of the E values for all the ions involved, modified by the permeability coefficient (P) for each ion. Because the concentration gradient and the electrical gradient can independently influence the movement of an ion across the membrane, it is possible that the electropotential difference could be large enough to produce diffusion of an ion even against a gradient in concentration. Therefore it is wrong to conclude that an ion has been transported solely by an active process simply because its intracellular concentration is greater than its extracellular concentration.

Table 6.5. *Nutrient transport equations*

1.	$J = P\Delta C$
2.	$J = \frac{D}{X}\Delta C$
3.	$E = \frac{RT}{ZF} \ln ao/ai$
4.	$V = V_{max} \frac{S}{K_s + S}$
5.	$V = \frac{V_{max}}{(1 + K_s/S)(1 + i/K_i)}$

Symbols: ao/ai = chemical potential difference between outside and inside of membrane; ΔC = concentration gradient across membrane; D = diffusion coefficient of molecule; E = Nernst potential; F = Faraday constant; i = inhibitor concentration; J = flux of molecules; K_i = half-saturation constant for inhibitor; K_s = half-saturation constant for substrate; P = membrane permeability coefficient; S = substrate (nutrient ion) concentration; T = temperature (Kelvin); V = initial uptake rate; V_{max} = maximum uptake rate at saturating substrate concentration; X = membrane thickness; Z = charge per ion (valence).

Since passive diffusion occurs without the expenditure of metabolic energy, metabolic inhibitors have no direct effect on diffusive flux. In addition, no carriers or binding sites are involved in diffusion and therefore it is nonsaturable.

6.3.3 Facilitated diffusion

Facilitated diffusion resembles passive diffusion in that transport occurs down an electrochemical gradient, but frequently the rate of transport by this process is faster. In facilitated diffusion, carriers or enzymes (permeases) are thought to bind the ion at the outer membrane surface and to cross the membrane to the inner surface where the bound ion is released. Facilitated diffusion exhibits properties similar to active transport: (1) it can be saturated, and transport data fit a Michaelis-Menten-like equation (see Sec. 6.3.4); (2) only specific ions are transported; and (3) it is susceptible to competitive and noncompetitive inhibition. However, in contrast to active transport mechanisms, any energy expenditure required for transport must be indirect.

6.3.4 Active transport

Active transport is the transfer of ions or molecules across a membrane at rates greater than by the combined rate of free diffusion and facilitated diffusion, against an electrochemical gradient. For this reason, active transport has also been termed "uphill" transport. Because external concentrations of inorganic nutrients are typically in the micromolar range and intracellular concentrations of these same nutrients are in the millimolar range, passive diffusion along an electrochemical gradient is unlikely to be important because of the large concentration difference between the inside and outside of the cell. Therefore, transport is generally considered to be active, but in some cases facilitated diffusion might be an important secondary process (Fig. 6.3). To conclusively demonstrate active transport as opposed to free or facilitated diffusion, the following criteria must be satisfied:

1. Active uptake is energy dependent, and a change in the uptake rate should occur after the addition of a metabolic inhibitor (e.g., dinitrophenol) or a change in temperature, since both of these factors influence energy production.
2. The active transport rate should exceed the combined rate of free and facilitated diffusion estimated from the electrochemical gradient, or there should be net ion movement against an electrochemical gradient.

Other properties of active transport, such as unidirectionality (in or out of the cell), selectivity of ions transported, and the saturation of the carrier system (exhibiting Michaelis-Menten kinetics), are not definitive criteria for active transport because they are also characteristic of facilitated diffusion. From the few electrochemical measurements that have been made on algae, primarily in the giant cells of the freshwater alga *Chara*, and by analogy also with vascular plants, transport in seaweeds is thought to be primarily active. Obtaining electrochemical measurements with the use of microelectrodes is difficult for most macroalgae because of the relatively small size of individual cells.

Active transport may also be coupled to the transport of other ions. Coupled transport may arise from the transfer of different ions at separate sites of the same carrier in opposite directions (antiport or countertransport) or the same direction (symport or cotransport). In microalgae, proton-linked cotransport of sugars and thiourea has been demonstrated (Syrett 1981). In many animal cells, transport generally takes place by cotransport of Na^+ ions rather than H^+ ions. Some preliminary evidence suggests that this is also true for transport by marine microalgae (Syrett 1981). This is not surprising, since seawater is high in Na^+ and low in H^+ ions.

6.4 Nutrient uptake kinetics

Nutrient ions generally enter plant cells by three mechanisms: passive diffusion, facilitated diffusion, and active transport (Fig. 6.3). If transport occurs solely by passive diffusion, the transport rate is directly propor-

Figure 6.4. Hypothetical plots of nutrient uptake rate (V) and concentration of the limiting nutrient (S), for (a) passive diffusion only, where V is directly proportional to S; (b) facilitated diffusion or active transport in which V_{max} in example 2 is half V_{max} of example 1, resulting in a concomitant decrease in the K_s; (c) linearized plot of the data in (a) to illustrate how the kinetic parameters V_{max} and K_s are determined graphically; (d) passive diffusion plus active transport (—), and active transport (- - -) with the passive diffusion component (— • —) subtracted.

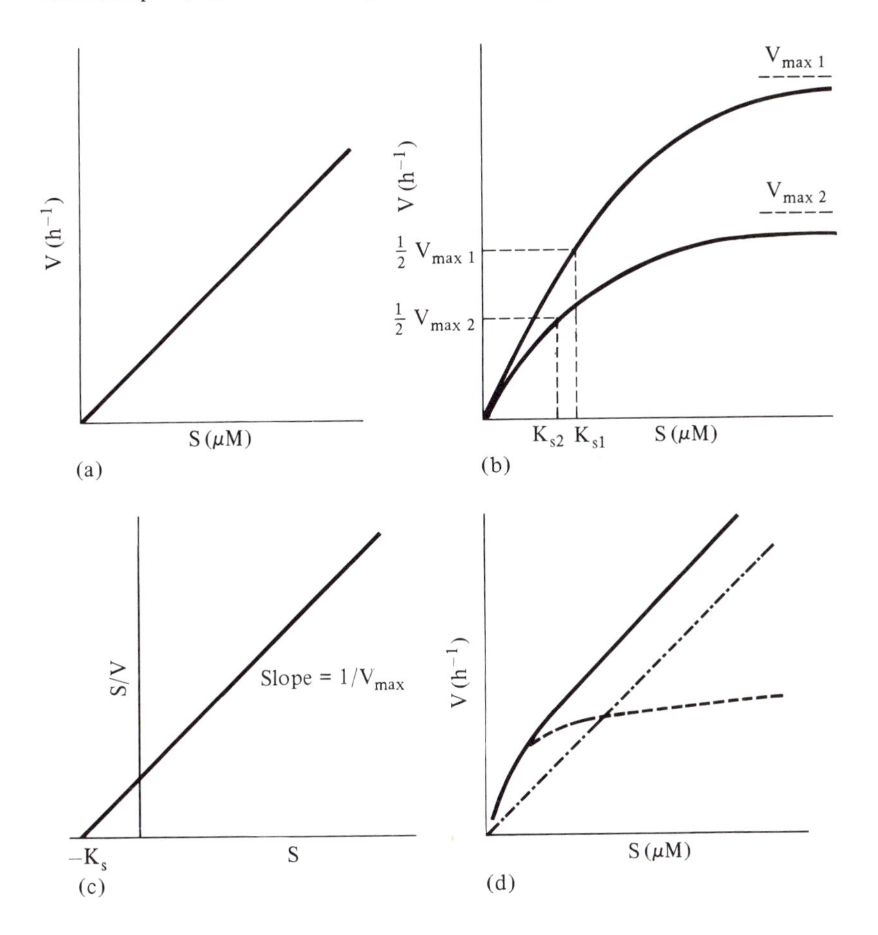

tional to the external concentration (Fig. 6.4a). In contrast, facilitated diffusion and active transport exhibit a saturation of the membrane carriers as the external concentration of the ion increases. The relationship between uptake rate of the ion and its external concentration is generally described by a rectangular hyperbola, similar to the Michaelis-Menten equation for enzyme kinetics (Fig. 6.4b). The equation is given in Table 6.5: equation 4. K_s (equivalent to K_M) is called the half-saturation constant and it is the substrate concentration at which the uptake rate is half its maximum. The lower the value of K_s, the higher is the affinity of the carrier site for the particular ion.

The transport capabilities of a particular macrophyte are generally described by the parameters V_{max} and K_s. Healey (1980) suggested that the slope of the initial part of the hyperbola (the linear portion) may be a more useful parameter for comparing the competitive ability of various species for a limiting nutrient; this is similar in concept to the use of the slope, α, in the P vs. I curve (Chapter 2). One of the problems in using K_s is that its value is not independent of V_{max} (i.e., when V_{max} decreases the value of K_s will also decrease even though the initial slope of the hyperbola remains the same) (Fig. 6.4b). The kinetic parameters may be estimated graphically from the rectangular hyperbola, but generally the Michaelis-Menten equation is rearranged to yield a straight line from which K_s and V_{max} can be calculated more accurately by linear regression (Fig. 6.4c) (Atkins & Nimmo 1980). The advantages and disadvantages of the three possible linear plots have been examined by Dowd & Riggs (1965); the S/V vs. S or the V/S vs. V plots are generally superior to the 1/S vs. 1/V plot.

Active uptake may not follow the simple saturation kinetics just described (Cornish-Bowden 1979). Studies with higher plants have revealed that the pattern of uptake may be biphasic or multiphasic in nature (Epstein 1972, Nissen 1974). In the case of biphasic kinetics, a plot of V vs. S reveals two rectangular hyperbolas, frequently referred to as the high and low affinity systems. At low substrate concentrations, the high affinity system operates, exhibiting a high degree of ion specificity and a low value for K_s, whereas at high concentrations the low affinity system is operative, exhibiting much less ion selectivity and a very high value for K_s. Others have refuted the evidence for biphasic uptake kinetics, claiming that it is not statistically significant (Borstlap 1981). To date there is no strong evidence for biphasic uptake kinetics in macroalgae. Deviation from saturation kinetics may also occur if active uptake and diffusion occur simultaneously. Diffusion is not likely to be important at low substrate concentrations, but at concentrations well above environmental levels diffusion may be significant, in which case the total uptake rate is composed of active uptake plus diffusion; the resultant uptake pattern is shown in Figure 6.4d.

6.4.1 Measurement of nutrient uptake rates

While the reasons for measuring nutrient uptake may vary, the methods discussed here will be primarily for ecological purposes (e.g., as an index of competitive ability) rather than for the elucidation of physiological mechanisms. The latter methods have been thoroughly reviewed by Lüttge & Pitman (1976) and Lüttge & Higinbotham (1979); the former have been discussed in detail by Harrison & Druehl (1982). There are three main techniques for measuring nutrient uptake rates: (1) radioactive isotope uptake, (2) stable isotope uptake, and (3) disappearance of nutrient from the medium measured colorimetrically (Harrison & Druehl 1982); the loss of a radioactive tracer from the culture medium can also be measured. One of the problems of using either radioactive or stable isotopes is that different parts of the thallus accumulate the isotope at different rates and samples from different areas of the thallus should be taken and averaged to obtain a whole thallus uptake rate (the most useful measurement for ecological purposes). In all three of these techniques very short incubation times of probably less than 10 to 15 min would yield an estimate of gross uptake rate (influx) while long incubation times (more than 6 h) would give rates that approximated net uptake, taking account of efflux of the nutrient from the thallus back into the medium.

Ideally, for autecological studies one would like to make uptake measurements in the field to avoid the difficulties of simulating in the laboratory important physical and chemical factors of the field. However, there are disadvantages to field experiments, such as rapidly varying natural conditions (e.g., nutrient concentrations, irradiance) and the difficulty in producing a significant number of simultaneous replicates. These problems are overcome under laboratory conditions, but every effort has to be made to produce realistic conditions. Important factors include saturating irradiance, light quality appropriate to the depth from which the specimen came, photoperiod (because, for example, nitrate uptake rates exhibit diel periodicity; Syrett 1981), water temperature, and concentrations of nutrients, trace metals, and vitamins.

Nutrient uptake rates of seaweeds are measured in the laboratory by incubating epiphyte-free tissue disks or, preferably, whole plants in filtered natural seawater to which nutrients (except the one under study), trace metals, and vitamins have been added at saturating levels. To eliminate the effect of rapid diffusion into the apparent free space, plants are frequently preincubated in saturating nutrients for a few minutes and then placed in appropriate experimental concentrations of the nutrient. (See Reed & Collins 1980 and Harrison & Druehl 1982 for further discussion.) There are two basic approaches to following disappearance of the nutrient from the culture medium. The first is to spike the culture with the nutrient of interest and follow the nutrient disappearance for several hours until nutrient exhaustion occurs; this method gives a time series of uptake rate (Fig. 6.5a). The second method involves the use of many containers with different concentrations of the nutrient and each with a different specimen of the species being studied. The incubation period in this case is usually short (10–60 min) but constant for all concentrations (Fig. 6.5b). If the uptake rate is constant (Fig. 6.5a), the choice of method is not important. However, if the uptake rate varies (Fig. 6.5c), as is frequently the case when ammonium is limiting (Probyn & Chapman 1982), a short incubation period must be used to correctly estimate the maximal uptake rate.

The reason why the uptake rate varies is that under nitrogen limitation, intracellular nitrogen pools may be low and the initial enhancement in uptake rate over the first 10 to 60 min may represent a pool-filling phase. As the pools fill, the decrease in uptake rate may be due to feedback inhibition. Therefore, this uptake rate does not represent the true transmembrane transport, which is free from feedback inhibition; this effect is discussed in more detail in Section 6.6.1. The method in which the nutrient is followed until it is depleted from the medium is not recommended for the estimation of V_{max} and K_s because the nutritional past history of the thallus is changing with time. Nevertheless, this method is useful in determining the assimilation rate, the rate at which intracellular nitrate or ammonium is incorporated into amino acids and proteins (Fig. 6.5c); this rate has been termed V_i (Conway et al. 1976).

While uptake determined from thalli in the incubation medium is generally attributed solely to the thalli, in some cases this may not be true because most thalli have epiphytes such as microscopic algae and bacteria on their surfaces. These epiphytes are not easily removed. Antibiotics such as streptomycin and penicillin have been employed to inhibit bacterial uptake (Harlin & Craigie 1978), but it is unlikely that there is an antibiotic concentration that completely inhibits the bacterial uptake of nutrients without also affecting the alga.

Nutrient uptake rates may be expressed in four commonly used units. Uptake may be normalized to area

Figure 6.5. Hypothetical time series of nutrient disappearance from the medium using the following methods: (a) perturbation method where saturated uptake is linear with time; (b) the multiple container–constant incubation time method; and (c) the perturbation method where saturated uptake is nonlinear with time. Phase 1 is the enhanced or rapid uptake, possibly due to filling of intracellular pool(s); phase 2 may represent an assimilation rate (also referred to as V_i); phase 3 is the final depletion of the limiting nutrient from the medium. (From Harrison & Druehl 1982, with permission of Walter de Gruyter & Co.)

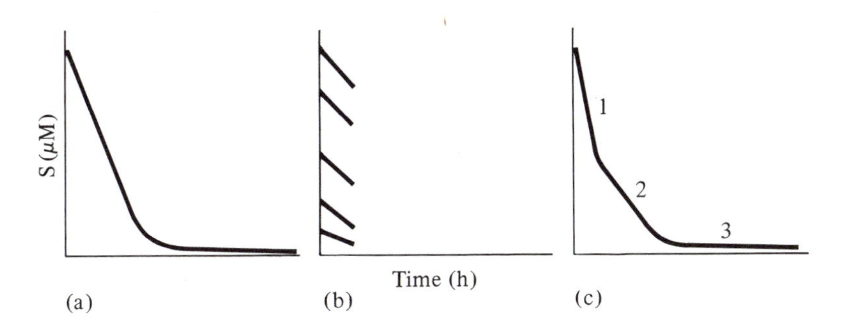

Figure 6.6. Nitrate and ammonium uptake rates for *Macrocystis pyrifera* mature blade disks as a function of (a) irradiance, (b) current speed. (From Wheeler 1982, with permission of Walter de Gruyter & Co.)

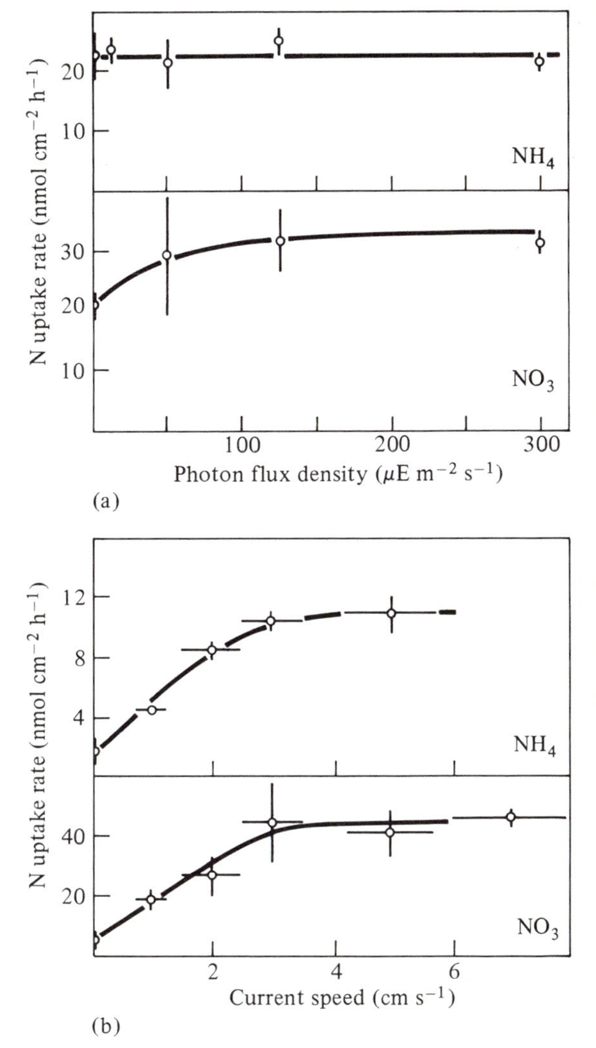

(μmol cm^{-2} h^{-1}), wet weight (μmol g wet wt^{-1} h^{-1}), dry weight (μmol g dry wt^{-1} h^{-1}), or, finally, nutrient content in the plant, which simplifies to a specific uptake rate (h^{-1}). If conversion factors are not given, nutrient uptake data in the literature that are expressed in different units cannot be accurately compared.

6.4.2 Factors affecting nutrient uptake rates

Factors affecting nutrient uptake rates can be divided into three broad categories: physical, chemical, and biological. Physical factors are discussed first.

The primary influence of light on nutrient uptake is indirect, through photosynthesis. Light can (1) provide energy for active transport (ATP production by photosynthesis); (2) produce carbon skeletons, which are necessary for the incorporation of nutrient ions into larger molecules; (3) provide energy for the production of charged ions, which establish Donnan potentials; and (4) increase the growth rate and thus increase nutrient uptake. Strong evidence of the effect of irradiance on uptake rates can be seen in plots of nitrate uptake against irradiance (Fig. 6.6a). The data fit a rectangular hyperbola for *Macrocystis* (Wheeler 1982) and also for many microalgal species (Syrett 1981). In contrast, ammonium uptake in *Macrocystis* was found to be independent of irradiance. Irradiance effects on the uptake of other ions have been summarized by Floc'h (1982). Photoperiod affects nitrate uptake, possibly due to diel periodicity shown by the nitrate reductase enzyme.

Temperature effects on active uptake and general cell metabolism approximate a Q_{10} = 2. In the case of simple diffusion, temperature has less effect and the Q_{10} value is 1.0 to 1.2. Several studies indicate that the effect of temperature on ion uptake is ion specific and dependent on the algal species. For example, a marked decrease in nitrate uptake rate was observed for *Laminaria longicruris* (Harlin & Craigie 1978) but not for *Fucus spiralis* (Topinka 1978) as the temperature was lowered.

Since temperature can fluctuate daily, it would be interesting to study how rapidly uptake rates respond to a change in temperature.

Water motion is another factor that is important in the movement of ions to the surface of the thallus (Wheeler & Neushul 1981; see Chap. 5). In areas of low turbulence or in unstirred laboratory cultures, transport across the boundary layer is primarily by diffusion. Thus it is the concentration of the nutrient in the medium that is the rate limiting step to nutrient uptake (Pasciak & Gavis 1974, Wheeler & Neushul 1981). Diffusion limited transport rates have been demonstrated for single-celled phytoplankters and the effect may be more pronounced for multicellular algae in quiescent waters, especially if the thallus is thick rather than filamentous (hence lower surface area: volume ratio). Wheeler (1980) and Gerard (1982a) demonstrated that the giant kelp *Macrocystis* encounters such transport limitation for carbon and nitrogen when the current over the fronds is less than 3 to 6 cm s^{-1} (Fig. 6.6b).

Exposure to air during a low tide frequently results in the loss of water from the thallus, depending on the season (Sec. 9.3.1). Recent studies have shown that mild desiccation (10–30% water loss) enhances nutrient uptake rates compared with hydrated plants in several intertidal seaweeds when they are submerged in nutrient-saturated seawater (Fig. 6.7) (Thomas & Turpin 1980, Thomas 1983). This enhanced uptake response occurred when growth was limited by that particular nutrient and when the thallus had been exposed to repeated periodic desiccation for several weeks. The relative degree of enhancement of uptake rate, the percentage of desiccation producing maximal uptake rates, and the tolerance to higher degrees of desiccation were positively related to tidal height (Thomas 1983).

Chemical factors such as the concentration of the nutrient being taken up and the ionic form of the element will affect uptake rates. For example, nitrogen in the form of ammonium is frequently taken up more rapidly than nitrate, urea, or amino acids (DeBoer 1981). Uptake rates can also be influenced by the concentration of competing ions in the medium. Ammonium may inhibit nitrate uptake by up to 50%. In contrast, *Gelidium*, *Macrocystis*, and *Laminaria* take up nitrate and ammonium at equal rates when they are supplied simultaneously (Bird 1976, Haines & Wheeler 1978, Harlin & Craigie 1978). Ca^{2+} and Na^{+}, as well as Mg^{2+} and K^{+}, are mutually antagonistic. Exogenous inhibitors (or pollutants) may affect membrane carriers by altering their activity or the rate of synthesis. Intracellular ion concentrations in the cytoplasm and vacuoles will also influence uptake rates. Wheeler & Srivastava (1984) found that the nitrate uptake capacity was inversely proportional to the intracellular nitrate concentration in *Macrocystis integrifolia*.

Biological factors that influence uptake rates include the type of tissue, the age of the plant, its nutritional past history, and interplant variability (e.g., Gerard 1982b). Uptake determinations on whole thalli are preferable for ecological measurements, but if the thallus is too large a portion must be used. Cutting the thallus to produce tissue segments may alter the uptake rate due to a wounding response, resulting in increased respiration (Hatcher 1977) or elimination of the translocation system or active sink region (Penot & Penot 1979). A comparison of excised sections of *Macrocystis* tissue with whole blades showed a marked decrease in uptake rate by the cut sections (Wheeler 1979).

Many seaweeds are perennials, and therefore natural populations often consist of different age classes. The nitrate uptake rate of three age classes of *Laminaria groenlandica* decreased with increasing age. The uptake rate (per g dry wt) of first-year plants was several times greater than for third-year plants (Harrison & Druehl 1982). Sharp differences in nutrient uptake abilities also occur between early life history stages and mature thalli of the same species. Ammonium uptake rates for *Fucus distichus* germlings were an order of magnitude higher than for the mature thalli (Thomas 1983). This sizable difference in uptake abilities is probably due to the large proportion of storage and support tissue in the adult

Figure 6.7. Nitrate uptake rate as a function of percentage of desiccation for *Fucus distichus*. Plants were desiccated to different degrees and then placed in medium containing 30 μM NO_3^-. The uptake was determined over a 30-min interval. (From Thomas & Turpin 1980, with permission of Walter de Gruyter & Co.)

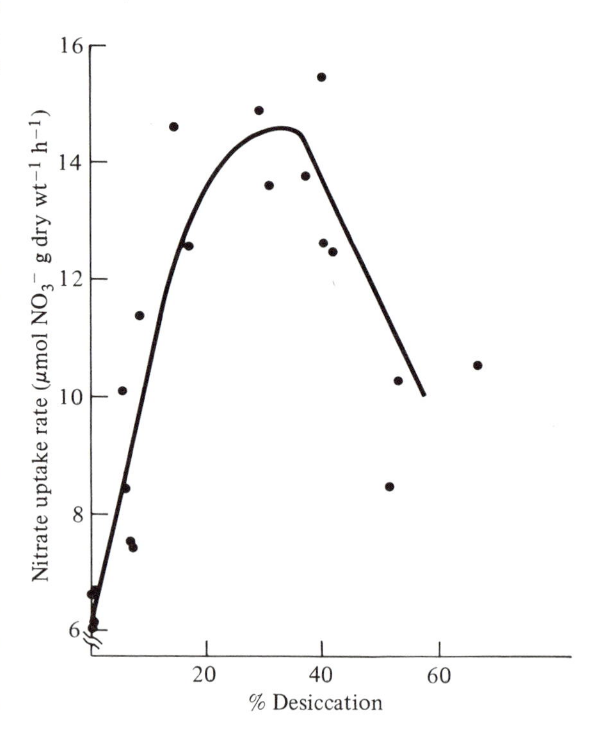

plants, which does not actively require nitrogen. Also interesting is that old fronds and stipes retain some ability to take up ammonium but entirely lose their capacity to take up nitrate. Young tissue, which is metabolically active, appears to need both nitrate and ammonium to meet its greater nitrogen requirements. The stipe demonstrated the lowest nitrogen uptake activity in *Fucus spiralis*, which is in keeping with its low metabolic activity (Topinka 1978). Schmitz & Srivastava (1979) observed the opposite trend of uptake with age of tissue in *Macrocystis*. They showed that mature regions took up and translocated phosphate to meristematic sink regions. Davison & Stewart (1983, 1984) obtained similar results for nitrogen with another kelp, *Laminaria digitata*. They demonstrated that 70% of the nitrogen demand of the intercalary meristem of this species was supplied by the transport of nitrogen, probably in the form of amino acids, assimilated by the mature blade.

In *Macrocystis*, Jackson (1977) deduced from indirect evidence that mature blades deeper in the water column may serve an important role in uptake of nitrogen from the relatively nutrient-rich water and its translocation (as amino acids) to the blades in the nutrient-poor surface water. Additional evidence from Gerard (1982c) supports this suggestion that at least some nitrogen can be moved from deeper, mature-senescent blades to shallower, growing blades.

The uptake rate of ammonium (as well as of most nutrients) is a function of the nutritional past history of the plant. When *Gracilaria foliifera* and *Agardhiella subulata (Neoagardhiella baileyi)* were grown under nitrogen-limiting conditions, the C/N ratio of the thalli was greater than 10 (by atoms) and the plants showed higher rates of ammonium uptake at a given ammonium concentration than plants that were not nitrogen limited (C/N ratio less than 10) (D'Elia & DeBoer 1978).

The effect of surface area:volume (SA : V) ratio on nutrient uptake was predicted by Littler & Littler (1980) on the basis of seaweed functional-form groups (see Fig. 2.28 and Table 9.3), and demonstrated by Rosenberg & Ramus (1984). The order of SA : V ratio and nitrogen uptake was *Ulva* > *Fucus* ≈ *Gracilaria* > *Codium*.

6.5 Uptake, assimilation, and metabolic roles of essential nutrients

There have been very few studies on nutrient metabolism in macroalgae compared with the number of studies on microalgae. Consequently, for many aspects of nutrient assimilation and metabolic roles, we refer to the research on phytoplankton to avoid leaving large gaps in certain areas, acknowledging that the situations may be rather different in seaweeds. Uptake refers to transport across the plasmalemma, while assimilation refers to a sequence of reactions in which inorganic ions are incorporated into organic cellular components. The key metabolic roles of the elements are summarized in Table 6.6 and are discussed in the following sections.

6.5.1 Nitrogen

Nitrogen uptake and assimilation in macroalgae have been recently reviewed by Hanisak (1983). In the few seaweeds that have been studied, the uptake rate of ammonium generally exceeds that for nitrate at environmental concentrations. At very high concentrations (greater than 30–50 μM) ammonium may be toxic to some seaweeds (Waite & Mitchell 1972). For some seaweeds (e.g., *Codium fragile*), ammonium has been reported to exhibit saturation kinetics, suggesting that a carrier is involved through facilitated diffusion or active transport (Hanisak & Harlin 1978). The actual mechanism of uptake of ammonium is unknown, even for microalgae (Syrett 1981). For other species of macroalgae, the ammonium uptake rate does not saturate as the ammonium concentration is increased, but instead increases linearly (Fig. 6.4d). This linear increase at high ammonium concentrations may represent a second transport mechanism, possibly a strong diffusion component. Further evidence that this might represent a diffusive component comes from the good hyperbolic fit (linear regression of transformed data in Fig. 6.4c) when the linear component is subtracted from the total uptake rates (Fig. 6.4d). However, definitive experiments have not been conducted to confirm this suggestion. Linear ammonium uptake rates over a range of ammonium concentrations have also been reported for several kelps, such as *Macrocystis pyrifera* (Haines & Wheeler 1978) and *Laminaria groenlandica* (Harrison & Druehl 1982), and for the red alga *Gracilaria verrucosa* (Thomas 1983).

Although no long-lived radioisotope of ammonium is available, the closest chemical homologue is available as ^{14}C-labeled methylamine (CH_3NH_2). This ammonium analogue has been used to study ammonium uptake kinetics in *Macrocystis pyrifera* (Wheeler 1979). However, if ammonium is present (e.g., only 1 μM), separate experiments must be performed to quantitatively assess the degree of inhibition of ammonium on methylamine uptake (Wheeler & McCarthy 1982).

Uptake of nitrate generally exhibits saturation kinetics (Harlin & Craigie 1978, DeBoer 1981). There are some reports, however, of a linear increase in nitrate uptake with increasing concentration for *Laminaria groenlandica* (Harrison & Druehl 1982) and *Gracilaria verrucosa* (Thomas 1983). Intracellular (cytoplasmic and/or vacuolar) pool concentrations that are up to three orders of magnitude greater than the surrounding seawater strongly suggest an unfavorable electrochemical gradient for transport; transport thus is likely to be primarily an active process. Even for microalgae, the actual mechanism of plasmalemma transport is not known (Syrett 1981).

Nitrite uptake has been studied in only a few cases. The maximal uptake rate of nitrite was similar to nitrate uptake but lower than ammonium uptake for *Codium fragile* (Hanisak & Harlin 1978).

While there are several reports of growth on dis-

Table 6.6. *Function and compounds of the essential elements in seaweeds*

Element	Probable functions	Examples of compounds
Nitrogen	Major metabolic importance as compounds	Amino acids, purines, pyrimidines, porphyrins, amino sugars, amines
Phosphorus	Structural, energy transfer	ATP, GTP, etc., nucleic acids, phospholipids, coenzymes including Co-A, phosphoenolpyruvate
Potassium	Osmotic regulation, pH control, protein conformation and stability	Probably occurs predominantly in the ionic form
Calcium	Structural, enzyme activation, ion transport	Calcium alginate, calcium carbonate
Magnesium	Photosynthetic pigments, enzyme activation, ion transport, ribosome stability	Chlorophyll
Sulfur	Active groups in enzymes and coenzymes, structural	Methionine, cystine, glutathione, agar, carrageenan, sulfolipids, coenzyme A
Iron	Active groups of porphyrin molecules and enzymes	Ferredoxin, cytochromes, nitrate reductase, nitrite reductase, catalase
Manganese	Electron transport in photosystem II, maintenance of chloroplast membrane structure	None
Copper	Electron transport in photosynthesis, enzymes	Plastocyanin, amine oxidase
Zinc	Enzymes, ribosome structure (?)	Carbonic anhydrase
Molybdenum	Nitrate reduction, ion absorption	Nitrate reductase
Sodium	Enzyme activation, water balance	Nitrate reductase
Chlorine	Photosystem II, secondary metabolites	Violacene
Boron	Regulation of carbon utilization (?), ribosome structure (?)	None
Cobalt	Component of vitamin B_{12}	B_{12}
Bromine[a]	Toxicity of antibiotic compounds (?)	Wide range of halogenated compounds, especially in Rhodophyceae
Iodine[a]		

[a]Possibly an essential element in some seaweeds.
Source: DeBoer (1981), with permission of Blackwell Scientific Publications.

solved organic nitrogen compounds such as urea and amino acids, there has been only one definitive study on the uptake kinetics of an amino acid, L-leucine. Schmitz & Riffarth (1980) examined the uptake of 17 different amino acids by the filamentous brown alga *Giffordia mitchelliae* and found the highest uptake rate for L-leucine. The uptake rate was observed to be light independent and carrier mediated, with a very low affinity (high K_s of 30–120 μM) for L-leucine and a low maximal uptake rate (0.03 μmol g dry wt^{-1} h^{-1}). They suggested that exogenous amino acids contribute probably less than 5% of the nitrogen demand of this alga. Urea is an excellent source of nitrogen for many seaweeds (Nasr et al. 1968, Probyn & Chapman 1982), but poor growth has been reported for others (DeBoer et al. 1978). Assessment of urea as a source of nitrogen in bacterized cultures is complicated, however, because the bacteria may break down the urea to ammonium, which can be used by the seaweed. Urea uptake is generally thought to be active, even though it should have a high rate of diffusion because of its small size.

Values obtained for nitrogen uptake kinetic "constants" are summarized in Table 6.7. The K_s values are generally high (2–13 μM), up to an order of magnitude higher than phytoplankton. V_{max} values range from 8 to 25 μmol g dry wt^{-1} h^{-1} for both nitrate and ammonium. Since specific uptake rates (where the rate is normalized to the nitrogen content of the thallus) are generally not reported for seaweeds, comparisons with phytoplankton are not possible.

Preliminary studies indicate that nitrogen assimilation in seaweeds and phytoplankton is reasonably similar and the main difference is the faster rate of the reactions in phytoplankton. Since no macroalgae are known to fix N_2, the inorganic sources for these plants are nitrate, nitrite, and ammonium. Ammonium is already reduced and can be directly incorporated into amino acids, but the oxidized ions must be first reduced intra-

Table 6.7. *Nitrogen uptake kinetic "constants" for seaweeds*

Alga	Temperature (C)	NO_3^-			NH_4^+		
		$K_s \pm$ SE μM	μmol h^{-1} g^{-1}dw	μmol $h^{-1}cm^{-2}$	$K_s \pm$ SE μM	μmol $h^{-1}g^{-1}$dw	μmol h^{-1} cm^{-2}
Fucus spiralis	5	6.6 ± 0.9	—	0.16	6.4 ± 2.0	—	0.18
	10	6.7 ± 0.8	—	0.20	5.4 ± 2.0	—	0.26
	15	7.8 ± 1.4	—	0.23	9.6 ± 2.6	—	0.35
	15 (dark)	12.8 ± 3.5	—	0.28	5.8 ± 1.8	—	0.29
Laminaria longicruris	15[a]	4.1	9.6	—	—	—	—
	10[b]	5.9	7.0	—	—	—	—
Macrocystis pyrifera	16	13.1 ± 1.6	30.5	—	5.3[c] ± 1.0	23.8[c]	—
	6–9	—	—	—	50[d]	23.6[d]	—
Iridaea cordata	—	—	—	—	2.5	5.5	—
Gracilaria foliifera	20	2.5 ± 0.5	9.7	—	1.6[c]	23.8[c]	—
Agardhiella subulata	20	2.4 ± 0.3	11.7	—	3.9[c]	15.9[c]	—
Hypnea musciformis	26	4.9 ± 3.9	28.5	—	16.6 ± 1.8	—	—
Enteromorpha spp.	15	16.6	129.4	—	—	—	—
Codium fragile	6	1.9 ± 0.5	2.8	—	1.5 ± 0.2	13.0	—
	24	7.6 ± 0.6	9.6	—	1.4 ± 0.2	28.0	—

[a]Summer tissue. [b]Winter tissue. [c]Mechanism 1 uptake (high affinity). [d]Uptake of methylamine, an analogue of NH_4^+.
Source: DeBoer (1981), with permission of Blackwell Scientific Publications.

Figure 6.8. Main features of nitrogen uptake and assimilation in a eukaryotic algal cell. (From Syrett 1981, reproduced by permission of the Minister of Supply and Services, Canada)

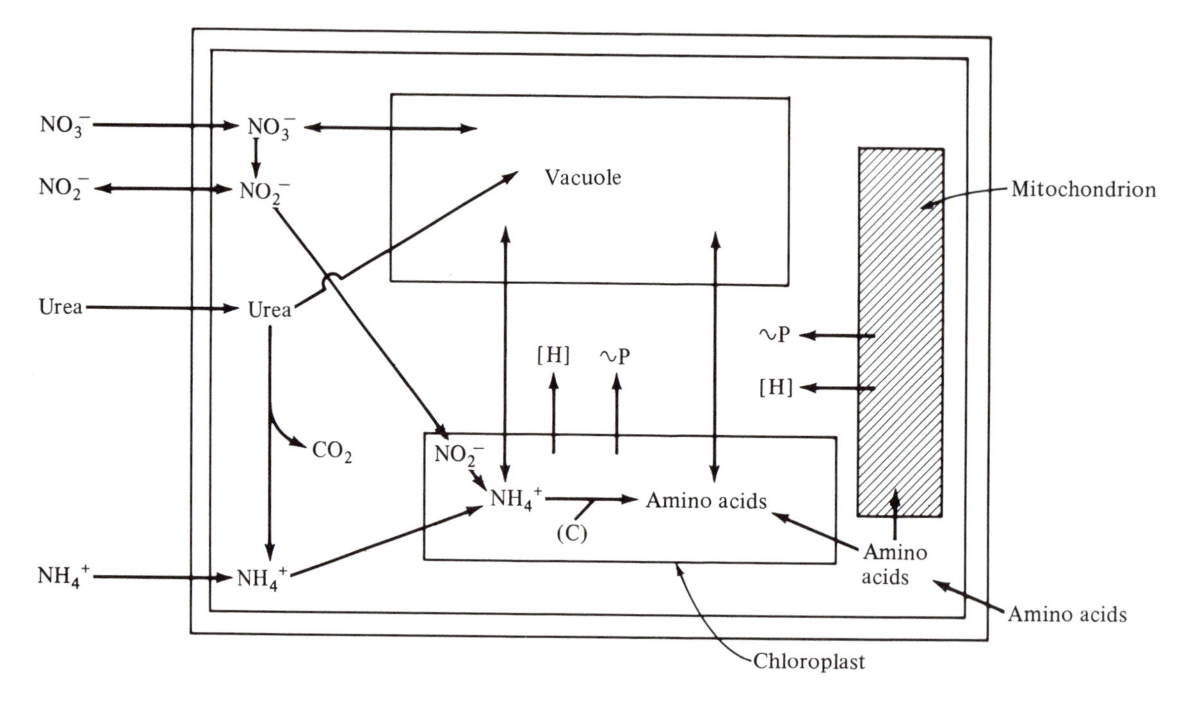

cellularly to ammonium. Eight electrons are necessary to reduce nitrate (oxidation state +5) to ammonium (oxidation state −3). The reduction occurs in two main steps.

The first step is the reduction of nitrate to nitrite, catalyzed by nitrate reductase (NaR) (Davison & Stewart 1984)

$$NO_3^- + NAD(P)H + H + \rightarrow NO_2^- + NAD(P) + H_2O$$

This enzyme has been isolated and purified from several species of microalgae (Syrett 1981). A few recent measurements indicate that it is also present in seaweeds (Weidner & Kiefer 1981, Wheeler & Weidner 1983). Nitrate reductase is a relatively large molecule (350,000 daltons) and has as cofactors molybdenum, heme, and flavin adenine dinucleotide (FAD). The electron donor is usually NADH, but there are some microalgae in which the donor is NADPH instead (Lee 1980, Syrett 1981). The enzyme is thought to occur in the cytoplasm, although preliminary evidence suggests that it may be associated with chloroplast membranes. A technique for measuring NaR activity has been worked out for microalgae (Eppley 1978) and adapted to seaweeds (Weidner & Kiefer 1981). Two problems in the measurement of NaR in seaweeds are the grinding of tough, often rubbery, thalli and the presence of phenolics, which tend to inactivate the enzyme. The problems, particularly pronounced in the Phaeophyta, have been overcome by adding polyvinyl pyrrolidone to bind the phenolics (e.g., Haxen & Lewis, 1981) or by purification steps to remove alginate, which also interferes with enzyme activity (Kerby & Evans 1983). Longitudinal and transverse profiles of nitrate reductase activity within the thallus of *Laminaria digitata* have been investigated by Davison & Stewart (1984). When plants were growing at their maximum rate, NaR activities were highest in the mature blades and decreased toward the meristem. This is consistent with the hypothesis that meristematic growth is maintained by translocation of organic nitrogen from mature tissue. The transverse profile of NaR activities in the stipe is similar to that of carbon-assimilation enzymes (Kremer 1980), with highest activity in the photosynthetic meristoderm. NaR activities were low in the stipe and holdfast.

An alternative to reducing nitrate to nitrite in the cytoplasm is to store nitrate in the vacuole (Fig. 6.8). Intracellular nitrate pool analysis indicates that a substantial amount of nitrate may accumulate in the cytoplasm/vacuole in some intertidal macroalgae, especially when NaR is relatively inactive (Thomas 1983). From earlier studies on phytoplankton, NaR had been thought to provide a good index of nitrate uptake, but more recent work shows a poor correlation between uptake and enzyme activity (Syrett 1981). It now appears that the enzyme activity may underestimate uptake by as much as an order of magnitude. Storage of nitrate in the vacuole, or a high nitrate efflux (excretion) may partially

Figure 6.9. Pathways of incorporation of ammonia. GDH = glutamate dehydrogenase; GOGAT = glutamate-oxoglutarate aminotransferase. (a and b based on Syrett 1981, reproduced by permission of the Minister of Supply and Services, Canada)

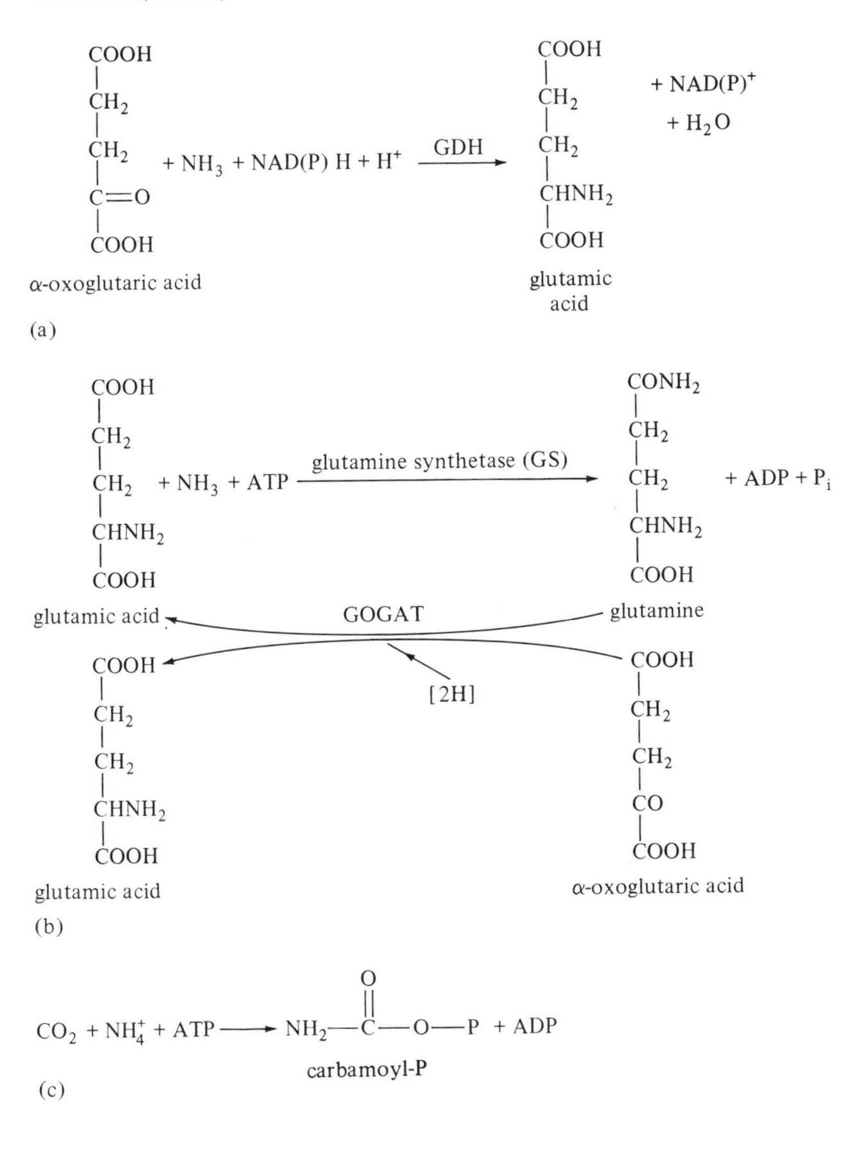

account for this poor correlation. If the same pattern holds true for seaweeds, NaR activity should be used only as a qualitative measure of nitrate utilization.

After nitrate is reduced to nitrite, nitrite is transported to the chloroplasts for reduction to ammonium (Fig. 6.8). This reduction is catalyzed by nitrite reductase (NiR), which is thought to be located in the chloroplasts, according to higher plant studies:

$$NO_2^- + 6Fe_{red} + 8H^+ \rightarrow NH_4^+ + 6Fe_{ox} + 2H_2O$$

(where Fe_{red} and Fe_{ox} are the reduced and oxidized forms, respectively, of ferredoxin). Although the six electrons are probably added sequentially in pairs, the two possible intermediates, nitric oxide and hydroxylamine, have not been detected. Both are toxic, and are presumed to remain enzyme bound (see Bidwell 1979). Nitrite reductase has an iron prosthetic group and uses ferredoxin as a cofactor to supply the electrons.

Ammonium incorporation into amino acids was thought at first to begin with the formation of glutamic acid via glutamate dehydrogenase (GDH) (Fig. 6.9a). However, in the early 1970s work with bacteria revealed an alternate pathway (Tempest et al. 1970), which was subsequently shown to operate in leaves of higher plants as well as in blue-green and green algae (Miflin & Lea 1977). In this pathway (Fig. 6.9b), glutamine rather than glutamate is the first product of ammonium assimilation. Glutamic acid is formed later by a second reaction in which the amide group of glutamine is transferred to α-oxoglutaric acid (α-ketoglutarate) from the Krebs cycle, forming two molecules of glutamate (Syrett 1981). The enzyme catalyzing the formation of glutamine is glutamine synthetase (GS) and the enzyme catalyzing the second reaction is glutamine-oxoglutarate aminotransferase (GOGAT) (also referred to by its older name, glutamate synthase).

Recent studies have shown that the GS-GOGAT pathway is the primary route for ammonium incorporation into amino acids in the fronds of *Macrocystis angustifolia* (Haxen & Lewis 1981), *Giffordia mitchelliae* (Schmitz & Riffarth 1980) and *Laminaria digitata* (Davison & Stewart 1984). The activity of GDH in *Macrocystis* is evidently not significant in the assimilation of ammonium produced from nitrate reduction, even if a large concentration of ammonium is created by inhibiting GS with a specific inhibitor, such as methionine sulfoxide (Haxen & Lewis 1981). An investigation of chloroplasts isolated from the siphonous green alga *Caulerpa simpliciuscula* showed that both pathways of ammonium assimilation (i.e., GS-GOGAT and GDH) are present simultaneously (McKenzie et al. 1979). GDH had an unusually low K_M for ammonium (0.4 to 0.7 mM), indicating that the enzyme could provide an alternative means of ammonium incorporation into amino acids in this species (Gayler & Morgan 1976). Other studies on phytoplankton suggest that under conditions of high external ammonium, the GDH pathway is operative; it has a low affinity for intracellular ammonium (K_M ca. 5–28 mM) and does not require ATP (Syrett 1981). Under low ammonium concentrations or in nitrate-grown cells, the GS-GOGAT pathway operates; it has a very high affinity for intracellular ammonium (K_M ca. 29μM for a marine diatom; Falkowski & Rivkin 1976) and requires ATP. This view is attractive but still lacks conclusive evidence.

Another, minor, pathway of ammonium assimilation is the carbamoyl phosphate pathway, in which ammonium is incorporated into citrulline and arginine and thence into pyrimidines (including thiamine) and biotin (Fig. 6.9c; see also Fig. 7.5). However, the ammonia for carbamoyl phosphate synthesis is derived from the amide group of asparagine or glutamine rather than from free ammonium (Beevers 1976).

The control of these various pathways is complex. There are three main controls: the interaction among various nitrogen compounds, the effect of light on the pathways, and the past nutritional history of the alga. As previously mentioned (Sec. 6.4.2), most algae take up ammonium rather than nitrate when both are present in the medium. Active NaR is not formed in the presence of ammonium, nor is the nitrate uptake system present. When active NaR and a nitrate uptake system are initially present, addition of ammonium can lead to a cessation of nitrate use in about one hour. The lag suggests that a product of ammonium assimilation (e.g., glutamine) rather than ammonium itself is the repressor of NaR (Syrett 1981). Low NaR activity has been reported when cells grown in ammonium are transferred to nitrogen-free medium. This observation has been explained by the formation of small amounts of nitrate, perhaps by intracellular nitrification (Spiller et al. 1976). There are at least three mechanisms by which NaR activity can disappear from cells. There are two reversible inactivation phenomena and one irreversible loss of the enzyme, perhaps caused by degradation. Inactivation appears to occur by reduction of the enzyme at the molybdenum site; nitrate provides some protection against this. Solomonson & Spehar (1977) suggested that NaR activity is regulated by cyanide, which is produced by an interaction between glyoxylate (from photorespiration) and hydroxylamine (from nitrate reduction).

Table 6.8. *Possible interactions of light with inorganic nitrogen metabolism of algae*

Photosynthetic (chloroplast) effects
Generation of reduced ferredoxin, which then:
Reduces NO_2^- (and N_2 and NO_3^- in blue-greens)
Reduces NAD(P)H and hence NO_3^- in eukaryotic algae
Drives GOGAT reaction of NH_4^+ assimilation
Activates/inactivates enzymes via thioredoxin
Generation of ATP through photophosphorylation, which then:
Is used to drive transport mechanisms for NO_3^-, NO_2^-, NH_4^+
Drives GS reaction of NH_4^+ assimilation
Stops the reoxidation of mitochondrial NADH by O_2, making this NADH available for NO_3^- reduction
Drives N_2 fixation in blue-greens
Photosynthetic fixation of CO_2 makes C acceptors available for NH_4^+ assimilation, thus removing feedback inhibition by organic N compounds of NO_3^- (NO_2^-) uptake
Other effects
Phytochrome (red light) effects?
Direct enzyme activation/inactivation by blue light possibly mediated through flavoproteins
Effects of light quality on protein synthesis

Source: Syrett (1981), reproduced with permission of the Minister of Supply and Services, Canada.

The effect of light on nitrate and ammonium assimilation in seaweeds has not been well studied. More extensive studies on phytoplankton have revealed that the interactions are complex (Table 6.8) and dependent on the metabolic state of the cells (Table 6.9). Both nitrate and ammonium assimilation can take place in darkness if enough carbon reserves are available. Because reduced ferredoxin is also required for NiR, there must be catabolic reactions that can lead to ferredoxin reduction.

Table 6.9. *Interaction of light and metabolic state in determining NH_4^+ or NO_3^- assimilation by green microalgae*

		Metabolic state of cells		
		Carbon-starved $-CO_2$	Normal growth $+CO_2$	Nitrogen-starved
Storage of C compounds		Nil	Very low	High
Rate of assimilation of				
NH_4^+	Light	0	+	++++
	Dark	0	0	+++
NO_3^-	Light	+[a]	+	++
	Dark	0	0	+[b]
NH_4^+ inhibition of NO_3^- uptake		0	+[c]	±[c]

[a]Ammonium accumulates. [b]Nitrite accumulates. [c]With dependence on NH_4^+ concentration.
Source: Syrett (1981), reproduced with permission of the Minister of Supply and Services, Canada.

Urea is usually assimilated by being first broken down into carbon dioxide and ammonium via the enzyme urease and then the free ammonium being incorporated into amino acids via the GS-GOGAT pathway. One study suggests that urea may be directly assimilated into organic nitrogen (Kitoh & Hori 1977). Purified urease contains nickel, and the development of urease activity in some marine diatoms has a strong dependence on nickel (Syrett 1981). There are some algae, mainly belonging to certain orders of the Chlorophyceae such as Volvocales, Chlorococcales, Chaetophorales, and Ulotrichales, which contain a urea carboxylase instead of urease (Syrett 1981).

Assimilation of nitrogen by nitrogen-deficient or nitrogen-limited cells is limited by the rate of protein synthesis, as suggested by the early work of Syrett (1956). This has been substantiated recently by evidence for the accumulation of internal pools of nitrate, ammonium, and free amino acids after the addition of nitrogen to nitrogen-limited cultures of phytoplankton (DeManche et al. 1979, Dortch 1982) and seaweeds (Haxen & Lewis 1981, Thomas 1983). Such pools would not accumulate if rates of protein synthesis were equal to or greater than rates of membrane transport and subsequent metabolism to amino acids. Several species of *Laminaria* accumulate nitrate, and tissue levels of nitrate represent a significant portion of the total nitrogen in the plant (Chapman & Craigie 1977). In many brown algae, however, internal nitrate levels are low and never account for more than 5% of the total tissue nitrogen (Buggeln 1978, Wheeler & North 1980, Asare & Harlin 1983). Thus, storage of nitrate nitrogen is not widespread in the brown algae. Amino acids, especially alanine, and proteins appear to form the major storage pools in *Gracilaria tikvahiae* (Bird et al. 1982) and *Macrocystis pyrifera* (Wheeler & North 1980).

Little is known about catabolism and turnover of cellular protein. Recycling of nitrogen may occur via the photorespiratory pathway (see Fig. 7.5) and by catabolism of specialized nitrogen storage compounds, such as guanine (Pettersen 1975) and arginine (Wheeler & Stephens 1977) in the Chlorophyceae. Pigments and associated proteins such as phycoerythrin may serve as nitrogen storage compounds in the Rhodophyceae (Gantt 1980; Bird et al. 1982). Although some enzymes, such as nitrate reductase, turn over constantly, this recycling of nitrogen is thought to be of minor importance.

The roles of nitrogen in cellular metabolism are well known (Table 6.7); it is a component of all amino acids (hence enzymes and other proteins) and nucleotides (including nucleic acids, FAD, FMN, coenzyme A, and ATP), as well as some pigments (chlorophylls, biliproteins, and cytochromes).

6.5.2 Phosphorus

Phosphorus is generally not considered to be a limiting nutrient in the marine environment, and there are almost no studies on phosphorus uptake kinetics in seaweeds, despite the fact that phosphorus concentrations are frequently near the limit of detection when nitrogen is exhausted from the seawater. Preliminary studies indicate that it appears to be taken up actively in *Porphyra* (Eppley 1958), and saturation kinetics have been obtained for the red alga *Agardhiella subulata*, yielding a V_{max} of 0.47 μmol g dry wt^{-1} h^{-1} and a K_s of 0.4 μM (DeBoer 1981).

The major form in which algal cells acquire phosphorus is as orthophosphate ions. Other sources are inorganic polyphosphates and organic phosphorus compounds. Sugar phosphates are reportedly taken up intact by bacteria (Rubin et al. 1977), but most eukaryotic algae require extracellular enzymatic hydrolysis to remove sugar before phosphate uptake (Nalewajko & Lean 1980). Polyphosphates may also require extracel-

Figure 6.10. Main features of phosphorus uptake and assimilation in a microalgal cell. DOP = dissolved organic phosphate; Pi = inorganic phosphate. (From Cembella et al. 1983, reprinted with permission from *Critical Reviews of Microbiology* 10:317–391, © 1983 CRC Press, Boca Raton, Fla.)

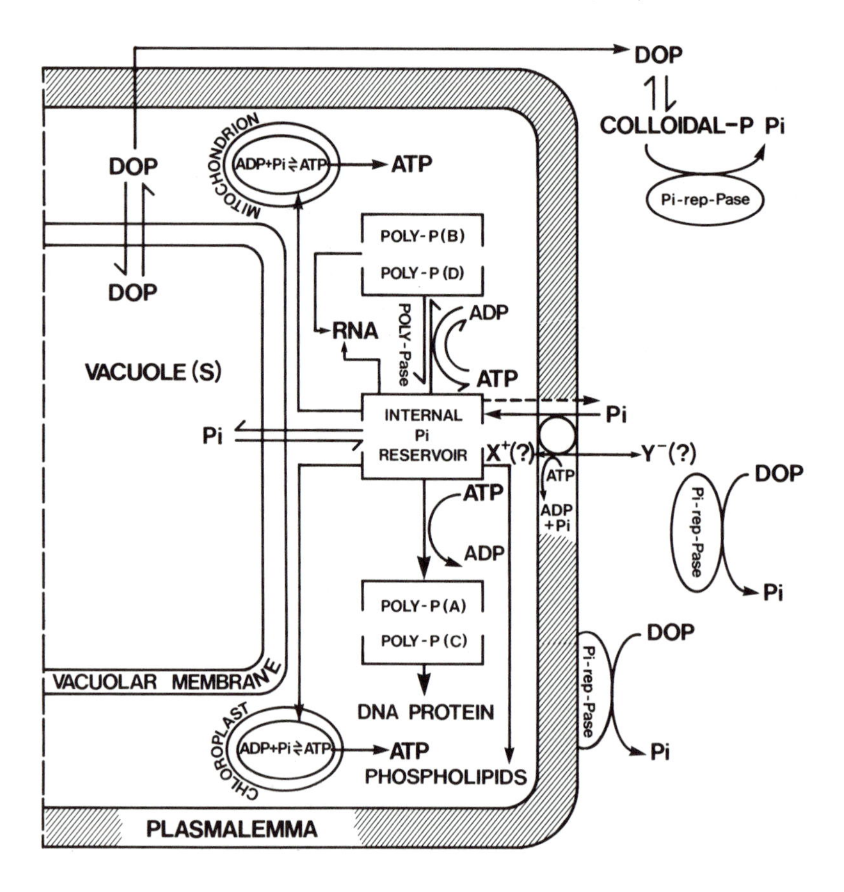

lular cleavage; the freshwater macroalga *Cladophora glomerata* breaks down pyrophosphate and triphosphate (both common in detergents) and takes up the phosphorus as orthophosphate (Lin 1977). Some seaweeds can use some organic forms of phosphate, such as glycerophosphate, by producing extracellular alkaline phosphatase (Walther & Fries 1976). The ability of cells to enzymatically cleave the ester linkage joining the phosphate group to the organic moiety is achieved by the activity of phosphomonoesterases (commonly called phosphatases) at the cell surface. Two groups of these enzymes have been distinguished on the basis of their pH optima, phosphate repressibility, and cellular location. Alkaline phosphatases are phosphate repressible, inducible, and generally located on the cell surface or released into the surrounding seawater. Acid phosphatases are phosphate irrepressible, constitutive, and generally found intracellularly in the cytoplasm. Both types may be found simultaneously in algal cells, with alkaline phosphatases aiding in uptake of organic phosphorus compounds and acid phosphatases playing a crucial role in cleavage and phosphate transfer reactions in metabolic pathways within the cell. The essential feature of phosphatases that allows them to participate efficiently in cellular metabolism is their ability to be alternately induced or repressed, depending on metabolic requirements. When external inorganic phosphate concentrations are high, the synthesis of alkaline phosphatase is repressed and cells exhibit little ability to use organic phosphorus compounds. Upon entry into phosphorus limitation, cells typically exhibit an increase in alkaline phosphatase activity. The magnitude of the increase depends on a species-specific response, the availability of organic phosphates, and the degree of phosphate limitation experienced by the cells (Healey 1973, Cembella et al. 1983). Generally, after the external inorganic phosphate has been exhausted, intracellular phosphorus from stored polyphosphates and orthophosphate is used up quickly, followed by increased alkaline phosphatase activity.

Inorganic phosphate transported across the plasmalemma enters a dynamic intracellular phosphate pool from which it is incorporated into phosphorylated metabolites or stored as luxury phosphorus in vacuoles or in polyphosphate vesicles in microalgae (Fig. 6.10). Some of the cytoplasmic phosphate pool may leak back out of the cell and reappear as external phosphate. Phosphorus-

Figure 6.11. Interconversion of inorganic and organic phosphate storage compounds. (From Cembella et al. 1983, reprinted with permission from *Critical Reviews of Microbiology* 10:317–391, © 1983 CRC Press, Boca Raton, Fla.)

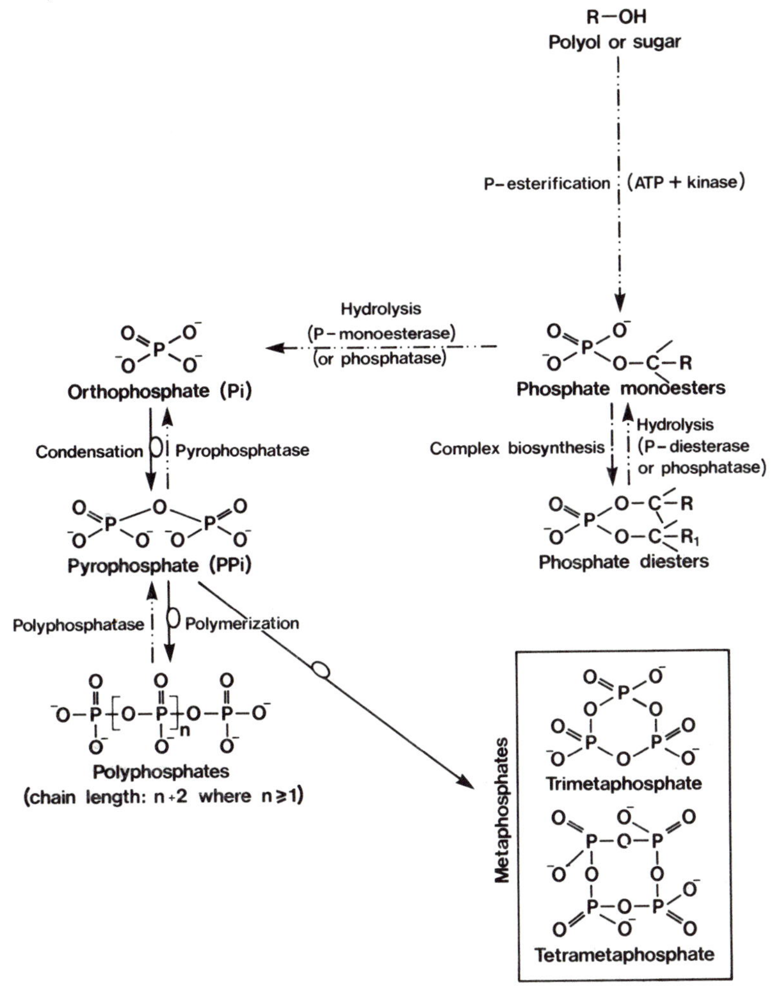

deficient algae possess the ability to incorporate phosphate extremely rapidly, and the amount taken up usually exceeds the actual requirements of the cell. The excess is built into polyphosphates by the action of polyphosphate kinase:

$$\text{ATP} + (\text{polyphosphate})_n \rightarrow \text{ADP} + (\text{polyphosphate})_{n+1}$$

(Kuhl 1974, Cembella et al. 1983). An important difference between phosphorus metabolism in vascular plants and algae is the formation of these polyphosphates; seaweeds known to form polyphosphates include species of *Acetabularia, Enteromorpha, Ceramium,* and *Ulothrix* (Kuhl 1962). The storage compounds are classified into cyclic and linear polyphosphates (Fig. 6.11). These two types cannot be easily separated by simple extraction procedures but can be divided into four categories (A, B, C, D; see Fig. 6.10) on the basis of sequential extraction techniques (Cembella et al. 1983). Intracellular polyphosphate acts as a noncompetitive inhibitor of phosphate uptake in phosphate-limited cultures of the freshwater unicellular green *Scenedesmus* (Rhee 1974). The kinetics could be described by an equation similar to that for enzyme kinetics under noncompetitive inhibition (Table 6.5: equation 5).

Phosphorus plays a key role in many biomolecules, such as nucleic acids, proteins, and phospholipids (the latter are important components of membranes). Its most important role, however, is in energy transfer through ATP and other high energy compounds in photosynthesis and respiration (Fig. 6.10), and in "priming" molecules for metabolic pathways.

6.5.3 Calcium and magnesium

Calcium deposits in the coralline red algae and a few calcareous green algae such as *Halimeda* are primarily extracellular or intercellular. Consequently, in these algae, calcium uptake is not metabolically con-

Table 6.10. *Site of formation and organization of the calcium carbonate deposits in calcifying marine algae*

	Group A	Group B	Group C
Site of deposit	Deposits formed on and within organic matrix within Golgi cisterna	Deposits formed in organic matrix of cell wall or mucilage	Deposits generally wholly extracellular
Organization	Extreme organization	Some organization near plasmalemma	No organization
Nature of deposit	Calcite	Calcite	Aragonite
Taxa	Coccolithophorids	Corallinaceae, some Cyanophyceae	Chlorophyta, Phaeophyta (*Padina*), Nemalionales, Peyssonneliaceae(?), some Cyanophyceae

Source: Modified from Borowitzka et al. (1974), with permission of Blackwell Scientific Publications.

trolled and calcium may remain bound to acidic wall polysaccharides. In the freshwater macrophyte *Cladophora glomerata*, calcium adsorption to cell walls was separated from internal transport (Sikes 1978). Intracellular calcium transport appeared to occur by active transport. Although strontium competed for calcium binding sites of acid polysaccharides, it did not inhibit internal transport of calcium, suggesting that the carrier for calcium may be specific.

Calcium is important to all organisms in the maintenance of cellular membranes. It is also a major component of the walls of several members of the Chlorophyceae, Rhodophyceae, and Phaeophyceae. Some of these algae, particularly the Rhodophyceae, contribute significantly to coral reef formation, and the mechanisms of uptake and deposition of calcium have been studied (recently reviewed by Darley 1974, Littler 1976, and Borowitzka 1977, 1982). Calcium is deposited as calcium carbonate, sometimes along with small amounts of magnesium and strontium carbonates. $CaCO_3$ occurs in two crystalline forms, calcite (hexagonal-rhombohedral crystals) and aragonite (orthorhombic), which never occur together in the same alga under natural conditions.

Studies of the structure, localization, and organization of the calcium carbonate deposits of algae suggest that there are at least three different mechanisms of calcification (Table 6.10), including the complex process in Coccolithophorids. In seaweeds (and some other macroalgae) there is a mechanism for calcite deposition and a mechanism for aragonite deposition. In red algae of the Corallinaceae, long calcite crystals are oriented at right angles to the plasmalemma, except in meristematic, genicular, and reproductive cells, which are noncalcified. The calcite crystals are in close association with the organic cell wall material, and some cell wall components probably act as a template for deposition and orientation of new crystals. Studies of the compartmentation of exchangeable Ca^{2+} in the thallus of *Amphiroa foliacea* indicate that there are at least two major organic calcium binding and exchange compartments, which are presumed to be the COO^- and $O–SO_2–O^-$ groups of the acidic polysaccharide wall compounds (Borowitzka 1979). Since aragonite is the normal crystal form of calcium carbonate precipitated from seawater, the organic wall material of the corallines is presumed to be responsible for deposition as calcite.

Little is known about the actual mechanism of calcification in these algae except that it is directly proportional to photosynthesis (carbonate uptake and hydroxyl ion extrusion, causing an increase in pH) and that deposition is stimulated by light and is highest in young tissue. There is some suggestion based on stable isotope ratios of $^{13}C:^{12}C$ and $^{18}O:^{16}O$ that the carbonate is derived from metabolic carbon dioxide. Inhibitory effects of orthophosphate on coralline calcification have been reported by Brown et al. (1977), although the actual mechanism is unknown. This explains the observation that when coralline algae are grown in phosphorus-enriched culture medium (30–150 μM PO_4^{3-}) they are only weakly calcified. These observations suggest that the growth of corallines could be inhibited in phosphate-polluted coastal waters.

Aragonite deposition has been thoroughly studied in the green alga *Halimeda*, and is reasonably well understood. Aragonite needles are deposited entirely outside the cell wall within the intercellular space, which is separated from the external seawater by the appressed tips of the utricles (Fig. 6.12). Passage of inorganic carbon and calcium ions into this intercellular space must be either through the cells or by diffusion over a long path through the cell walls. During photosynthesis CO_2 is taken up from the intercellular space, resulting in an increase in intercellular pH and CO_3^{2-} concentration, with subsequent deposition of aragonite. This pH-induced deposition hypothesis for calcification in *Halimeda* is not applicable to all aragonite depositors, for in the brown alga *Padina* there are no intercellular spaces and aragonite is precipitated in concentric bands on the

Figure 6.12. Schematic representation of the postulated ion fluxes affecting calcium carbonate precipitation in *Halimeda*. A black dot at the plasmalemma indicates that the flux is postulated to be active. (From Borowitzka 1977, with permission of *Oceanography and Marine Biology: an Annual Review*)

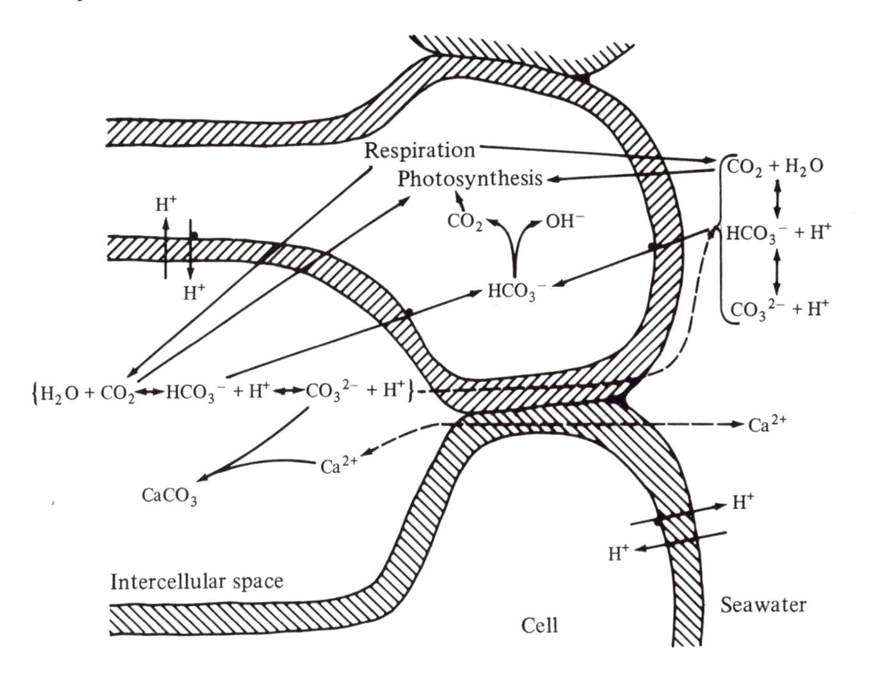

outer surface of the thallus. Moreover, there are other seaweeds with apparently suitable morphology (intercellular spaces) that do not calcify (e.g., *Enteromorpha*) (Borowitzka et al. 1974, Borowitzka 1977).

Calcium, along with other divalent cations, has been shown to activate adenosine triphosphatase in three calcareous algae (Okazaki 1977). Activation by calcium was stronger than by other divalent cations, which in fact competed with Ca^{2+}. Another role of calcium is in the development of cellular polarity in fucoid eggs (see Sec. 10.3.1).

Magnesium is an essential cofactor or activator in many reactions, such as nitrate reduction, sulfate reduction, and phosphate transfers (except phosphorylases). It is also important in several carboxylation and decarboxylation reactions, including the first step of carbon fixation, where the enzyme ribulose-1,5-bisphosphate (RuBP) carboxylase attaches CO_2 to RuBP. Magnesium also activates enzymes involved in nucleic acid synthesis, and binds together the subunits of ribosomes. There are several means by which magnesium may act (Bidwell 1979): (1) it may link enzyme and substrate together, as, for example, in reactions involving phosphate transfer from ATP; (2) it may alter the equilibrium constant of a reaction by binding with the product, as in certain kinase reactions; (3) it may act by complexing with an enzyme inhibitor; (4) it can form metalloporphyrins, such as chlorophyll; and (5) it can play a role in binding charged polysaccharide chains to one another, since it is a divalent cation.

6.5.4 Potassium and sulfur

Sodium, potassium, and chloride ions are never likely to be limiting to macroalgal growth in the marine environment (in contrast to nitrogen or possibly phosphorus). However, interest in their uptake rates is associated with understanding osmoregulatory processes. Radioisotope equilibration technqiues have been used to study influx and efflux of K^+, Na^+ and Cl^- in *Porphyra purpurea* (Reed & Collins 1980). The initial rapid uptake of these ions was due to extracellular adsorption. Plasmalemma transport showed saturation kinetics with cells discriminating in favor of K^+ and Cl^- and against Na^+. Hence, there was an active uptake and a passive loss of K^+ and Cl^- and an active efflux of Na^+.

The role of potassium in ionic relations is nonspecific, as it is only one of several monovalent cations involved. Potassium has a more specific role as an enzyme activator (O'Kelley 1974); many protein synthesis enzymes do not act efficiently in the absence of K^+, but the way in which K^+ binds to the enzymes and affects them is not well understood. It is known to bind ionically to pyruvate kinase, which is essential in respiration and carbohydrate metabolism (Bidwell 1979). While rubidium may in some cases or to a certain extent substitute for potassium, West & Pitman (1967) have shown that the rate of uptake of $^{86}Rb^+$ by *Ulva lactuca* and *Chaetomorpha darwinii* is very much slower than uptake of $^{42}K^+$. More recently, the kinetics of ^{86}Rb exchange have been used to study intracellular compartmentation of K^+ in *Porphyra* (Reed & Collins 1981).

Sulfate uptake rates in *Fucus serratus* showed Michaelis-Menten-type saturation kinetics (K_S = 6.9 × 10^{-5} M), but Coughlan (1977) also noted inhibition by selenate, molybdate, tungstate, and especially chromate. The kinetics of this inhibition were not worked out. In the economically important red alga *Chondrus crispus*, sulfate uptake was multiphasic, which is similar to the pattern in higher plants (Jackson & McCandless 1982). The value of K_S was about 3 mM. Sulfate uptake by the unicellular marine red alga *Rhodella maculata* was biphasic with a K_S of 22 mM for the low affinity system and 63 μM for the high affinity system (Millard & Evans 1982).

Most algae can supply all of their sulfur requirements by reducing sulfate, the most abundant form of sulfur (25 mM) in the aerobic marine environment (O'Kelley 1974). For this reason few studies have been conducted on the use of other forms of sulfur such as sulfite and organic sulfur-containing compounds. Sulfur nutrition and utilization in algae has been reviewed by O'Kelley (1974), Schiff (1980, 1983), and Raven (1980). Before sulfate can be incorporated into various compounds, it must be activated, since it is a relatively unreactive compound (Schiff & Hodson 1970). The enzyme ATP sulfurylase catalyzes the substitution of SO_4^{2-} for two of the phosphate groups of ATP, to form adenosine-5′-phosphosulfate (APS) (Fig. 6.13). APS can have another phosphate added from another ATP to form adenosine-3′-phosphate-5′-phosphosulfate (PAPS), which is believed to be the starting point for sulfate ester formation in many systems and for sulfate reduction. (Like nitrogen, sulfur is incorporated into proteins in its most reduced form.) Although the majority of the studies on algal sulfate reduction have been done on freshwater unicells, a sulfite (SO_3^{2-}) reductase has been demonstrated in several marine red algae (e.g., *Porphyra* spp.) and green algae (O'Kelley 1974). The significance of an enzyme with the ability to reduce sulfite was questioned by Schiff & Hodson (1970), who pointed out that sulfite is extremely reactive and therefore may react with any nonspecific reductase; they believe that thiosulfate ($S_2O_3^{2-}$) is the intermediate rather than sulfite. Coughlan (1977) recently showed that the SO_4^{2-} → APS → PAPS activating system is also operative in *Fucus serratus*, but that the bulk of the sulfate is attached to fucoidan without further reduction.

Much of a cell's sulfur is incorporated into proteins. Two sulfur-containing amino acids, cysteine and methionine, are very important in maintaining the three-dimensional configurations of proteins through sulfur bridges. Sulfur is also part of the biologically important molecules biotin and thiamine (Fig. 6.11), and of coenzyme A. However, many algae, especially seaweeds, produce commercially valuable sulfated polysaccharides, which are important in thallus rigidity (e.g., carrageenan in red algae) and adhesion (e.g., fucoidan in brown algae) (see Secs. 7.5.2 and 10.3.1). The uptake of sulfate by *Fucus serratus* has been studied in this connection (Coughlan 1977). Other sulfur-containing compounds in seaweeds include, in red algae, taurine and its derivatives (sulfur at the sulfite level of reduction) (O'Kelly 1974, Ragan 1981). Some species of *Desmarestia* have so much sulfuric acid in their vacuoles that the pH is close to 1; this may serve as a grazer deterrent (Sec. 9.4.1). Crystalline sulfur has been found in *Ceramium rubrum* and shown to be responsible for the toxicity of this alga to the bacterium *Bacillus subtilis*; *C. rubrum* is unusual in having high free sulfur content (Ikawa et al. 1973).

6.5.5 Iron

To date there is only one study of iron uptake by seaweeds. Saturation kinetics were exhibited for iron uptake by *Macrocystis pyrifera* (Manley 1981). Labeled iron exchanged more slowly from the free space than did other divalent cations. The slow exchangeability of iron may reflect the higher affinity of the cell wall and intercellular constituents for ferric ion. Manley (1981) also found that when bathophenanthroline disulfonate (BPDS) was added to the culture medium, the inhibition of iron uptake was immediate and drastic. Since BPDS has a high affinity and specificity for chelating the ferrous ion, there may be a reduction of Fe^{3+} to Fe^{2+} before iron is taken up. This is similar to phytoplankton (Anderson & Morel 1982) and higher plants (Brown 1978). After the iron is taken up by *Macrocystis* it enters the sieve tubes, where it reaches 11 times the external concentration. Manley postulated that the iron was chelated by some organic compound and translocated to juvenile fronds. This iron chelation is similar to higher plants in which ferrous ion is oxidized to ferric in the xylem and chelated with citrate (Brown 1978).

Recent experimentation suggests that phytoplankton take up iron mainly as the ferrous ion (Fe^{2+}), but ferric ions (Fe^{3+}) may be taken up also (Anderson & Morel 1980, 1982). Ferrous ions were previously thought to be present in very low amounts, but recent evidence suggests that they can be formed by (1) oxidation rates ($Fe^{2+} \rightarrow Fe^{3+}$) that are 150 times slower than previous estimates (Andseron & Morel 1982); (2) chelators that can chemically or photochemically reduce Fe^{3+} to Fe^{2+} (Huntsman & Sunda 1980); and (3) direct photoreduction of Fe^{3+}. Other suggested iron sources have been iron oxides, which, often in association with colloids, adsorb onto the cell surface and subsequently dissolve, releasing iron for transport into the cell. However, since iron oxide dissolution kinetics are not well understood, the possible importance of this process is unknown.

In special cases when bacteria and algae have been grown under iron limitation, they have produced extracellular chelators called siderophores (Neilands 1973, 1981). Siderophores are low molecular weight organics that coordinate and solubilize the ferric ion by forming a highly stable complex that may be taken up by the

Figure 6.13. Sulfate activation. APS = adenosine phosphosulfate; PAPS = phosphoadenosine phosphosulfate. (From Schiff & Hodson 1970, with permission of the New York Academy of Sciences)

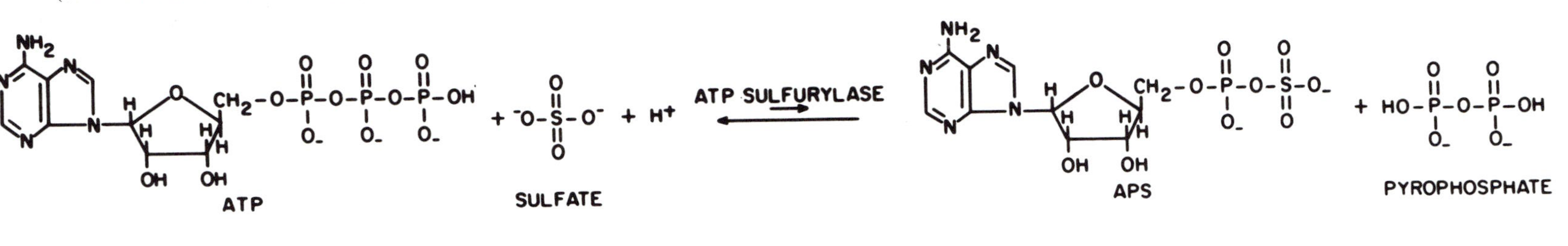

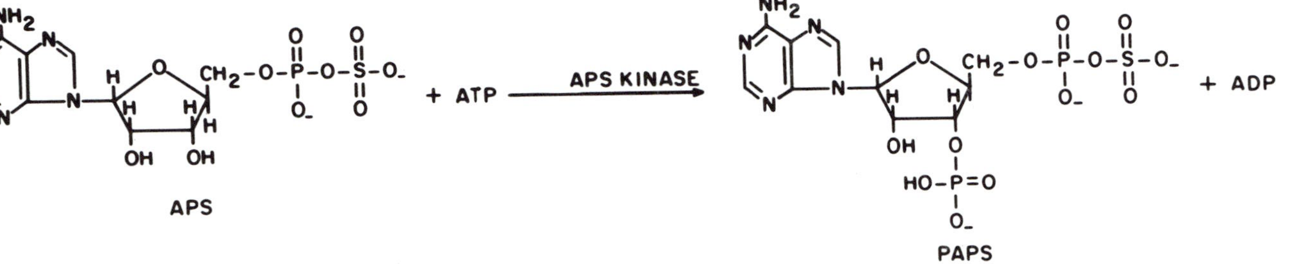

cell. Siderophore production is induced under iron-limiting conditions and repressed when the iron concentration is high. There are two main types of siderophores. The catechol type binds iron via three catechol groups, while the hydroxamate type has three hydroxamate groups, which form a ferric complex. There is evidence that both prokaryotic (Huntsman & Sunda 1980) and eukaryotic (Trick et al. 1983) algae produce siderophores under low iron concentrations. While the actual mechanism of iron uptake is unknown, one possibility is that the siderophore forms a complex with the ferric ion. This complex may be taken across the membrane or reduced at the cell surface, liberating the ferrous ion with concomitant recycling or extracellular release of the siderophore. Because the complexing strength of the hydroxamic acids is much greater for the ferric ion than for the ferrous ion, the reduction of the ferric ion complex provides a means of releasing the complexed iron and freeing the ligand to pick up more ferric ion. Other possible mechanisms are discussed by Neilands (1973).

Iron plays a parallel role to magnesium in chlorophyll, in being at the center of the cytochrome molecules, which transfer electrons in the respiratory chain and in photosynthesis. The importance of iron in this connection lies in its ability to change valence between Fe^{2+} and Fe^{3+}, but iron is also present in a number of oxidizing enzymes (such as catalase) in which it does not change valence (Bidwell 1979). While iron is not part of the chlorophyll molecule, it is required as a cofactor in the synthesis of chlorophylls, at least in *Euglena* (O'Kelley 1974). Another important iron-containing molecule, involved in photosynthesis and other electron transfers, is ferredoxin, a protein that also contains labile sulfur (in addition to the sulfur in amino acids) (Lehninger 1975).

6.5.6 Trace metals

Zinc, cesium, strontium, cobalt, molybdenum, and rubidium uptake kinetics have been studied (Gutknecht 1965, Penot & Videau 1975, Floc'h 1982). Uptake of zinc by *Ascophyllum nodosum* (Skipnes et al. 1975) and cobalt in *Laurencia corallopsis* (Bunt 1970) appears to be by active transport. A simple exchange process involving intracellular polysaccharides is the main uptake mechanism for strontium (Skipnes et al. 1975), and in some algal species it is also the uptake mechanism for zinc (Gutknecht 1963, 1965).

The effects of Cu, Zn, Mn, and Co on the growth of gametophytes of *Macrocystis pyrifera* have been examined in chemically defined medium (Kuwabara 1982). His results indicated that toxic copper and zinc ion concentrations, together with cobalt and manganese deficiencies, may be among the factors controlling the growth of some marine macrophytes in deep seawater off southern California.

The principal roles of Mn, Cu, Zn, and Mo are as enzyme cofactors (Table 6.6). Manganese plays a vital role in the oxygen-evolving system of photosynthesis and is a cofactor in several Krebs cycle enzymes (Bidwell 1979). Copper is present in plastocyanin, one of the photosynthetic electron transfer molecules, and is a cofactor in some enzyme reactions (Bidwell 1979). Zinc, in higher plants, is an activator of several important dehydrogenases and is involved in protein synthetic enzymes. It is essential to algae and probably plays similar roles in them (O'Kelley 1974). McLachlan (1977) could not demonstrate a requirement for Zn, Cu, or Mn by embryos of *Fucus edentatus* in defined medium, but he concluded that probably these elements were present as contamination in sufficient concentrations to satisfy nutritional needs. Zinc, at an optimum concentration of 0.5 nM, was shown by Noda & Horiguchi (1977) to be required by *Porphyra tenera*; without it, chlorophyll and phycobilin production was hindered and high molecular weight protein content decreased. Molybdenum is most important, in algae as well as in other plants, in nitrate reduction. Like iron, molybdenum can participate in redox reactions by its ability to change valence, in this case between Mo^{5+} and Mo^{6+}. Nasr & Bekheet (1970) reported that addition of trace amounts of ammonium molybdate increased the dry weight of *Ulva lactuca, Dictyota dichotoma,* and *Pterocladia capillacea*; presumably much of the increase can be attributed to molybdenum stimulation of nitrate reductase, since there was only a trace of ammonium in the culture medium.

Fries (1982a) has reported that additions of selenium increase the growth of *Fucus spiralis* and the red alga *Goniotrichum alsidii*. She found it necessary to add 0.01 μM Se to artificial seawater to obtain normal growth. Vanadium, at a concentration of about 10 $\mu g\ L^{-1}$, is required for maximal growth of some macroalgae (Fries 1982b).

6.5.7 Vitamins

The main role of vitamins in algae appears to be as enzyme cofactors (Swift 1980). Thiamine, for example, as thiamine pyrophosphate, is a cofactor in the decarboxylation of pyruvic acid and other α-keto acids. Biotin is a cofactor in carboxylation reactions.

6.6. Long-distance transport (translocation)

Many large kelps are similar to vascular plants in that movement of inorganic and organic compounds occurs by translocation. The movement of organic compounds and the anatomical features of translocating tissues will be discussed in Section 7.3. This section focuses only on long-distance transport of inorganic ions.

Various tracers (^{32}P, ^{86}Rb, ^{35}S, ^{99}Mo, ^{45}Ca, ^{36}Cl) have been used to show the movement of mineral elements in the thallus of *Laminaria digitata* (Floc'h 1982). Phosphorus, sulfur, and rubidium show pronounced long-distance transport, whereas chloride, molybdenum, and calcium do not seem to move. Nitrate was found in the sieve tube sap of *Macrocystis pyrifera* but no translocation measurements were made (Manley 1983).

There is a high demand for phosphorus in growing or meristematic regions. In all ^{32}P translocation studies dealing with the Fucales or Laminariales, ^{32}P movement has been from the older tissues toward the younger growing regions (Fig. 6.14). In *Cystoseira baccata* the movement of phosphorus was also toward the meristems, which in the Fucales occur at the tips of the branches (Fig. 6.14). Therefore, a source-to-sink relationship exists, similar to that observed in vascular plants. In *Laminaria hyperborea* the older parts of the blade serve as a source of phosphate for the meristematic regions, in much the same way as has been shown for carbon assimilation (Floc'h 1982). Indirect evidence of phosphorus translocation is provided, since there is no significant difference in ^{32}P uptake by young and old tissues. The older tissues thus probably translocate unused phosphorus (they have a low P requirement since they are not growing) to new, actively growing tissues.

The intensity of the labeling in autoradiographs suggests that the midrib of many Fucales is the main pathway of mineral transport (Fig. 6.14). Most of the translocation occurs through the medulla, with some secondary lateral transport from the medulla to the meristoderm (meristematic epidermis) in the stipe. Since ^{32}P has been found in the sieve tube sap of *Macrocystis* and shows the same velocity and directionality of transport as ^{14}C, it most probably moves through the sieve elements (see Section 7.5).

Since phosphate translocation frequently occurs against a concentration gradient and is related to algal metabolism, the mechanism of transport is certainly not by simple diffusion. At present there is no general agreement on the mechanism of phosphate transport or on whether it is transported in inorganic form or organically bound (Floc'h & Penot 1978, Floc'h 1982). A few hours after inorganic phosphorus was taken up it was incorporated into organic compounds, especially hexose monophosphates. Later in the experiment the ratio of inorganic phosphate to phosphate ester remained constant in the conducting zone (medulla), whereas it increased considerably in the region of uptake as well as in the sink region (Floc'h & Penot 1978). This suggests that inorganic phosphate could be one of the translocation forms. On the other hand, Schmitz & Srivastava (1979) reported ^{32}P-labeled hexose monophosphates in the fed region (blade) as well as in the sieve sap of *Macrocystis integrifolia*. This suggests that phosphorus may be translocated in an organically bound form, but it does not exclude the possibility that inorganic phosphorus may be translocated, followed by phosphorylation in the sieve tube sap. This latter possibility is similar to what occurs in higher plants (Bidwell 1979). Nitrogen shows a similar pattern to phosphorus, with the highest demand by the meristematic regions. This demand cannot be met by uptake rates, and recent studies on *Laminaria digitata* demonstrated that up to 70% of the nitrogen demand by the meristem is met by translocation from the mature blade, probably in the form of amino acids.

6.7. **Growth kinetics**

6.7.1. Theory

The classical principles of microbial growth kinetics derived by Monod (1942) for growth limited by a single substrate are based on the assumption that the formation of new biomass is simply related to uptake of the substrate. Thus, the specific growth rate is related directly to the concentration of extracellular substrate, according to equation 1 in Table 6.11. Note that this equation, generally referred to as the Monod equation, is almost identical to the equation describing the relationship between uptake rate and nutrient concentration (Table 6.5: equation 4); both equations describe a rectangular hyperbola. In Monod's experiments, the growth-limiting substrate was glucose (carbon), which was metabolized almost immediately after uptake by bacterial cells. In this special case, growth was proportional to the external concentration of the limiting substrate because glucose was not stored, and thus the yield (number of cells formed per unit of limiting substrate) remained constant.

In only a few cases for microalgae has the relationship between steady-state growth rate and external nutrient concentration been described by the Monod equation (Rhee 1980). Deviations from the equation appear to be due to increased cell mortality at low dilution rates and the ability of cells to store nutrients such as nitrogen and especially phosphorus. In many studies (Rhee 1980, 1982), external nutrient concentrations have been undetectable over a wide range of growth rates. In these cases, growth rate was related to the intracellular nutrient concentration or the cell quota, q, according to equation 2 in Table 6.11 (the Droop equation). The subsistence cell quota, q_o, is the minimum concentration of the limiting nutrient per cell required before growth can proceed. When the specific growth rate (μ) is plotted against q, the Droop equation describes a rectangular hyperbola with a threshold, q_o, and an asymptote equal to the maximum specific growth rate (μ'_m) (Fig. 6.15). In Droop's model, growth is empirically related to the total intracellular substrate concentration. From a biochemical point of view, growth rate is likely to be related to a certain intracellular pool (Rhee 1980). This internal nutrient pool is used during growth, but it is constantly replenished by the concentration-dependent uptake of the external nutrient. At steady state, net uptake is in equilibrium with the internal pool and therefore growth rate is related to the total intracellular concentration (i.e., the cell quota), assuming that the size of the internal nutrient pool is directly proportional to the cell quota.

In using either equation, one should consider several assumptions or criteria. The Monod equation is useful when carbon is the limiting nutrient because the cell

Figure 6.14. Translocation of phosphorus. Autoradiographs of brown algae labeled with ^{32}P for 3 h, showing point of uptake and direction of migration (arrows). (a) *Cystoseira baccata* showing the whole thallus and the autoradiograph after isotope translocation; (b) autoradiograph for *Laminaria setchellii*. (From Floc'h 1982, with permission of Walter de Gruyter & Co.)

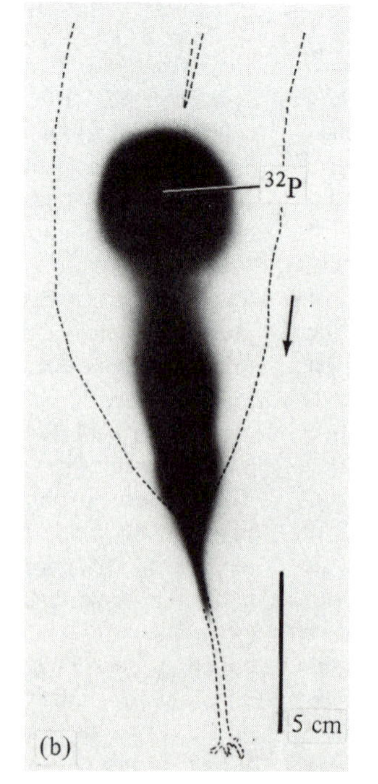

Table 6.11. *Algal growth equations*

1.	$\mu = \mu_m \dfrac{S}{K_s + S}$
2.	$\mu = \mu'_m(1 - q_o/q)$
3.	$\mu = \dfrac{100\ [\ln (N_t / N_o)]}{t}$

Symbols: K_s = half-saturation constant for growth; N_o = initial biomass; N_t = biomass on day t; q = cell quota (amount of nutrient per cell); q_o = subsistence cell quota; S = substrate concentration; t = time in days; μ = specific growth rate; μ_m = true maximum specific growth rate; μ'_m = specific growth rate at infinite S.

Figure 6.15. Relationship between dilution rate (*D*, which is equivalent to the steady-state growth rate) and cell quota (*Q*) of the phytoflagellate *Monochrysis lutheri* in a vitamin B_{12}-limited chemostat. (From Droop 1968, Vitamin B_{12} and marine ecology. *J. Mar. Biol. Assoc. U.K.*, vol. 48, pp. 689–733, with permission of Cambridge University Press)

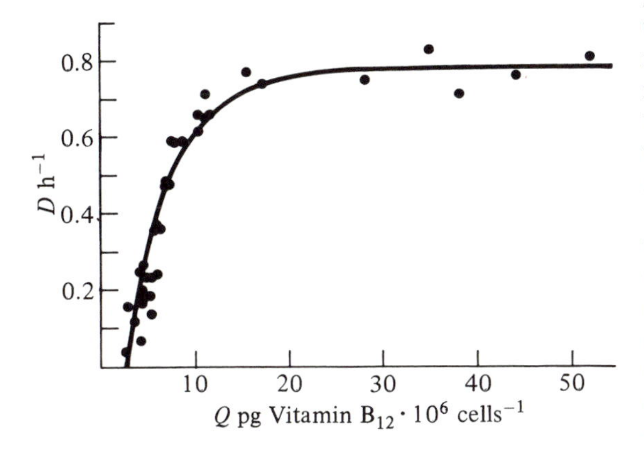

quota (reciprocal of cell yield) is constant with varying growth rate. However, for other nutrients, especially phosphorus (Rhee 1980, 1982, Cembella et al. 1983), cell quota increases sharply as growth rate increases. This increase in cell quota allows growth to continue for several generations under phosphorus-limiting conditions, by mobilization of phosphorus storage reserves. Under these conditions, growth rate is not related to external phosphate concentration, but rather to phosphorus cell quota. In this case and similarly for nitrogen, the Droop equation is recommended for determining growth kinetic parameters. Further comparison between the Monod and Droop equations are found in Goldman & McCarthy (1978), Cembella et al. (1983), and Droop (1983). The theory for growth kinetics just described has been developed from numerous experiments on bacteria and phytoplankton. No carefully executed experiments have been conducted on macroalgae.

6.7.2 Measurement of growth kinetics

Various means are available for measuring rate of growth of seaweeds (Chapman 1973a). Nondestructive methods include changes in wet weight, surface area or (in kelps) the movement of holes punched in the meristematic region at the base of the blade. Sampling for changes in dry weight, or plant carbon or nitrogen, is destructive to the plant. From the time series of changes in any one of the above parameters, the specific growth rate (μ) is calculated as the percentage increase in the parameter per day. For example, the specific growth rate could be calculated from a daily increase in fresh weight according to equation 3 in Table 6.11, assuming steady-state exponential growth (DeBoer et al. 1978). For ecological or field purposes, nondestructive sampling is preferable since the growth of the same plant can be followed over time and related to the limiting nutrient concentration in the ambient water.

The most accurate way of determining the relationship between external nutrient concentration and growth rate in seaweeds is to use continuous flow cultures. In these, an attempt is made to keep the external nutrient concentration constant by using a high dilution rate (flow rate/container volume) and a high container volume/plant biomass ratio. Then growth rate is measured at a series of external nutrient concentrations when steady state is achieved. During the approach to steady state, a reasonably constant biomass must also be maintained by harvesting at frequent intervals or by increasing the inflowing nutrient concentration to compensate for the increase in biomass. In phytoplankton continuous cultures, this adjustment of biomass is achieved automatically because cells are removed with the outflowing medium. Steady state can also be approximated by using semicontinuous cultures, where changes in nutrient concentration in the medium are minimized by frequent medium changes. The time required to reach steady state depends primarily on the growth rate and culture conditions. DeBoer et al. (1978) have arbitrarily chosen the criterion that the biomass must increase by an order of magnitude. For example, if an alga is growing at a rate of 10% per day, steady state will be reached in 23 days (DeBoer 1981). Similarly, many phytoplankton physiologists use the guideline of 10 generation times to steady state.

Uptake rate is also a good approximation of growth rate when nutrients do not limit growth or when steady state growth occurs under nutrient-limiting conditions. However, nutrient uptake rates may greatly exceed growth rates when nutrients are added to an algal culture growing under nutrient-limiting conditions. The half-saturation constant, K_s, determined during transient conditions of

Figure 6.16. Growth rates (± 1 S.E.) of the red seaweed *Agardhiella subulata* as a function of residual N concentration for various N enrichment sources. (From DeBoer et al. 1978, with permission of *Journal of Phycology*)

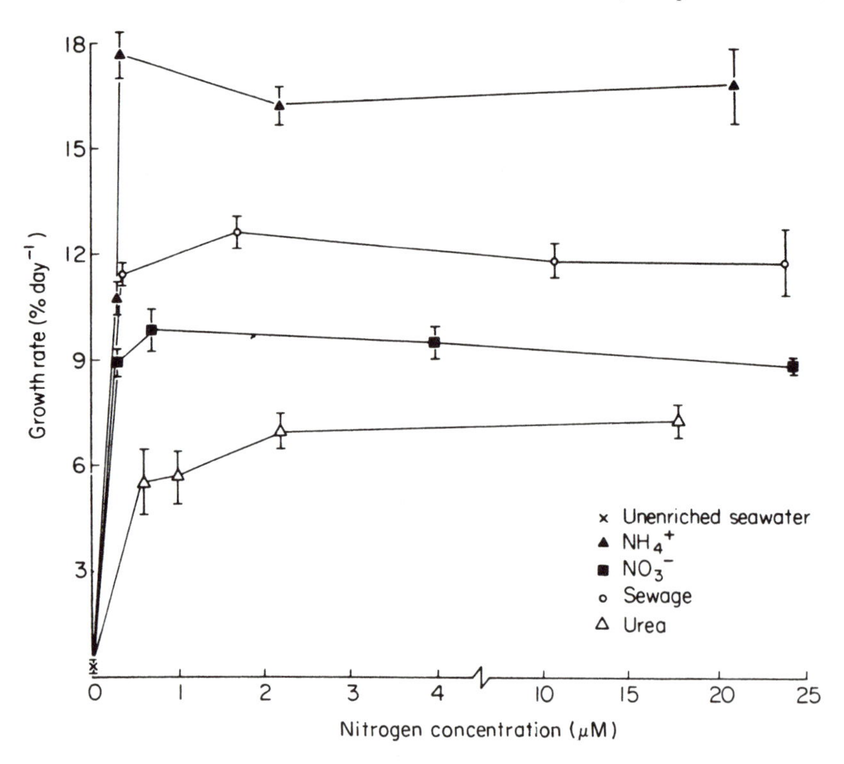

short-term uptake experiments, may also be greater than for growth.

6.7.3 Growth rates

In contrast to the numerous phytoplankton studies on the relationship between growth rate and ambient nutrient concentration, only a few studies have been conducted on seaweeds. DeBoer et al. (1978) found that the nitrogen growth kinetics of two red algae, *Agardhiella subulata* and *Gracilaria foliifera*, followed typical growth saturation curves (Fig. 6.16). The values of K_s ranged from 0.2 to 0.4 μM for various nitrogen sources, and growth rate was saturated as low as 1 μM NH_4^+ or NO_3^+. Similarly, the annual brown alga *Chordaria flagelliformis* had low K_s values (0.2 to 0.5 μM) for the three nitrogen substrates, nitrate, ammonium, and urea (Probyn & Chapman 1983). These low values are in contrast to other studies where, for example, growth rate was saturated at 10 μM nitrate for *Laminaria saccharina* (Chapman et al. 1978; Wheeler & Weidner 1983) and at 30 μM nitrate or ammonium for the estuarine green alga *Cladophora albida* (Gordon et al. 1981). These latter seaweeds would be poor competitors against phytoplankton, where growth rate is saturated at 1 μM nitrogen or even less.

Studies with phytoplankton have recently shown that while growth rate is often related to the concentration of nutrients in the external medium, growth rate can be estimated more accurately from the concentration of nutrients within cells (Droop 1968, 1973, Rhee 1980). This basic principle is related to the common agricultural practice of plant tissue analysis, where the critical tissue concentration is that which just gives saturating growth. Higher or lower concentrations indicate nutrient reserves or deficiency, respectively. The techique has now been applied to aquatic vascular plants (Gerloff & Krombholz 1966) and seaweeds (Hanisak 1979, DeBoer 1981). The growth of *Codium fragile* was more directly related to the amount of nitrogen present in its thallus than to the nitrogen in the surrounding water (Fig. 6.17). Even though internal nitrogen values ranged from 0.9 to 4.8%, growth rate remained constant at nitrogen concentrations in excess of ca. 2%. While tissue nitrogen generally reaches saturating values with increasing concentrations of external nitrogen, phosphorus does not saturate as readily, if at all (Fig. 6.18). Therefore, the critical tissue phosphorus concentration is more difficult to determine accurately.

6.8 **Effects of nutrient supply**

6.8.1 Chemical composition

Extensive analysis of the chemical composition of marine plankton has revealed that the ratio of carbon, nitrogen, and phosphorus is 106C:16N:1P (by atoms).

Figure 6.17. Relationship between growth and internal nitrogen concentration of *Codium fragile*, growth measured as increase in dry weight after 21 days. (From Hanisak 1979, with permission of Springer-Verlag)

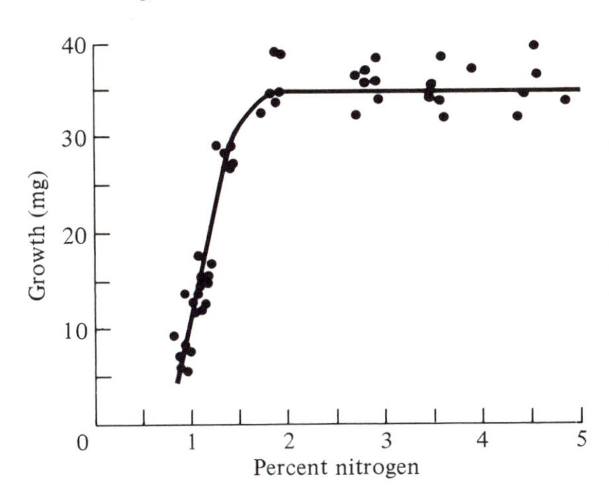

Figure 6.18. Concentrations of total nutrient in *Cladophora* tissue as a function of the nutrient concentration supplied in complete medium. (a) Phosphorus, supplied at 0–6 μM, N at 375 μM; (b) nitrogen, supplied at 0–375 μM, P at 12 μM. Each point is the mean ± S.E. of three replicates. (From Gordon et al. 1981, with permission of Walter de Gruyter & Co.)

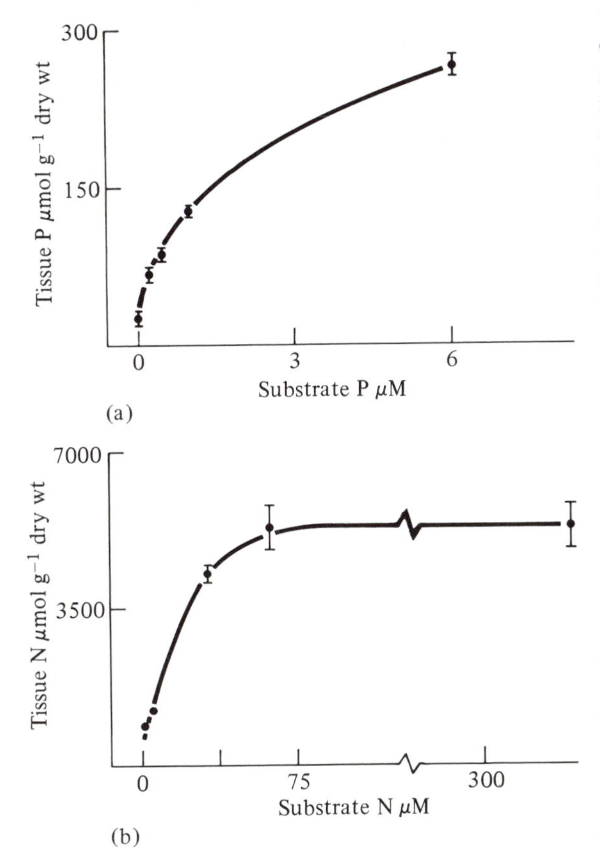

This is commonly referred to as the Redfield ratio. Decomposition of this organic matter occurs in the same ratio. However, Atkinson & Smith (1983) have recently shown that benthic marine macroalgae and seagrasses are much more depleted in P and less in N, relative to C, than phytoplankton. The median ratio for seaweeds was about 550:30:1. An important ramification of these observations is that the amounts of nutrients required to support a particular level of net production is much lower for macroalgae than it is for phytoplankton. The high C:N:P ratio in seaweeds is thought to be due to their large amount of structural and storage carbon. The average carbohydrate and protein content of seaweeds has been estimated at about 80% and 15%, respectively, of the ash-free dry weight (Atkinson & Smith 1983). In contrast, the average carbohydrate and protein content of phytoplankton is 35% and 50%, respectively (Parsons et al. 1977). The large amounts of carbon in seaweeds is in keeping with the observed taxonomic differences in the C:N ratio. Niell (1976) found higher ratios in the Phaeophyceae than in either the Chlorophyceae or Rhodophyceae.

Deviations from the Redfield ratio have frequently been used to infer which nutrient is limiting the growth of phytoplankton (Goldman et al. 1979). Phytoplankton deprived of P during growth typically have N:P atomic ratios greater than 30:1, whereas phytoplankton deprived of N during growth have N:P atomic ratios less than 10:1. Similarly C:N and C:P atomic ratios are dependent on growth conditions. Recent studies have shown that seaweeds respond in a similar manner. The C:N atomic ratio is generally higher when plants are grown under nitrogen limitation because of a decrease in proteins and an increase in carbohydrates. For example, in *Ulva lactuca*, the concentrations of β-alanine and asparagine can decrease 20-fold under nitrogen starvation (Nasr et al. 1968). Neish & Shacklock (1971) demonstrated that *Chondrus crispus* has a higher carrageenan content in unenriched seawater than in N-enriched medium and a similar effect is seen in agar content of *Gracilaria foliifera* (DeBoer 1979). Elevated transient uptake rates of ammonium have also been used to indicate nitrogen limitation in *Gracilaria foliifera* and *Agardhiella subulata* when C:N atomic ratios rise above 10 (D'Elia & DeBoer 1978).

The chemical composition of many temperate seaweeds varies seasonally, due primarily to the onset of nitrogen limitation in the coastal waters in summer. Wheeler & Srivastava (1983) found that tissue nitrate (as ethanol soluble nitrate) and total nitrate paralleled the ambient nitrate levels and showed summer minima and winter maxima (from 0 to 70 μmol g fresh wt^{-1} for nitrate and from 0.9 to 2.9% of dry wt for total nitrogen) in *Macrocystis integrifolia*. In contrast, Wheeler & North (1981) found that neither nitrate nor ammonium accumulated in the tissue of *M. pyrifera*; free amino acids accounted for a major portion of the soluble nitrogen.

Juvenile *M. pyrifera* sporophytes do not appear to store nitrogen (Wheeler & North 1980). In culture experiments, Chapman et al. (1978) have shown for *Laminaria saccharina* that internal concentrations of nitrate increased at substrate concentrations above 10 μM NO_3^- and reached concentrations that were several thousand times higher than the surrounding medium.

Other higher molecular weight compounds may be involved in nitrogen accumulation and storage. The dipeptide L-citrullinyl-L-arginine has been observed to accumulate to high concentrations when *Chondrus crispus* is supplied with nitrate or ammonium at low temperatures (Laycock et al. 1981). Much of this reserve can be readily mobilized for growth of the plant during higher temperatures, increased irradiance, and low levels of external nitrogen, which are common during late spring and summer months. Consequently, rapid growth rates, sustained by declining nitrogen reserves, persisted well after the disappearance of ambient nitrogen. The accumulation of soluble nitrogen reserves appears to be optimized under conditions of low temperature and reduced light (Rosenberg & Ramus 1982). Under these conditions the rate of accumulation exceeds the requirements for growth. Pigments, or more likely the proteins associated with them, may also serve a secondary role of nitrogen storage (Perry et al. 1981, Smith et al. 1983). There are several observations of marked decreases in pigment content under nitrogen deficiency (DeBoer 1981). Chlorophyll and phycoerythrin concentrations in *Gracilaria foliifera, Agardhiella subulata,* and *Ceramium rubrum* were strongly influenced by concentrations of inorganic nitrogen in the medium (DeBoer & Ryther 1977).

6.8.2 Development, morphology, and reproduction

A consequence of multicellularity is a decreased surface:volume ratio of cells. Some seaweeds appear to respond to nutrient deficiency by the production of hairs from the thallus surface, akin to the role of root hairs in vascular plants. Whoriskey & DeBoer (unpublished) observed that under low nitrogen concentrations and moderate agitation, hair cell formation was enhanced in *Hypnea musciformis, Gracilaria* spp., *Agardhiella subulata,* and *Ceramium rubrum*. Interestingly, cytoplasmic streaming occurred in hairs on the apical (meristematic) regions of the thalli, but not in hairs in the lower part of the thallus. A greater development of apical hairs was observed in *Fucus spiralis* germlings grown in low nutrient concentrations (Schonbeck & Norton 1979c); these tufts could also be seen on germlings in the field except when nutrients were at the highest concentration. The whorls of *Acetabularia* may also serve to increase absorptive surface area (Gibor 1973, Adamich et al. 1975). Adamich et al. found an inverse relationship between whorl development and the concentration of nitrate in the medium.

Cladosiphon zosterae is a specific epiphyte on eelgrass, *Zostera marina*. Ammonium or urea in the medium induced a hairless, compact morphology of the discoid stage of the epiphyte compared to the morphology in nitrate medium (Lockhart 1979). In the microscopic stages of the life history of the kelp *Lessonia nigrescens*, irradiance and nitrate and phosphate concentrations influence the course of development of gametophytes and the attainment of fertility (Hoffmann & Santelices 1982). The interacting effects are summarized in Figure 6.19. Under nitrogen limitation the gametophytes did not survive, while under phosphorus limitation they were multicellular but did not show the usual sexual differentiation.

Embryos of *Fucus edentatus* are morphologically distinct when they are grown under nitrogen deficiency (Fig. 6.20d). Generally, the nitrogen-deficient embryos were smaller, tapered, with a single long primary rhizoid and reduced secondary rhizoid development in contrast to those grown in complete medium (Fig. 6.20a) (McLachlan 1977). Growth of the embryos in phosphorus- or iron-deficient medium resulted in much smaller embryos with normal morphology. Omission of bromine from the medium resulted in considerably reduced growth (Fig. 6.20b), while in medium lacking boron the embryos became moribund (Fig. 6.20c). The concentration of boron has also been reported to influence the development and reproduction of *Ulva lactuca* and *Dictyota dichotoma* (Nasr & Bekheet 1970). Iodine is required for vegetative growth and normal formation and maturation of plurilocular sporangia of *Ectocarpus siliculosus* (Woolery & Lewin 1973) and also appears to be required for the development and reproduction of both crusts and blades of *Petalonia fascia* (Hsiao 1969).

Nutrient availability is known to influence reproduction in microalgae (Drebes 1977) and there are a few reports for macroalgae. Nitrogen depletion enhanced gamete formation in *Ulva fasciata*, whereas higher nitrogen favored vegetative growth and asexual reproduction (Mohsen et al. 1974). Abundant zoospore formation was observed after ammonium was added to the medium of *Ulva lactuca* (Nasr et al. 1968). Nitrogen deficiency suppresses the reproduction of many species (McLachlan 1982), but the source of nitrogen may also control reproduction, as in species of *Acetabularia* (Adamich et al. 1975). More information is needed on effects of nutrients on reproduction, especially nutrients other than nitrogen.

6.8.3 Growth rate and distribution

Growth and productivity of seaweeds is controlled by environmental factors such as irradiance, temperature, nutrient availability, and water movement. Marked seasonal fluctuations in nutrient availability (especially nitrogen) occur, which affect growth rate. This has been most thoroughly studied in the kelp *Laminaria longicruris* in Nova Scotia (Hatcher et al. 1977, Gagné et al.

Figure 6.19. Life cycle of *Lessonia nigrescens* showing the influence and interaction of light intensity and culture media on development and fertility. Gametophyte stages: 1 = multicellular, vegetative; 2 = multicellular, no sexual differentiation; 3 = few celled, fertile; 4 = few celled, vegetative; 5 = multicellular, fertile. (From Hoffmann & Santelices 1982, with permission of Elsevier Biomedical Press)

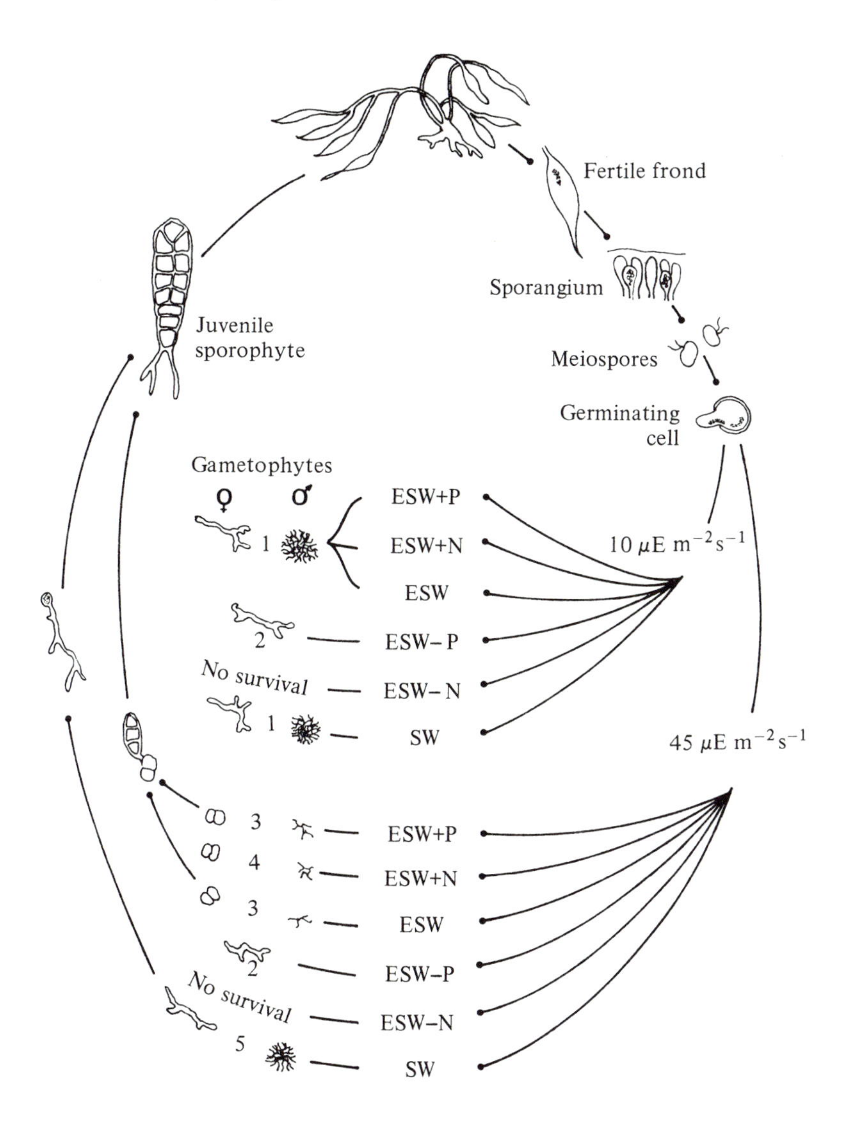

1982). In these kelp beds there are two main limiting factors: light and nitrogen availability. Nitrogen is present year-round at Centreville, in the southwest part of the province, due to upwelling, but is limiting for eight months of the year at Boutlier's Point, St. Margaret's Bay (near Halifax). The interaction of light and nitrogen availability determines the seasonality of kelp growth. At Boutlier's Point the plants grow mainly during the period of nitrogen availability, in winter and early spring. By also building up internal nitrogen reserves they are able to prolong their rapid growth for at least two months and to take advantage of improved light conditions during the spring. The bulky thallus is able to store substantial quantities of both inorganic and organic nitrogen during late autumn and winter for use in growth in late spring and early summer, when nitrogen is becoming limiting but light conditions are improving (Fig. 6.21). During the summer when irradiance is high, plants at Boutlier's Point store carbohydrate as laminaran. These carbon reserves are then remobilized and used in conjunction with nitrogen (high ambient concentrations) to produce amino acids and proteins for growth in early winter. At Centreville, where nitrogen does not become limiting, plants do not build up laminaran reserves, and kelp growth rate follows irradiance, greatest in summer (Fig. 6.21). Without carbohydrate reserves, growth rate declines during late autumn and early winter as irradiance falls.

Figure 6.20. Fourteen-day-old embryos of *Fucus evanescens* grown in various media. (a) Complete medium; (b) medium minus bromine; (c) medium minus boron; (d) medium minus nitrogen. (From McLachlan 1977, with permission of Blackwell Scientific Publications)

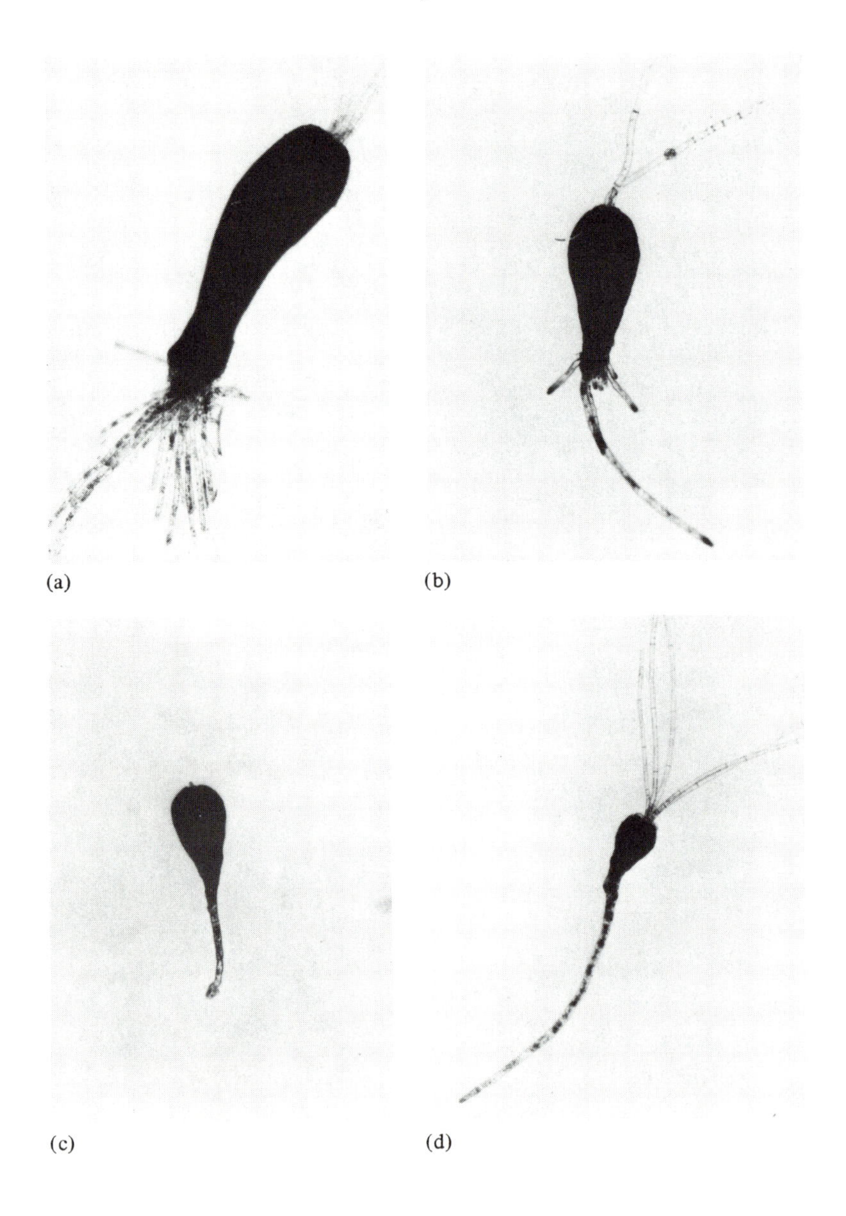

Seasonality of growth has been studied in other seaweeds, such as *Gracilaria foliifera* and *Ulva* spp. (Rosenberg & Ramus 1982). During late winter, both these species accumulated substantial soluble nitrogen reserves that were depleted during the spring–summer growth period. Species may survive during the summer by using intermittent peaks in nutrient concentration as they occur. In contrast to the kelps, neither of these species started the winter with a significant store of reserve carbohydrate. On the other hand, an annual, *Chordaria flagelliformis*, has been shown to maintain higher growth rates during the summer even when ambient nitrogen is exhausted (Probyn & Chapman 1982). Measurement of K_s values for *Chordaria* indicated that they were very low (0.2–0.5 μM) for nitrate, ammonium, and urea; therefore, this annual effectively scavenges nitrogen from seawater (Probyn & Chapman 1983). The comparatively small intracellular nitrogen pool (0.1–0.4% of dry wt) indicates that in *Chordaria*, typical of many opportunistic species, newly absorbed nitrogen is directed into growth rather than storage.

Another suggested source of nutrients, especially nitrate, is the submarine discharge of ground water in coastal areas (Johannes 1980). Since nitrate in ground

Figure 6.21. Growth, internal nitrogen reserves, and laminaran contents of blades of *Laminaria longicruris* at two sites in Nova Scotia with contrasting light and nitrogen environments. (From Gagné et al. 1982, with permission of Springer-Verlag)

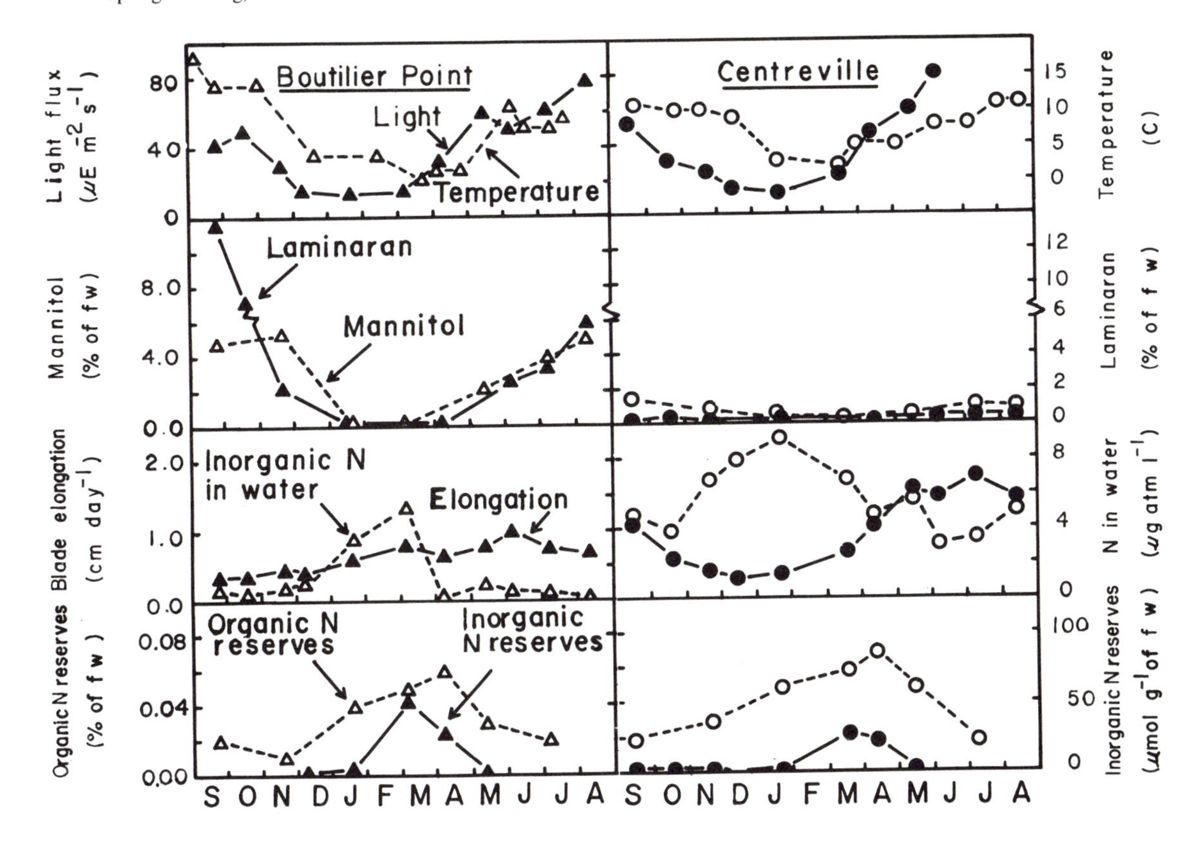

water is very high (50–120 μM), a small amount of discharge could significantly enrich nitrogen-impoverished coastal waters in the summer. Regenerated nutrients from the sediments could also be an important source of nitrogen enrichment. Martens et al. (1978) showed that in the top meter of sediments from Long Island Sound, ammonium concentrations may reach 10 mM and reactive phosphate more than 1.0 mM. If the circulation of seawater is such as to entrain some fraction of the interstitial sediment water into the surface layers, nutrient enrichment of the water column results from these regenerated nutrients. Smetacek et al. (1976) have demonstrated this point in association with high-salinity seawater intrusions in the Kiel Bight. The authors showed that an intrusion of high-salinity water could displace low-salinity interstitial water and that the resulting input of ammonium, phosphate, and silicate to the water column was approximately 10 times the concentration prior to the intrusion. In Narragansett Bay, Rhode Island, up to 80% of the nutrients entering the bay at certain times of the year are estimated to come from the sediments. Transport of nutrients into the water column was reported to occur as a result of two mechanisms that were approximately equal in magnitude. One resulted from the activity of the burrowing benthic organisms and the other was by diffusion (Nixon et al. 1976).

6.9 **Synopsis**

Seaweeds require various mineral ions and up to three vitamins for growth. Certain elements are essential for growth while others may be taken up even though they are not apparently required. There are a few elements, such as nitrogen, phosphorus, iron, and possibly some trace metals (e.g., cobalt and manganese) that may limit the rate of growth of some seaweeds at certain times of the year because of their low concentrations in seawater. To date nitrogen is the element most frequently observed to limit seaweed growth.

Elements are taken up as ions (charged particles) that diffuse to the cell surface, through the cell wall to the plasmalemma. Some ions, especially cations, may not reach the plasmalemma because they become adsorbed to certain chemical components of the cell wall. Ions may pass through the plasmalemma passively, either by passive diffusion or facilitated diffusion. Nonelectrolytes (uncharged particles) generally diffuse through the membrane at a rate that is proportional to their solubility in lipid and inversely proportional to their molecular size.

Ions are usually transported by an active process requiring energy. The uptake rate of most ions is related to the ion concentration in the seawater by a rectangular hyperbola, with the term V_{max} denoting the maximum uptake rate and K_s the concentration at which $V = ½V_{max}$. Nutrient uptake rates can be measured by isotope accumulation in the plant tissue or by the disappearance of the nutrient from the medium.

Various forms of dissolved inorganic nitrogen (nitrate, nitrite, ammonium) and dissolved organic nitrogen (urea and amino acids) are the main sources of nitrogen for seaweeds. Ammonium is generally taken up in preference to all other nitrogen sources. The uptake rate of nitrogen and other ions is influenced by light, temperature, water motion, desiccation, and the ionic form of the element. Biological factors that influence uptake include the type of tissue, the age of the plant, its nutritional past history and interplant variability. After ions are taken up, some (e.g., N, P, S, Rb) may be translocated to other tissues within the plant.

After nitrate and ammonium are taken up they are usually used to synthesize amino acids and proteins, although the ions may be stored in the cytoplasm and vacuoles. Nitrate is reduced intracellularly to ammonium with the aid of two enzymes, nitrate and nitrite reductases. Ammonium is incorporated into amino acids via two main pathways. In the glutamine synthetase pathway, glutamine is the first product. The second pathway, which is thought to be of secondary importance, is the glutamate dehydrogenase pathway, which results in formation of glutamic acid.

Seaweeds can take up phosphorus as orthophosphate ions or obtain phosphate from organic compounds through extracellular cleavage using the enzyme alkaline phosphatase. The most important role of phosphorus is in energy transfer through ATP and other high energy compounds in photosynthesis and respiration.

Calcium ions are used in maintenance of membranes and in cross-linking cell wall polysaccharides. Calcium carbonate is deposited in the walls of certain seaweeds. Magnesium is an essential cofactor in many cellular reactions. Other important nutrients are K, S, Fe, Cu, Mn, Zn, and Co, and the vitamins B_{12}, thiamine, and biotin.

The relationship between external nutrient concentration and growth rate can be described by a rectangular hyperbola in which growth rate approaches a maximum at higher nutrient concentrations. Uptake rate is a good approximation of growth rate only when nutrients do not limit growth, or when steady state growth occurs under nutrient limiting conditions. The half-saturation constants (K_s) for growth are generally much higher for seaweeds than for phytoplankton, indicating that phytoplankton have a higher affinity for nitrogen when the concentration is low. However, seaweeds, particularly kelps, can store large quantities of nutrients when the external concentration is high. These cellular reserves are used during periods of low nutrients, and in this way the kelps are able to prolong their period of rapid growth for up to two months after nitrogen in the water is virtually depleted.

7 Carbon metabolism

Because the element carbon plays such a central role in metabolism of plants, it is treated apart from the other nutrients. This chapter examines the means by which seaweeds obtain carbon, from processes ranging from photosynthesis to heterotrophy, and the fate of fixed carbon in respiration, transport, storage, and utilization in characteristic algal compounds, including polysaccharides and secondary metabolites.

7.1 Carbon sources

7.1.1 Inorganic carbon

Seaweeds are photoauxotrophic, except for parasitic species that are at least partially heterotrophic. The use of the term auxotrophic, rather than autotrophic, reflects the plants' needs for small quantities of vitamins (see Secs. 6.2.3, 6.6.8) and, in some cases, external growth factors. The principal source of carbon for photoauxotrophs is inorganic C, which is available as CO_2 and HCO_3^- in the sea.

Carbon dioxide from the atmosphere dissolves in water to form carbonic acid, and, at pH greater than 8.0, bicarbonate ion is formed directly (Borowitzka 1977). These forms and carbonate are in dynamic equilibrium (Fig. 7.1). Because of this equilibrium, uptake of CO_2 or HCO_3^- by seaweeds causes more CO_2 to dissolve or carbonate to dissociate (the latter from shells or rock), so that the supply of inorganic carbon in the sea is virtually limitless. (Carbon limitation may occur in brackish waters, where there is less HCO_3^- – see Sec. 4.3.3 – and in aquaculture systems when algal biomass is high.) The relative proportions of the forms of carbon depend on pH and salinity, as shown in Figure 7.2, and also temperature (Kalle 1972, Broecker 1974). In seawater of pH 8 and salinity 35 ppt, about 90% of the inorganic carbon occurs as HCO_3^-; there is relatively little free CO_2. Absolute values are approximately 10 μM CO_2 and more than 2 mM HCO_3^-; this compares with 300 ppm (ca. 13 μM) CO_2 in air (Kremer 1981a). CO_2 solubility in seawater is inversely proportional to temperature, yet a reduction in water temperature reduces the CO_2 concentration because it shifts the carbonate equilibrium toward HCO_3^-.

Apparently all seaweeds can use CO_2, which diffuses readily across the cell membrane (Raven 1974), but not surprisingly some seaweeds (and perhaps many that have not yet been tested) can use HCO_3^-. Although CO_2 concentrations are similar in air and seawater, CO_2 diffusion rate is very much slower in water, so that seaweeds that can use HCO_3^- should have an advantage over those that cannot. However, since RuBP carboxylase uses CO_2 as a substrate, HCO_3^- must be dehydrated. An enzyme, carbonic anhydrase (carbonate dehydratase), which greatly speeds up the interconversion of CO_2 and HCO_3^-, has been found but not localized in seaweeds. In unicellular algae and higher plants it has been located variously within or around the plastids or in the cell wall (Kimpel et al. 1983). This enzyme, which has a turnover number of 36 million molecules of CO_2 per minute in human blood (Lehninger 1975), may well eliminate the photosynthetic rate differences between CO_2-

Figure 7.1. The carbonate equilibrium.

$$CO_2 + H_2O \rightleftharpoons H_2CO_3 \xrightleftharpoons{pK' = 6.0} HCO_3^- + H^+ \xrightleftharpoons{pK' = 9.1} CO_3^{2-} + H^+$$

$$CO_2 + OH^- \rightleftharpoons HCO_3^- \rightleftharpoons CO_3^{2-} + H^+$$

Figure 7.2. Percentage distribution of different forms of inorganic carbon in seawater as a function of pH at three different salinities. (Modified from Kalle 1945, *Der Stoffhaushalt des Meeres*, with permission of Akademische Verlagsgesellschaft, Geest und Portig KG)

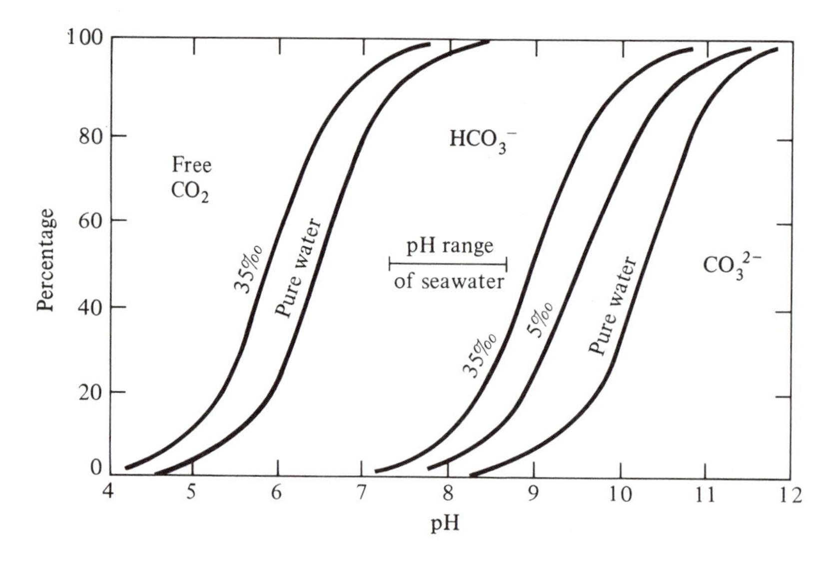

and HCO_3^--users. Bicarbonate users, such as *Laminaria*, can be made to photosynthesize in air at normal (submerged) rates if they are provided with CO_2 concentrations of several thousand ppm (Lloyd et al. 1981). Kremer's (1981a) survey of CO_2 versus HCO_3^- use in seaweeds shows no trends, but data are still scarce. This question might usefully be investigated in terms of Littler's functional/form types (Sec. 2.9).

Shell-boring algae, such as the conchocelis stage of *Porphyra* and the gomontia stage of *Monostroma*, live in particularly favorable carbon environments. *Porphyra tenera* can satisfy most of its carbon and calcium requirements from the shells, presumably by acidifying $CaCO_3$ to give Ca^{2+} and HCO_3^- (Ogata 1971).

7.1.2 Heterotrophy in seaweeds?

Are any photoauxotrophic seaweeds able to use organic carbon sources? This question was posed by Wilce (1967) with regard to arctic seaweeds in low light environments. Wilce's hypothesis has so far generated much speculation but relatively little experimental work. Organic carbon utilization by photosynthetic phytoplankters is well established (Neilson & Lewin 1974), but phytoplankters have a great advantage over seaweeds in their surface:volume ratios. Bacteria and fungi in the water and on seaweed surfaces are excellent scavengers of organic carbon and are expected to easily outcompete seaweeds for organic carbon. Moreover, since inorganic carbon is so abundant there is little reason to expect that seaweeds would need to compete for organic carbon, except if irradiance were very low. Some recent evidence shows that, under certain experimental conditions (usually including organic carbon concentrations well above those in the sea), some seaweeds can take up and use organic molecules, including acetate (Gemmill & Galloway 1974), hexoses (Turner & Evans 1977), and amino acids (Schmitz & Riffarth 1980) (reviewed by Kremer 1981a). Gemmill & Galloway's work on *Ulva lactuca* showed that acetate assimilation was light dependent and also required Krebs cycle activity; this also proved that uptake was not due to epiphytic microorganisms. Photoassimilation of acetate was enhanced at low CO_2 concentrations. When *Ulva* grows in polluted areas, this mechanism may have some ecological significance. Critical experiments have yet to be performed on the abilities of seaweeds to obtain organic carbon from natural concentrations and in competition with the microbial flora.

The argument for facultative heterotrophy in arctic sublittoral seaweeds, as set out by Wilce (1967), was based on the following premises: (1) seaweeds have been dredged from as deep as 100 m where there is virtually no light even in summer; (2) during the winter there is a long period of continuous darkness and during the summer, when daylight is available, the water is ice-covered, turbid, or both, so that there would be little light even in more moderate depths; (3) thus seaweeds in the arctic would barely have adequate photosynthesis for maintenance, let alone for storage for winter. Until recently these arguments could be countered only on the basis of equally indirect evidence (e.g., Drew 1977). In 1980, however, Chapman & Lindley reported year-round irradiance measurements at Igloolik, a site 69° N, along with measurements of *Laminaria solidungula* growth. The lower limit of the kelp was 20 m, and experiments demonstrated that the light available at that depth was adequate for the growth of the plant. Growth of plants in the field was greatest in winter and early spring under thick ice cover; this correlated with nitrogen availability.

In summer, plants could not grow because of nitrogen starvation, but they photosynthesized and stored carbohydrate, which was used the following winter. Chapman & Lindley also found that *L. solidungula* did not use glucose in darkness. The question of how deep seaweeds grow in the arctic remains unanswered. (Chapman & Lindley did not venture below 23 m, being 2500 km from the nearest decompression unit!) However, dredging records are suspect since there is no assurance that the plants had grown at the depth from which they were collected. Although Chapman & Lindley's study does not eliminate the possibility of heterotrophy, it does at least clearly show that this kelp can grow photoauxotrophically at the natural depth limit of the population.

7.1.3 Parasitic algae and symbiotic fungi

The nutrition of parasitic algae is by definition at least partly heterotrophic. Various colorless or weakly pigmented algae, particularly in the Rhodophyta, grow on other seaweeds and are assumed to receive organic carbon from their hosts. Obligate epiphytes, particularly *Polysiphonia lanosa*, which has penetrating rhizoids, have sometimes also been thought to be partially parasitic. *P. lanosa* has now been shown to be at most auxotrophic (Harlin & Craigie 1975, Turner & Evans 1977). Some colorless pustules that had been classified as parasitic algae have now been shown to be galls of host tissue, presumed to be caused by bacteria (e.g., "Lobocolax" – McBride et al. 1974; see also Evans et al. 1978).

The best-known parasitic red alga is *Harveyella mirabilis*, an alloparasite of (i.e., not related to) *Odonthalia* and *Rhodomela*, which has been thoroughly studied by Goff (reviewed in her 1982 paper) and Kremer (1983). *Harveyella* cells penetrate among the host cells via grazing wounds and cause proliferation of host cells to form pustules, over which the cortex of *Harveyella* forms its reproductive structures. Contiguous cells of the host thallus and isolated host cells in the pustule of *Harveyella* export assimilates to *Harveyella* cells, but the secondary pit connections formed between parasite and host are plugged, probably precluding translocation through them. Goff (1979b) proposed that translocation could take place across the cell wall matrix, but the exact pathway and the composition of the translocated material have not yet been determined in this association. *Harveyella* relies completely on its host for carbon, which it receives as digeneaside and perhaps some amino acids (Fig. 7.3.) (Kremer 1983). *Gracilaria verrucosa* transfers carbon as floridoside to its nonpigmented alloparasite *Holmsella pachyderma* (Evans et al. 1973), while *Polysiphonia lanosa* exports mannoglycerate to its weakly pigmented alloparasite *Choreocolax polysiphoniae* (Callow et al. 1979). The adelphoparasite *Janczewskia gardneri* is moderately pigmented as an adult, and Court (1980) found no translocation to it from its closely related host, *Laurencia spectabilis*. Clearly, each host–parasite relationship is unique and must be assessed separately. Three possible mechanisms have been suggested by Goff (1979b) to explain how *Harveyella* induces its host to

Figure 7.3. Nutritional relationship between *Rhodomela confervoides* and the parasite *Harveyella mirabilis*. Daily carbon budget for the host (a portion 200 mg fresh wt) bearing an average-size parasite (0.05 mg fresh wt). DOM = uptake of dissolved organic matter. (From Kremer 1983, with permission of Springer-Verlag.)

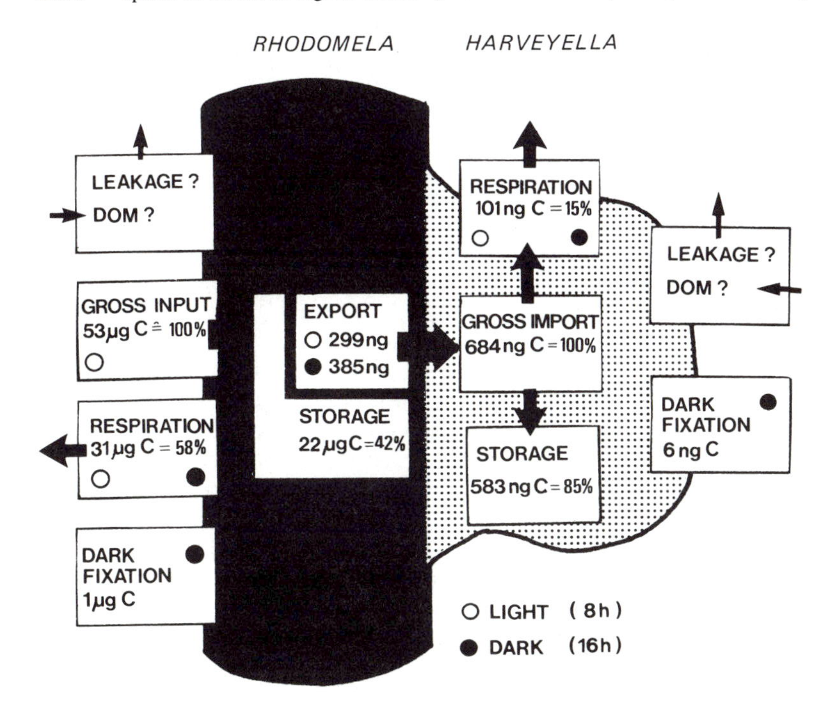

release organic carbon: (1) changing the permeability of the host plasmalemma; (2) modifying the host's cell wall polysaccharide synthesis so that compounds normally deposited in the wall remain soluble; and (3) mechanically or enzymatically damaging the host cell wall, leading to lysis and release of metabolites.

A few seaweeds are known to be characteristically associated with fungi in relationships that are most likely symbiotic. The relationships include *Blidingia minima* var. *vexata* (*Ulva vexata*) + *Turgidosculum ulvae*, *Prasiola borealis* + *Guignardia alaskana*, and *Apophlaea sinclairii* + *Mycosphaerella apophlaeae*, about which very little is known (Kohlmeyer & Kohlmeyer 1972, Hawkes 1983). In addition there are associations between the ascomycete *Mycosphaerella ascophylli* and two fucoids, *Ascophyllum nodosum* and *Pelvetia canaliculata*. Specimens of these fucoids over 5 mm long are invariably infected with *Mycosphaerella*, but specimens less than one year old lack the fungus (Kohlmeyer & Kohlmeyer 1972, Kingham & Evans 1977). The fungus grows throughout the host thallus, without penetrating the cells or forming haustoria. Fungal fruiting bodies (ascocarps) are formed chiefly (although not exclusively) in receptacles of the hosts, where they appear to the naked eye as small black dots (Fig. 7.4). The fungus also occurs in vegetatively reproducing salt marsh *Ascophyllum* (J. Higgins, personal communication). The fungus undoubtedly obtains carbon from the host; it has recently been isolated into axenic culture and found able to utilize laminaran and mannitol, but not alginic acid (nor, incidentally, D-xylose or D-fructose) (Fries 1979). *M. ascophylli* is similar to saprophytic marine fungi in requiring a pH of 7 to 8 and high NaCl concentration, but differs in having a lower temperature optimum and a requirement for the vitamins thiamine and biotin. The presence of the fungus in *Pelvetia canaliculata* changes the ratio of mannitol:volemitol from 1.7:1.0 (in plants less than 5 mm long and in plants treated with fungicides) to 0.8:1.0 (Kingham & Evans 1977). (Volemitol is restricted to *P. canaliculata* from Europe; it is not found in other species of *Pelvetia*, nor in *Ascophyllum* – Kremer 1979a.) The compounds, if any, that the fucoids get from the ascomycete are unknown. The nature of the cues involved in the simultaneous fruiting of the fungus and the alga also merit investigation. Kohlmeyer & Kohlmeyer (1972) considered whether this association and others like it could be classified as lichens. They concluded that the associations are not lichens, because lichens have a thallus that is not morphologically like either the phycobiont (algal) or mycobiont (fungal) thallus and because lichen phycobionts reproduce only asexually. The morphologies of the fucoid–*Mycosphaerella* associations are dominated by the sexually reproducing algae. Moreover, no characteristic secondary lichen substances are formed in these associations or the three similar ones mentioned above. The Kohlmeyers proposed the term "mycophycobiosis" to describe these associations. They also cautioned people studying the biochemistry of these algae to be aware that compounds extracted from the association may be coming from the fungus.

7.2 Anabolic and catabolic pathways

Much of the basic biochemistry of metabolism in seaweeds is assumed to be similar to that in other organisms, particularly where common pathways have been found in bacteria, higher plants, and animals. The basic outlines of glycolysis and the Krebs cycle, as discussed in Section 7.2.1, oxidative phosphorylation, and the syntheses of amino acids, proteins, and nucleic acids are probably universal. However, considerable variety in the details of some of these "universal" pathways is now being found by molecular biologists. For example, the transfer RNA that reads the initiation codon in protein synthesis in the cytoplasm of eukaryotes has slightly different structure and behavior from that found in prokaryotes, chloroplasts, and mitochondria (see Goodwin & Mercer 1983). The assumption of similarity must be made guardedly. In this section, respiration will be briefly reviewed and some aspects of carbon fixation that are unique to seaweeds will be described.

7.2.1 Respiration

The carbon dioxide that is liberated in respiration comes largely from the oxidation of sugars in the energy-yielding pathways of glycolysis (in the cytosol) and the Krebs cycle (in mitochondria). These pathways are common to virtually all organisms and are treated extensively in any biochemistry text, but a brief account is useful here (Fig. 7.5). Sugars are primed for the reaction sequence by phosphorylation at the expense of ATP. There is one oxidation during glycolysis and also some substrate-level phosphorylation of ADP, but no CO_2 is released. During the oxidation (of glyceraldehyde-3-phosphate [GAP] to 1,3-diphosphoglycerate), NAD^+ is reduced to NADH. The NADH, as well as that from the oxidations in the Krebs cycle, is reoxidized via the respiratory electron transport chain, a sequence of oxidation–reduction reactions (akin to those in photosynthetic electron transport) coupled to phosphorylation of ADP. The electron acceptor at the end of the chain is oxygen, which is reduced to water: this is where the oxygen consumption takes place. The compound at the end of glycolysis is pyruvate. This enters the mitochondrion and is oxidatively decarboxylated in a complex, regulated reaction to give acetyl-coenzyme A (see Fig. 7.9). During this reaction CO_2 is given off and NADH is formed.

For each hexose that enters glycolysis, two acetyl-CoA are formed, each containing two of the original six carbon atoms. These two carbons are added to oxaloacetic acid (OAA) at the start of the Krebs cycle and sequentially removed in two further oxidative decarboxylations. OAA is regenerated by a series of reductions. For each hexose that is oxidized, 6 CO_2 are given off

Figure 7.4. The *Mycosphaerella-Ascophyllum* association. (a) Branch tip of *Ascophyllum* with receptacles, showing conceptacles (*) as circular areas and ascocarps of the fungus as black dots (**). (b) Vertical section through an ascocarp, showing two-celled ascospores in one ascus. (From Kohlmeyer & Kohlmeyer 1968, with permission of J. Cramer)

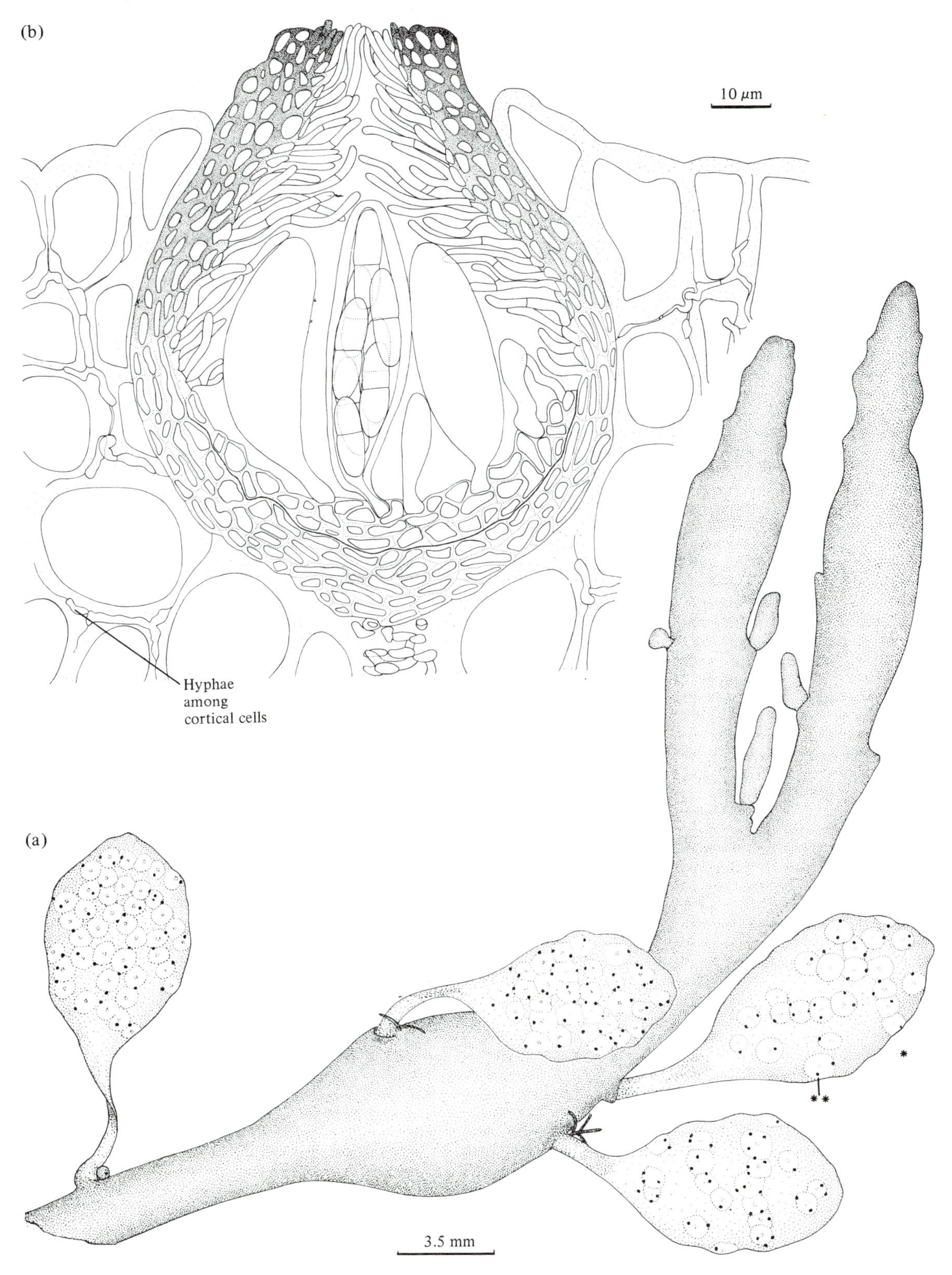

Figure 7.5. Summary of algal metabolism. Enzymes: (1) RuBP carboxylase; (2) PEP carboxylase or carboxykinase; (3) carbamoyl-phosphate synthetase. (Based on Raven 1974 and other sources, with permission of Blackwell Scientific Publications)

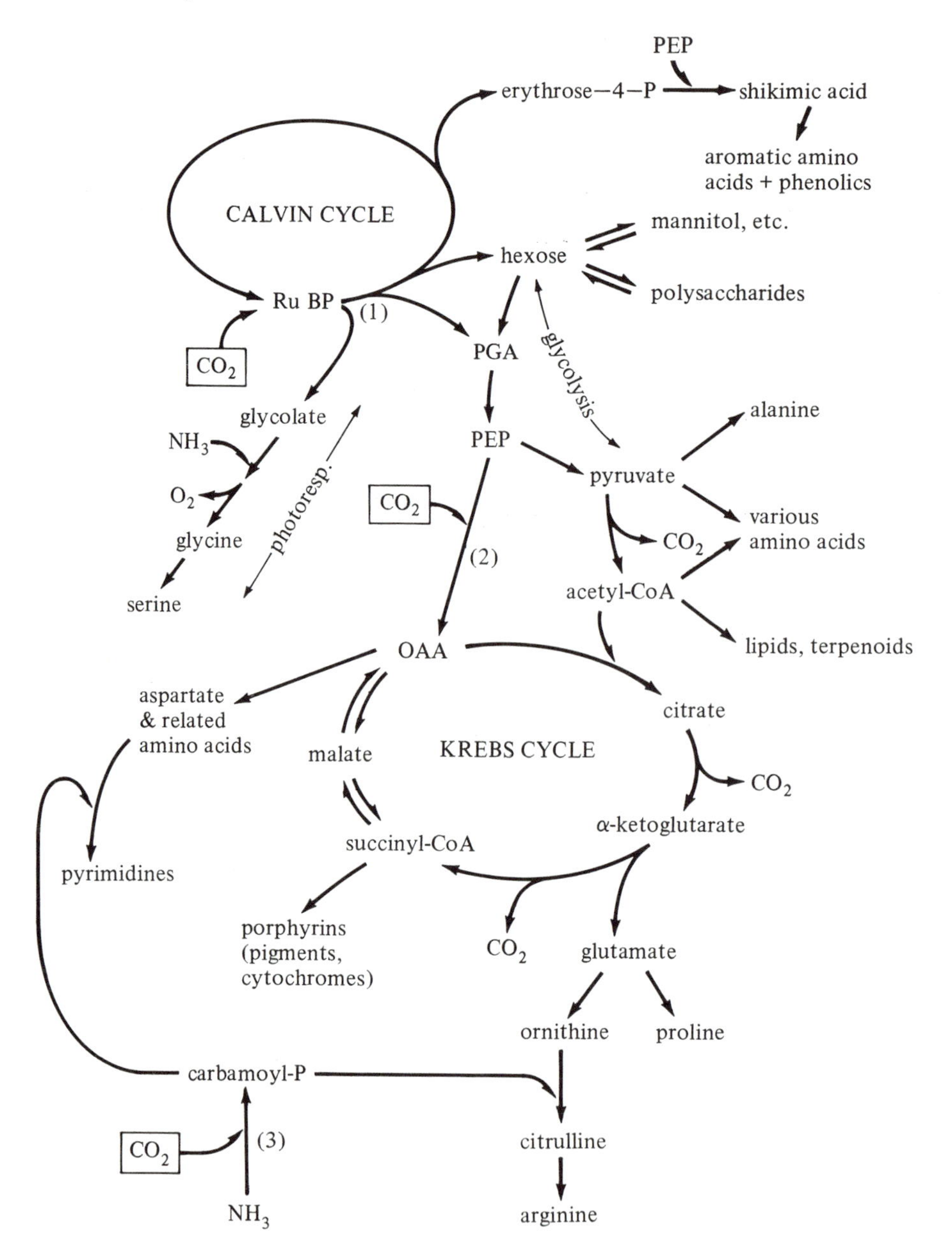

and 12 pairs of electrons are passed on to a total of 6 O_2. Glycolysis and the Krebs cycle provide energy and reducing power (NADH) for cell metabolism, but, during the light, photosynthesis can also provide these. Glycolysis and the Krebs cycle are also the source of carbon skeletons for synthesis of other cell components (Fig. 7.5). A useful summary of the energy requirements of various metabolic processes has been compiled by Raven (1982).

7.2.2 The Calvin cycle

The reductive pentose phosphate pathway, or Calvin cycle, was first demonstrated in a freshwater green alga, *Chlorella* (see Calvin 1962). It has subsequently been found in every other photosynthetic plant (including the blue-green algae) that has been investigated (Kremer 1981a). It operates only in the light with a direct supply of ATP and NADPH from photosynthetic electron transport. All the enzymes of the Calvin cycle are located in the plastids. Mannitol synthesis in the brown algae is also located in the plastids, but sucrose synthesis in the green algae may take place in the cytoplasm, as it does in higher plants. The key enzyme in the Calvin cycle is ribulose-1,5-bisphosphate (RuBP) carboxylase, which is in the pyrenoids in those species that have them (e.g., *Pilayella littoralis* – Kerby & Evans 1978, *Caulerpa simpliciuscula* – Wright & Grant 1978). This enzyme has been studied widely in higher plants (Jensen & Bahr 1977), where it has been found to consist of 16 subunits of two different polypeptides and to need Mg^{2+} and CO_2 for activation. The larger subunits are made in the plastid, coded by chloroplast DNA. The smaller subunits are made in the cytoplasm; a signal segment attached to the N-terminal end permits passage across the plastid membranes and is cleaved off before the subunits are assembled (Goodwin & Mercer 1983). Akazawa & Osmond (1976) found that this enzyme from the green alga *Halimeda cylindracea* has a similar construction. RuBP carboxylase from a diatom, studied by Estep et al. (1978), was found to have some properties different from the higher plant enzyme: it was stimulated by aspartate and malate, as well as requiring the two cofactors for activation.

In the Calvin cycle, RuBP carboxylase inserts CO_2 into RuBP below the keto (C = O) group, cleaving RuBP into two molecules of 3-phosphoglycerate (PGA) (Fig. 7.6). PGA is reduced, using NADPH and ATP, to two other C_3 compounds, glyceraldehyde-3-phosphate (GAP) and dihydroxyacetone phosphate (DHAP). From then on, two series of reactions take place: GAP and DHAP can be combined to give fructose-1,6-bisphosphate (FBP), and these three compounds can be interconverted to regenerate RuBP (Fig. 7.6). The net effect of the Calvin cycle is: $6\ CO_2 + 18\ ATP + 12\ NADPH \rightarrow$ hexose $+ 18\ ADP + 18\ PO_4^{3-} + 12\ NADP$. Following this common fixation into PGA and FBP, carbon accumulates in various low molecular weight compounds that are characteristic of the class to which the alga belongs (Craigie 1974, Raven 1974, Kremer 1981a).

Sucrose is the principal low molecular weight product in the majority of the Chlorophyceae examined so far, as in higher plants. Smaller amounts of some amino acids, including alanine and glycine, accumulate. Sucrose, a disaccharide of glucose and fructose, is derived from FBP, whereas the amino acids come from PGA either directly or via pyruvate (Fig. 7.5). There are scattered (and sometimes conflicting) reports of green algae producing glucose and fructose rather than sucrose, or producing considerable quantities of fructose alone (Kremer 1981a). *Caulerpa simpliciuscula* deposits its hexoses largely in β-linked glucans and sugar monophosphates rather than in sucrose and starch (Howard et al. 1975).

The Phaeophyceae are noted for their production of the sugar alcohol, mannitol. One species, *Pelvetia canaliculata*, also produces a C_7 alcohol, volemitol. Mannitol is formed by reduction of fructose-6-phosphate (Fig. 7.6) (Ikawa et al. 1972). Volemitol probably arises in a similar manner from sedoheptulose-7-phosphate (Kremer 1977). Glycine is the major amino acid formed during light-dependent carbon fixation in the brown algae.

There are three groups of Rhodophyceae on the basis of their low molecular weight photoassimilates (Fig. 7.7) (Kremer 1981a). In the genus *Bostrichia* (Ceramiales, Rhodomelaceae) only, two sugar alcohols, dulcitol and sorbitol, are formed. In all other Ceramiales, as far as is known, digeneaside is formed. In all other orders the principal product is floridoside. Floridoside consists of glycerol plus galactose, whereas digeneaside consists of glyceric acid plus mannose. Principal amino acids in the red algae are glycine, serine, and aspartate. The enzymatic steps by which these characteristic carbohydrates are made have yet to be worked out, but one may expect that galactose and mannose come from fructose-6-phosphate by epimerization, and that they are primed by being esterified to a nucleotide phosphate (UTP, GTP) before being coupled to glycerol or glyceric acid (compare with isofloridoside formation, Fig. 4.4). The glycerol/glyceric acid is presumably derived from one of the C_3 compounds early in the Calvin cycle (or glycolysis), such as PGA, GAP, or DHAP. The pathway of dulcitol and sorbitol formation may parallel mannitol formation.

7.2.3 Photorespiration

Photorespiration is a result of the oxygenase activity of RuBP carboxylase, in which RuBP is split with the introduction of O_2 rather than CO_2 (Sec. 2.6.3). One of the early products is the C_2 acid glycolate (see Fig. 7.5), which, in angiosperms, is oxidized through a pathway involving peroxisomes and mitochondria, in such a way that three-fourths of the carbon in glycolate is salvaged and used in the Calvin cycle (Bidwell 1979). The pathway of glycolate metabolism in algae is not fully known yet, but it appears to be different from that in angiosperms in that glycolate is oxidized by a dehydrogenase rather than an oxidase; most algae do not have peroxisomes, which are the site of glycolate oxidase in angiosperms.

7.2.4 Light-independent carbon fixation

If seaweeds are supplied with labeled carbon in the dark, they fix it into various characteristic products that are not necessarily the same as those derived from the Calvin cycle. Red and green macroalgae in darkness form only small amounts of amino acids. Recent evidence on light-independent carbon fixation in *Gracilaria verrucosa* (Bird et al. 1980) suggests that CO_2 is fixed along with NH_4^+ via a urea–ornithine cycle (perhaps via carbamoyl phosphate synthetase – see Fig. 7.5). This provides a mechanism for rapid uptake of ammonium by nitrogen-starved plants at the expense of stored carbon.

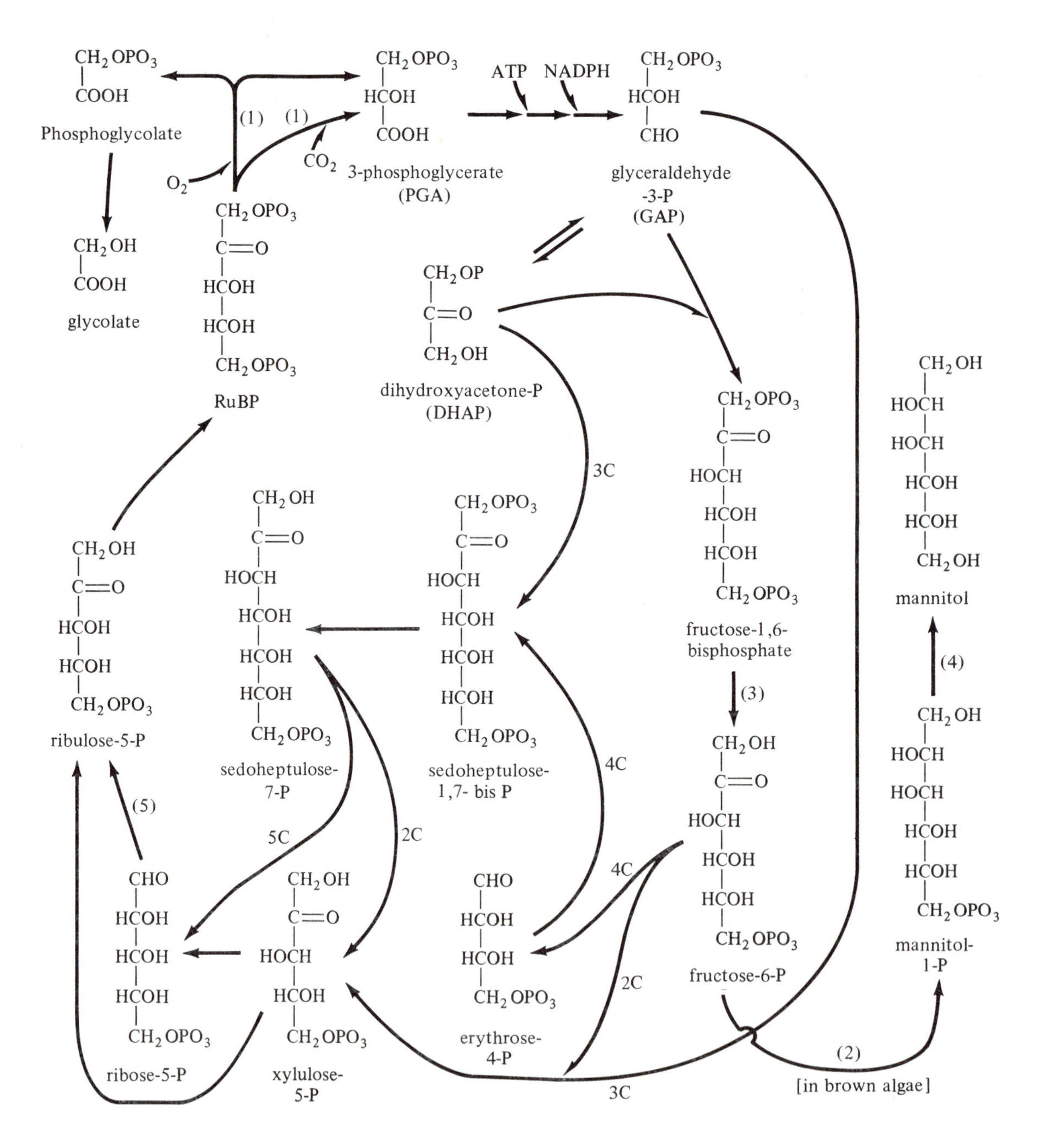

Figure 7.6. The Calvin cycle, also showing pathways of photorespiration and mannitol synthesis. Enzymes: (1) RuBP carboxylase/oxygenase; (2) mannitol-1-P dehydrogenase; (3) fructose-1,6-bisphosphate phosphohydrolase; (4) mannitol-1-phosphate phosphohydrolase; (5) ribose-5-phosphate isomerase. Unbracketed numbers beside arrows indicate number of carbon atoms in fragments transferred. (For example, fructose-6-P [lower right] is split into a 4C piece [erythrose-4-P] and a 2C piece, the latter attached to GAP to give xylulose-5-P). (Modified from G.R. Noggle & G.J. Fritz, *Introductory Plant Physiology* © 1976. Reprinted with permission of Prentice-Hall Inc., Englewood Cliffs, N.J.)

Figure 7.7. Some low molecular weight carbohydrates from brown and red seaweeds.

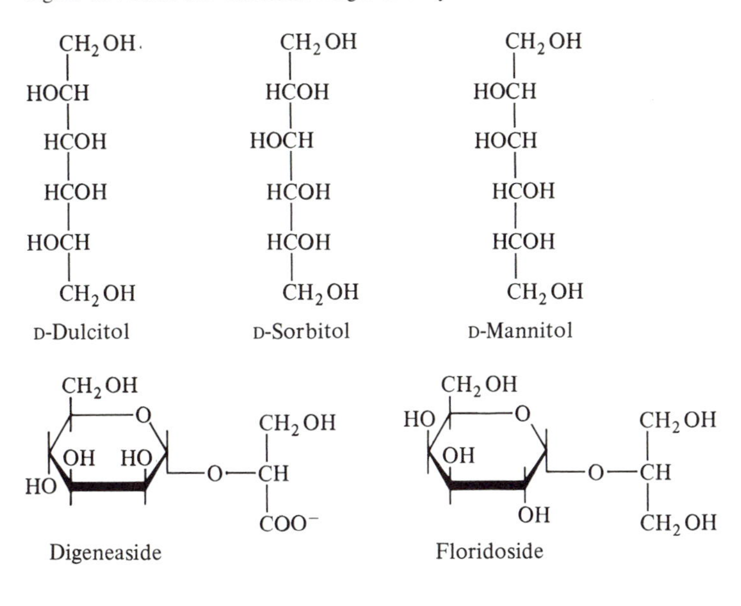

Brown algae, with considerably higher rates of light-independent fixation, do not form mannitol in darkness, but rather form malate, aspartate, citrate and alanine. Although present throughout the brown algae, dark carbon fixation occurs to a significant extent only in young tissues of kelps and fucoids, where it can account for over 20% of the total carbon fixed (Kremer 1981a). It takes place in light as well as in darkness, through β-carboxylation of phosphoenolpyruvate (PEP), that is, by addition of CO_2 to the β-carbon of PEP, which yields initially oxaloacetic acid (OAA) (Fig. 7.8). This reaction, which also generates ATP, is carried out by PEP carboxykinase, probably in the cytoplasm. PEP carboxykinase is stimulated by Mn^{2+} and ADP (Kremer 1981b, Kerby & Evans 1983). A similar reaction, but without the involvement of ATP, occurs via PEP carboxylase in some red and green algae, but activities are low. PEP carboxylase, which is the β-carboxylating enzyme in higher plants, uses HCO_3^- as a substrate, but PEP carboxykinase, like RuBP carboxylase, uses CO_2, which, however, can be rapidly formed by carbonic anhydrase from the abundant HCO_3^- in seawater. Oxaloacetic acid is rapidly converted to malate, citrate, and the other stable compounds just mentioned.

The role of light-independent carbon fixation in seaweeds is enigmatic. In higher plants it is found in species from dry, high-irradiance habitats. In C_4 plants it serves to maintain a flow of carbon to the Calvin cycle even when CO_2 flux is restricted because stomata are nearly closed. In Crassulacean acid metabolism (CAM) species, it is used to fix carbon at night for Calvin cycle activity during the day with stomata closed. In both of these cases there is no net CO_2 fixation because the C_4 acid formed by β-carboxylation subsequently releases CO_2, which is refixed in the usual way by RuBP carboxylase (Bidwell 1979). The process is energetically expensive, but this is not important in high-irradiance environments. However, the habitat of kelps and fucoids is quite different.

The significance of β-carboxylation in adult kelps and fucoids is seen during utilization of stored mannitol. Mannitol is remobilized from mature or old tissue to provide energy and carbon skeletons for growth in the meristems. PEP carboxykinase permits salvage of CO_2 lost when pyruvate is converted to acetyl-CoA in glycolysis (Fig. 7.8). There is no energy consumed in this reaction, since ATP is generated from PEP whether PEP is used to fix CO_2 or is converted to pyruvate in glycolysis (Kremer 1981b). Since 1 mole of mannitol yields 2 moles of PEP and thus can refix 2 moles of CO_2, there can be *net* carbon fixation even in darkness. This is an important ecological advantage, particularly since kelp growth is often tied to nitrogen availability, which is high in winter when irradiance is low. This mechanism is also of use to juvenile plants during reuse of recently manufactured mannitol. One should expect kelp and fucoid meristems, because of their high PEP carboxykinase activities, to manifest little photorespiratory loss of CO_2. A more general role of β-carboxylation is as an anaplerotic pathway providing carbon skeletons for amino acid synthesis (see Fig. 7.5).

7.3 **Translocation**

In the simplest seaweeds each cell is virtually independent of the others for its nutrition. However, many seaweeds contain nonpigmented cells in their me-

Figure 7.8. Utilization of PEP via PEP carboxykinase (1) in light-independent carbon fixation, or via pyruvate kinase (2) in glycolysis. (Modified from Kremer 1981b, with permission of Springer-Verlag)

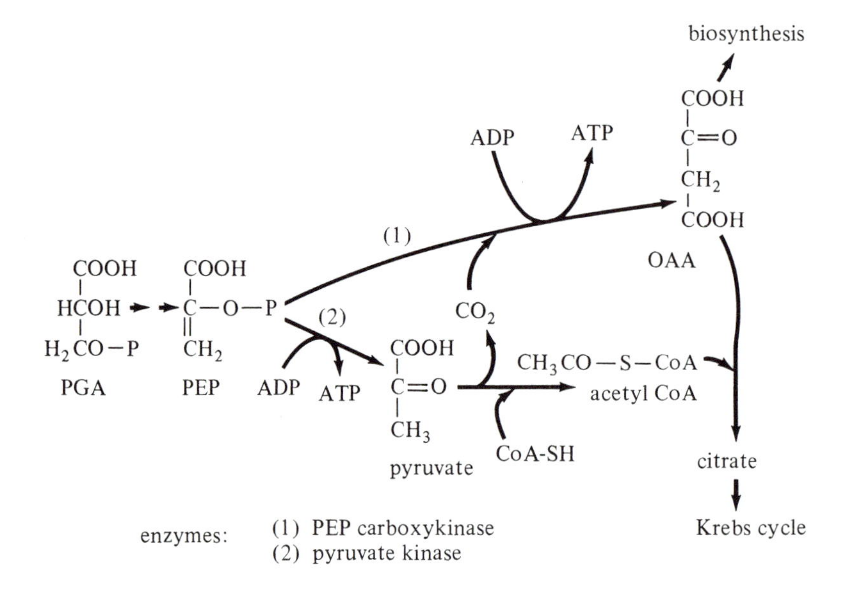

dulla that are evidently supplied with photoassimilates by the pigmented cortical or epidermal cells. Parasitic algae receive organic carbon via short-distance translocation from their hosts. Moreover, in those algae with an apicobasal gradient there is clearly movement of growth-regulating substances within the thallus. Such short-distance transport might take place through plasmodesmata, which, in some green and perhaps most brown algae, traverse the cross walls and join the cytoplasm into a continuous symplast (see Fig. 2.9d). Another possible route is across the cell wall, but one would expect this to be a more difficult route since the molecules would have to cross two cell membranes. In this case molecules could also diffuse out of the thallus while between cells, since the cell walls and intercellular spaces form a continuum, the apoplast, which is in direct contact with the seawater. Plasmodesmata are not found in those green algae that divide by furrowing (Stewart et al. 1973), nor in the red algae.

The Florideophycidae and some of the Bangiophycidae (and one green seaweed, *Smithsoniella earleae* – Brawley & Sears 1982) have septal plugs (pit connections) between the cells; plugs also form between cells of red algal parasites and their hosts. The red algal plugs consist of an acid polysaccharide–protein complex. The plasmalemma is continuous between cells and is often also continuous with a membrane over the plug. If the plug is completely surrounded by plasmalemma it is extracellular and there is no connection between cells (Brawley & Wetherbee 1981). Possibly, however, the plugs are not always isolated by a membrane, or possibly they can nevertheless provide a preferential pathway for transport: at least they might prevent leakage into the apoplast. In at least one freshwater red alga, *Batrachospermum sirodotii* (Aghajanian & Hommersand 1978), the plugs can break down to permit cytoplasmic continuity.

Long-distance translocation evolved in higher plants as an adaptation to habitats where light and CO_2 are available in air, whereas water and minerals are available chiefly in the soil. Since the whole outer surface of a seaweed is photosynthetic and involved in nutrient absorption, there is no need of translocation for exchange of materials between different regions of the plants. However, translocation can also serve to redistribute photoassimilates from mature (i.e., nongrowing), strongly photosynthetic areas to rapidly growing regions. This role is useful only where there is a localized growing region and a relatively large or distant mature region. Kelps have such a structure. Although there have been reports of translocation of photoassimilates in two red algae (Hartmann & Eschrich 1969) and in *Sargassum* (Titlyanov & Peshekhodko 1973), the only well-established cases of translocation of photoassimilates are in the sporophytes of the Laminariales (recently reviewed by Schmitz 1981 and Buggeln 1983). Moss (1983) has recently reinvestigated the anatomy of Fucales medullary filaments and shown that they can be considered sieve elements. Directional transport of ^{14}C-labeled assimilates through these filaments has now been shown by Diouris & Floc'h (1984). Transport of mineral ions is well established for Fucales and Laminariales (see Sec. 6.6).

Translocation in virtually all Laminariales takes place through the sieve elements, as has been shown by radioautography (Steinbiss & Schmitz 1973). The sieve elements lie in a ring between the cortex and medulla

Figure 7.9. Longitudinal sections through sieve tubes with sieve plates of (a) *Laminaria groenlandica* and (b) *Macrocystis integrifolia*. Mitochondria can be seen in both (arrows); plastids (P) and vacuoles (V) are indicated in *Laminaria*. Note the many narrow pores through the sieve plate of *Laminaria* compared with the few large pores in *Macrocystis*. Scales: 5 µm. (a from Schmitz & Srivastava 1974, with permission of Wissenschaftliche Verlagsgesellschaft mbH., Stuttgart; b from Schmitz 1981, with permission of Blackwell Scientific Publications)

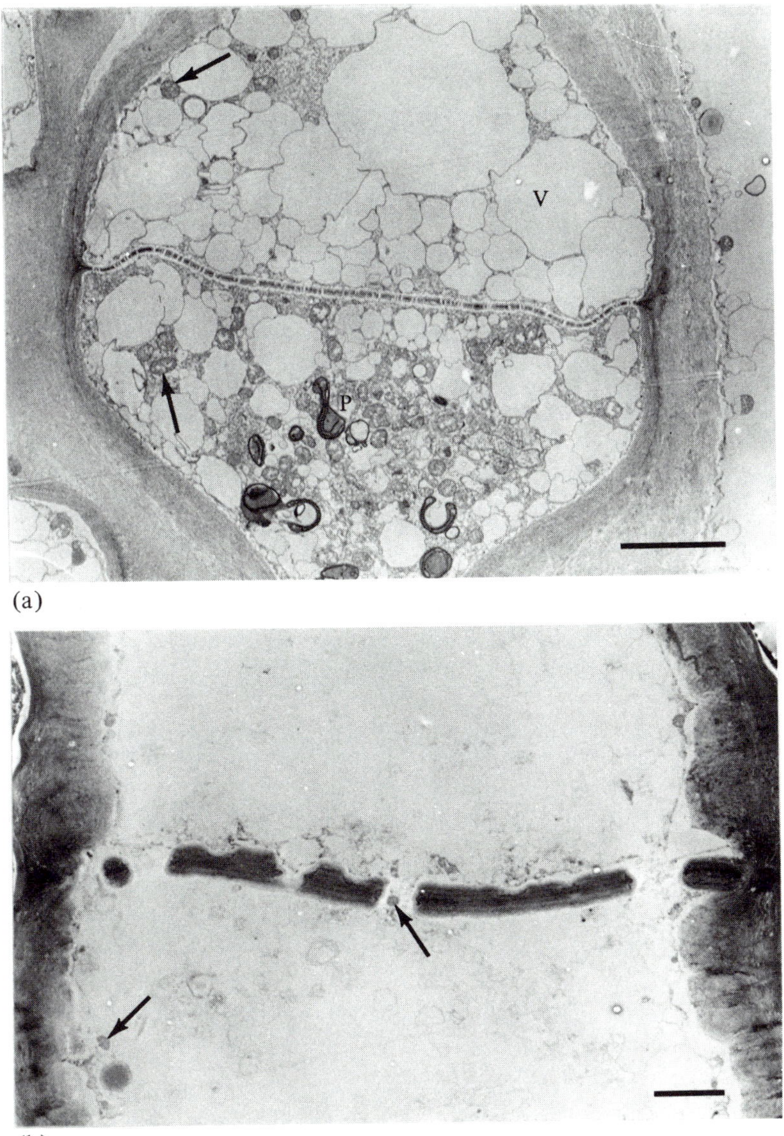

throughout the stipe and blades. Longitudinal files of sieve elements branch and interconnect. Although translocation had been suspected ever since kelp sieve elements were first discovered in the late nineteenth century, only circumstantial evidence had accumulated until 1963 when Parker used ^{14}C and fluorescein dye to demonstrate translocation in *Macrocystis pyrifera*. The structure of sieve elements and, in particular, the characteristically perforated sieve plates on their end walls show a trend from the smaller, simpler kelps (e.g., *Laminaria*) to the largest, most complex (*Macrocystis*). The pores become less numerous but larger in larger kelps, thus more effective in transport. The sieve elements of *Laminaria* are filled with cytoplasm, organelles, and numerous small vacuoles, whereas those in *Macrocystis* more closely resemble vascular plant sieve elements in having a peripheral layer of enucleate cytoplasm and a very large central vacuole or lumen (Fig. 7.9). The pores in the sieve plates are lined with callose (a β-1,3-linked glucan), which, as in vascular plants, can be deposited to occlude the pores. This mechanism prevents great loss of sieve tube sap in case of injury and may also regulate routes of translocation.

The only kelp known to differ from the description just given is *Saccorhiza dermatodea* (Emerson et al. 1982). Here translocation takes place in the medulla, through highly elongated cells (solenocysts) that are cross-connected by smaller cells (allelocysts). The ultrastructure of solenocysts is similar to the sieve elements of other genera. *Chorda* may be another exception. Titlyanov & Peshekhodko (1973) claimed to have found translocation in this genus, but the ultrastructure of the "trumpet-hyphae" has not been studied. This genus is in a separate family from all other Laminariales.

Kelps translocate the same materials that they make in photosynthesis; there is apparently no selectivity in sieve tube loading. Mannitol and amino acids (chiefly alanine, glutamic acid, and aspartic acid) each account for about half of the exported carbon. These materials move at velocities ranging from less than 0.10 m h^{-1} in *Laminaria* to about 0.70 m h^{-1} in *Macrocystis*, the speed generally correlating with the size of the sieve plate pores. The *rate* of translocation depends on not only the velocity but also the amount of material that is moving and hence on concentration of solutes and the cross-sectional area of transporting sieve elements. Rate is much harder to determine than velocity, but some values have been calculated, ranging from 1 g dry wt h^{-1} cm^{-2} in *Alaria* to 5 to 10 g h^{-1} cm^{-2} in *Macrocystis*. Rates in most vascular plants range from 0.2 to 6.0 g h^{-1} cm^{-2}, although some translocate at a very much higher rate (Schmitz 1981).

Photoassimilates follow a source-to-sink pattern of translocation. Sources include mature tissue; sinks include intercalary meristems and, to a lesser extent, sporophylls and haptera. In species with only one blade the pattern is simple: mature distal tissue exports and meristematic tissue at the blade–stipe junction imports. The pattern becomes complicated in *Macrocystis*, which has numerous blades in various stages of growth and

Figure 7.10. Translocation patterns in *Macrocystis*. (a) Import (→) and export (←) patterns, and direction of exported assimilates in *M. integrifolia* in spring, with much upward translocation, and in autumn, with more transport downwards. The numbers indicate distances in meters from the apex at which changes take place. (b) Translocation between successive generations of fronds (x°, x′, x″, etc.) of *M. pyrifera*. Only the apical blade is shown on each frond. Older fronds support the young fronds closest to them until the young fronds reach a certain size. (a from Lobban 1978c, with permission of *Journal of Phycology*; b from Lobban 1978b, with permission of the American Society of Plant Physiologists)

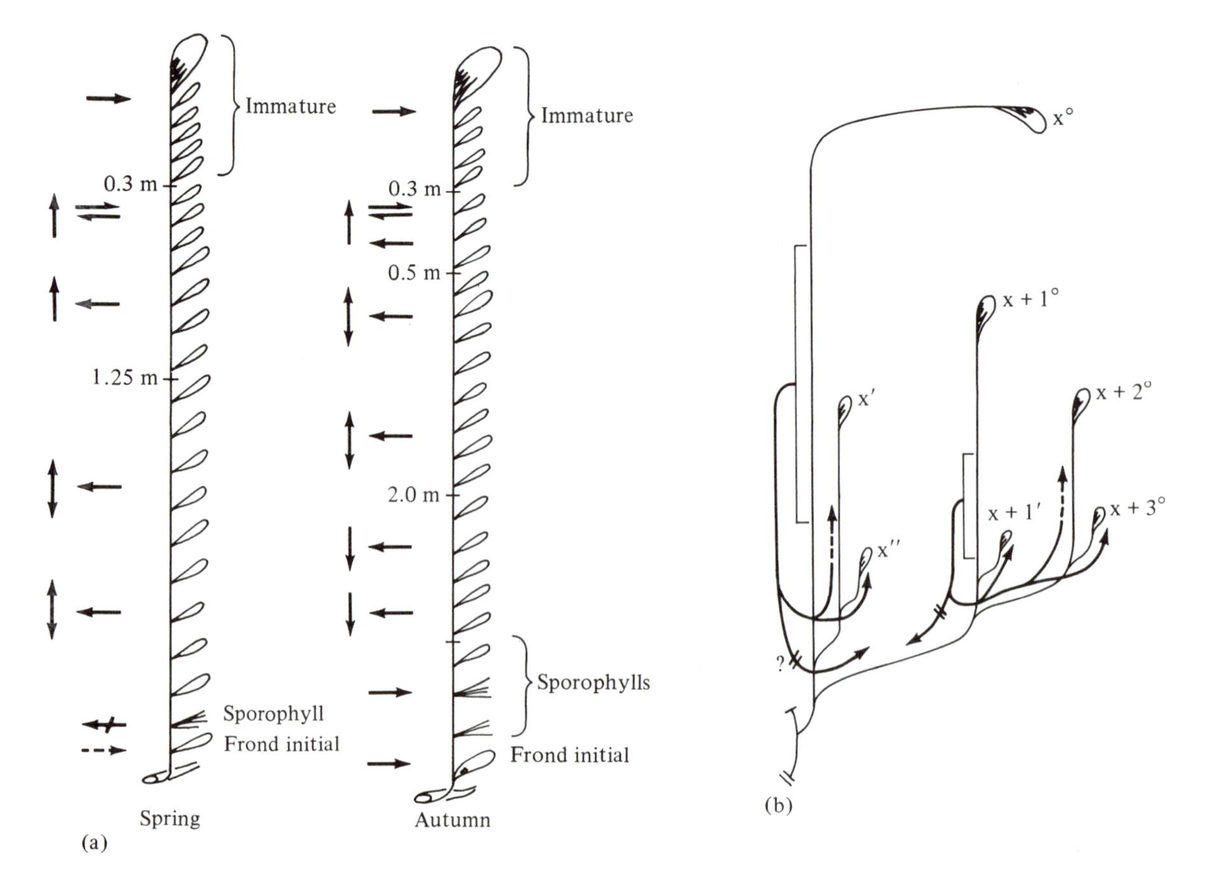

maturity as well as young fronds developing from old fronds. In this genus, the changing import-export pattern of a blade as it matures is very similar to the pattern found in dicotyledonous angiosperms (Fig. 7.10). Young blades only import, but as they near full size export begins. Initially export is upward, to the meristem from which the blade developed; later it is also downward to fronds developing from the base of the parent frond. In *M. integrifolia*, which lives in a seasonally variable habitat, downward export in autumn also appears to carry photosynthates to the base of the plant for storage (Lobban 1978b, c). When growth stops in kelps, translocation also stops (Lüning et al. 1973, Lobban 1978c).

The mechanism of translocation in kelps has not yet been established; this is hardly surprising in view of the controversy over mechanisms in vascular plants, which have been far more intensively studied. The structure of *Macrocystis* sieve elements (Fig. 7.9b) would be consistent with a mass flow (Münch) mechanism, whereby sieve tube loading in the source causes an osmotic influx of water, which in turn pushes the solutes to the sink, where assimilates are unloaded. Schmitz (1981) has argued in favor of this mechanism. The structure of the *Laminaria* sieve tubes (Fig. 7.9a) does not seem to offer the open pathway required by the mass flow hypothesis, and a different mechanism may be operating there. However, recent work on vascular plants suggests that such a vesiculate structure is not inconsistent with pressure flow (Buggeln 1983). The data on translocation in kelps have been critically evaluated with regard to the Münch hypothesis and found to be generally consistent with it (Buggeln 1983).

The ecological significance of translocation is that it allows more rapid growth of localized meristems. This is especially important in plants like *Macrocystis*, which may be attached in deep water where the new frond initials are shaded by both the water column and the surface canopy of blades. In populations of *M. pyrifera* growing in stratified water where the surface layer is poor in nutrients, translocation may also serve to carry nitrogen (as amino acids) to the surface canopy (Wheeler & North 1981). In perennial species of *Laminaria*, translocation serves to move stored carbon from the mature blade to the meristem when new growth begins; this start of new growth is triggered by photoperiod in *L. hyperborea* (Lüning, unpublished) and by nitrogen availability in several other species (e.g., *L. longicruris* – Gagné et al. 1982).

7.4 **Storage and structural polysaccharides**

7.4.1 Storage polymers

Carbon may be stored in monomeric compounds such as mannitol, but much is stored in polymers. One advantage of polymers is that they have smaller effects on osmotic potential than the same amount of carbon in monomeric form. A number of characteristic storage polysaccharides are found in red, brown, and green seaweeds, but most are structurally similar in being branched and unbranched chains of glucose units (glucans) (Craigie 1974, McCandless 1981). Most green algae, like higher plants, store starch, a mixture of branched molecules (amylopectin) and unbranched molecules (amylose) (Fig. 7.11). Amylose consists of α-D-glucose units linked $1 \rightarrow 4$; amylopectin also has $\alpha(1 \rightarrow 6)$ branch points. Amylose is insoluble in water, forming micelles in which the molecules are helically coiled, whereas amylopectin is soluble. Dasycladales, such as *Acetabularia*, do not always store starch but often store inulin, a fructose polymer (Percival 1979). Red algae store primarily floridean starch, a branched glucan similar to amylopectin except for having a few $\alpha(1 \rightarrow 3)$ branch points. Recent evidence from McCracken & Cain (1981) shows that primitive red algae, including five marine species, also have amylose in their starch. Brown algae store laminaran, which, like starch, comprises a branched, soluble molecule and an unbranched, insoluble molecule. (These compounds are called simply soluble and insoluble laminaran.) The glucose in laminaran is in the β form, however, and the links are $\beta(1 \rightarrow 3)$ and $\beta(1 \rightarrow 6)$ (Fig. 7.11c). Furthermore, some laminaran molecules, called M-chains, have a mannitol molecule attached to the reducing (C-1) end.

Some algal polysaccharides are known to be storage compounds and others to be structural, but the functions of many are not yet understood (Percival 1979). Knowledge of polysaccharides, particularly their native conformation, is still fragmentary. Polysaccharides are still described in terms of their solubility characteristics and component units. Rees (1977) has outlined the principles that determine the shapes of polysaccharides, and has summarized what is known about the functional significance of the shapes.

Storage compounds show quantitative changes correlated with season (really with growth), plant part, and reproductive condition. These changes have been particularly well documented in commercially valuable kelps, fucoids, and Gigartinales. Studies such as those by Black (1949, 1950) and Jensen & Haug (1956) on *Laminaria* spp. and fucoids were primarily concerned with fluctuations in the valuable wall matrix polysaccharides and iodine, but also documented changes in mannitol and laminaran contents. More recently the relationships among growth, storage, and sometimes nitrogen availability have been worked out for *Hypnea musciformis* (Durako & Dawes 1980), *Eucheuma* spp. (Dawes et al. 1974), and *Laminaria longicruris* (Sec. 6.8.3) (Chapman & Craigie 1977, Gagné et al. 1982). Build-up of carbohydrate in matrix polysaccharides during periods of low growth or nitrogen starvation has also been shown, for instance, in *Chondrus crispus* (Neish et al. 1977) and *Eucheuma* (Dawes et al. 1977). Although most of the few known cases of such storage seem to be the result of conditions that permit photo-

Figure 7.11. Algal storage polysaccharides. (a) amylopectin; (b) amylose; (c) two types of laminaran chains – M, with mannitol attached to the reducing end; G, with glucose at the reducing end. (a, b from Lehninger, 1975, *Biochemistry*, 2nd ed., © 1975 Worth Publishers, New York; c reprinted with permission from Percival & McDowell 1967, *Chemistry & Enzymology of Marine Algal Polysaccharides*, Copyright 1967 Academic Press Inc. [London], Ltd.)

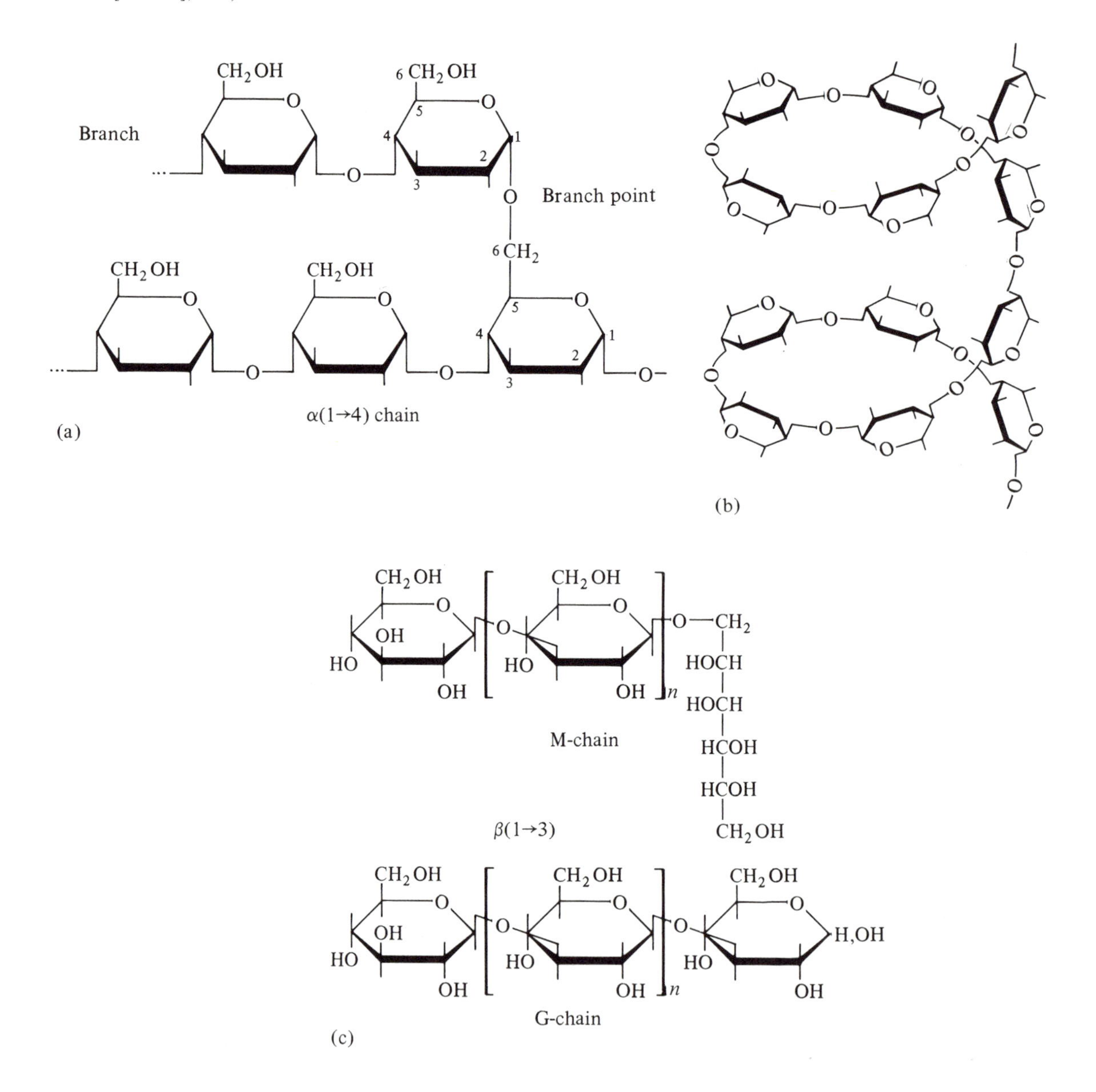

synthesis but not growth, there is no evidence that seaweeds can retrieve this extracellular carbon. There is the possibility, not yet documented, that the larger kelps, such as *Macrocystis* and *Pterygophora*, may transport materials to storage in overwintering portions of the thallus. This kind of storage would require not only a translocation system but also a triggering mechanism; if it occurs at all it must be rare among the algae. Another special case of storage is the accumulation of carbohydrate reserves by spores and gametes. Red algal spores accumulate floridean starch; red algal spermatia, being colorless, must also rely on reserves (Brawley & Wetherbee 1981). Other spores and gametes are capable of photosynthesis (McLachlan & Bidwell 1978, Kremer & Markham 1979), but the extent to which these cells may depend on stored carbon has not been determined.

7.4.2 Structural polymers

The structural polysaccharides are those that constitute the cell walls; some, such as cellulose, are fibrillar and often uncharged, whereas others, including the commercially important gums, are charged due to sulfate or acid groups and show no structure under the electron microscope. Many are very variable, so that terms such as carrageenan cover a range of similar but not identical molecules.

The fibrillar molecules include cellulose, xylan, and mannan. Cellulose, a $\beta(1 \rightarrow 4)$-linked, unbranched glucan, is common in higher plants and was once thought to be the main structural component in algal cell walls as well. Now it is known to be replaced in some species by xylan or mannan; alginate in the brown algae may also play this role. Moreover, there is a biochemical alternation of generations in some species (including *Acetabularia, Derbesia*, and *Urospora*), in which the walls of the sporophyte contain mannans whereas the walls of the gametophyte (cysts of *Acetabularia*) contain xylans and/or glucans (McCandless 1981). Cellulose molecules lie parallel to one another, forming ribbons or fibrils that can be readily seen in the electron microscope. Algal xylans can be either $\beta(1 \rightarrow 4)$-linked, forming flat celluloselike ribbons, or $\beta(1 \rightarrow 3)$-linked, forming hollow, usually triple helices. $(1 \rightarrow 4)$ xylans have been found in several genera of Caulerpales, the $(1 \rightarrow 3)$ xylans in Bangiophycideae; other red algae have been found to have both types. Mannans form short rods rather than fibrils, which is thought to be significant in the elasticity of *Porphyra* walls (see Chap. 4). The mannans consist of $\beta(1 \rightarrow 4)$-linked mannose residues.

The cell wall matrix polysaccharides present a much more complex picture than either storage or fibrillar polysaccharides. In general, these polymers, like amylose, form helices that are aggregated in various ways in the gel state (Fig. 7.12) (Rees 1975). Each class produces a range of characteristic compounds.

Chlorophyceae produce highly complex sulfated heteropolysaccharides, each molecule of which is made up of several different residues. The major sugars are glucuronic acid, xylose, rhamnose, arabinose, and galactose, made up in several combinations, as shown in Table 7.1. The quantities of each residue vary, and there is as yet no information on their arrangement. Considerably less is known about these green seaweed polysaccharides than about those from reds and browns, perhaps because of their complexity and because none has yet found commercial application.

Matrix polysaccharides of the brown algae were once thought to be relatively simple: alginic acid con-

Figure 7.12. Some mechanisms by which a carbohydrate chain may bind to another of similar or different structure with the formation of a gel network. Lower series of drawings, from left to right, correspond to: isolated double helix, e.g., ι-carrageenan; aggregated double helices, e.g., agarose; "egg carton," e.g., calcium polyguluronate; mixed aggregate, e.g., agarose-galactomannan. (From Rees 1975, with permission of the ajuthor)

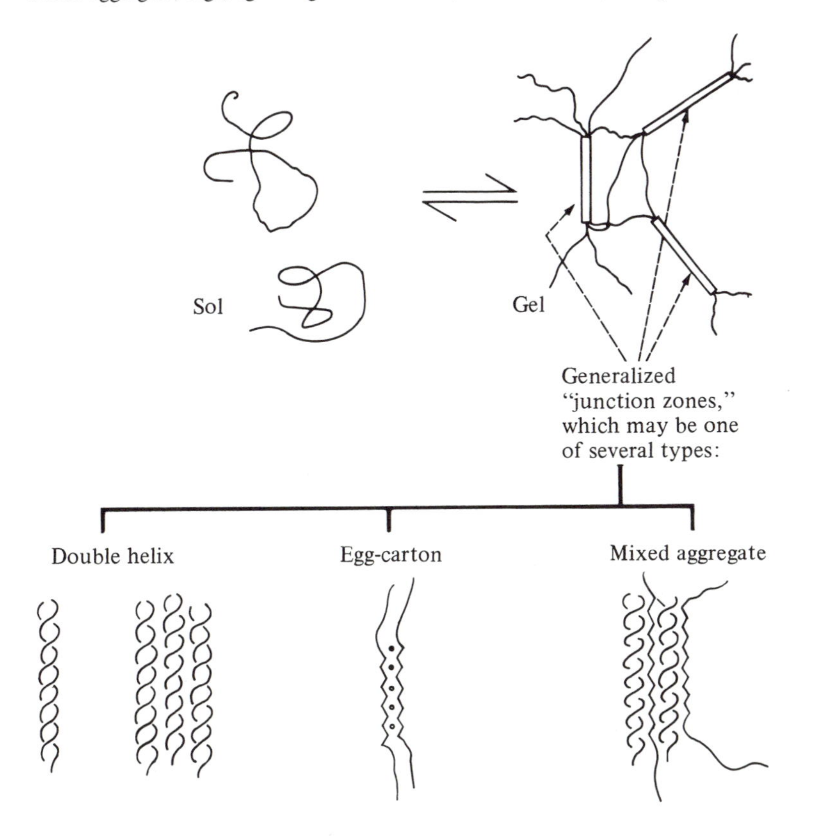

Table 7.1. *Structural polysaccharides in three groups of green seaweeds*

Species	Polysaccharide
Ulva lactuca *Acrosiphonia centralis* *Enteromorpha compressa* *Urospora penicilliformis* *Urospora wormskioldii* *Codiolum pusillum*	Sulfated glucuronoxylorhamnans
Acetabularia crenulata	Sulfated glucuronoxylorhamnogalactan
Chaetomorpha capillaris *Chaetomorpha linum* *Cladophora rupestris* *Caulerpa filiformis* *Codium fragile*	Sulfated xyloarabinogalactans

Source: Percival (1979), with permission of the British Phycological Society.

sisting of mannuronic and guluronic acids, and fucoidan consisting of fucose (Percival & McDowell 1967). Recent evidence indicates that "fucoidan" covers a very wide range of compounds, the simplest being nearly pure fucan (e.g., in *Fucus distichus*). The most complex (e.g., in *Ascophyllum nodosum*) are heteropolymers containing fucose, xylose, galactose, and glucuronic acid, in which the uronic acid may form a backbone and the neutral sugars extensive branches (Larsen et al. 1970, McCandless & Craigie 1979). Even fucan is complex, having $\alpha(1 \rightarrow 2)$- and $\alpha(1 \rightarrow 4)$-links between fucose units as well as $\alpha(1 \rightarrow 3)$-branches and varying degrees of sulfation on the remaining hydroxyl groups (Fig. 7.13a). Alginic acid consists of two uronic acids (sugars with a carboxyl group on C-6): D-mannuronic acid (M) and L-guluronic acid (G) (Fig. 7.13b). Pure chains exist, but in many chains both acids occur. The residues are not in random sequence; rather, they occur in blocks as $(\text{-M-})_n$, $(\text{-G-})_n$, and $(\text{-MG-})_n$.

The bulk of red algal matrix polysaccharides are galactans in which $\alpha(1 \rightarrow 3)$- and $\beta(1 \rightarrow 4)$-links alternate. The variety in the polymers comes from sulfation, pyruvation, or methylation of some of the hydroxyl groups and from the formation of an anhydride bridge between C-3 and C-6 (Fig. 7.14). The commercially most important groups of red algal polysaccharides are the agars and the carrageenans. Agars consist of alternating β-D-galactose and α-L-galactose with relatively little sulfation. The best commercial agar, neutral agarose, is virtually free of sulfate. Some more highly sulfated polymers with the agar structure do not gel and indeed are not referred to as agars (e.g., funoran from *Gloiopeltis* spp.). In carrageenans, β-D-galactose alternates with α-D- (not -L-) galactose and there is much more sulfation. The sulfate groups project from the outside of the polymer helix; this polyelectrolyte surface makes the molecule more soluble (Rees 1975). Conversion of a 6-sulfate group to an anhydride bridge, as in ι- and κ-carrageenan, yields a stronger gel.

Classically, two carrageenan fractions, λ and κ, were distinguished on the basis of their solubility in KCl. Currently, two categories are recognized on the basis of the presence or absence of sulfate on C-4 of the β-galactose residues (Fig. 7.14) (McCandless & Craigie 1979). κ- and ι-carrageenans, which gel, have the extra sulfate, whereas nongelling forms, including λ-, ξ-, and π-carrageenans, do not. κ-carrageenan, the stronger gel, is found predominantly in the gametophytes and λ-carrageenan in the tetrasporophytes of *Chondrus crispus* and some other Gigartinaceae and Phyllophoraceae (McCandless et al. 1973, McCandless 1981). The significance of this, in terms of the physiology of the plants and in view of the spectrum of carrageenans that actually exists, is not yet clear. The type of carrageenan appears to be associated with reproductive phase and not with ploidy level, as seen in the results of van der Meer et al. (1983) from sporogenous male mutants of *Chondrus crispus*. Other Gigartinales, such as *Eucheuma*, produce only one kind of carrageenan in both phases (Dawes 1979).

The roles of the matrix polysaccharides are still the subject of controversy. Gelling polysaccharides no doubt aid in rigidity of the cell wall, while perhaps providing a certain amount of elasticity necessary in the aquatic environment. The conversion of galactose-6-sulfate to the 3,6-anhydride results in a stiffer gel because it takes a "kink" out of the chain, allowing more extensive double helix formation and thus a more compact gel (Percival 1979). The strength of other gels is increased by binding Ca^{2+} or other divalent cations that can cross-link polymer chains. The properties of the polymers that tend to increase divalent ion binding are

Figure 7.13. Brown algal cell wall matrix polysaccharides. (a) fucan, showing overall structure of part of a chain, and details of three kinds of linkage; (b) portions of alginic acid: left, polymannuronic acid; right, polyguluronic acid. (a reprinted with permission from Percival & McDowell 1967, *Chemistry & Enzymology of Marine Algal Polysaccharides*, Copyright 1967 Academic Press Inc. [London], Ltd.; b from Mackie & Preston 1974, with permission of Blackwell Scientific Publications)

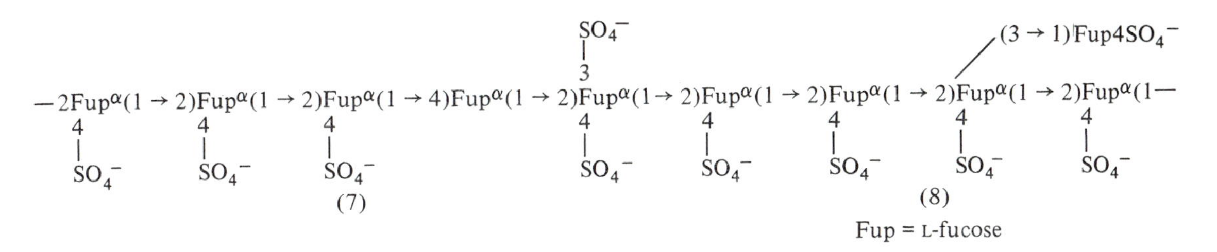

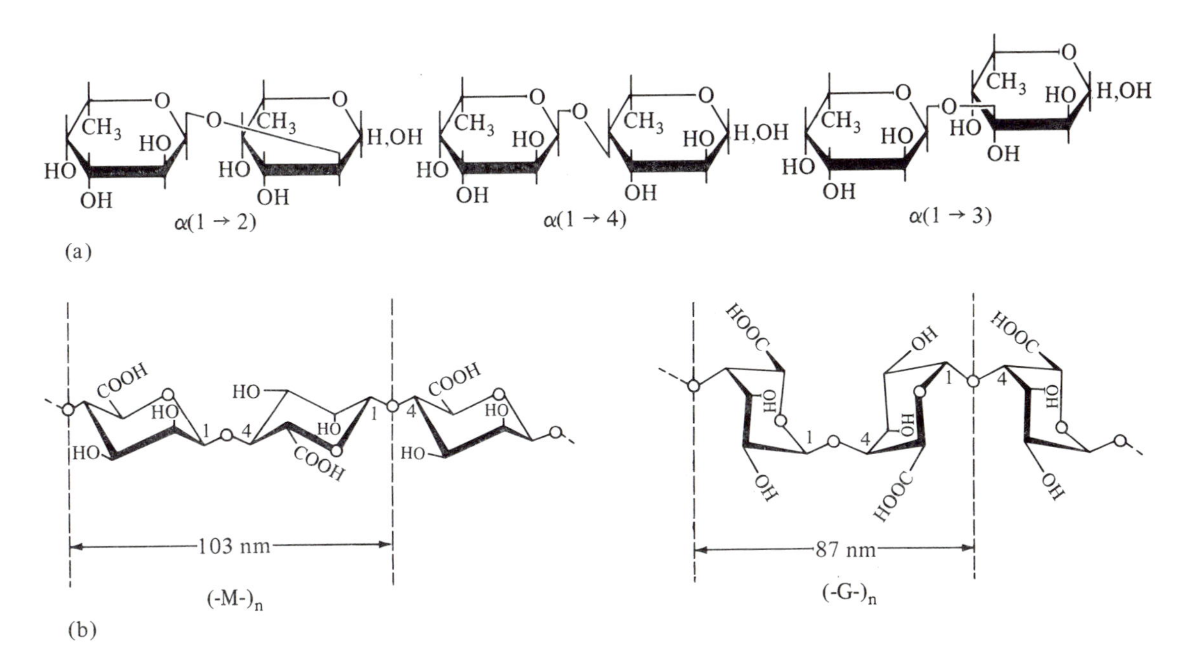

diverse, however. Sulfation of fucoidan is necessary if *Fucus* embryos are to adhere to the substrate, suggesting a correlation between sulfation and divalent ion binding. The sulfate ester is negatively charged and would be expected to bind cations. However, agars gel better with less sulfation and, according to Haug (1976), sulfate esters in *Ulva lactuca* polysaccharides interfere with calcium binding. In the latter instance, borate complexes with rhamnose residues in the polymers, and Ca^{2+} stabilizes that complex (Fig. 7.15a). Sulfate prevents complexing with borate, resulting in a weaker gel.

The strengths of alginates again depend on Ca^{2+} binding, with guluronic acid having a much greater affinity for Ca^{2+} than has mannuronic acid. This effect is not related to sulfate, since neither acid is sulfated, but apparently depends on the more zigzag conformation of polyguluronic acid, which allows Ca^{2+} to fit into the spaces like eggs in an egg carton (Fig. 7.15b). Polyguluronic acid is thus the most rigid form of alginic acid, polymannuronic acid the most lax, with molecules of alternating residues intermediate in strength. While a change in sulfation, or isomerization between uronic acids, could alter wall strength in seaweeds (as might, for example, be useful for spore release), there is so far only a little evidence that seaweeds can make such changes once the polymers are outside the cells (see Sec. 7.4.3).

What are the roles of the nongelling matrix polysaccharides? There is some evidence that they are hygroscopic and may therefore provide protection for the cells against desiccation. Quillet & de Lestang-Brémond (1978) found evidence that fucoidan and λ-carrageenan bond to gel-forming but less hygroscopic polymers such as alginate and ι-carrageenan, respectively, to give hygroscopic gels. Species of fucoids from low to high water show increasing quantities of fucan, but Schonbeck & Norton (1979a) found that *Fucus spiralis*, not *Pelvetia*, was the most hygroscopic fucoid at higher humidities, while there were no differences among various fucoids

Figure 7.14. Red algal cell wall matrix polysaccharides. (a) agar; (b) κ- and λ-carrageenans showing differences in sulfation, and the C-4 of β-D-galactose used in classifying carrageenans.

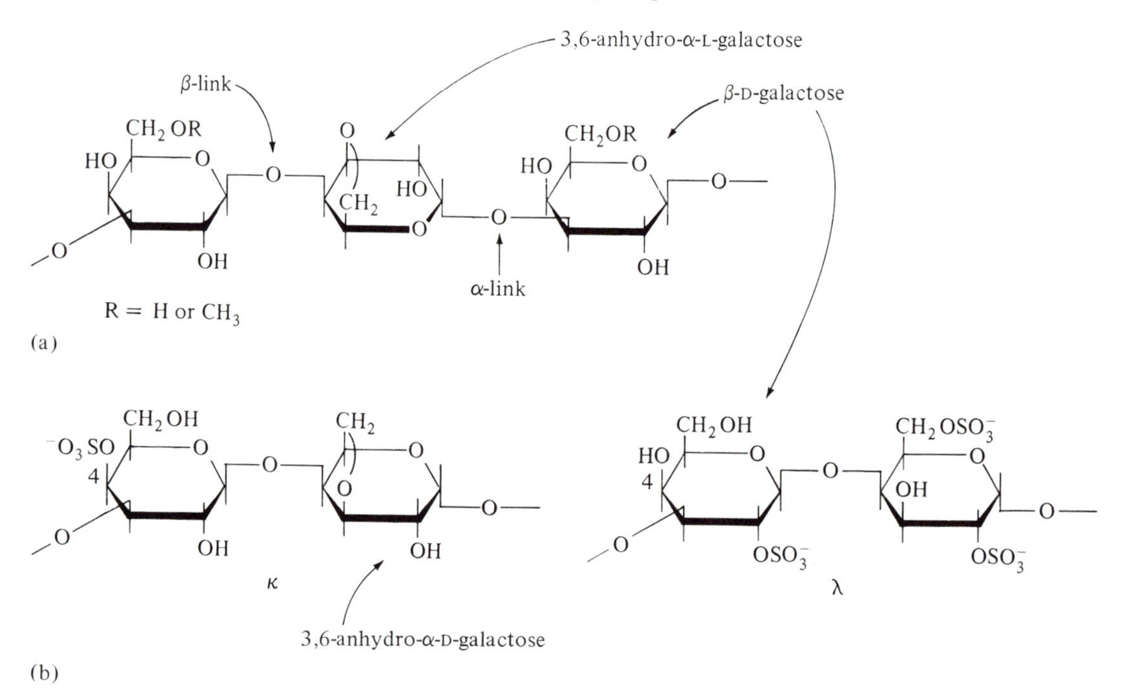

Figure 7.15. Calcium binding (a) to a rhamnose–borate complex from *Ulva* (also showing free rhamnose-2-sulfate); (b) between two chains of polyguluronic acid (the egg-carton model); oxygen atoms shaded black are involved in binding to Ca^{2+}. (a from Percival 1979, with permission of the British Phycological Society; b from Rees 1975, with permission of the author)

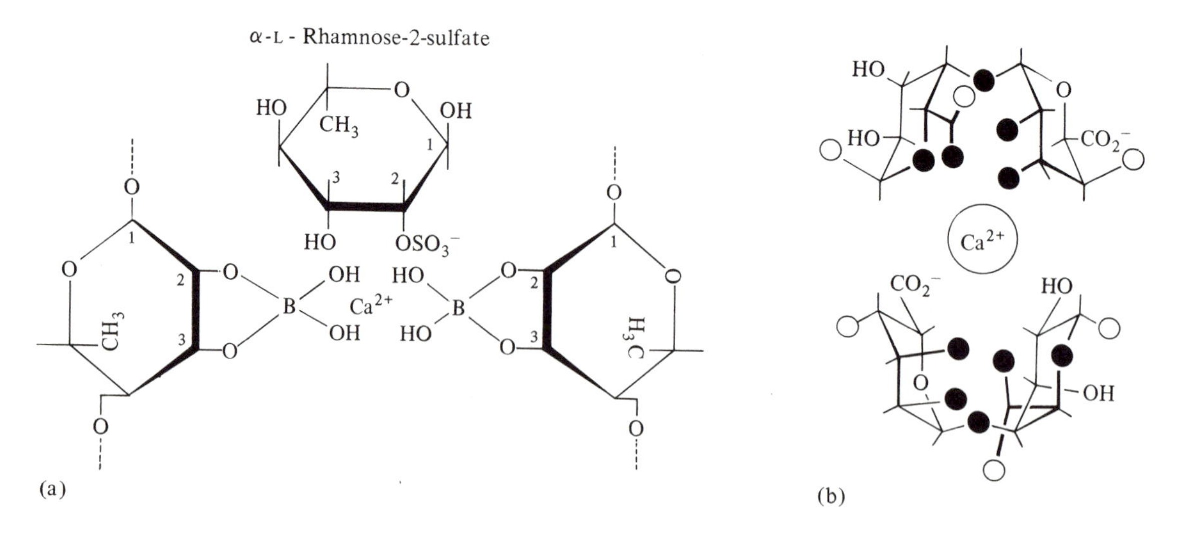

under drier conditions. Besides possibly providing desiccation resistance, these molecules may also serve as a kind of ion exchange material. Seaweed cell walls certainly adsorb cations readily, but there is no evidence that these ions can be subsequently released into the cells. In view of the very limited knowledge of their structure, it is not surprising that the roles of the matrix polysaccharides are not well understood.

7.4.3 Polysaccharide synthesis

The various cell wall and storage polysaccharides are synthesized at several sites in the cell. Cell wall

Figure 7.16. Elongation of a glucan chain. A glucose residue is transferred from ADP-glucose to the reducing end of the chain.

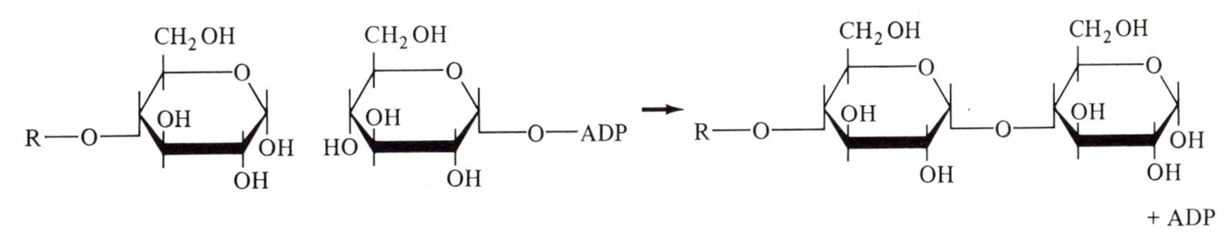

materials are probably made in Golgi vesicles, which pass to the outside of the cell by reverse pinocytosis. The enzyme UDP-galactosyltransferase, which can transfer galactose to fucoidan, has been located in Golgi bodies of *Fucus serratus* (Coughlan & Evans 1978). Starch granules in the cytoplasm of the red alga *Serraticardia maxima* contain the starch-synthesizing enzyme ADP-glucose: α-1,4-glucan α-4-glucosyltransferase (Nagashima et al. 1971). In other algae, starch is stored, and presumably made, in the chloroplasts.

Polysaccharide synthesis seems generally to involve the addition of a nucleotide diphosphate-linked monomer to a primer or existing chain, as shown in Figure 7.16 (Turvey 1978). The route by which mannose is incorporated into alginic acid was postulated by Lin & Hassid (1966) to be D-mannose → D-mannose-6-P → D-mannose-1-P → GDP-D-mannose → GDP-D-mannuronic acid → alginate. Each different bond type in a polymer is made by a different enzyme. Thus amylopectin synthesis requires one enzyme to make the (1 → 4) links, another to make the (1 → 6) branch points (Haug & Larsen 1974). Complexity seems to be added after polymer synthesis. Alginic acid is made initially as polymannuronic acid, then residues are converted to guluronic acid by epimerization at C-5. Addition of sulfate and methyl groups and the formation of anhydride bridges in various polysaccharides also take place after polymerization (Percival 1979). Wong & Craigie (1978) partially characterized from *Chondrus crispus* an enzyme (also known from other red algae) that forms the 3,6-anhydride bridge on galactose residues sulfated at C-6, with the release of the sulfate. The structure of the resulting κ-carrageenan is shown in Figure 7.14. There is evidence from another red alga, *Catenella opuntia*, for extracellular turnover of sulfate ester on λ-carrageenan after deposition. The sulfate appears to be carried by a methylated cytidine monophosphate (de Lestang-Brémond & Quillet 1981, Quillet & de Lestang-Brémond 1981).

Formation of microfibrils in cell walls involves an extracellular crystallization step after the polymers have been secreted. The orientation of the microfibrils is controlled by microtubules just below the plasmalemma. Experimental separation of the three events (polymerization, orientation, crystallization) has been achieved through the use of inhibitors. Colchicine is well known as a microtubule disrupting agent (much used in studies of mitosis); cycloheximide, which inhibits protein synthesis by 80S ribosomes, inhibits polyglucan synthesis; and two carbohydrate binding dyes, Congo red and Calcofluor white ST/M2R, inhibit crystallization of the glucan chains into microfibrils (Herth 1980, Quader 1981).

7.5 Secondary metabolites

Seaweeds provide a rich source of interesting organic molecules that are not part of the major metabolic pathways (i.e., pathways of protein, pigment, lipid, and nucleotide synthesis) (Ragan 1981). Most of these compounds are merely curiosities or of interest as potential drugs for human use. For some there are at least indications of their usefulness to the plants in which they occur. Some compounds, however, may be of no use at present to the plants that produce them. Over the course of evolution, gene mutations have led to the synthesis of many new compounds. If a new compound is not harmful to the plant, in terms of fitness, the pathway will not be selected against. Because environments change, a compound that now may provide no benefit to the species may do so in the future – or may have in the past. Bell (1980) recently discussed possible roles of secondary metabolites in plants. For most of these compounds the biosynthetic pathways are unknown.

The best-known groups of secondary metabolites are those thought to be grazer or epiphyte deterrents, or antibiotics important in wound healing and probably in disease resistance. Brown algae have phenolic compounds in physodes (Figs. 2.9d, 9.23) and excrete these compounds (Ragan & Jensen 1979). The sporophylls of *Alaria marginata* have higher concentrations of phenolics than does the vegetative blade and are consumed by snails at a lower rate than vegetative tissue (Steinberg 1984). Phenolics probably originate from the shikimic acid pathway, which is also part of the biosynthesis of the aromatic amino acid phenylalanine (Fig.7.5) (see Harborne 1980). Simple phenols, such as phloroglucinol (Fig. 7.17a), polymerize within the physodes through a variety of linkage types, and may reach molecular weights of 10^5 (Ragan & Craigie 1976, Ragan 1976, 1981). In addition to their demonstrated role as toxins, polyphenols are also important in heavy metal binding and may con-

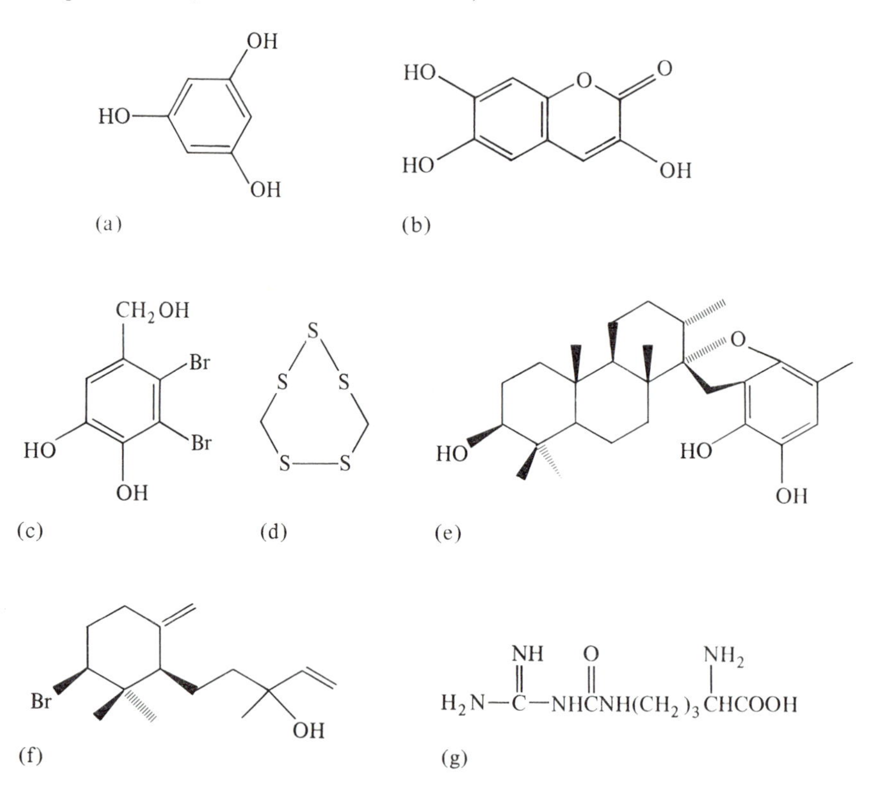

Figure 7.17. Some seaweed secondary metabolites. (a) phloroglucinol; (b) hydroxycoumarin; (c) lanosol; (d) lenthionine; (e) stypotriol; (f) β-snyderol; (g) gigartinine. (b from Menzel et al. 1983, with permission of Walter de Gruyter & Co.; f from Young et al. 1980, with permission of *Journal of Phycology*; the rest from Ragan 1981, with permission of Blackwell Scientific Publications)

ceivably be of value to seaweeds in polluted areas (Ragan et al. 1979). Coumarins (e.g., Fig. 7.17b), a group of phenolic compounds, have been found in some siphonous green algae (Menzel et al. 1983). There, as in higher plants where they are common, they may serve protective roles in (1) controlling epiphytes, (2) defense against microbial infection, and (3) grazer deterrence. They may also take part in polymerizing and thus solidifying the wound plug. The great diversity of halogenated compounds found in seaweeds, especially in red algae, may also play antibiotic roles; certainly some show antibacterial activity in vitro (Table 9.5) (Fenical 1975). The halogenated compounds include a different group of phenolics, not associated with physodes, numerous terpenoids, and other types of structures (Fig. 7.17c; Table 9.5). Also among the probable antibiotics are some sulfur-containing compounds such as lenthionine (Fig. 7.17d). Fish deterrents, including udoteal (Fig. 9.17) and stypotriol (Fig. 7.17e), are unhalogenated terpenoids, and often have reactive groups such as aldehydes (−CHO) and acetate ester (−OAc). However, avrainvilleol is brominated (Sun et al. 1983).

Little is known about the cellular (or extracellular) location of these compounds, and such observations as there are shed little light on the roles of the compounds. In red algae, brominated compounds have been detected in chloroplasts of *Lenormandia prolifera* (von Hofsten et al. 1977) and in intercellular granules in *Thysanocladia densa* (Pallaghy et al. 1983). Some cytoplasmic inclusions of *Laurencia snyderae*, called "corps en cerise," contain snyderol (Fig. 7.17f) (Young et al. 1980).

Some unusual nonprotein amino acids found in seaweeds have a potential role as nitrogen storage compounds. One such compound is gigartinine (Fig. 7.17g), isolated from *Chondrus crispus* by Laycock & Craigie (1977).

7.6 **Synopsis**

Carbon, the major nutrient element of all plants, is available in seawater chiefly as HCO_3^-. There are small amounts of CO_2, the proportion depending on pH and salinity. Carbon is fixed into organic compounds in the "dark" reactions of photosynthesis. Parasitic algae rely on carbon from their hosts, but other seaweeds have little or no ability to use exogenous organic carbon, particularly at environmental concentrations. Most seaweeds are thus photoauxotrophic and are not facultatively heterotrophic. The relationships between some seaweeds and symbiotic fungi pose interesting nutritional and hormonal questions.

CO_2 is fixed by RuBP carboxylase in the Calvin cycle. HCO_3^- is converted to CO_2 with the aid of carbonic anhydrase. In the brown algae, especially in young tissue of kelps and fucoids, light-independent carbon fixation occurs, in light and dark, via PEP carboxykinase, which also uses free CO_2. RuBP carboxylase also acts as an oxygenase, leading to the photorespiratory or glycolate cycle.

Various principal products of photosynthetic carbon fixation are found among the seaweeds. Green algae form chiefly sucrose and starch; brown algae form mannitol and laminaran; red algae form several low molecular weight compounds and floridean starch. In addition, all groups accumulate some amino acids.

In the kelps, the low molecular weight compounds can be translocated from mature regions to meristematic regions. The driving force is probably a water potential gradient set up by the loading of organic molecules into the sieve elements in the source and their unloading in the sink.

Seaweed cell walls have a fibrillar layer, consisting of cellulose, mannan, or xylan. A variety of characteristic mucilaginous polysaccharides is also found in seaweed walls. These include some of commercial value, such as agars, carrageenans, and alginates. Chemical changes in these polymers can make them weaker or stiffer gels. The native structures of these compounds are not well known. Different stages in the life history may have different wall polysaccharides.

Numerous secondary metabolites occur in seaweeds. Some of these, notably phenolics and halogenated compounds, are active as antibiotics or grazer or epiphyte deterrents.

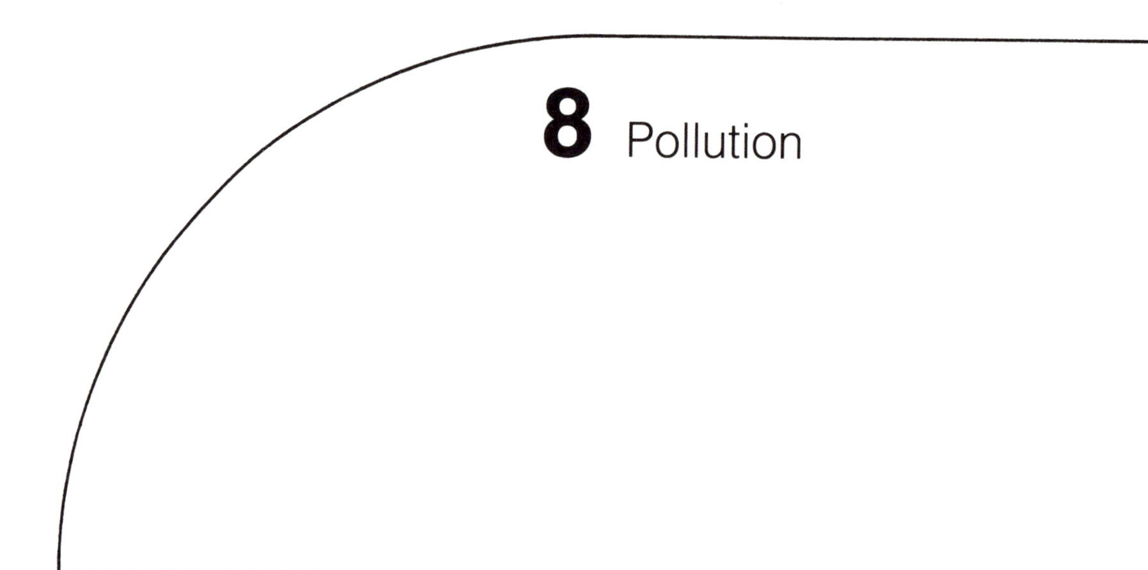

8 Pollution

8.1 Introduction

Concern over marine pollution has developed comparatively recently, with several important events, such as the world's first major oil spill (106,000 metric tons) by the supertanker *Torrey Canyon*, accelerating public concern in the early 1960s. Although there is no precise definition of the term *pollution*, one general definition is a stress caused by human activities on the natural environment, resulting in unfavorable alteration of an ecosystem. Other definitions, referring to the introduction of a substance into the environment by humans, are more restrictive since they do not include thermal pollution. The term ''unfavorable'' in the definition involves human value judgments; therefore it is common to see disagreement among scientists and politicians on whether certain events are examples of pollution (Rosenberg et al. 1981).

The adverse effects of pollutants on aquatic organisms are generally identified from their acute and lethal impacts. Mortality can be readily recognized and quantified (recall comments on strain in Sec. 3.3.3). However, there is considerable difficulty in determining sublethal effects and whether the responses observed in the laboratory can be extrapolated to the more natural and varied conditions in the sea. The sublethal response of an organism can be categorized according to the effect on the organism's (1) biochemistry/physiology; (2) morphology; (3) behavior; and (4) genetics/reproduction. Physiologists and ecologists debate how to measure responses to sublethal concentrations of a pollutant and whether laboratory bioassays give meaningful results. In reality, *both* laboratory and field measurements are necessary. To bridge the gap between the two areas, some laboratory facilities are scaled up and taken into the field to conduct experiments. For example, large plastic enclosures have been used to capture part of the water column to observe effects of a pollutant on the community (Grice & Reeve 1982).

Figure 8.1. Hypothetical relation of concentration of pollutant to response of marine organisms, showing some significant points and regions on the curve. (From Waldichuk 1979, with permission of The Royal Society, London)

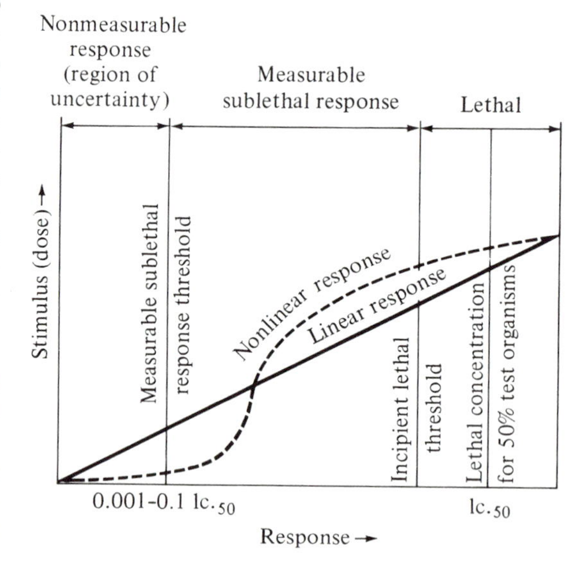

The general stimulus–response relationship of biological assays on the effect of pollutants on aquatic organisms is schematically represented in Figure 8.1. The relationship is frequently nonlinear with a threshold response. The dose–response relationship is often characterized by the parameter $l.c._{50}$, the lethal concentration for 50% of the test organisms. Bioassays are particularly useful in assessing toxicity of mixtures of substances such as pulpmill effluents. At present it is not possible to chemically estimate or anticipate the biological consequences of chemical transformation, complexation, and interaction of contaminants on their toxicity, because only living systems can integrate the effects of those variables that are biologically important (Stebbing 1979). Bioassay techniques have recently been assessed and the following three test organisms recommended: oyster larvae, sea urchin larvae, and microalgae (Stebbing 1979). Stebbing also recommended that manipulative techniques be used in conjunction with bioassays to aid in the categorization of contaminants (Table 8.1).

The Marine Pollution Subcommittee of the British National Committee on Oceanic Research has recently summarized past deficiencies and future needs in pollution research in the laboratory and field (Cole 1979b). First, in the laboratory care must be taken in experimental design and operation of bioassay tests. The physicochemical aspects of the pollutant, such as solubility, adsorption, and chemical complexation and speciation should be taken into account, because for these reasons only a small portion of the pollutant may be biologically available (i.e., available to be taken up by an organism) (Burton 1979). Total concentration of a contaminant may give little indication of its toxicity. The choice of a test organism is often based on the ease of handling and culturing; frequently the most resistant stage (adult) is used for testing. A proper evaluation should take into account the full life cycle, and certainly tests with the most sensitive stage should be conducted. For application to survival of populations, the response must ultimately be related to a healthy progression through the full life cycle, including successful reproduction. In many cases, several species should be tested (Cairns 1983). The recovery process should also be examined in experiments; the possibility that recovery may occur after damage by pollutants is often ignored and rarely assessed. Short-term laboratory bioassays are particularly limited, and both short- and long-term bioassays may be invalid in the absence of suspended particulate matter, which is known to profoundly influence the effects of many pollutants on seaweeds.

Because of the simplicity of laboratory experimental conditions in comparison with the complexity of the marine environment, physiologists should also make observations directly in the field. This could be done, for example, by attaching macrophytes to artificial substrates and taking them to polluted areas. Long-term monitoring of community structural changes in the field is required for a minimum of several years to assess effects at low pollutant concentrations. Remote sensing using infrared photography may be used to survey large areas of macrophytes rapidly and document major changes over several years. In the field the ecologist must try to sort out the responses caused by natural spatial and temporal variability from effects attributable to pollution (Hawkins & Hartnoll 1983). Since the bulk of most long-lived pollutants often ends up in the sediments, the mechanisms and dynamics of uptake and release of pollutants from sediments and their transfer to biota need further investigation.

Table 8.1. *Manipulative techniques that might be used in conjunction with bioassays to aid the identification of toxic contaminants*

Manipulation	Technique
Removal of organics	Activated charcoal
Breakdown of organics	UV photo-oxidation
Removal of divalent metals	Ion exchange resin
Binding of metals	Chelating agents (NTA, EDTA)
Removal of chlorinated hydrocarbons	Ion exchange resin
Removal of PCBs and DDT	Membrane filtration

Source: Stebbing et al. (1980), with permission of Conseil International pour l'Exploration de la Mer.

Examples of five categories of marine pollution will be discussed in this chapter: thermal pollution; heavy metals such as mercury, lead, cadmium, zinc, and copper; eutrophication (excessive nutrients such as nitrogen or phosphorus); organic wastes such as herbicides and pesticides; and hydrocarbons, particularly oil.

8.2 Thermal pollution

Some industries and most power plants use water for cooling and discharge heated waste water into the aquatic environment. The effect on macrophytes can be either deleterious or beneficial, depending on the geographic location, season, and species involved. Unlike higher plants, which may be exposed to a wide temperature range (ca. 50 C), seaweeds generally exist in a much narrower range (ca. 10–25 C). Whether macrophytes can survive the increased water temperature depends on how close they are to their upper limit of temperature tolerance (see Sec. 3.3.3). This temperature tolerance is not constant for a species and may depend on other environmental factors such as light, salinity, nutrients, and pollutants (Laws 1981). For example, during the summer in temperate shallow bays or tropical areas, the ambient water temperature may already be near the upper range of tolerable temperatures. Therefore, an increase of only a few degrees may quickly produce sublethal or lethal conditions for some species. Extensive reductions in macrophytes were noted in power plant discharge areas of semitropical Biscayne Bay, Florida (Wood & Zieman 1969). On the other hand, in most temperate coastal areas with a high water exchange, point increases in temperature will usually result in increased growth rates and primary productivity.

Many northern European countries now recommend that when a new power plant is built that is cooled by seawater, the temperature increase that results after mixing is not to exceed 2 C and in the summer the

Figure 8.2. Graphic model of potential impact of elevated temperatures on populations of *Ascophyllum nodosum*. PSN = photosynthesis; P/R = primary productivity. (From Vadas et al. 1978, with permission of the Technical Information Center, U.S. Department of Energy)

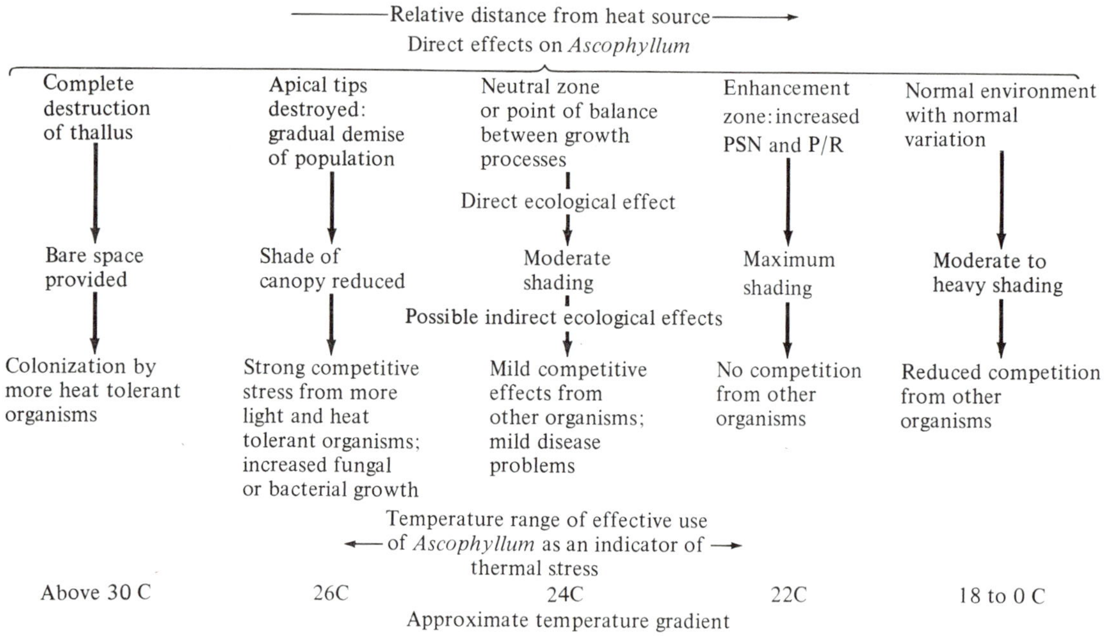

——— Relative distance from heat source ———→

Direct effects on *Ascophyllum*

Complete destruction of thallus	Apical tips destroyed: gradual demise of population	Neutral zone or point of balance between growth processes	Enhancement zone: increased PSN and P/R	Normal environment with normal variation
		Direct ecological effect		
Bare space provided	Shade of canopy reduced	Moderate shading	Maximum shading	Moderate to heavy shading
		Possible indirect ecological effects		
Colonization by more heat tolerant organisms	Strong competitive stress from more light and heat tolerant organisms; increased fungal or bacterial growth	Mild competitive effects from other organisms; mild disease problems	No competition from other organisms	Reduced competition from other organisms

←— Temperature range of effective use of *Ascophyllum* as an indicator of thermal stress —→

Above 30 C	26C	24C	22C	18 to 0 C

Approximate temperature gradient

temperature of the mixed water is not to rise above 26 C. In spite of this recommendation, the species composition of an area may change, even in the absence of thermal damage to individuals, since the increase in temperature may affect species competition. A change in algal species composition may also affect the species composition of herbivores and animals higher up the food web. Other deleterious effects could occur when there is an abrupt plant shutdown, which could create a cold shock for many macrophytes. In addition, chlorine or copper are periodically introduced into the cooling water system to reduce fouling, creating a temporary toxic environment in the vicinity of the discharge.

There is little published information on the direct effects of thermal effluents on macrophytes. Some symptoms of thermal stress include frond hardening, bleaching, or darkening and cell plasmolysis. Adult *Macrocystis* plants displayed substantial tissue deterioration after surface temperatures reached 20 C or more for several weeks (North 1979). Three disorders have been observed: black rot, tumorlike swellings, and stipe rot. Black rot refers to a darkening of the blades that usually appears first at the tips, then spreads toward the base. On the Atlantic coast, a prolonged but intermittent thermal stress affected the growing (apical) tips of *Ascophyllum nodosum* (Fig. 8.2) (Vadas et al. 1978). Significant declines occurred in percentage of cover, biomass, growth, and survival of *Ascophyllum*, but basal sections of these plants survived, although weakly attached to the substrate. Apical meristems were not initiated in the second spring after the onset of the thermal discharge, but in later years the population recovered fully. However, the cover of *Fucus vesiculosus* decreased when the thermal discharge began and it never reestablished.

The temperature curve for growth is the result of numerous metabolic interactions, some of which have different temperature optima (Cairns et al. 1975). Tolerance limits in the field are lower because of interaction with other factors and the integration of long-term effects. For example, *Macrocystis pyrifera* displays a temperature optimum for photosynthesis between 20 and 25 C, yet plants deteriorate when ocean temperatures reach 20 C (North 1979). Hence, field results are not always predictable from laboratory measurements.

Ocean Thermal Energy Conversion (OTEC) is a method of converting solar energy in tropical waters into electrical energy using the temperature difference between the warm surface water and the colder water at 600 m depth. Surface water is pumped through a heat exchanger to evaporate a working fluid (freon). The vapors then spin a turbine and electricity is generated. The OTEC program has the opposite effect of power plants on the environment, because colder water is pumped from depth to the surface. Two important environmental changes occur: surface temperatures are cooled from near 30 C to 20 to 25 C, and nutrient concentrations are increased as the nutrient-rich deep water mixes with the nutrient-impoverished surface water. Preliminary studies

Table 8.2. *Classification of elements according to toxicity and availability*

Noncritical	Toxic but very insoluble or very rare	Very toxic and relatively accessible
Na C F	Ti Ga Hf La	Be As Au Co
K P Li	Zr Os W Rh	Se Hg Ni Te
Mg Fe Rb	Nb Ir Ta Ru	Tl Cu Pd Pb
Ca S Sr	Re Ba	Zn Ag Sb Sn
H Cl Al		Cd Bi Pt
O Br Si N		

Source: Wood (1974), *Science*, vol. 183, pp. 1049–1052, Copyright © 1974 by the American Association for the Advancement of Science.

have shown that this process is beneficial; two red algae, *Laurencia poitei* and *Gracilaria ferox*, and the brown *Sargassum fluitans* all exhibited faster growth rates (Thorhaug & Marcus 1981).

8.3 Heavy metals

The term ''heavy metal'' has been generally used to describe those metals having atomic numbers greater than iron (59) or having a density greater than 5 g mL^{-1} (Sorentino 1979). From the standpoint of environmental pollution, metals may be classified according to the following three criteria: (1) noncritical; (2) toxic but very insoluble or rare; (3) very toxic and relatively accessible (Table 8.2) (Wood 1974). Some heavy metals in category 3, however, such as manganese, iron, copper, and zinc, are essential micronutrients and are frequently referred to as trace metals (Sec. 6.5.6). They may limit algal growth if their concentration is too low and can be toxic at higher concentrations; frequently the optimum concentration range for growth is narrow (Fig. 8.3). Other heavy metals in category 3, such as Hg or Pb, are not required for growth and they may become toxic to algae at very low concentrations (e.g., 10–50 μg L^{-1}).

Metals in minerals and rocks are generally harmless and only become potentially toxic when they dissolve in water. They enter the environment by natural weathering of rocks, leaching of soils and vegetation, and volcanic activity. Some of the highest mercury levels are found not in coastal waters but in the deep sea, near mid-ocean ridges, due to submarine volcanic activity. Therefore, in the assessment of marine pollution, a distinction must be made between natural sources and those caused by human activities. Humans contribute metals to the environment by mining and smelting, combustion of fossil fuels, and industrial waste disposal, from processing and manufacturing. Most of the metal load is transported by water in a dissolved or particulate state, and most of it reaches the oceans via rivers or land runoff. Also, rainwater carries significant cadmium, copper, zinc, and, especially, lead from the atmosphere to the oceans. These metals in the atmosphere come from the burning of fossil fuels. Metals in sediments may be reduced or oxidized primarily by bacteria and released into the overlying water.

Figure 8.3. The oligodynamic dose relation between population parameters and concentration of metabolite or pollutant. (From Perkins 1979, with permission of The Royal Society, London)

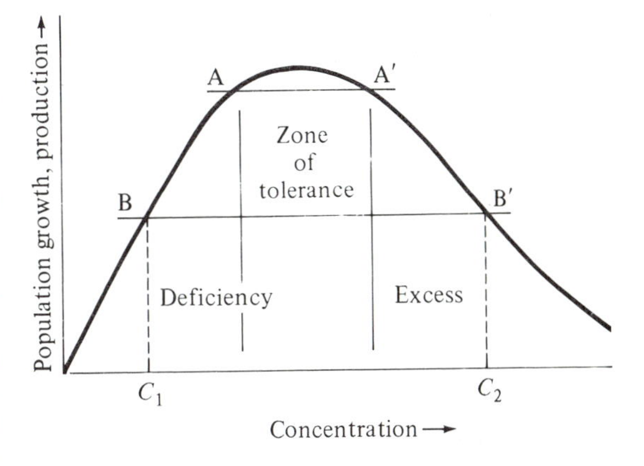

Metals in an aquatic environment may exist in dissolved or particulate forms. They may be dissolved as free hydrated ions; complex ions, chelated with inorganic ligands such as OH^-, Cl^-, or CO_3^{2-}; or complexed with organic ligands such as amines, humic and fulvic acids, and proteins. Particulate forms may be found as colloids or aggregates (e.g., hydrated oxides); adsorbed onto particles; precipitated as metal coatings onto particles; incorporated into organic particles such as algae; and held in the structural lattice in crystalline detrital particles (Beijer & Jernelöv 1979). The physical and chemical forms of metals in seawater are controlled by environmental variables such as pH, redox potential, ionic strength, salinity, alkalinity, presence of organic and particulate matter, biological activity, and the intrinsic properties of the metal. Changes in these variables result in transformation of the chemical form and contribute to the availability, accumulation, and toxicity of the element to aquatic organisms.

In coastal waters, the concentration of heavy metals decreases with distance from river mouths. This is not only the result of dilution but also of salting out of high molecular weight fractions and flocculation of inorganic matter as salinity increases. Metals may adsorb to these newly formed particles and sink to the sediments. On the other hand, some metals previously attached to particles in the river may be displaced by chloride ions and become available for uptake by algae and other organisms. Further information on the biogeochemistry of metals can be found in Förstner & Wittman (1979), Förstner (1980), and Laws (1981).

8.3.1 Uptake and accumulation

Metals are taken up both passively and actively by algae. Some, such as Pb and Sr, may be passively adsorbed by charged polysaccharides in the cell wall and intercellular matrix (Sec. 6.3.1) (Morris & Bale 1975, Eide et al. 1980). Other metals (e.g., Zn, Cd) are taken up actively against large intracellular concentration gradients (Eide et al. 1980).

Macrophytes concentrate metal ions from seawater, and variations in the concentration of the metal in the thallus are often taken to reflect metal concentrations in the surrounding seawater. On this basis, macroalgae, and especially the Phaeophyceae, frequently have been used as indicators of trace metal pollution (Morris & Bale 1975, Phillips 1977). The rationale for using seaweeds as indicators of metal contamination is based on three main reasons (Luoma et al. 1982). The first is that metal concentrations in solution are often near analytical detection limits and may be variable with time. Seaweeds concentrate metals from solution and integrate short-term temporal fluctuations in metal concentrations. Second, empirical methods for distinguishing the biologically available fraction of total dissolved metal levels have not been developed for natural systems. By definition, seaweeds will accumulate only those metals that are biologically available. Finally, since plants do not take up particulate bound metals (as animals do), plants will accumulate metals only from solution.

While these reasons have been used in the past to argue for using seaweeds as indicators of metal pollution, their use has probably led to erroneous results in view of what is known now. For example, light and nitrogen availability affect uptake of Fe, Mn, Zn, Cd, and Rb by *Ulva fasciata* in outdoor continuous cultures (Rice & LaPointe 1981). Light exerted a greater overall effect than nutrient level. Concentrations of Cd and Rb decreased and Mn increased as the specific growth rate increased, suggesting metabolic regulation of these three metals. Algal nitrogen content controlled the uptake of zinc and iron. Other investigators have found that the age of the frond has an important bearing on the levels of heavy metal accumulation, with older parts being more retentive (Bryan & Hummerstone 1973). Other factors that can influence metal accumulation include the position of algae on the shore (Bryan & Hummerstone 1973), the season of the year (Burdon-Jones et al. 1982), and the presence of other pollutants in the surrounding water (Bryan 1969).

Seaweeds thus may *not* accurately reflect metal concentrations in the surrounding water. Several examples illustrate this important point. A population of *Fucus* growing near a metal-polluted estuary showed the same seasonal pattern of trace metal concentration changes as another population growing in unpolluted waters in the same region (Fuge & James 1973). Foster (1976) found that concentrations of Cd, Pb, Cr, and Ni in polluted *Fucus vesiculosus* and *Ascophyllum nodosum* from polluted areas were lower than in control plants even though the concentrations of these metals were higher in the water. He suggested that the reason why these two seaweeds did not reflect the ambient metal concentrations was that the high dissolved concentrations of Cu and Zn resulted in these elements being the predominant occupants of the uptake binding sites to such an extent that accumulation of Cd, Pb, Cr, and Ni was inhibited.

Since the amount of metals found in seaweeds varies considerably, the concentration factor will vary also. The concentration factor is calculated as μg of the metal per g dry wt of seaweed divided by $\mu g\ mL^{-1}$ of the dissolved metal in seawater. Examples of concentration factors for four metals in two brown seaweeds are given in Table 8.3 (Rao & Tipnis 1967, Foster 1976). The concentration factors of these elements range from 10^3 to 10^4, but chromium reaches a factor of 10^6 (Saenko et al. 1976).

If macrophytes are to be considered good indicators of metal pollution they must be able to release metal ions as well as take them up, especially if metal concentrations in the water fluctuate over the long term. Several studies have examined metal release in situ by transplanting whole plants from polluted to unpolluted areas (Myklestad et al. 1978, Eide et al. 1980). Zinc and cadmium showed some release from older tissue, whereas very little lead was released; mercury release was intermediate. Newly grown tissue attained the same heavy metal composition as the native plants, suggesting that only young portions should be used to assess ambient metal concentrations because of the retention of metals by the older portions.

With recent improvements in the analytical capability to measure trace elements, direct measurement of trace metal concentrations in the water are preferred to indirect measurements via bioaccumulation in seaweeds. Nevertheless, biological availability of these concentrations must be assessed by using bioassay organisms.

8.3.2 Mechanisms of tolerance to toxicity

Few studies on tolerance to metal toxicity have been conducted on seaweeds. Therefore, some investigations using microorganisms (Iverson & Brinckman 1978, Gadd & Griffiths 1978) and higher plants (Foy et al. 1978) will be discussed to illustrate basic principles that may apply to macrophytes.

Table 8.3. *Concentrations of selected heavy metals in seawater and in brown algae and concentration factors*

		Fucus vesiculosus		*Ascophyllum nodosum*	
Heavy metal	Seawater (μg L^{-1})	(ppm)	conc. factor ($\times 10^3$)	(ppm)	conc. factor ($\times 10^3$)
Zn	11.3	116	10	149	13
Cu	1.4	9	6.4	12	8.6
Mn	5.3	103	19	21	3.9
Ni	1.2	8	6.8	5.5	4.6

Concentration factor = ppm dried seaweed per μg mL^{-1} of dissolved metal in seawater.
Source: Modified from Foster (1976), with permission of Applied Science Publishers.

Algae may produce extracellular compounds or compounds in or on the cell wall that bind to certain metal ions, rendering them nontoxic. Detoxification of metal ions in the culture medium or at the cell surface is referred to as an exclusion mechanism since the metal ions do not cross the cell membrane and consequently the uptake of the metal is prevented. This mechanism has been invoked to partially account for the high copper tolerance of some multicellular freshwater algae such as *Mougeotia, Microspora*, and *Hormidium*, where more than half the copper associated with the cells accumulated in the cell walls (Francke & Hillebrand 1980). Carrageenan from *Eucheuma striatum* has been shown to effectively bind heavy metals such as cadmium, lead, and strontium through an exchange mechanism (Veroy et al. 1980). The results of this study demonstrate that metal binding capacity was correlated with the degree of sulfation of the carrageenan. Hall et al. (1979) found that tolerant strains of the ship-fouling alga *Ectocarpus siliculosus* accumulated less copper in comparison with intolerant strains, but the tolerant strain did not always produce greater amounts of extracellular products. Moreover, the extracellular products of the tolerant strain did not confer tolerance on the intolerant strains. Membrane and intracellular changes are believed to account for tolerance in the strain of *E. siliculosus*. Further studies on this strain revealed that it was also more tolerant of zinc and lead (Hall 1980).

Another exclusion mechanism, not yet observed for macrophytes, is the adsorption or detoxification of the metal ion by surface-living microorganisms. Possibly epiphytes such as diatoms or bacteria may take up and sequester the metal ion before it reaches the membrane surface of a seaweed.

If no exclusion mechanism is operating, the metal ion will enter the cytoplasm. A number of detoxification mechanisms are possible inside the cell. Intracellular precipitation of copper has been observed within vacuoles and nuclei of the freshwater green *Scenedesmus acutiformis* when it was grown at high copper concentrations (Silverberg et al. 1976), whereas in *Porphyra umbilicalis* the nucleus was the site for intracellular bound cadmium (McLean & Williamson 1977).

Once metals are inside the cell, they may, as a result of biological action, undergo changes in valence and/or conversion to organometallic compounds. Both processes can be considered detoxification mechanisms, since volatilization and removal of the metal may result. Transformations involving changes of valency have been studied mainly with mercury. Several types of bacteria and yeast can reduce cationic mercury (Hg^{2+}) to the elemental state (Hg^0) that volatilizes from the medium (Gadd & Griffiths 1978). Certain metals (e.g., Hg, Pb, Cd, Sn) are transformed into organometallic compounds by methylation. Although products of methylation may be more toxic to some animals than the free ion, they are often volatile and can be released into the atmosphere.

In addition to physiological tolerance to heavy metals, resistance to metals is likely to be genetically controlled. Copper tolerance in the ship-fouling alga *Enteromorpha compressa* appears to be genetically determined, because progeny from the ship-fouling plants were also very copper tolerant even though they had not been previously exposed to copper (Reed & Moffat 1983). The rate of accumulation of copper in ship-fouling thalli was equal to the rate for nonfouling thalli, suggesting that tolerance may be due primarily to internal detoxification, rather than to an exclusion mechanism.

8.3.3 Effects of metals on algal metabolism

The order of metal toxicity to algae varies with the algal species and experimental conditions, but generally it is Hg > Cu > Cd > Ag > Pb > Zn (Rice et al. 1973, Rai et al. 1981).

Mercury, the most toxic metal, interacts with enzyme systems and inhibits their functions, especially enzymes with reactive sulfyhdryl (–SH) groups. The toxic effects of mercury on algae generally include (1) cessation of growth in extreme cases; (2) inhibition of photosynthesis; (3) reduction in chlorophyll content; and (4) increase in cell permeability and loss of potassium ions from the cell (Rai et al. 1981). There are two studies on physiological effects of mercury on marine macroalgae. Hopkin & Kain (1978) studied how different life history stages of *Laminaria hyperborea* were affected and found that the growth of gametophytes was most sensitive. Respiration rates of the sporophyte increased only at the highest concentrations of mercury. The effect of Hg on the increase in length of five intertidal Fucales showed that exposure to an average concentration of 100 to 200 μg Hg L^{-1} for 10 days gave a 50% reduction in growth rate (Strömgren 1980b). Even at 5 to 9 μg Hg L^{-1}, a reduction in growth was seen in adults of *Fucus spiralis*.

Copper, even though an essential micronutrient, is the second most toxic metal, and copper sulfate has been used to control nuisance algae in freshwaters. Copper toxicity is dependent on the ionic activity (concentration of free Cu^{2+}) and not the total copper concentration (Sunda & Guillard 1976). Toxicity effects have been shown to pass through several stages (Sorentino 1979). First, copper affects the permeability of the plasmalemma, causing loss of K^+ from the cell and changes in cell volume. Next, Cu^{2+} may be transported to the cytoplasm and then to the chloroplasts, where it inhibits photosynthesis by uncoupling electron transport to $NADP^+$. As the ionic concentration increases, copper is bound to chloroplast membranes and other cell proteins, causing degradation of chlorophyll and other pigments. At still higher concentrations, copper produces irreversible damage to chloroplast lamellae, preventing photosynthesis and eventually causing death.

The effects of copper on macrophytes have been the most extensively studied of all the metals, although studies on physiological effects are still lacking. Toxic effects on phytoplankton have been summarized by Davies (1978), Hodson et al. (1979), and Lewis & Cave (1982). Studies on *Laminaria hyperborea* showed that copper had patterns similar to mercury in that gametophyte growth was more sensitive than sporophyte growth, but copper was less toxic than mercury (Hopkin & Kain 1978). When the lowest toxic concentration of copper for sporophytes of *L. hyperborea* was compared to concentrations of copper typically found in the ocean, the ratio was 3.3. This low ratio is in contrast to ratios of 200 for mercury and up to 2000 for cadmium, suggesting that copper concentrations in the ocean need only be increased by a small factor before becoming toxic to this species. When the effect of copper concentrations on the growth of four intertidal fucoids was examined, copper was somewhat more toxic than mercury and far more toxic than zinc, lead, or cadmium (Strömgren 1980a, b). A 50% reduction in growth took place when copper ranged from 60 to 80 μg L^{-1} total copper, with *Pelvetia canaliculata* and *Fucus spiralis* being the most sensitive species. Fielding & Russell (1976) showed that species in mixed culture gave a different response to copper than when grown in unialgal culture. For example, for unknown reasons *Ectocarpus* grew better when grown with *Erythrotrichia* than by itself at the same copper concentration. These investigators warned that results from unialgal cultures may be misleading because of species interactions (see also Ch. 9.5.2).

Cadmium is a serious pollutant for plants and animals, particularly in coastal waters near industrial areas where concentrations may rise from normal levels of about 0.1 μg L^{-1} to several μg L^{-1}. Some physiological investigations have been conducted on phytoplankton (Davies 1978, Simpson 1981) but there is little research on macrophytes. Markham et al. (1980) reported on Cd uptake and its effects on growth, pigment content, and carbon assimilation in *Ulva lactuca* and *Laminaria saccharina*. Growth rate of *Laminaria* sporophytes was reduced by 50% at 2000 μg L^{-1} Cd (but this is several orders of magnitude higher than most polluted areas). However, when plants exposed to different concentrations of cadmium for 6 days were measured for growth after a further 8 days in unpolluted seawater, the concentration causing a growth rate reduction of 50% was 900 μg L^{-1} Cd. Markham et al. (1980) concluded that long-term effects are more serious than is immediately evident, and that exposure time is important in determining the extent of the effects. They found that Cd continued to accumulate over their 6-day experiment, with slower-growing plants and regions of the thalli (e.g., stipe, holdfast) accumulating relatively more cadmium. At concentrations over 2300 μg L^{-1} Cd the blades showed a sharp loss of pigment in the distal region. At sublethal concentrations (<2000 μg L^{-1}), a sharp reduction in photosynthetic and growth rates was observed. Enzymes that are involved in primary metabolism were extracted and Cd added to them in vitro (Kremer & Markham 1982). Generally, the activities of RuBP carboxylase, PEP carboxykinase, and mannitol-1-phosphate dehydrogenase were not affected. However, in vivo uptake and incorporation of ^{14}C-leucine was drastically reduced in Cd-treated plants. Kremer & Markham concluded from these observations that cadmium inhibits one or more steps in protein synthesis and thus leads to enzyme deficiencies and a series of secondary effects. In tests with five intertidal Fucales, Cd enhanced the growth of *Pelvetia canaliculata* and *Ascophyllum nodosum*, even at concentrations up to 1000 μg L^{-1}, whereas growth of both *Fucus spiralis* and *F. serratus* was inhibited at 450 μg L^{-1}. The reasons for the enhanced growth are unknown, but possibly Cd displaced essential trace elements from particles, making them available to the plants, or reduced the toxic effects of other metals.

There has been little research on less toxic heavy metals such as lead and zinc in seaweeds. One preliminary report assessed the effects of lead on four small, finely branched red algae: *Platythamnion pectinatum, Platysiphonia decumbens, Pleonosporium squarrulosum*, and *Tiffaniella synderae* (Stewart 1977). Significant reductions in growth occurred only at unrealistically high lead concentrations (10 mg L^{-1} Pb as $PbCl_2$). Even though zinc is generally considered to be actively taken up by seaweeds (Skipnes et al. 1975) (actually entering the cell, as opposed to being adsorbed onto the surface), it has a relatively low toxic effect. Strömgren (1979) found 5 to 10 mg L^{-1} of Zn was required for a 50% reduction in growth of five intertidal Fucales. In contrast, Cu and Hg toxicity occurred at one hundredth and Cd and Pb at one fifth this concentration.

Some of the heavy metals in seawater are radioactive, and normal seawater has a radioactivity level of approximately 0.32 nCi L^{-1}. *Fucus* and *Porphyra* normally have a radioactivity of 5 to 15 nCi kg^{-1} wet wt, while the level in mollusks and fish is 1 to 3 nCi kg^{-1} (Gerlach 1982). The level of radioactivity may be even higher in certain areas due to fallout or wastes from nuclear recycling plants. In the vicinity of Selafield (formerly Windscale) on the Irish Sea, the additional radioactivity originating from the reprocessing plant results in contamination of *Porphyra*, especially with ^{106}Ruthenium, at levels of radioactivity up to 340 nCi kg^{-1} (about 35 times normal). For residents of south Wales, "laver bread" made from *Porphyra* is a specialty, and by regularly consuming it they can expose themselves to considerable radiation (up to about 20% of the permissable radioactivity dose) (Hetherington 1976).

8.3.4 Factors affecting metal toxicity

One of the most important factors that determine the biological availability of a metal is its physicochemical state. Adsorption to particles in the water or complexation with dissolved organics generally reduces toxicity. Because the form in which the metal exists is difficult or often impossible to characterize, most studies measure the total concentration of the metal, which does not correlate well with toxicity. This may explain why two studies examining the same *total* concentration of the metal on a particular alga may obtain very different results.

The pH and redox potential can have a considerable effect on the availability and thus the toxicity of heavy metals. In general, at a low pH, metals exist as free cations, but at an alkaline pH like seawater they tend to precipitate as insoluble hydroxides, oxides, carbonates, or phosphates.

The interaction of salinity and temperature with toxicity is not always clear. Usually the heavy metal content of seawater is lower than that of freshwater. An increase in temperature has resulted in an increase in toxicity in some cases and a reduction in other instances (Rai et al. 1981). The increased toxicity at higher temperature may be explained by increases in the energy demand, which results in enhanced respiration of the organism, but the decreased toxicity at high temperature has not been satisfactorily explained (Förstner & Wittman 1979).

The concentration of certain nutrients, such as phosphorus, may reduce toxicity because of the formation of insoluble phosphates. Large additions of nitrate have been found to reduce cadmium toxicity in a marine diatom, *Thalassiosira fluviatilis*, for unexplained reasons (Li 1978).

Algae growing in temperate coastal areas in summer may endure the double stress of nitrogen limitation and the presence of a pollutant. The possibility of increased sensitivity to the pollutant under this condition of double stress has not been extensively examined. In a study using the marine diatom *Skeletonema costatum*, Cloutier-Mantha & Harrison (1980) found that the growth of nitrogen-limited cells was not any more sensitive to mercury than nitrogen-saturated cells. Nevertheless, the nitrogen-limited cells had a significantly reduced ability to take up NH_4^+ when it was added to the medium (i.e., had a higher K_s) when they were previously exposed to Hg. Thus, mercury pollution may decrease the ability of a species to utilize the limiting nutrient during periods of seasonal nutrient limitation and decrease its chances of surviving.

The effect of algal extracellular products on metal toxicity has been studied in the laboratory for phytoplankton (Davies 1978, Lewis & Cave 1982) and has been shown to reduce toxicity when the culture density is high. The importance of extracellular products in the natural environment is not clear, since dilution effects are considerable. However, abnormally high concentrations of dissolved organics from dispersed sewage occur near sewage outfalls and probably aid in reducing metal toxicity in these areas (Sec. 8.6.1).

One interesting area of environmental research that has not received adequate consideration is the impact of other pollutants on the toxicity of heavy metals. For example, the presence of 2,4-D decreases the toxicity of nickel and aluminum in a marine phytoplankter, and copper decreases the toxicity of the herbicide Paraquat to freshwater phytoplankters (Rai et al. 1981). Metal-metal antagonism is also known. Selenium may relieve mercury toxicity (Rai et al. 1981) and manganese or iron may reduce copper toxicity in various microorganisms (Gadd & Griffiths 1978, Lewis & Cave 1982). Significant antagonistic effects appeared with exposure to Cu + Zn and with Hg + Zn in measurements of increase in length of *Ascophyllum nodosum* (Strömgren 1980c). When two highly toxic metals such as Cu and Hg are added simultaneously, generally the toxic effects are additive. There are few good examples in plants of synergism between metals (i.e., where the total effect is greater than the sum of the effects of the individual metals), perhaps because of the lack of attention that this general area has received (Gadd & Griffiths 1978).

8.3.5 Ecological aspects

The fact that metal concentrations in marine organisms are typically several orders of magnitude higher than concentrations of the same metal in seawater has led to the suggestion that metals are accumulated in

Table 8.4. *Cadmium concentrations in water (μg Cd L^{-1}), seaweeds, and shore animals (mg Cd kg^{-1}) at four collecting stations on the southern side of the Severn estuary and Bristol Channel*

Location	Distance from Avonmouth (km)	Seawater	*Fucus*	*Patella*	*Thais*
Portishead	4	5.8	220	550	—
Brean	25	2.0	50	200	425
Minehead	60	1.0	20	50	270
Lynmouth	80	0.5	30	50	65

Source: Butterworth et al. (1972). Reprinted by permission from *Marine Pollution Bulletin*, vol. 3, pp. 72–74.

higher concentrations in higher trophic levels of the food chain due to biological magnification. Comparison of concentrations of various metals in phytoplankton and zooplankton with those in seawater values shows that the metal concentrations in the plankton are indeed about 1000 times higher (Martin & Knauer 1973). However, the concentrations of only Cu, Zn, and Pb are substantially higher in zooplankton than phytoplankton. For Mn, Ag, Cd, and Hg the differences are small. Studies with mercury in anchovies and other animals showed that bioaccumulation varies with the tissues sampled, with liver having the highest levels (Knauer & Martin 1972).

Support for the biological magnification hypothesis in natural areas comes from the fact that the tuna and swordfish, both top-level carnivores, have mercury concentrations that are several orders of magnitude higher than phytoplankton (Laws 1981). In heavily polluted areas, examples of bioaccumulation also exist. For example, coastal waters along the southern shore of the Bristol Channel, England, have very high concentrations of Cd, Zn, and Pb (Butterworth et al. 1972). Analysis of the seawater and a simple food chain showed that Cd and Zn are accumulated up the food chain. Table 8.4 shows that cadmium is found in relatively low levels in *Fucus* (the primary producer), at higher concentrations in the limpet *Patella* (herbivore), and in greatest concentrations in the carnivorous dog whelk *Thais*. The same process could also occur for metals.

In the laboratory, for unknown reasons the transfer of zinc and iron from *Fucus serratus* to *Littorina obtusata* did not result in accumulation in the snail (Young 1975). Similarly, when the abalone *Haliotis* sp. was fed on a lead-treated brown alga, *Egregia laevigata*, little bioaccumulation occurred (Stewart & Schulz-Blades 1976).

Although the elevated metal concentrations in animals in some examples suggest biological magnification, it has been argued that the same effects may be produced by very different mechanisms. From studies on DDT, Hamelink et al. (1971) suggested that elevated levels in animals may be the result of direct uptake of the pollutant from the water and by differences in pollutant exchange equilibria between water and different classes of organisms.

8.4 **Oil**

Petroleum, or crude oil, is an extremely complex mixture of hydrocarbons with some additional compounds containing oxygen, sulfur, nitrogen, and metals such as Ni, V, Fe, and Cu. The main components of oil may be classified into three broad categories (older names given in parentheses). The alkanes (paraffins) are saturated straight chain or branched hydrocarbons, whose general composition is C_nH_{2n+2} (Fig. 8.4a–c). These aliphatic compounds make up about 20% of crude oil and are very common in gasoline and fuel oils. The presence of only saturated bonds makes reactions with alkanes difficult, hence they are very resistant to degradation. Low molecular weight ($<C_6$) alkanes are generally gases (e.g., methane, ethane, and propane), while high molecular weight aliphatics ($>C_{18}$) are solids (e.g., waxes). Cycloalkanes (cycloparaffins or naphthenes) are similar to alkanes but the carbon chain is joined in a ring. These compounds have the general formula C_nH_{2n} and account for about 50% of crude oil, the most prevalent being cyclopentane and cyclohexane (Fig. 8.4d–f). Frequently, alkyl groups (e.g., $-CH_3$) are substituted on the cycloalkane ring, forming compounds such as methylcyclohexane.

Aromatics are the third major group of compounds; these contain one or more benzene rings. Aromatics are commonly found in crude oil or produced during refining and include benzene, toluene, naphthalene, and phenol (Fig. 8.4g–i). Aromatics usually constitute less than 20% of crude oil, but they are very toxic to plants and animals. Other hydrocarbons, such as alkenes, occur in crude oil in much smaller amounts. Alkenes (olefins) are unsaturated chain compounds possessing double or triple bonds, but without the regular arrangement found in the benzene ring. Examples include ethylene and acetylene, which are produced during refining.

Figure 8.4. Some hydrocarbons from crude oil.

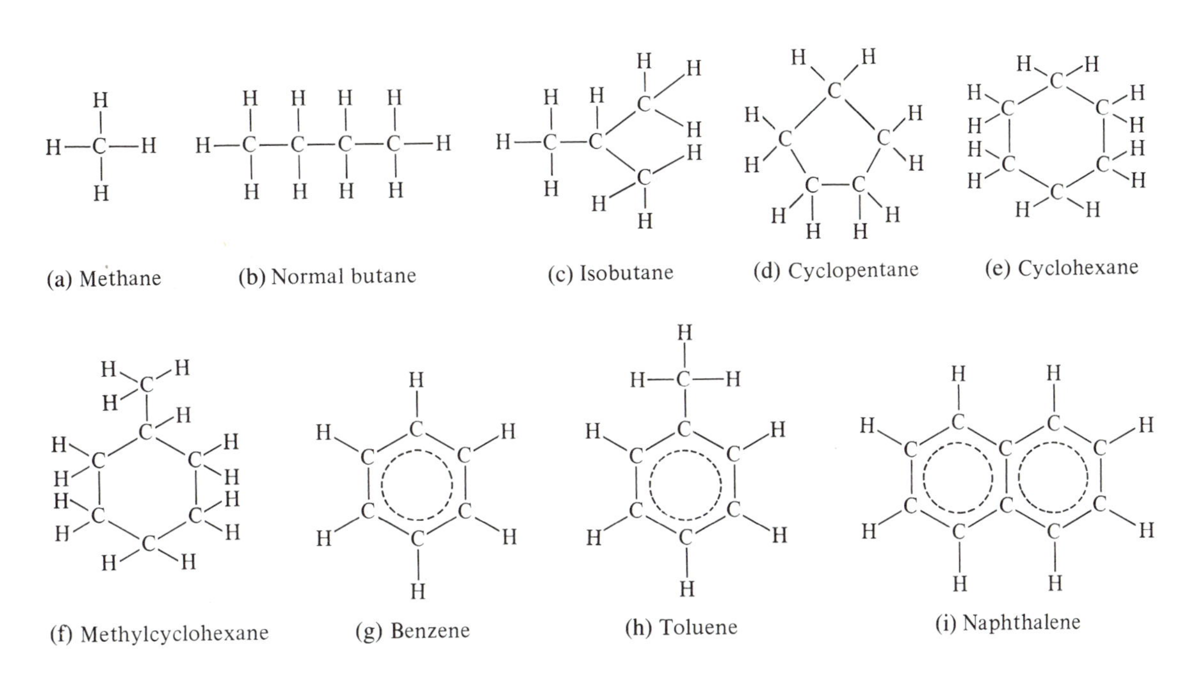

8.4.1 Inputs and fate of oil

Estimates have been made by the U.S. National Academy of Sciences of the magnitudes of various sources and discharges into the marine environment up to 1970 (Fig. 8.5). Tanker accidents and oil well blowouts, which make the newspaper headlines, contribute only a small percentage of the total input but can be devastating in local areas. Most of the oil comes from discharge of waste oil from industrial and municipal sources and routine operations of ships, refineries, and other oil-processing facilities. Examination of the important source categories reveals that there will probably be no reduction in the input of oil into the ocean until there is a significant decline in the use of oil. The greatest input of oil is in coastal areas, which are often the most biologically productive, rather than in the open ocean.

The main physical, chemical, and biological processes governing the fate of oil in the ocean are summarized in Figure 8.6 and Table 8.5 and have been reviewed by Lee (1980). The fate of the oil depends on the type spilled and where it is spilled. The source of the crude oil determines its unique characteristics (often denoted by place of origin, e.g., Nigerian or Kuwait crude oil). Many refined petroleum products are spilled, including gasoline, kerosene, fuel oils (Nos. 2, 3, 4, etc.), and lubricating oils.

Most oil spills immediately form a surface slick, a thin boundary layer between the seawater and the atmosphere. Light oils spread faster than heavy oils and may form films as thin as 0.1 μm. Many of the hydrocarbons are volatile and begin to evaporate immediately. At the end of 24 h, half of the compounds with up to 14 carbon atoms have vaporized, and after three weeks, only half of the hydrocarbons shorter than C_{17} have evaporated. Vaporization continues slowly, leaving tarlike lumps; it is the most important natural factor removing oil from the water surface. Refined products such as gasoline and kerosene may disappear almost completely, whereas viscous crudes may lose less than 25% by evaporation (Table 8.5) (Bishop 1983).

Some polar hydrocarbons dissolve in seawater. As a guideline, the quantities dissolving are about 10 mg L^{-1} for C_6 compounds, 1 mg L^{-1} for C_8, and 0.01 mg L^{-1} for C_{12}. While the time scale of dissolution is similar to weathering, the magnitude of the process is only about one fifth as important (Table 8.5).

When the sea surface is agitated by wind, the oil may absorb water up to 50% of its weight and form brown masses called "chocolate mousse." Besides the water-in-oil emulsions, oil-in-water emulsions (dispersions) form, especially under the influence of added chemicals (dispersants). While emulsification and dispersion give the impression that the oil has disappeared from the surface, it actually exists as tiny droplets and its potentially poisonous effects persist. The toxicity of this dispersion is reduced, however, because the lighter fractions such as the aromatics and aliphatics have largely evaporated.

On a larger time scale, photochemical oxidation may contribute to the weathering of oil. Through the actions of atmospheric oxygen and solar radiation, the proportions of oxygenated compounds in the slick in-

Figure 8.5. Sources of petroleum going into the oceans, millions of metric tons per annum (mta). (From Geyer 1980, with permission of Elsevier Science Publishers)

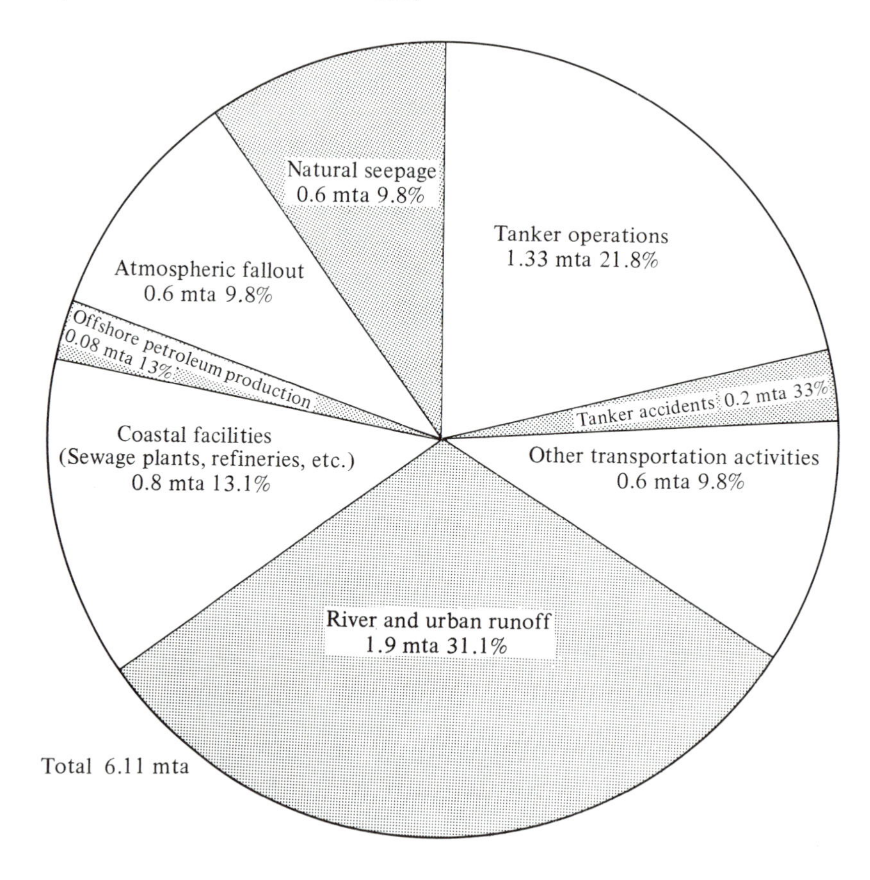

crease. For example, aromatics and alkyl-substituted cycloalkanes tend to be oxidized more rapidly and form soluble and insoluble (tars) compounds.

Microbial degradation only begins to take place after the oil at the surface has aged and lost part of its highly volatile, poisonous components by vaporization. At least 90 strains of marine bacteria and fungi and a few algae are capable of biodegrading some components of petroleum. However, oil-decomposing bacteria increase in number slowly after an oil spill. In some cases, their growth is restricted because there is not enough nitrogen or phosphorus in the water and it is necessary to supplement the low quantities normally present in oil. Moreover, many major oil spills occur in temperate waters in winter, when temperature restricts bacterial growth rate. Nontoxic dispersants enhance biodegradation by greatly increasing the surface area of the oil. Normal alkanes are the most easily degraded, whereas aromatics, cycloalkanes, and branched alkanes are more difficult. Further aspects of microbial degradation of oil are discussed by Carlberg (1980), Stafford et al. (1982), and Gundlach et al. (1983).

Beaching of the oil may occur if a spill occurs near shore and the wind is in the right direction. This oil may adhere to rocks, plants, and animals, or be worked into the sediments if a dispersant is used. Penetration into the interstitial system between sand grains results in very slow degradation rates, often due to lack of oxygen in the interstitial water. As a result, oil may persist in sediments for years.

Tar lump formation and sinking are the final stages of weathering. Oil may adsorb to particles, which sink, or it may be consumed by filter-feeding plankton such as copepods and become incorporated into fecal pellets, which also sink. Weathered oil may form lumps (usually about the size of peas), which can coalesce and become large enough to form a substratum for sedentary animals such as goose-neck barnacles, again causing the lump to sink.

The *Amoco Cadiz* oil spill (223,000 metric tons), off the north coast of France, is the largest (nearly twice the size of the *Torrey Canyon* spill) and the best-studied tanker spill in history. Evaporation and stranding on shore accounted for 60% of the oil spilled. After three years, most of the obvious effects had gone but high hydrocarbon concentrations remained in estuaries and marshes that initially received large amounts of oil (Gundlach et al. 1983).

Figure 8.6. The transport of spilled oil in the marine environment. (Reprinted by permission from R.A. Horne 1978, *The Chemistry of Our Environment*, Copyright 1978 John Wiley & Sons, Ltd.)

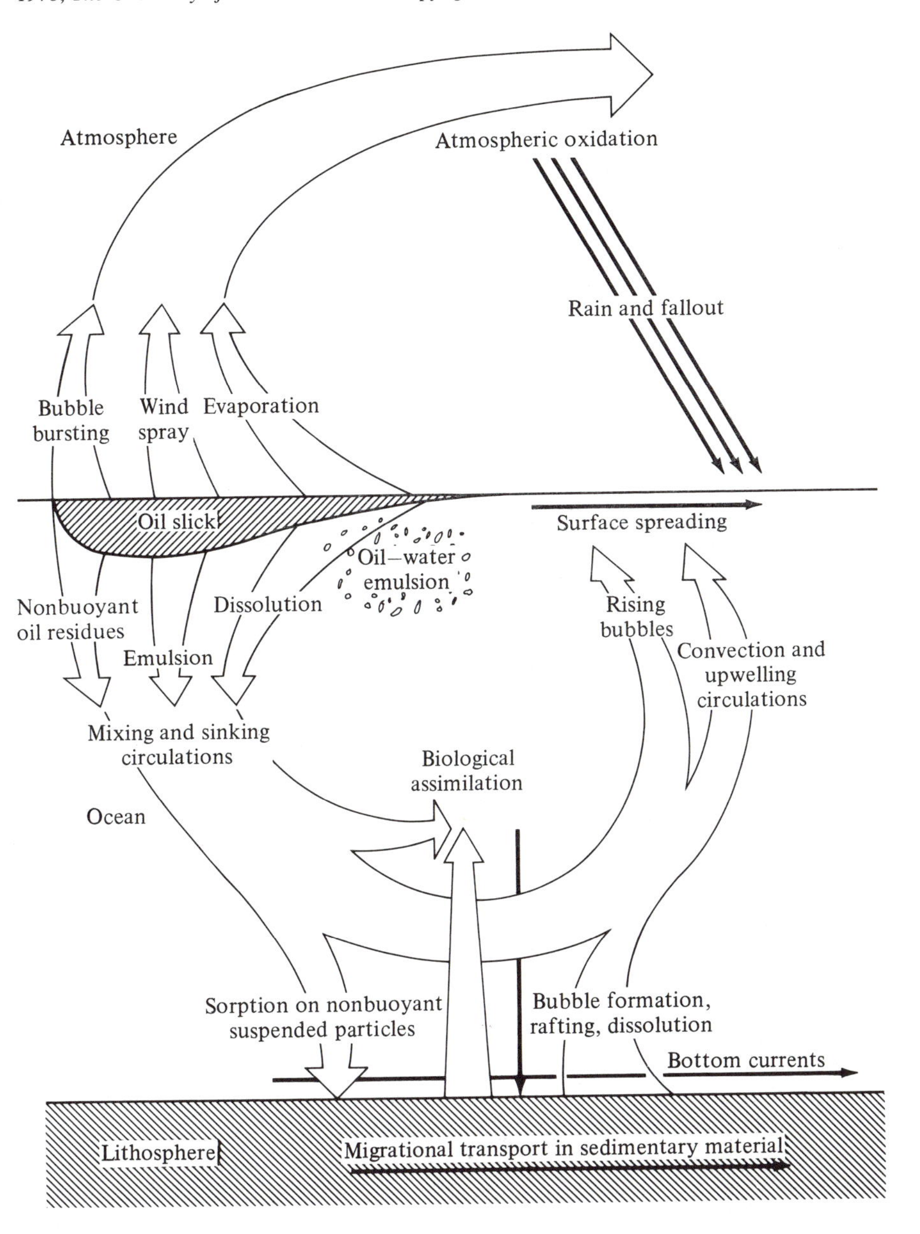

8.4.2 Effects of oil on algal metabolism

The toxic effects of oil fall into two categories: those associated with coating of the organism and those due to uptake of hydrocarbons and the subsequent disruption of cellular metabolism. In the case of coating, there is a reduction in diffusion of CO_2 into the plant and some reduction of light penetration. Schramm (1972) observed that in *Porphyra umbilicalis, Fucus vesiculosus*, and *Laminaria digitata*, reduction in photosynthetic rate correlated with the thickness of the oil layer. He also found that during exposure to the air the oil reduced water loss from the blades via desiccation, allowing photosynthesis to occur for longer than normal but at a reduced rate. In severe cases of oiling, breakage of kelp fronds has been observed due to overweighting because of the oil adhering to the frond. The loss of thalli by this mechanism is associated primarily with the high molecular weight, water insoluble hydrocarbons (Nelson-Smith 1972).

The second category, disruption of cell metabolism, has been examined primarily by monitoring changes in photosynthesis, respiration, growth, pigment content,

Table 8.5. *Pathways for the environmental fate of crude oil*

Pathway	Time scale, days	Percentage of initial oil
Evaporation	1–10	25
Solution	1–10	5
Photochemical degradation	10–100	5
Biodegradation	50–500	30
Disintegration and sinking	100–1,000	15
Residue	>100	20
Total		100

Source: Reproduced with permission from Butler et al. (1976), *Sources, Effects and Sinks of Hydrocarbons in the Aquatic Environment*, pp. 287–297, © 1976 by the American Institute of Biological Sciences.

morphology, and ultrastructure. Bioassay investigations have revealed variable effects, depending on the physical and chemical properties of the oils and components tested (e.g., whole oils vs. refined products), the parameters being measured, and the test species employed. There are problems in interpreting many of the earlier published results (Vandermeulen & Ahern 1976), especially if crude oil was used, because its composition was not known and even now it cannot be easily determined. The major problems are the lack of details about the manner of preparation of the oil extract, how old the extract was before it was applied (hence what fraction of the volatiles had been lost), and total or differential losses of hydrocarbons, especially during long-term experiments (Vandermeulen & Ahern 1976). For these reasons many investigators have chosen to work with individual components of oil, whose composition is at least understood and measurable, and then they extrapolate back to the original oil.

The penetration of oil depends on the covering on the thallus. The brown algae in particular are thought to be largely protected from oil damage by the presence of the mucilaginous coating. This is assumed to be what "saved" the *Macrocystis* beds off Santa Barbara after an oil well blowout (Mitchell et al. 1970). The compounds that penetrate the thallus most easily, and hence are most toxic, are the lower molecular weight, lipophilic compounds such as aromatics. The least toxic components, and the least water soluble, are the long-chain alkanes. Intermediate in toxicity are the cycloalkanes, followed by olefins. The aromatics and other toxic hydrocarbons appear to exert their toxic effect by entering the lipophilic layer of the cell membrane, disrupting its spacing. As a result, the membrane ceases to properly control the transport of ions in and out of the cell.

Disruption of cellular metabolism has usually been measured through changes in the rate of photosynthesis or respiration. North et al. (1965) observed complete inhibition of photosynthesis of young blades of *Macrocystis* after 3 days of exposure to a 1% emulsion of diesel oil in seawater. As little as 10 to 100 ppm of unspecified fuel oils reduced photosynthesis by 50% during 4-day exposures (Clendenning & North 1960). More detailed studies conducted recently showed that photosynthetic rate reductions vary with the type of crude oil, its concentration, the length of exposure, method of preparation of the oil–seawater mixture, the irradiance, and the algal species (Hsiao et al. 1978). *Laminaria saccharina* was more sensitive than the red alga *Phyllophora truncata*, and Venezuela crude was more toxic than three Canadian crudes. Other experiments with arctic seaweeds also showed that *Laminaria saccharina*, plus *Cladophora stimpsonii* and *Ulva fenestrata*, were very sensitive to only 7 ppm of Prudhoe Bay crude oil (Shiels et al. 1973). They found that effects were most acute at high irradiance. In experiments with the green seaweed *Acrosiphonia sonderi*, inhibition of photosynthesis by a crude oil extract increased during the first 4 h of incubation, presumably because of the toxicity of volatile aromatics such as benzene and naphthalene. In all the studies to date, the actual mechanism of inhibition has not been investigated. This is mainly because of the difficulty in separating the toxicity effect from the purely mechanical effect of coating (smothering) the thallus and the reduction in light reaching the plant. Bleaching is commonly observed among red algae and is probably caused by the breakdown of phycoerythrin by kerosene-related compounds. Lipid-soluble pigments such as chlorophylls may be leached out of cells by oil (O'Brien & Dixon 1976).

Several dioicous brown algae, including *Ectocarpus* and *Fucus*, secrete olefinic hydrocarbons into seawater as gamete attractants (see Sec. 10.4.3). The possibility that petroleum hydrocarbons could confound recognition of the attractant fucoserratene by *Fucus* spermatozoids was investigated by Derenbach & Gereck (1980). They found that a combination, rather than a single compound, of petroleum hydrocarbons attracted spermatozoids, but at concentrations about 100 times that at which fucoserratene is active.

The reproductive stages of *Fucus edentatus* and

Laminaria saccharina are particularly sensitive to oil, especially during gamete or spore release (Steele & Hanisak 1979). Concentrations as low as 2 $\mu g\ L^{-1}$ of Willamar crude or several fuel oils blocked fertilization in *Fucus*, apparently because of toxic effects on the sperm. *Laminaria* spores did not germinate above 20 $\mu g\ L^{-1}$. Male gametophytes were more sensitive to oils than female gametophytes, because in both *Fucus* and *Laminaria* they are smaller than females and hence have a higher surface-to-volume ratio, possess fewer stored reserves, and respire at a higher rate, creating a greater energy demand. Sporophyte development in both species was inhibited at higher concentrations of 200 $\mu g\ L^{-1}$. Hopkin & Kain (1978) found the opposite response to phenol: zoospores and gametophytes were more tolerant than sporophytes of *Laminaria hyperborea*.

The effects of oil contamination on algal respiration are not well known due to the paucity of experimental evidence. The respiration rate of *Laminaria hyperborea* was inhibited by 100 ppm phenol immediately after the addition of the pollutant (Hopkin & Kain 1978). To compare respiration and growth effects, Hopkin & Kain calculated the ratio of the minimum concentration of the pollutant reducing frond respiration to the minimum reducing sporophyte growth in culture. The ratios were phenol 1.3, mercury 100, zinc 500, and copper 2500. Since all ratios were >1, growth of plants in culture was more sensitive than tissue disks in the respirometer. Interference by oil with respiration could occur in a number of processes such as gas diffusion, glycolysis, and oxidative phosphorylation. Mechanical blockage of gas diffusion is thought to be less pronounced for oxygen than for carbon dioxide (Schramm 1972). Other physiological mechanisms explaining the inhibition of respiration have not been examined in macrophytes, but some studies of higher plants have been reviewed by Baker (1970).

Inhibition of algal DNA and RNA activity has been reported upon exposure to high concentrations of crude oil (Davavin et al. 1975). A 24-h exposure to emulsified oil–seawater mixtures (100–10,000 ppm) resulted in decreased DNA in the red algae *Grateloupia dichotoma* and *Polysiphonia opaca* and significantly reduced DNA- and RNA-specific activity in the green alga *Ulva lactuca*. Given the fundamental importance of nucleic acids for reproduction and protein synthesis, this work is a start in understanding possible mechanisms whereby resistance to damage by oil may be conferred on tolerant species.

8.4.3 Ecological aspects

For the last 20 years, attempts have been made to quantify the effects of large oil spills in different parts of the world on various flora and fauna. Conclusions drawn from these studies have varied considerably, ranging from minimal effects to severe damage. The assessment has varied depending on the ecosystem studied and the community or population observed. Gundlach & Hayes (1978) constructed an "oil spill index," in which different ecosystems were ranked according to their vulnerability. Rocky exposed cliffs are the least vulnerable, while salt marshes and mangroves are extremely vulnerable (Table 8.6). Communities also have been ranked, with birds and benthic subtidal communities being most vulnerable, plankton and benthic rocky intertidal communities only slightly vulnerable.

Another important factor that will determine the magnitude of the impact is the location of the spill relative to the shore. The impact is full scale if the spill occurs close to the beach and is quickly washed on shore. The least impact occurs if the oil does not reach the shore for several days, giving time for many of the toxic volatile compounds to evaporate. Other factors that affect the ecological impact are the type of oil, the dosage, water temperature, weather conditions, prior exposure of the area to oil, presence of other pollutants, and type of remedial action (e.g., use of dispersants).

The ecological impact of oil on several specific seaweed habitats can be examined against the general background presented above. On rocky shores there may be a slight short-term impact, but no significant long-term effect on the macrophyte community has been observed (Nelson 1982, Gundlach et al. 1983). Rocky intertidal areas that have been cleaned with detergents after an oil spill show recolonization rates comparable to rates on control plots. The first macroalgae to recolonize the Cornwall shore after the *Torrey Canyon* spill in 1967 were *Ulva* and *Enteromorpha*. They quickly covered the entire area because herbivores that usually graze on them (e.g., limpets and periwinkles) had been killed by the oil. Limpets increased in number the next year, and 7 years after the spill the macrophyte–herbivore community had returned to normal (Gerlach 1982). Similar observations were made on the Somerset coast of England, where oil was reported to not even adhere to *Fucus spiralis*, and the percentage of cover of this alga increased from 50% to 100% after the spill (Crothers 1983). No significant effects of the *Amoco Cadiz* spill were observed for *Laminaria*, *Fucus*, or *Ascophyllum* (Gundlach et al. 1983). Some of the damage to corallines, such as loss of pigments, appears to be partially or wholly caused by the dispersant BP 1002 and its toxic aromatic solvent (Boney 1970). A number of nontoxic dispersants, such as Corexit, are now available; therefore, toxic effects attributable to the dispersant should no longer be a problem.

When dispersants have not been used to aid in the oil cleanup algal growth is generally less affected. In the San Francisco Bay oil spill of 1971, caused by the collision of two tankers carrying Bunker C fuel oil, prespill algal densities were restored in 2 years. Upper-shore algae, *Endocladia muricata* and *Gigartina cristata*, were coated with oil, but growth in the following summer appeared to be normal. Other algae, such as *Halosaccion*

Table 8.6. *Expected impact of oil spills on marine habitat types and clean-up recommendations*

Exposed rocky cliffs	Under high wave energy, oil spill clean-up is usually unnecessary
Exposed rocky platforms	Wave action causes a rapid dissipation of oil, generally within weeks. In most cases clean-up is not necessary
Flat, fine sand beaches	Due to close packing of the sediment, oil penetration is restricted. Oil usually forms a thin surface layer that can be efficiently scraped off. Clean-up should concentrate on the high tide mark; lower beach levels are rapidly cleared of oil by wave action
Medium- to coarse-grained beaches	Oil forms thick oil-sediment layers and mixes down to 1 m deep with the sediment. Clean-up damages the beach and should concentrate on the high water level
Exposed tidal flats	Oil does not penetrate in compacted sediment surface, but biological damage results. Clean-up only if oil contamination is heavy
Mixed sand and gravel beaches	Oil penetration and burial occur rapidly; oil persists and has a long-term impact
Gravel beaches	Oil penetrates deeply and is buried. Removal of oiled gravel is likely to cause future erosion of the beach
Sheltered rocky coast	The lack of wave action enables oil to stick to rock surfaces and tidal pools. Severe biological damage. Clean-up may cause more damage than if the oil is left untreated
Sheltered tidal flats	Long-term biological damage. Removal of the oil nearly impossible without causing further damage. Clean-up only if the tidal flat is very heavily oiled
Salt marshes and mangroves	Long-term deleterious effects. Oil may continue to exist for 10 years or more

Source: Gerlach (1982), reprinted with permission of Springer-Verlag.

glandiforme, Enteromorpha intestinalis, Urospora penicilliformis, and *Ralfsia pacifica*, were more dense than normal, possibly because of a reduction in grazers.

If weathering eliminates the more volatile toxic components before the oil reaches the shore, the effects of heavy oil deposition on intertidal floras appears to be largely physical, with injury due to smothering and adsorption of oil. Most seriously affected by oil coating are algae growing between neap and spring high tide marks, especially those near spring high tide, where oil may be stranded for a long time. Many high intertidal species of Rhodophyceae and Phaeophyceae become oleophilic as their surfaces dry out (O'Brien & Dixon 1976). This strong adsorptive capacity for oil has been documented for *Ascophyllum nodosum, Fucus* spp., *Pelvetia canaliculata, Gigartina stellata* and *Gelidium crinale*. Such algae can become severely overweighted by adsorbed oil and subject to breakage by waves. For algae with annual basal regrowth, loss of distal blades may be no more debilitating to the plant than losses during a winter storm (Nelson-Smith 1972). However, the loss of too many photosynthetic blades during the growing season when metabolic products are stored, could impair a plant's regenerative ability (O'Brien & Dixon 1976).

No clear patterns emerge in the relationship between systematics and the susceptibility of intertidal algae to oil. Several studies indicate that Cyanophyceae are particularly resistant to oil (O'Brien & Dixon 1976). Species of Chlorophyceae in particular have a remarkable ability to invade areas where other species have been eliminated. The spread of green algae is often due to the die-off of herbivores, which are more susceptible to oil than the algae. Early observations suggested that filamentous red algae and corallines were most susceptible to oil and oil–emulsifier blends, possibly because of the destruction of phycoerythrin (Nelson-Smith 1972), but this suggestion requires further confirmation.

Salt marshes (Sanders et al. 1980) and coral reefs (Loya & Rinkevich 1980) are the most severely affected by oil spills. Upon the death of corals, rapid colonization of algae on skeletons of the dead corals may be enhanced by oil pollution. In an oil-polluted reef at Eilat (Red Sea), up to 50% of the surfaces of the coral *Stylophora pistillata* were covered by the brown alga *Lobophora variegata* (*Pocockiella variegata*) (Loya & Rinkevich 1980).

8.5 Synthetic organic chemicals

8.5.1 Herbicides

Among herbicides, phenoxycarboxylic acid derivatives (2,4-D, 2,4,5-T, 2,4,5-TB, etc.) outweigh all other compounds in tonnage produced. These weed killers are used in high concentrations (mg L^{-1}) in freshwater streams and ponds to control nuisance vascular plants such as *Elodea*. Two other compounds, Paraquat® and Diquat®, are widely used because both disappear rapidly from the water and do not appear to be released from the sediments where they tend to concentrate (Duursma & Marchand 1974, Hurlbert 1975). After a herbicide treatment, growth of freshwater phytoplankton may be temporarily depressed and then flourish, usually because of the nutrients released from the macrophyte kill and subsequent decomposition (Kohler & Labus 1983).

Although herbicides have not been directly used in the marine environment, they can enter estuarine areas through river discharge or runoff. Studies on herbicide effects on marine algae have been conducted primarily in the laboratory. The sporeling growth of five species of red macroalgae, *Antithamnion plumula, Plumaria elegans, Callithamnion tetricum, Nemalion multifidum*, and *Brongniartella byssoides*, was inhibited by 3-amino-1,2,4-triazole (3AT or amitrole) at about 10 mg L^{-1} (Boney 1963). Short-term immersions in culture medium containing 3AT reduced the growth of sporelings, whereas protracted contact with 3AT resulted in chlorosis. Boney also found that growth inhibition was more pronounced with sporelings of intertidal algae than with sublittoral species.

Paraquat and 3AT were also tested on the settlement, germination, and growth of *Enteromorpha* (Moss & Woodhead 1975). Zygotes were able to develop into filaments in 7 mg L^{-1} Paraquat but germination was deferred at higher concentrations. Increased resistance of zygotes occurred when they settled in clumps on the substratum. Green filaments of *Enteromorpha* were more susceptible than ungerminated zygotes. *Enteromorpha* was more sensitive to 3AT than to Paraquat. The herbicide atrazine was lethal at 0.01 mg L^{-1} to young sporophytes of *Laminaria hyperborea* (Hopkin & Kain 1978). Three other herbicides tested by Hopkin & Kain, 2,4-D, Dalapon®, and MCPA, were nontoxic to the sporophytes at the highest concentration tested (>100 mg L^{-1}). To compare atrazine to toxicity of metals, Hopkin & Kain calculated on a molarity basis the minimum concentration causing a detrimental effect (Table 8.7); atrazine was seen to be even more toxic than copper or mercury.

Table 8.7. *Molarity of pollutants that have a minimum concentration causing a detectable effect on* Laminaria hyperborea

Pollutant	Molarity causing minimum detectable effect ($\times$ 10^{-6})
Atrazine	46
Copper	79
Mercury	250
Cadmium	890
DOBS 055	2,800
SDBS	2,900
Zinc	7,700

Source: Reprinted with permission from Hopkin & Kain (1978), *Estuarine & Coastal Marine Science*, vol. 7, pp. 531–553. Copyright © 1978 Academic Press Inc. (London) Ltd.

Growth and photosynthesis of marine phytoplankton are adversely affected by herbicide concentrations between 10 and 500 mg L^{-1}, depending on the compound (Duursma & Marchand 1974). Phytoplankton are most sensitive to the triazine group of herbicides, while Diquat and Paraquat are least effective (Kohler & Labus 1983).

8.5.2 Insecticides

Pesticides may be classified according to their target species or their chemical properties. In the former case, there are insecticides, herbicides, fungicides, and rodenticides. On the basis of chemical nature, two general categories of insecticides are delimited: the organic phosphate compounds, (e.g., parathion), which are the more degradable; and chlorinated hydrocarbons (e.g., DDT, dieldrin, endrin), which are the more persistent. Of these two categories, Ukeles (1962) found that organochlorine insecticides were the more toxic group to five species of phytoplankton.

DDT came into use as a pesticide in 1942 and was used extensively throughout the world until 1972, when its use in the United States was banned except under special circumstances. Although impact on freshwater communities has been documented (e.g., Rudd 1964), no effects on marine macrophytes have been reported. The concentration of DDT and its breakdown products in the open ocean is about 0.002 μg L^{-1} (Duursma & Marchand 1974). Reduction of phytoplankton photosynthesis occurs at about 10 μg L^{-1} or higher, depending on the species (Wurster 1968). To achieve such high concentrations in tests, DDT must be first dissolved in ethanol, since its solubility in seawater is only 1.2 μg L^{-1}. Therefore, DDT seems unlikely to affect phytoplankton in nature, and it may also have virtually no effect on macrophytes, although experiments are needed to test this (Duursma & Marchand 1974, Hurlbert 1975, Laws 1981). Estimates indicate that zooplankton are 100 times more sensitive and fish 10 times more sensitive than phytoplankton. In the freshwater environment, insecticide treatment reduced benthic herbivore populations, which was followed by increases in benthic filamentous algae such as *Zygnema* and *Mougeotia* (Hurlbert 1975). Epiphytes such as benthic diatoms also increased because of the susceptibility of the herbivores to DDT.

The rate of degradation of DDT in the sea varies from more than a few days to several months (Reutergårdh 1980, Laws 1981). However, its degradation components DDD and DDE may be as toxic as DDT itself. Much of the DDT in the ocean may exist inside or adsorbed onto plankton or other particulates. It is especially concentrated in the fatty tissues of animals, where its half-life may be considerably longer. The best-documented effect of DDT is the decrease in bird eggshell thickness, but whether DDT exhibits food chain magnification or amplification remains controversial (Laws 1981).

8.5.3 Industrial chemicals: PCBs

Polychlorinated biphenyls (PCBs) are complex mixtures of chlorine-substituted biphenyls. They have been marketed under the trade name Aroclor® 1242 and 1254, with the last two digits signifying the average percentage of chlorine by weight in the mixture. PCBs are exceptionally stable compounds (destruction by burning requires temperatures >1300 C) and they are toxic to most organisms. For these reasons and others, PCBs are no longer manufactured in the United States. Although they are being phased out, a few specialized uses remain; they are still used as dielectric fluids in capacitors, as plasticizers in waxes, in transformer fluids and hydraulic fluids, in lubricants, and in heat transfer fluids.

PCB concentrations in the upper 200 m of the North Atlantic Ocean have been reported to be about 20 ng L^{-1}, but near industrial areas they may be as high as 320 ng L^{-1} (Peakall 1975). Concern over these concentrations is warranted because at 100 ng L^{-1} PCBs affect growth of phytoplankton communities in continuous cultures (Fisher et al. 1974). Furthermore, Fisher & Wurster (1973) showed that phytoplankton living at suboptimal temperatures were even more sensitive to PCBs. Further studies showed that there was no loss of photosynthesis per unit chlorophyll *a*, but nevertheless, carbon fixation was reduced because chlorophyll *a* per cell was reduced. Generally, the higher the degree of chlorination, the higher the toxicity (e.g., Aroclor 1254 is more toxic than 1231). The estimates of toxicity relative to the concentration of PCBs in the water become even more alarming because of the recent finding that the large amount of PCBs adsorbed onto particles can be taken up by phytoplankton when they make contact with the particle (Harding & Phillips 1978). PCBs initially associated with microparticulates were rapidly transferred to four species of marine diatoms. The transferred PCBs can inhibit photosynthesis at a site on the electron transport chain, close to PSII (Sinclair et al. 1977). This research demonstrates that particle-bound PCBs are of great biological importance, especially in coastal and estuarine areas where suspended loads can be very high.

The effects of PCBs on natural phytoplankton communities have been studied outdoors in large controlled experimental enclosures or mesocosms (Iseki et al. 1981). Following the addition of 50 μg L^{-1} PCBs (over 200 times natural concentrations) to the enclosure, primary productivity was reduced by 30%, settling velocity of the particulate matter was accelerated, zooplankton standing stocks were reduced, and decomposition activity of sedimented matter by bacteria was reduced by 90%.

There are as yet no studies on the effects of PCBs on marine macrophytes. Given the marked effects on photosynthesis, respiration, and growth of phytoplankton, investigations are certainly warranted. Effects on invertebrates and vertebrates, especially birds, have been relatively well studied (Duursma & Marchand 1974, Peakall 1975, Reutergårdh 1980).

8.5.4 Antifouling compounds: triphenyltin

Organotins, particularly the trialkyltin compounds such as triphenyltin ($SnPh_3Cl$) and tributyltin, have been widely used as biocides in antifouling compositions. Although they are highly effective against *Enteromorpha* species, they are less effective in controlling *Ectocarpus* and the microfouling film that generally precedes settlement by macroalgae (Millner & Evans 1981). This microfouling film generally comprises bacteria, benthic diatoms, and filaments of green algae such as *Ulothrix pseudoflacca*.

Although there are a number of organic and inorganic tin compounds, the trialkyltin compounds have been found to be the most toxic to fouling algae (Wong et al. 1982). For this reason and the economic aspects associated with fouling, a number of extensive physiological and biochemical studies have been conducted (Evans & Christie 1970, Millner & Evans 1980, 1981).

The photosynthetic apparatus of zoospores and vegetative tissue of *Ulothrix* was relatively insensitive to triphenyltin compared with those of *Enteromorpha intestinalis* (Millner & Evans 1980). However, respiration in zoospores and vegetative tissue of both species was equally affected. The fact that respiration is more affected than photosynthesis suggests that the specific binding site might be in the mitochondria. Trialkyltins act as energy transfer inhibitors in animal mitochondria (Gould 1976). Questions remaining to be answered include, Why is *Ulothrix* more resistant to organotins than is *Enteromorpha*, even though *Ulothrix* takes up organotins more rapidly?

8.6 Complex wastes and eutrophication

8.6.1 Eutrophication

Sewage is classified as a complex waste since it contains inorganic nutrients (N and P in particular), organics, chlorine (from chlorination), and some heavy metals. In this section, the focus will be on the inorganic nutrients. The sewage is usually delivered to water by means of a pipe, often with holes in it to create a wider dispersal area. Before the sewage is released several treatments are possible. In primary treatment, the sewage is screened to remove large particulates and then passed to settling chambers where particles settle out. In secondary treatment, the liquid sewage is put into another tank where it is aerated to encourage bacterial growth and aerobic oxidation of the dissolved organics. This process removes a large portion of the organics from the sewage. In tertiary treatment, nutrients such as N and P are removed by chemical treatment (e.g., precipitation of phosphate by alum) or biological treatment (e.g., growing phytoplankton to remove nutrients) (Ryther et al. 1972). Tertiary treatment is generally not used because it is so expensive. On a worldwide basis, over 90% of the sewage from coastal areas enters the sea untreated (Cole 1979a). In developed countries where sewage is treated, the sludge that settles out in primary treatment is loaded into ships and dumped further out to sea.

The dumping of nutritive organic wastes into coastal areas with low amounts of water exchange may stimulate growth of algae to the point where excessive amounts of phytoplankton and/or macrophytes in the water create biological, aesthetic, or recreational problems. Above-normal plant growth and biomass production in response to added nutrients is termed eutrophication. The water body affected is said to be eutrophic or hypertrophic (Gerlach 1982). The increased nutrient input may be the result of land runoff, river inflow, or sewage. The latter source is the most important and the least studied. When eutrophication is extensive, the large amount of algal biomass (both phytoplankton and macrophytes) soon begins to decay and seriously depletes the oxygen concentration for animals. In extreme cases, where water exchange in a bay is very slow, this high biological oxygen demand (BOD) may result in an extensive fish kill.

Most of the studies on the response of macrophytes to eutrophication have been concerned with sewage outfalls, probably because they are convenient to study. They represent a small, defined area with a gradient in nutrient concentration away from the outfall. Generally, the studies have been from an ecological point of view, examining changes in community structure and diversity (Borowitzka 1972, Munda 1974). The best-studied sewage outfall is on San Clemente Island, off southern California. This is a low-volume outfall, producing only 95,000 L per day of untreated domestic sewage. Littler & Murray (1975) found 17 fewer species of macrophytes and less cover near the outfall than in a nearby control area. The outfall flora was less diverse and showed a reduction in community stratification (spatial heterogeneity) due to the absence of *Egregia laevigata, Halidrys dioica, Sargassum agardhianum*, and the sea-grass *Phyllospadix torreyi*. These were replaced in the mid-intertidal near the outfall by a low turf of blue-green algae, *Ulva californica, Gelidium pusillum*, and small *Pterocladia capillacea*, and in the lower intertidal by *Serpulorbis squamigerus* covered with *Corallina officinalis* var. *chilensis*. Littler & Murray suggested that sewage favors rapid colonizers and more sewage-tolerant organisms. Macrophytes near the outfall exhibited relatively higher net productivity, smaller growth forms, simpler and shorter life histories, and most were components of early successional stages.

Further studies were undertaken at this site to experimentally test whether algal communities that are characteristic of sewage-stressed habitats showed high resiliency. The measure was their ability to recover quickly after removal of all biota from some quadrats on the rocky shore. Murray & Littler (1978) found that blue-green algae, filamentous Ectocarpaceae, and colonial diatoms were the dominant forms in the early successional stages in the cleared (denuded) areas in both the sewage and control plots. The outfall plots showed rapid recovery by algae such as *Ulva californica, Gelidium pusillum*, and *Pseudolithoderma nigrum*, which have a capacity for rapid recruitment. The algal communities in the unpolluted (control) denuded areas did not fully recover, even after 30 months.

The studies by Littler and colleagues on the impact of sewage on macrophytes forms an excellent example of the combination of field and laboratory studies; they progressed from a community field study, to experimental manipulation (denuded plots) in the field, and to studies on environmental physiology of important species in the laboratory. Kindig & Littler (1980) studied the response of 10 macrophytes to untreated, primary, secondary, and secondary chlorinated sewage effluent during long-term culture studies in the laboratory. *Bossiella orbigniana* and *Corallina officinalis* var. *chilensis* exhibited an increased photosynthetic rate when exposed to primary treated sewage, and in long-term cultures growth was enhanced. Chlorination of effluent produced only a short-term reduction in growth for the first week of culturing. Three populations of *C. officinalis* with differing pollution histories (preexposure to pollution) showed a tolerance to sewage corresponding to the extent of the previous exposure. This finding indicates that this species may be able to adapt physiologically to sewage stress and suggests that considerable caution must be exercised in the selection of benthic algae as biological indicators of pollution (Burrows 1971). The results of the study by Littler and colleagues confirms earlier reports that coralline algae are extremely tolerant of high concentrations of sewage (Dawson 1959). Downstream and inshore from a domestic sewage outfall in Laguna Beach, California, Dawson (1959) observed that 90% of the algal biomass was composed of the corallines *Bossiella* and *Corallina*.

Excessive growth of green seaweeds in response to sewage effluents is becoming an increasingly common phenomenon in sheltered marine bays (Fig. 8.7) (Perkins & Abbott 1972, Reise 1983). Excessive growth of *Enteromorpha* spp. on the tidal flats of the Wadden Sea during summer was attributed by Reise (1983) to eutrophication by adjacent sewage effluents. Mats were first composed primarily of *Enteromorpha*, but later other algae such as *Ulva* spp., *Cladophora* spp., *Chaetomorpha* spp., and *Porphyra* spp. appeared as secondary components. Mats of these algae cover wide areas of sheltered flats, and in sandy flats the strands of *Enteromorpha* become anchored in the feeding tunnels of the abundant polychaete *Arenicola marina*, enabling the algae to resist displacement by tidal currents. Storms are able to dislocate the algal mats (5–20 cm deep), and the sand flats

Figure 8.7. Eutrophication: *Codium vermilara* drifts onto the beach in Puerto Madryn, Argentina, which lies at the head of a nearly enclosed shallow bay, and is removed by truck.

are usually covered for no more than one month. The sediments under the mats become anoxic. This condition is tolerated by polychaetes, but the more sensitive *Turbellaria* decreases in abundance and species richness.

The distribution of littoral algae in the inner part of Oslofjord in Norway has been studied over the last 40 years, and *Ascophyllum nodosum* was observed to be a dominant alga in the area before 1940. More than 20 years ago a large increase in the sewage load occurred. Many species, such as *Rhodochorton purpureum, Phyllophora broadiaei, Spermothamnion repens*, and *Ascophyllum nodosum*, have disappeared or become rare (Rueness 1973). Rueness cleared plots in the inner part of Oslofjord near the sewage outfall and in a control area to observe recolonization. In addition, rocks from the control area to which *A. nodosum* was attached were transplanted to the sewage-stressed area. Regrowth was much faster in the inner fjord than the control area. In the inner fjord, the dominant recolonizing species was *Enteromorpha compressa*, followed by *Fucus spiralis*. No *Ascophyllum* germlings were observed. In the cleared area in the control plots, regrowth proceeded more slowly and green algae were less predominant. The number of species that recolonized was also greater, and after 6 months regrowth was primarily dominated by a dense stand of *Porphyra purpurea* in the cleared control area. The *Ascophyllum* transplants into the sewage-stressed area were heavily infested with epiphytes and frequently overgrown by *Enteromorpha* spp., *Ulva lactuca, Ceramium strictum*, and small mussels (*Mytilus*). Rueness concluded that the increased competition for substrate and the shading effect of the *Enteromorpha* carpet reduced the chances of *Ascophyllum* germlings becoming established near the sewage outfall.

The green alga *Cladophora* cf. *albida* is an acknowledged symptom of increased eutrophication in Peel Inlet in Western Australia (Gordon et al. 1981). Rivers flowing into the inlet provide large inputs of nitrogen and phosphorus, presumably from agricultural runoff and sewage. This increase in nutrients has resulted in the formation of thick algal beds (10–100 mm deep) that accumulate in shallow waters and decompose. The resulting deterioration of the previously clean beaches is a concern for recreational usage, and commercial fishery may be threatened.

Even in oligotrophic areas of the ocean, significant effects from a sewage outfall have been observed (Laws 1981). Kaneohe Bay is a subtropical embayment in the Hawaiian Islands. By 1972, about 4×10^6 L per day of sewage was being emptied into it. This sewage affected the coral reef community in two basic ways. First, a reduction in water clarity was caused by increased phytoplankton growth. This reduced the amount of available light for the symbiotic zooxanthellae living in the hermotypic corals and thus resulted in reduced coral growth. Second, the sewage discharge also stimulated the growth of the green alga *Dictyosphaeria cavernosa*, commonly known as the bubble alga. This alga usually establishes itself within a coral head at the base of the frond and then grows outward, eventually enveloping the coral head and killing the coral. This alga is not found beyond the sewage-stressed area and therefore appears to grow in response to the elevated nutrient concentrations caused by the sewage. The sewage was recently diverted from the bay and the recovery of the community is now being documented (Laws & Redalje 1982). Although inorganic nutrient concentrations have begun to revert to presewage levels, the system will take some time to fully stabilize because of the slow release of nutrients from plankton that have sunk out and accumulated in the sediments during the sewage discharge period.

Ammonium concentration in discharged sewage may be very high (up to 2200 μM). However, the maximum value of ammonium found in the surface waters over the White's Point sewage outfall off Los Angeles was 35 μM, and more frequently the concentrations were 5 to 10 μM. This represents a dilution of about 100-fold compared with the discharged concentration. Ammonium concentrations of 10 to 30 μM are not toxic to phytoplankton, but Thomas et al. (1980) found that at concentrations of 200 μM the growth of two dinoflagellates was inhibited, although three diatoms tested were not. Macrophytes from the Chlorophyceae seem to be more tolerant of sewage toxicity than many phytoplankton species. *Enteromorpha linza* grew well in full-strength sewage effluent even though the ammonium concentration was 500 μM (Chan et al. 1982). *Enteromorpha compressa* appears to be more sensitive, showing inhibition of photosynthesis when exposed to about 75 μM NH_4^+. The inhibition of growth that has been observed near sewage outfalls is probably not due to too high ammonium levels per se, but rather to other inhibiting factors such as heavy metals (Hershelman et al. 1981) or chlorinated compounds such as chloramine in chlorinated sewage (MacIsaac et al. 1979, Thomas et al. 1980).

In addition to direct inhibitory effects of sewage on macrophytes, secondary effects may account for macrophyte decline in progressively eutrophicated freshwaters. There is recent evidence that the decline of macrophytes is due to increased growth and shading by epiphytes and filamentous algae associated with beds of (vascular) macrophytes as well as increased turbidity of surface layers because of phytoplankton growth (Phillips et al. 1978).

Other components of concern in sewage are detergents, which may cause oxygen depletion because of the organic load, and sewage sludge and its disposal. Since pollutants such as heavy metals and possibly PCBs are greatly concentrated in sludge, it is generally dumped further out to sea, but in the United States attempts are being made to ban ocean dumping. Detergents and surfactants in sewage are also considered pollutants. Anionic detergents account for the bulk of the detergents in household sewage, and three of these, sodium lauryl ether sulfate, sodium dodecyl benzene sulfonate (SDBS), and DOBS 055, were tested on *Laminaria hyperborea* (Hopkin & Kain 1978). Both SDBS and DOBS 055 reduced the growth of sporophytes at concentrations of 1 to 10 mg L^{-1}, and gametophyte germination was also inhibited by SDBS. The toxicity of anionic detergents is intermediate between nonionic detergents (the least toxic) and cationic ones, based on tests with phytoplankton (Duursma & Marchand 1974, Kohler & Labus 1983). Toxicity of detergents and surfactants is attributed to disruption of cellular and intracellular membranes. Indeed, detergents such as Triton X-100 are used in research to help extract cell components.

The effects of chlorine on algal photosynthesis have been documented in the laboratory but no effects of chlorination of waste water on algae have been observed in the field. Chlorine is highly reactive and rapidly forms a number of compounds. It may form highly toxic chlorinated organic compounds, initiate the production of the strong biocide hypobromite, and react with ammonium to produce chloramines, which are particularly toxic to larval zooplankton (Bishop 1983). At 10 ppm, chlorine irreversibly inhibits photosynthetic activity of phytoplankton (Eppley et al. 1976). However, field experiments have shown that there is no evidence of deleterious effects of chlorine on phytoplankton photosynthesis in waters receiving chlorinated sewage wastes off southern California (Thomas et al. 1974). This is probably attributable to the jet diffusion system that is used, which provides immediate dilution by over 100-fold. Tests of chlorine toxicity to marine macrophytes have not been conducted.

Industrial waste disposal into the oceans in the United States is being phased out because the wastes contain many compounds that are extremely toxic. All dumping was scheduled to cease by the end of 1981, but this deadline has been postponed because of a lack of suitable disposal alternatives. Elevated concentrations of trace metals in surficial sediments near the Los Angeles County outfall resulted in contamination factors (median outfall/ median baseline) greater than 20 for Ag, Cd, Cu, and Hg (Hershelman et al. 1981). These elevated metal concentrations could have significant effects on small macrophytes that remain close to the sediment/ substrate.

8.6.2 Pulp mill effluent

Different wood-processing systems have different wastes, depending on the quality of the final product. Two methods of making pulp from coniferous trees used in Canada and the United States are the kraft and sulfite processes. In the kraft process, wood chips are initially digested in an alkaline solution of sodium sulfide and sodium hydroxide. This is a cleaner process since most of the digesting chemicals are recovered before the effluent is discharged. In the sulfite process, digestion occurs with an acidic calcium bisulfite solution, and much less of the digestive solution is reclaimed. Wood-processing industries require large quantities of water (200,000 L per metric ton of pulp) and therefore release large quantities of effluent containing such toxic compounds as hydrogen sulfide, methyl mercaptans (giving most of the smell), resins, fatty acid soaps, and sodium thiosulfate (Carefoot 1977). In addition, the effluent contains large amounts of waste organic matter such as lignins, which color the water brown, and wood fibers, which blanket the sediments in the area of the discharge, creating a high biological oxygen demand and possibly anaerobic conditions. Both the lignins and fibers severely reduce light penetration into the water.

There is only one study of the effects of pulpmill effluent on seaweeds (Hellenbrand 1978). Under normal field conditions, *Chondrus crispus, Ascophyllum nodosum*, and *Fucus vesiculosus* were not adversely affected by treated kraft mill effluent. In fact, productivity of all seaweeds increased, probably because of the nutrients in the effluent.

Laboratory experiments on the effects of six different pulpmill effluents on marine phytoplankton were conducted by Stockner & Costello (1976). They found that some species required a preadaptation period before the cultures resumed exponential growth in relatively high concentrations (20–30%) of the kraft mill effluent. A green flagellate, *Dunaliella tertiolecta*, exhibited exponential growth even in 90% kraft effluent, which was the most toxic effluent of the six types tested. Their results suggest that in marine waters that receive effluent without a drastic pH change, phytoplankton may not be seriously affected except when effluent concentrations exceed 30 to 40%. If the area receives effluent from a sulfite process, then lower concentrations (ca. 10%) may produce some inhibition of growth. In actual field experiments, Stockner & Cliff (1976) found that light attenuation by the effluent, especially in the 400 to 500-nm region, was the major cause of the reduction in daily rates of primary production. The tea-colored effluent would also reduce light and primary productivity of some macrophytes in the area, but this has not been tested.

The storage of logs in booms while they are waiting to be processed through the pulp mill may destroy local macrophyte beds due primarily to the lack of light penetration.

8.7 **Synopsis**

Pollution includes human additions of deleterious materials and energy into the environment. The effects of pollutants on macrophytes can be lethal (acute) or sublethal. The effects are assessed with bioassay experiments that should be conducted in the laboratory *and* field. The physicochemical aspects such as solubility, adsorption, and chemical complexation and speciation are extremely important in quantifying the effects of a pollutant. On the other hand, *total* concentration of a contaminant may give little indication of its toxicity. The choice of bioassay organism, its life history stage, and the potential for long-term recovery are also important in pollution assessments. There are obvious limitations to laboratory bioassays, especially because they do not contain suspended particulates that are known to radically reduce the toxicity of pollutants such as heavy metals through adsorption in estuarine areas.

Thermal pollution originating primarily from the cooling water of power plants is stimulatory if the water temperature does not rise above the optimal temperature for growth of a species. Thermal stress on seaweeds has occurred in areas off southern California. Symptoms include frond hardening, bleaching, or darkening and cellular plasmolysis in *Macrocystis*.

Increased nutrient supply, especially nitrogen, near sewage outfalls has generally resulted in changes in community structure and diversity and increased epiphytism on macrophytes. Macrophytes near the outfall tend to show relatively higher net primary productivity, smaller growth forms, and simpler and shorter life histories; most were components of early successional stages.

Heavy metal toxicity is influenced by the type of metal ion, the amount of particulates in the water, and the algal species. Generally, the order of metal toxicity for seaweeds is $Hg > Cu > Cd > Ag > Pb > Zn$. Since metal toxicity usually occurs only when the metal exists as a free ion, adsorption of the ion onto particles may be a very significant detoxification process in some environments. Macrophytes show several mechanisms to detoxify the metal or increase their tolerance. Extracellularly, metals may be detoxified by binding to algal extracellular products. Exclusion of the metal ion may occur at the cell wall by binding to cell wall polysaccharides, or at the cell membrane by changes in transport properties of the membrane. Intracellularly, metals may undergo changes in valence, or be converted into nontoxic organometallic compounds. Intracellular precipitation within vacuoles and nuclei has been observed for copper in phytoplankton. If significant detoxification does not occur, the metal ion may inhibit the function of algal enzyme systems, which elicits the following responses: cessation of growth, inhibition of photosynthesis, reduction of chlorophyll content, and increase in cell permeability and loss of K^+ from the cell. Macrophytes tend to concentrate many heavy metals to several orders of magnitude above ambient seawater concentrations. The tendency for further bioaccumulation along the food chain is less clear because of the variation among seaweeds and animals, and it is also dependent on the type of pollutant or even the kind of metal.

Petroleum is an extremely complex mixture of hydrocarbons, including alkanes, cycloalkanes, and aromatics. Oil may reduce photosynthesis and growth in macrophytes by preventing gas exchange, disrupting chloroplast membranes, destroying chlorophyll, and altering cell permeability. In some cases, penetration of oil is reduced by the mucilaginous coating, especially on some brown seaweeds. The components that penetrate the thallus most easily and hence are the most toxic are the lower molecular weight, volatile, lipophilic compounds, including the aromatics. Alkanes are least toxic. In the laboratory, the concentrations at which oil is toxic depend on the type of oil, how the extract was prepared, when it was used, water temperature, and presence of other pollutants or dispersants. Additional factors in the field that influence oil toxicity are proximity of the spill to the shore and weather conditions, especially wind. Rocky intertidal areas suffer slight, short-term harm from oil spills, whereas the impact on salt marshes and coral reefs is severe and longer term. Weathering of oil occurs by a number of processes, the most important of which is evaporation of the most toxic compounds. Herbivores are often more susceptible to oil than macrophytes, and an increase in ephemeral algal biomass is often a response to the reduced grazing pressure.

The effects of synthetic organic chemicals such as insecticides, herbicides, industrial chemicals and antifouling compounds on macrophytes have received little attention. Likewise, complex wastes such as pulp mill effluent and domestic sewage have been largely ignored.

Most of the ecosystems that have been studied to date have shown remarkable ability to recover when the source of the pollutant has been removed. Most of the effects that have been discussed have been local and confined to coastal areas, where point sources of the pollutant occur. In many cases, the animals were more sensitive than the macrophytes, resulting in a decrease in grazing and an increase in some species of seaweed. Other changes were at the community structure level, where the pollutant rendered one species less competitive than another.

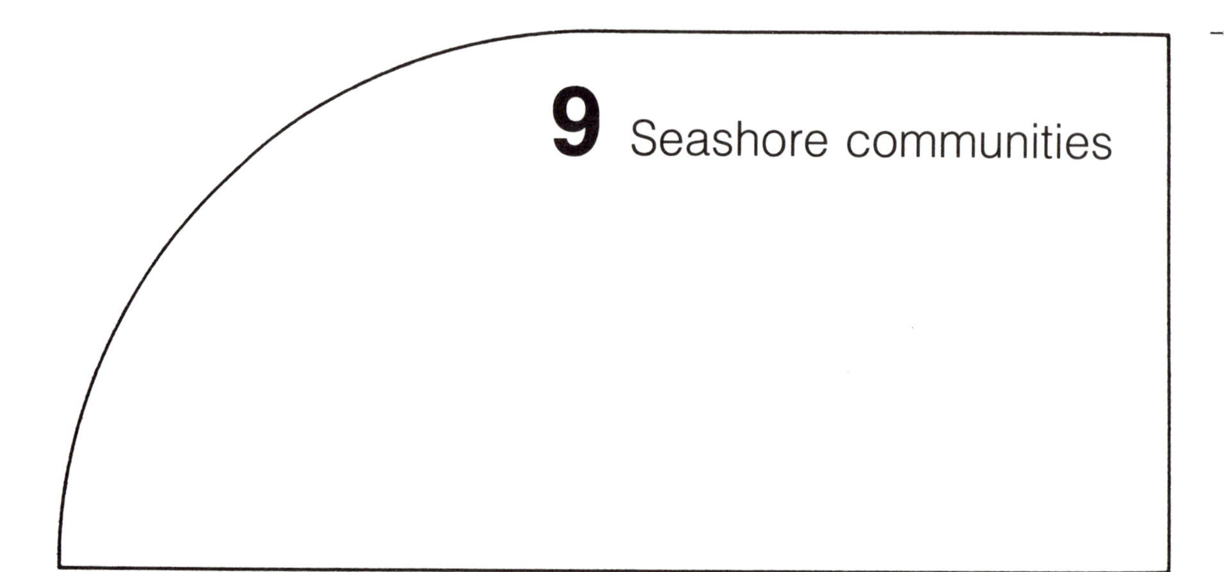

9 Seashore communities

In this chapter, zonation and some other aspects of community biology are considered against the background of environmental factors already introduced as well as others, including desiccation, competition, antibiosis, and factor interactions.

9.1 **Zonation patterns**

Almost all marine shores experience tides, although the tidal amplitude varies greatly from place to place and on some shores the change in water level is slight. The pattern of high and low waters also varies from place to place (Fig. 9.1), depending on the interaction between the tide waves and the standing waves caused by water slopping back and forth in the ocean basins (for details see texts such as Gross 1982, Thurman 1978, or Carefoot 1977). More important to seaweed ecology and physiology are the changes at one place. Tide waves are generated by both the moon and the sun; those caused by the sun are only 47% as powerful as lunar tides and of a different period (12 h, compared with 12 h 25 min). At new and full moon the two gravitational forces are more or less aligned, giving a larger tidal amplitude (spring tides) than when the forces are perpendicular (neap tides). Thus there is a monthly progression from neap tides to spring tides. Further, the times of high and low water change during the lunar month and sometimes from season to season; this is important because desiccation stress is increased when summer low tides occur during the day.

Tidal cycles are of three types (Fig. 9.1). Diurnal tides have one high and one low per day; this is an unusual type, occurring in parts of the Gulf of Mexico. Semidiurnal tides rise and fall twice a day, with successive highs and lows more or less equal in height; this type is common along open Atlantic Ocean coasts. Mixed tides occur twice a day but have clearly unequal highs and lows (Fig. 9.1b). Mixed tides are characteristic of Pacific and Indian Ocean coasts, as well as in smaller basins such as the Caribbean Sea and the Gulf of St. Lawrence (Gross 1982). In addition there are storm tides, with irregular periods, usually of several days, caused by barometric pressure changes and winds.

The height of the waterline is measured with reference to a standard level, the chart datum or zero tide. The definition of the level varies from country to country, and even from coast to coast in the United States. The numerical differences between the defined levels are small, except where there is a very large tidal range. For the Atlantic coast of the United States zero tide is defined as mean low water, whereas on the Pacific coast it is mean *lower* low water. This reflects the characteristic differences in the tidal cycle type. Other countries refer to mean low water springs, lowest normal tides (Canada), or lowest possible low water (Natl. Ocean. Atmos. Admin. 1983).

Two features of intertidal seaweed vegetation are readily apparent: distinct bands of particular species or associations run parallel to the shoreline and there are variations in the flora over short horizontal distances. Such patterns, though less obvious, also occur subtidally. Despite the extent of global shorelines there are general patterns upon which local variations are superimposed. A vertical strip of the shore may logically be divided into zones determined by the organisms present: a *Fucus* zone, a barnacle zone, and so forth. Alternatively, it may be divided according to tide levels, either in general terms (high, mid-, and low intertidal), or on the basis of critical tide levels.

9.1.1 Biological schemes

The Stephensons' worldwide survey of seashores enabled them to perceive the basic patterns underlying the variations in flora and fauna and to present a working hypothesis of "the universal features of zonation be-

Figure 9.1. Physical division of the seashore. Daily (first order) critical tide levels (CTLs) in diurnal, mixed, and semidiurnal tide regimes during a neap tidal cycle. Stippling indicates submergence; arrows indicate duration of continuous exposure or submergence immediately above and below CTLs. CTLs divide the intertidal region into four or five levels. (From Swinbanks 1982, reprinted with permission of Elsevier Biomedical Press)

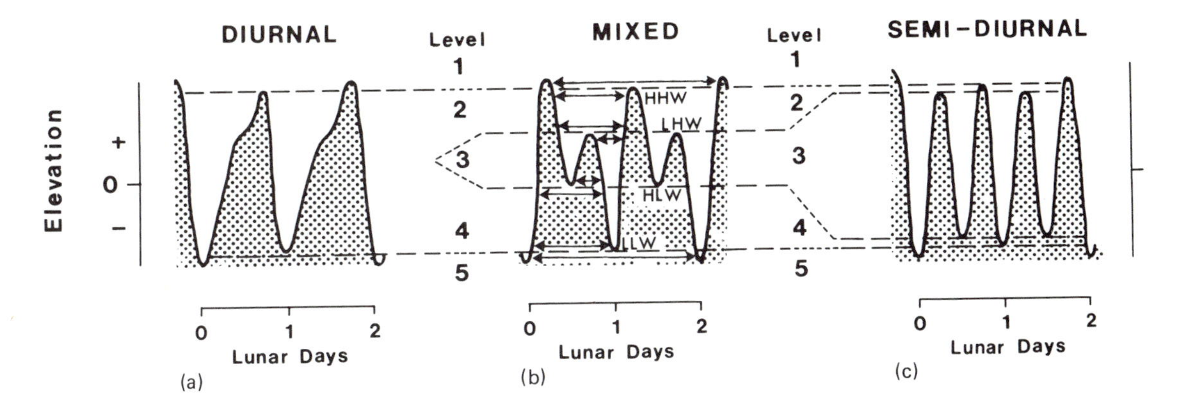

tween tide-marks on rocky coasts'' (as they titled their 1949 paper). Their now-familiar scheme, which is based on biological features, is shown in Fig. 9.2a.

In the Stephensons' scheme, the bulk of the intertidal zone is called the *littoral zone*, extending between extreme high water of spring tides (EHWS) and extreme low water of spring tides (ELWS), as taken from published tide tables. A *supralittoral fringe* straddles EHWS and is defined by limits of certain organisms (Fig. 9.2a). An *infralittoral fringe* occupies the lowermost littoral zone, from ELWS to the upper limit of kelps or kelplike seaweeds. The *supralittoral zone* is a region influenced by spray and sea air, and not usually occupied by seaweeds or marine animals. Below ELWS is the *infralittoral zone*, which is the upper part of the subtidal zone; it also can be divided (discussed subsequently). Pérès (1982a, b) uses the same terminology as the Stephensons.

The principal modification of this scheme, made by Lewis (1964), accounts for the effects of wave action in broadening and extending the vertical height of zones (Fig. 9.2b). Since zone boundaries are determined partly by duration of exposure to the atmosphere (see Sec. 9.3.2), the effect of wave action is to decrease the duration of exposure of a spot on the shore. The important criterion is not the theoretical exposure time determined from tide tables, but the actual exposure time. Lewis also extended the term littoral zone to include all of the supralittoral fringe, and his term is thus equivalent to the word ''intertidal'' as used in this book. He divided the littoral zone into a eulittoral zone and a littoral fringe (Fig. 9.2b). Both Lewis and the Stephensons point to three universal zones. They are, from upper to lower, (1) a ''black'' zone of littorinid snails, *Verrucaria* lichens, and blue-green algae; (2) a zone of barnacles, limpets, and assorted algae; and (3) a zone of laminarians, corals, or equivalent organisms.

Two examples of actual zonations from shores not studied by the Stephensons or Lewis are presented to show how in general this scheme is universal and how in detail there can be differences. An example of local variation is also included. Many additional examples can be found in the books by Lewis (1964) and Stephenson & Stephenson (1972) and in the review by Pérès (1982b).

Example 1: Three Saints Bay, Kodiak Island, Alaska (57° N) (Nybakken 1969). As part of a quantitative study, Nybakken presented diagrams showing the vertical distribution of the major organisms (Fig. 9.3) for comparison with Stephenson & Stephenson's universal scheme. On the vertical rock faces (Fig. 9.3a) the zonation fit the universal scheme fairly well. There were (1) a lichen zone above mean higher high water (there were no littorinids on these steep faces); (2) a barnacle zone, the lower part of which was dominated by *Fucus*; and (3) a zone of laminarians and the red alga *Odonthalia floccosa*. On other transects, however, where lichens were absent, there was effectively no supralittoral fringe. In the universal scheme the upper edges of the midlittoral zone and the supralittoral fringe are defined by the upper limits of, respectively, barnacles and littorinids. However, in Three Saints Bay, *Littorina* species were densest in the barnacle zone (East reef: Fig. 9.3c) or *Fucus* zone (beaches: Fig. 9.3b) and never extended above the barnacles. On soft bottoms the kelp zone was replaced by an equivalent eelgrass zone (Fig. 9.3b, c).

Example 2: Puerto Pardelas, Golfo Nuevo, Chubut, Argentina (Olivier et al. 1966). This part of the Argentine coast, 42° S latitude, presents very different intertidal communities, which can nevertheless be fitted to the universal scheme (Fig. 9.4). There are (1) a zone of blue-green algae in the supralittoral fringe merging into a limpet-dominated region in the uppermost part of the midlittoral; no littorinids; (2) a zone dominated by

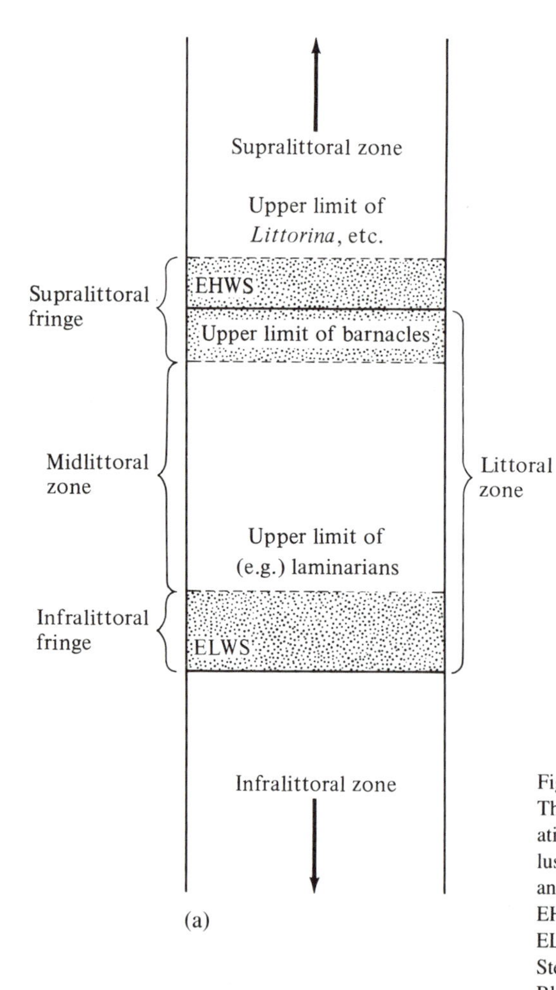

mussels and assorted algae, which merges directly or via a band of coralline algae into (3) a zone of *Codium* and associated algae that extends down from low water of spring tides. The *Codium* zone is the equivalent of the kelp zone.

Descriptions of subtidal zonation are fewer, owing to the need to use SCUBA equipment. Pérès (1982a, b) has proposed a universal scheme in which the depth occupied by multicellular algae is divided into an upper, infralittoral zone, extending from about low water of neap tides to the lower limit of photophilic algae, and a lower, circalittoral zone dominated by corallines. Two examples will illustrate subtidal zonation. On the wave-sheltered islands in the Strait of Georgia, between Vancouver Island and the mainland (Neushul 1965, Lindstrom & Foreman, 1978), the uppermost zone of the sublittoral is an extension of the infralittoral fringe, and hence it is dominated by kelp and associated understory and turf algae. This community extends 8 to 10 m below low

Figure 9.2. Biological division of the seashore. (a) The Stephensons' universal scheme for intertidal zonation. (b) Lewis's scheme for intertidal zonation, illustrating the effect of wave exposure in broadening and raising the zones (toward the left of the diagram). EHWS = extreme high water of spring tides; ELWS = extreme low water of spring tides. (a from Stephenson & Stephenson 1949, with permission of Blackwell Scientific Publications; b from Lewis 1964, with permission of the author)

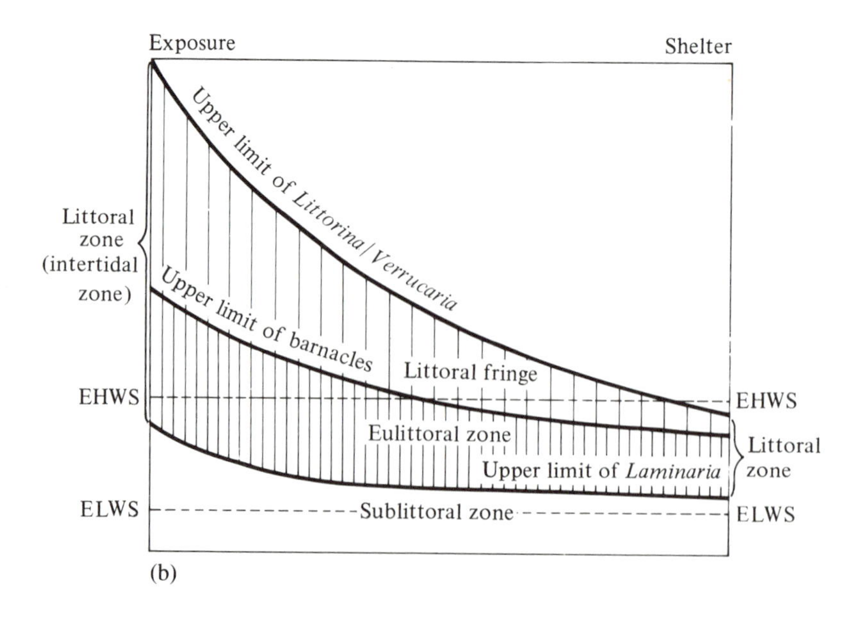

water. Beyond this is a deep sublittoral community consisting of red algae plus scattered *Agarum fimbriatum*. The transitions between zones are not as distinct as in the intertidal. A distinctive seaweed community was found on sandy bottoms. On West Island, South Australia (Shepherd & Womersley 1970), three zones were seen in high-energy areas (Fig. 9.5a), but in more sheltered areas the lowermost zone was absent. The composition of the communities and the depth ranges of the zones depend greatly on the amount of surge, as is shown for a nearby site in Fig. 5.17. In rough areas the zone limits are deeper (Fig. 9.5b). The mid- and lower sublittoral zones were found to be characterized, respectively, by brown and red algae. The upper zone, again an extension of the sublittoral fringe, had a turf community with some *Durvillaea* or *Cystophora* (equivalent to the kelps in the foregoing example).

There are difficulties inherent in defining zones by the organisms found in them. Floras and faunas change geographically. Furthermore, while a topographically uniform shore may have a uniform distribution of organisms, a jagged shoreline of irregular rocks and boulders is likely to present a very confusing pattern of organisms, with zones breaking down into patches. T.A. and Anne Stephenson, from their broad experience with intertidal zonation, expressed the problem this way: "It is essential to realize that all the different facets of the rock have their own appropriate zonation . . . One must expect one zonation on rocky faces directed landward, another on faces directed seaward . . . The zonation on a thoroughly broken shore has to be described in terms of the prevailing arrangement of species on all the rocky facets of the same type at the same level . . . [Probably] the best instrument for appreciating these interrupted zonations is the suitably trained human eye" (Stephenson & Stephenson 1972, pp. 15–16).

There is not only variability in space; time is also important. Aside from seasonal and successional changes in the vegetation, the timing of harsh conditions with respect to settling and of the clearing of space with respect to reproduction of potential settlers add to the heterogeneity in the communities (see, for example, Archambault & Bourget 1983, Dethier 1984). The patchiness in the pattern is not static. Vertical limits of species vary from year to year (Fig. 9.6), perhaps dependent on variations in emersion-submersion histories. Relative abundances and distributions of species nearly equal as competitors change from time to time, and longer-term changes are also known to occur (Lewis 1980). Understanding such long-term changes requires long-term observations, while understanding of the short-term changes still requires much work on how various factors affect the critical stages of the life cycles of the plants. Observations of the presence of a plant at a given spot can be interpreted to mean that conditions there have been suitable for its growth since it settled. Absence, on the other hand, only indicates that at some time conditions were unsuitable or that reproductive bodies of the species were unable to reach the spot. Conditions might have been unsuitable always or only at some brief time in the past if the plant was ever present.

9.1.2 Physical schemes

The shoreline communities can be studied in reference to physical parameters rather than the biological parameters just described. The concept of *critical tide levels* (CTLs) was introduced in one form by Colman in 1933, but Doty's (1946) modified version is most widely recognized. Doty tried to equate changes in vegetation with CTLs, which has been criticized by those who stress the importance of biological factors in determining zonation (e.g., Chapman 1973c). CTLs are discussed later in relation to zonation causes (Sec. 9.3.2); for the present they are introduced simply as a frame of reference within which zonation patterns can be described (as has been proposed by Swinbanks [1982] in his recent revision of Doty's scheme). Critical tide levels are levels in the intertidal zone at which there are marked increases in the duration of exposure or submergence. They occur at crests and troughs in daily, monthly, annual, or even longer-term tidal cycles. In a mixed tidal cycle (Fig. 9.1b), successive high and low waters are of different heights, so that there are several approximate doublings of duration of emersion; for example, from just below to just above lower high water (LHW), below-to-above higher high water (HHW). Thus on any given day the shore can be divided into five tidal zones. Shores with diurnal or semidiurnal tides have fewer zones (Fig. 9.1a, c). These diurnal zones are produced by so-called first-order CTLs. The monthly progression from spring to neap tides gives second-order CTLs. Swinbanks's scheme is summarized in Fig. 9.7. The major shortcoming of this physical scheme is that it depends on water levels being those predicted: it does not allow for wave action or storm tides.

Glynn (1965) determined the effective height of one wave-exposed high intertidal community (see Fig. 3.2a) to be 0.24 m lower than the predicted height. More recently, Druehl & Green (1982) attempted to correlate vertical distribution with actual submergence-emergence durations. Using a "surf-sensor" devised to record submergence events, they examined the submergence histories of three contiguous areas (within 50 m of each other) with differing slopes in a wave-exposed location, and the relations between seaweed vertical distributions and corrected exposure times. Actual submersion-emersion curves differed from predicted tidal curves in two ways. First, the intertidal range was greater and the lowest emerged level was higher than predicted; a gently shelving transect was closest to predicted and a rocky point the most different (Fig. 9.8a). This compares with the general observation depicted in Figure 9.2b. Second,

Figure 9.3. Zonation on Kodiak Island, Alaska. Composite diagrams from several transects of zonation on (a) vertical rock faces; (b) beach transects (cobble substratum from just above mean lower low water, MLLW, to mean higher high water, MHHW; sand below).

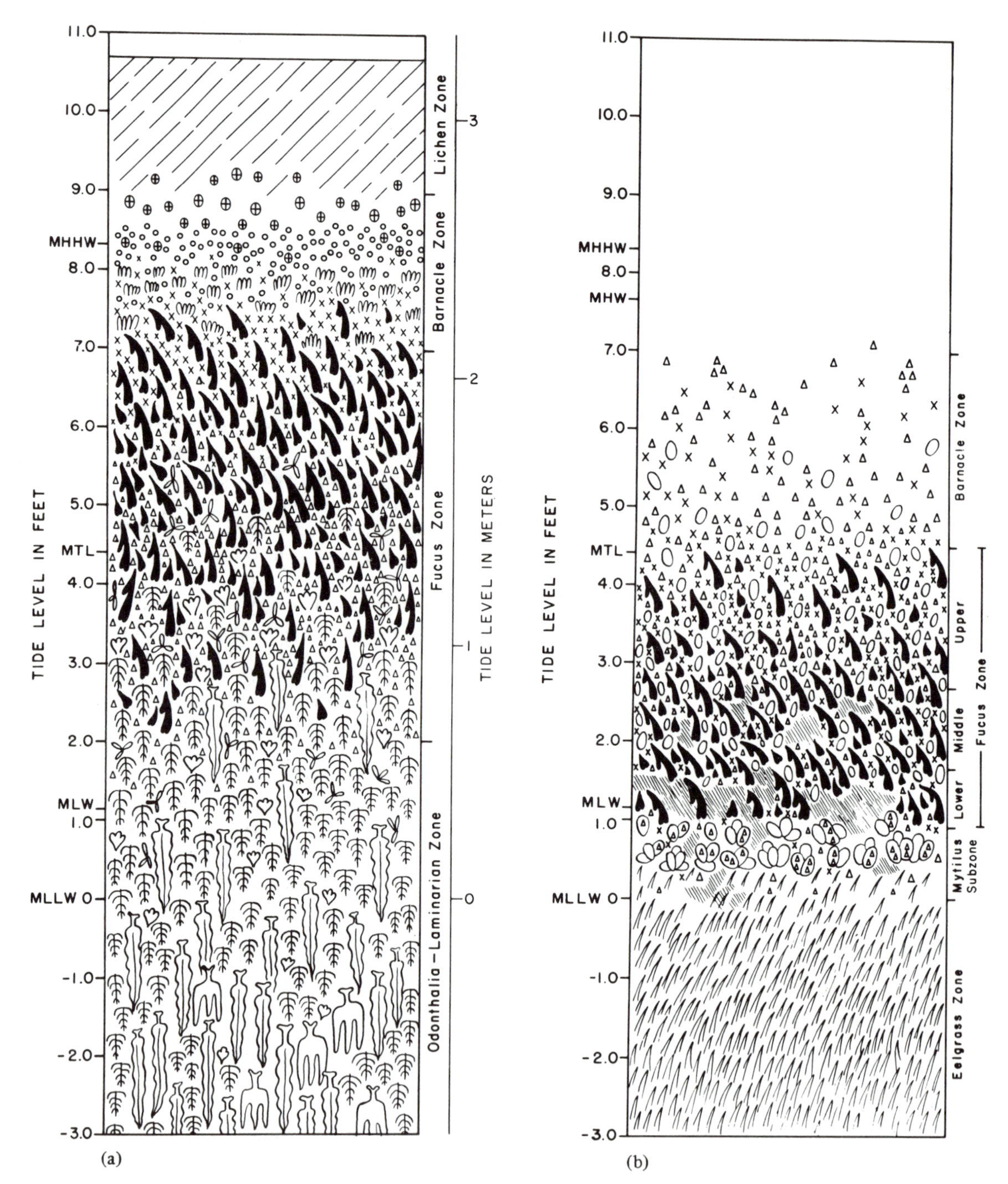

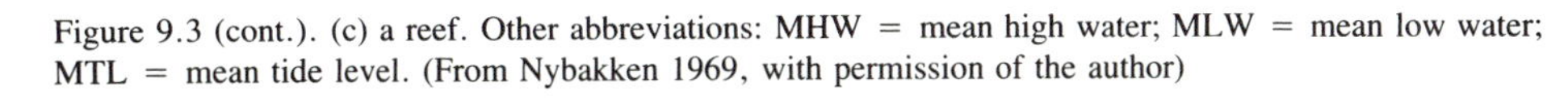
Figure 9.3 (cont.). (c) a reef. Other abbreviations: MHW = mean high water; MLW = mean low water; MTL = mean tide level. (From Nybakken 1969, with permission of the author)

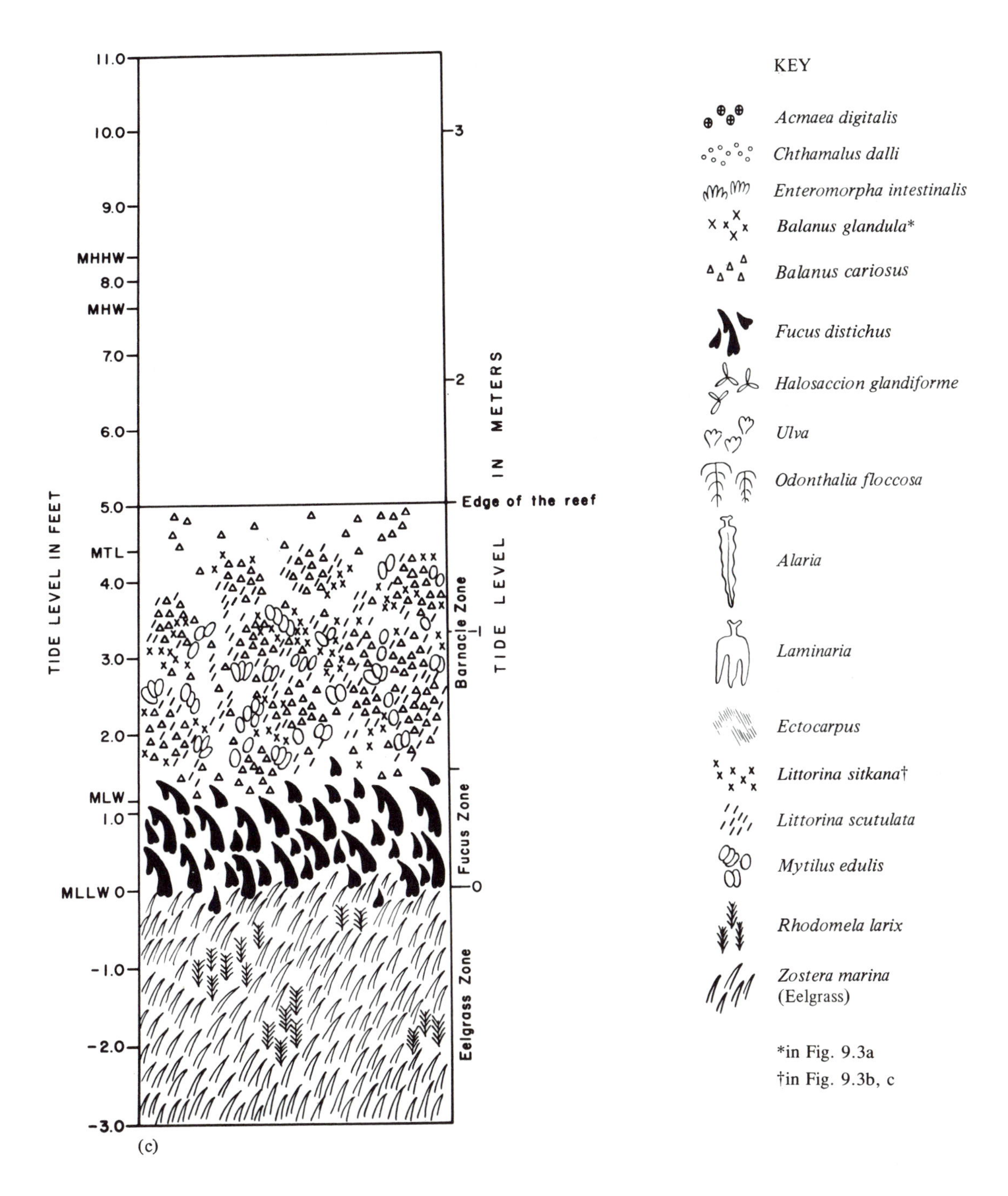

Figure 9.4. Profile of zonation on the rocky platforms at Colombo Beach, Chubut, Argentina. *Lyngbya* community = *Lyngbya aestuarii*, *Oscillatoria viridis*, and other blue-green algae; *Pachysiphonaria* community = *Pachysiphonaria lessoni* (a limpet), plus the red alga *Hildenbrandia lecanillieri*. (After Olivier et al. 1966)

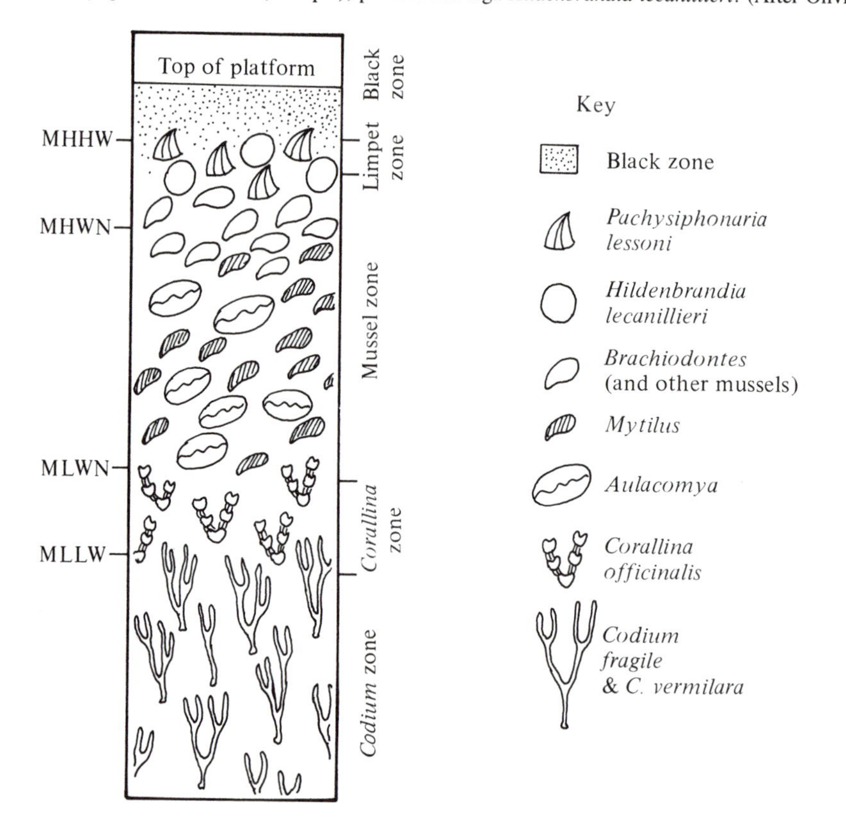

the harmonic pattern of the tidal curves was disrupted, again the most on the rocky point (Fig. 9.8b, c). Thus waves tend to blur the positions of CTLs.

9.2 Vegetation analysis and population dynamics

9.2.1 Vegetation analysis

Zonation descriptions of the type just given provide only a superficial view of the whereabouts of the most conspicuous and dominant organisms. Diagrams such as those in Figure 9.3 can hardly give even the crudest impression of abundances, neither relative abundance of species nor changes in abundance of one species as a function of tide level or of time. To properly understand population and community structure and the controlling forces, quantitative data and methods of vegetation analysis are needed. The abundance of an organism may be measured in several ways. The percentage of the substrate covered by the species or association is an appropriate measure where the vegetation is dense and uniform, for example, with algal turf or encrusting species. The number of individuals may be counted if they are clearly separate. The biomass (usually dry weight) of the population may be measured. Each method is difficult or impossible with some species: for example, one cannot count numbers of individuals in a turf – or even in an *Ascophyllum* bed – because each of the overlapping plants sends up numerous erect shoots; on the other hand, it is hard to determine the biomass of encrusting species or to ascertain percentage of cover of widely separated individuals.

The basis of any quantitative study of an area is a sound sampling procedure (Russell & Fielding 1981). The samples must be representative of the population as a whole and must, among other things, reflect heterogeneity of the population. Samples are most commonly chosen by placing quadrats (square frames enclosing a defined area) in the study area, employing some means of obtaining the randomness required for statistical analysis. The vegetation in the quadrat can be counted or photographed, which allows assessment of the same area on subsequent dates, or can be collected for one-time biomass and species determination. The achievement of randomness in placing the quadrat is difficult because the terrain is usually irregular and the vegetation patchy. "Putting [quadrats] in 'representative' or 'typical' places is not random sampling" (Green 1979). As an alternative to mathematically random sampling, many people man-

Figure 9.5. Subtidal zonation on West Island, South Australia. (a) Profile of zonation at a very wave-exposed point, showing three zones; (b) diagram showing changes in depths of the three zones from exposed to sheltered areas. (From Shepherd & Womersley 1970, with permission of The Royal Society of South Australia Inc.)

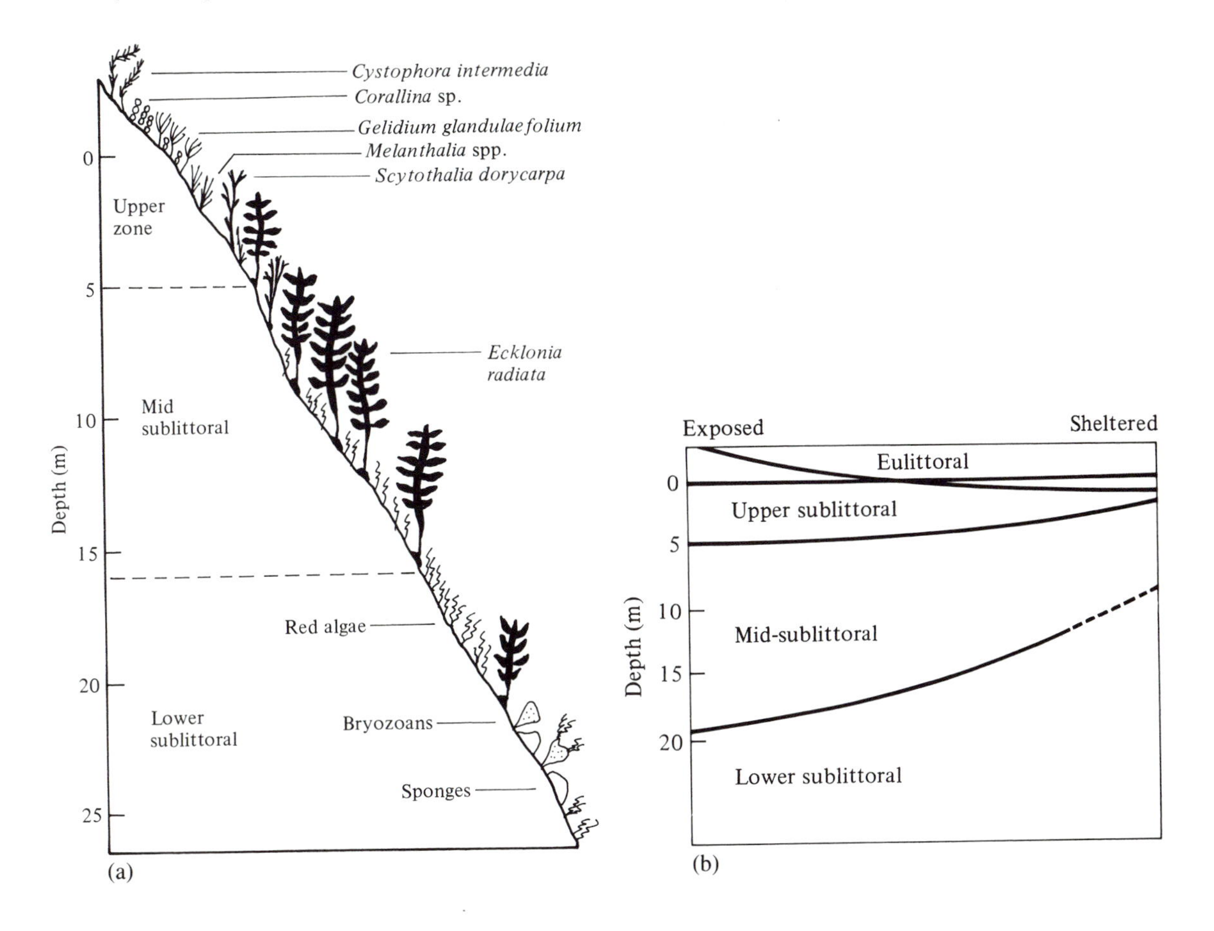

Figure 9.6. Year-to-year changes in upper and lower limits of two intertidal kelp species on three transects at an exposed site on the west coast of Vancouver Island, B.C. A gently shelving platform, a rocky point, and a narrow channel are compared (see also Fig. 9.8). (From Druehl & Green 1982, with permission of Inter-Research)

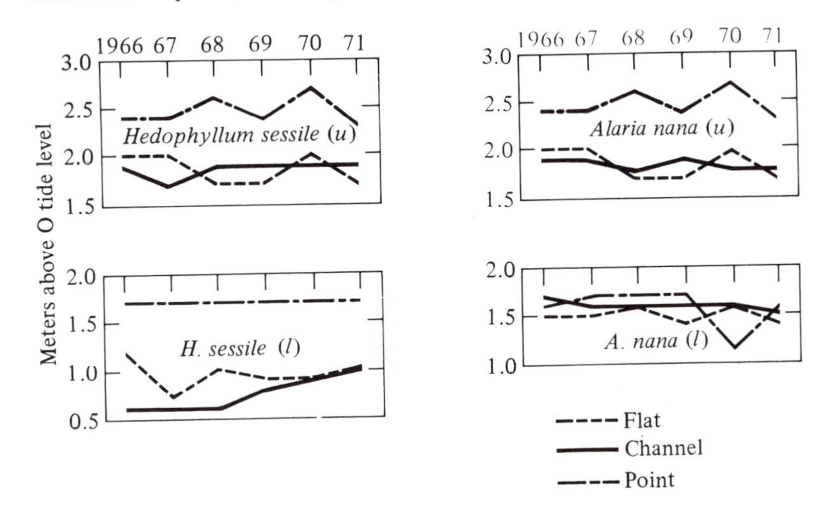

Figure 9.7. Curves of maximum continuous exposure and submergence with respect to tidal elevation, which illustrate the idealized sequence of first- and second-order CTLs for all types of tide. The CTLs used in this scheme are labeled for mixed tide (as in Fig. 9.1b); the stepped curve rising to the left indicates increasing exposure, while that rising to the right indicates increasing submergence. Abbreviations for tide levels: EHHW = extreme higher high water; ELLW = extreme lower low water; HLLW = highest lower low water; HSLLW = highest spring lower low water; LHHW = lowest higher high water; LLHW = lowest lower high water; LSHHW = lowest spring higher high water. (From Swinbanks 1982, with permission of Elsevier Biomedical Press)

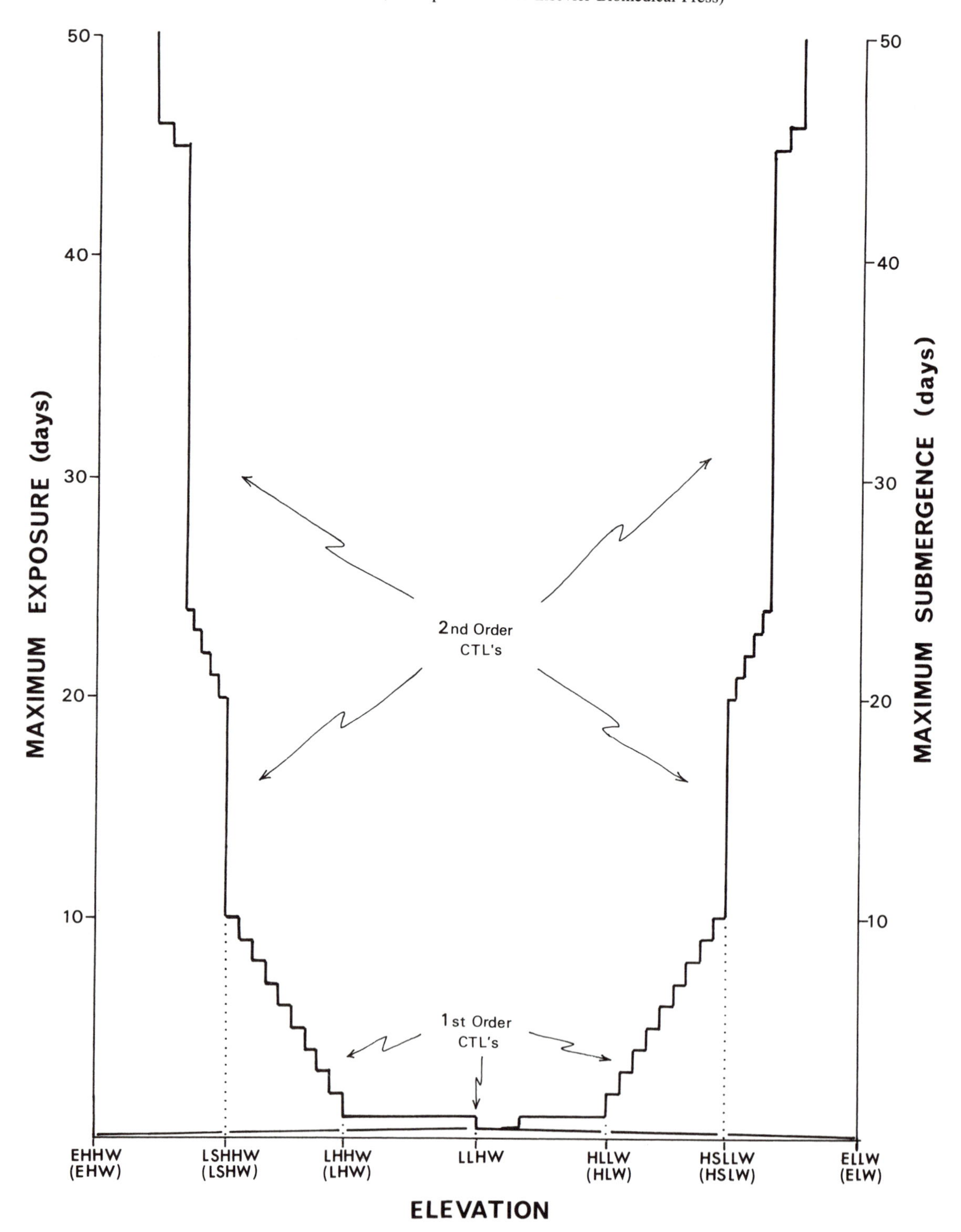

Figure 9.8. Submergence-emergence data from a site on Vancouver Island, British Columbia. (a) Measured accumulated time submerged as a function of elevation above zero tide for a rocky point (P), a channel (C), and a gently shelving rock face (F), all within 50 m of one another, compared with data from 6-min tidal predictions (U). (b) Predicted tide heights over a lunar cycle. (c) Actual tide heights and wave heights at the rocky point. The wave height data are derived from twice-daily observations. (From Druehl & Green 1982, with permission of Inter-Research)

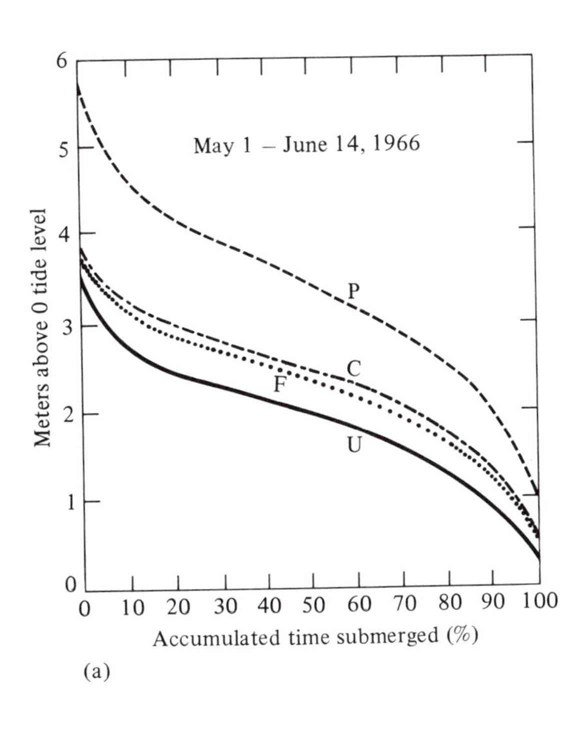

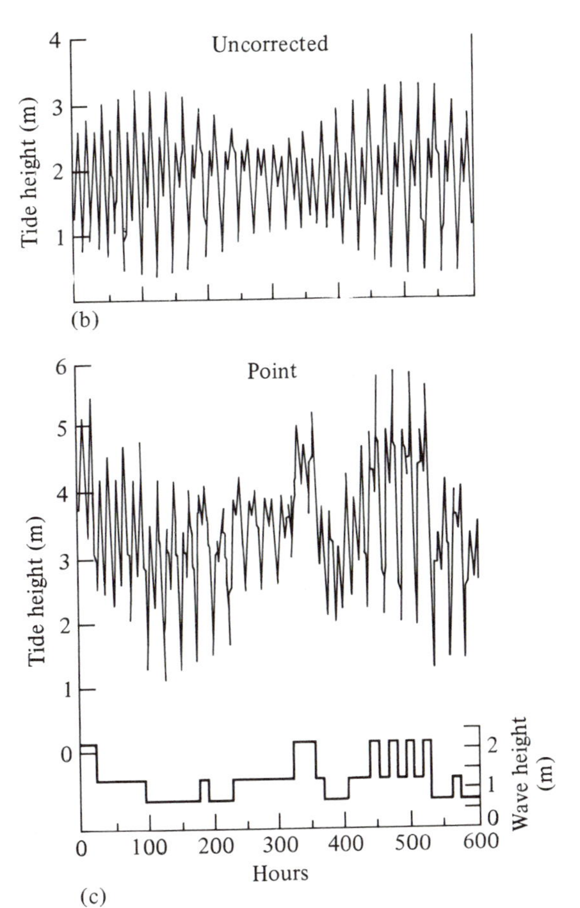

age *unbiased* sampling at fixed intervals along lines (transects) stretched horizontally or vertically across the shore (see Russell & Fielding 1981). Another practical consideration is the size of the quadrat, which, through preliminary work, must be made appropriate to the population or community to be studied and to the type of substrate.

Seaweeds are by no means uniformly distributed over the areas they occupy, as shown in Figure 9.9, where the percentages of cover by *Fucus* on five beach transects (depicted in Fig. 9.3b) are compared. Great variation between transects in one area is also evident. Community analysis can nevertheless show that the apparent discontinuities are real, for example, the breaks between the eulittoral and the littoral and sublittoral fringes (Bolton 1981).

For analysis of community structure, a combination of species incidence (presence or absence) and abundance may be used. Several methods of statistical analysis are available and have been used with varying success by different authors (see Lindstrom & Foreman 1978, Russell 1980, John et al. 1980, and Russell & Fielding 1981 for details, and Underwood 1981 for a review). For example, a method recommended by Russell (1980) produced no discrete groupings when used by Lindstrom & Foreman (1978) on their quadrats. The latter authors, after trying six methods on their data, concluded "that there is no single best method . . . and the use of several techniques is recommended since the results of one may aid in interpreting the results of another." Each method requires certain kinds of data, makes certain assumptions, and provides certain kinds of information. Once the requirements have been fulfilled it is wise to test the null-hypothesis (that "nothing is going on"), since these analyses can produce the appearance of pattern even from completely random data (Green

Figure 9.9. Changes in percentage of cover over the intertidal zone of dominant species on beach transects on Kodiak Island, Alaska (see Fig. 9.3b). (From Nybakken 1969, with permission of the author)

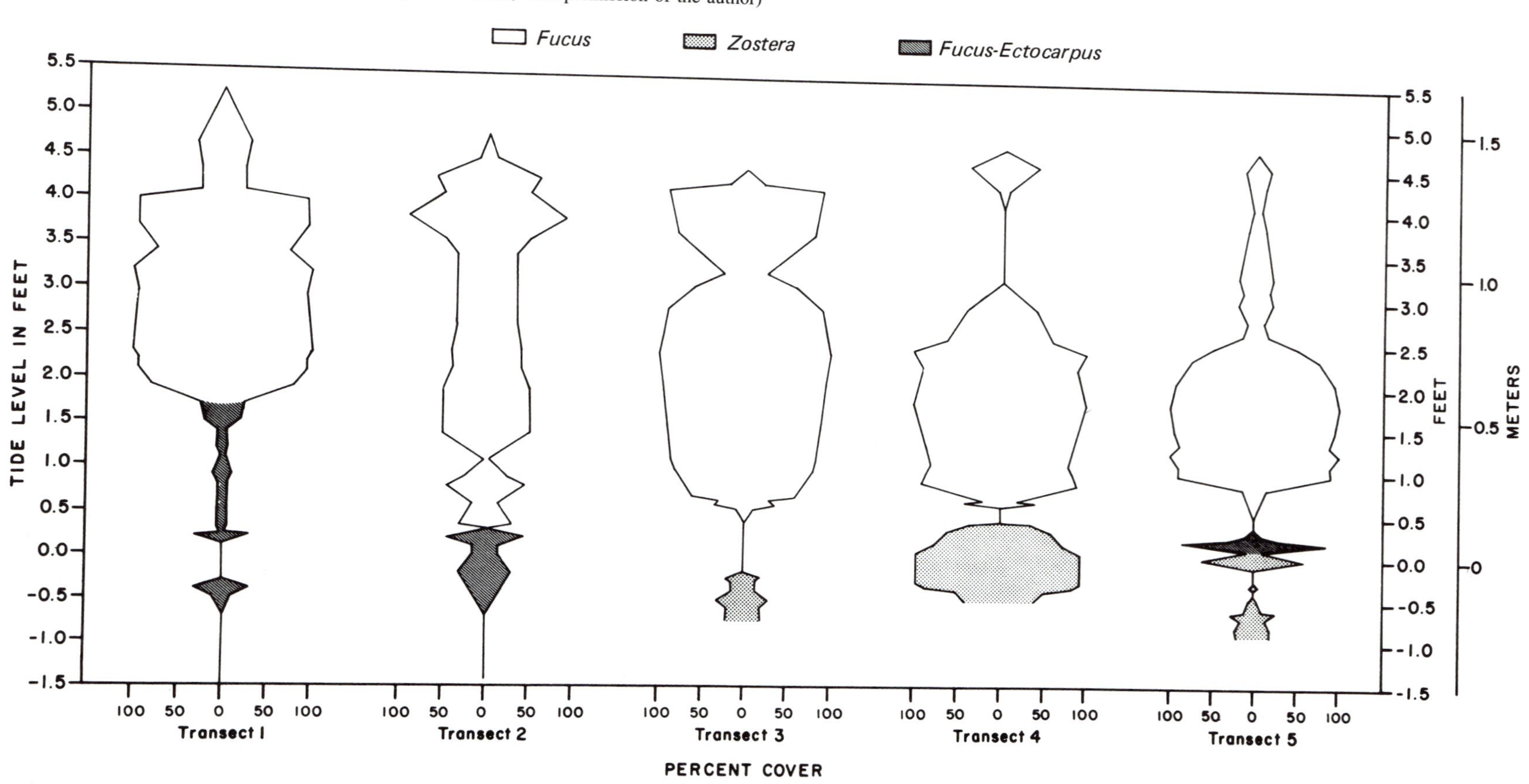

1979). Choosing a suitable method can be difficult: as Green points out in his 10 principles of statistical analysis, an unexpected or undesired result is not a valid reason for rejecting the method and hunting for a "better" one, while on the other hand there is no reason to use statistical analysis to prove the obvious. The use of these methods, while a refinement on qualitative observations, should not become an end in itself but should be the means to approach the second stage of ecology, the study of temporal changes (Lewis 1980).

9.2.2 Population dynamics

In seaweed vegetation analysis little attention has been paid to population dynamics as distinct from biomass or productivity studies. Population dynamics is the study of changes in the *numbers of organisms* in populations and of the factors influencing them (Harper 1977, Russell & Fielding 1981). Difficulties arise in determining numbers in most seaweed populations, although for some studies fronds can be considered as the functional units, obviating this difficulty (Cousens & Hutchings 1983).

The size of a population that can be supported by a resource-limited environment is called the carrying capacity, *K*, of the environment. The term "size" refers to the number of individuals, although the concept can still be applied if we substitute biomass. Marine organisms tend to produce vast numbers of offspring that gradually die off during the year, so that from year to year the population remains more or less constant at the carrying capacity (Figs. 9.10, 9.11) (Levinton 1982). Starting from a cleared environment, the increase in number of individuals over time at that site essentially follows a sigmoidal logistic curve. Most of the factors that cause the number of individuals to decrease can be grouped together as competition, including intraspecific and interspecific competition and predation. In addition, factors such as disease are important in stressed and especially monospecific stands (a problem of practical interest in aquaculture).

Population growth also includes emigration and immigration. Reproductive cells are generally shed into the water column (*Postelsia* is an exception). Some of the cells immigrate back into an existing population, others may found a new population. Algal spores and zygotes tend to settle indiscriminately all over the shore, except inasmuch as some zoids are phototactic and can migrate upward or downward. Some cells or germlings in the intertidal region are subsequently killed by adverse physical conditions. For example, *Costaria costata* and other kelps in British Columbia develop in the intertidal during spring, when low tides are in the early hours. In early summer, warmer weather and daytime low tides combine to produce temperature-desiccation stress lethal to individuals above the sublittoral fringe (Druehl & Duncan, unpublished). The situation for the red alga *Gastroclonium coulteri* in central California is similar, and Hodgson (1981) showed that desiccation, rather than light or temperature stress, was the damaging factor.

In theory, if a group of newly settled individuals (a cohort) is followed over time, the future changes in the population can be predicted by determining the number of individuals remaining at a given time, their life expectancy, and the number of offspring they can be expected to produce. In practice it is often impossible to determine the initial size of the cohort and the age-specific fecundity (Levinton 1982). Attempts to construct life-tables, showing recruitment and survivorship of individuals, have been made by Rosenthal et al. (1974) for *Macrocystis pyrifera* and by Gunnill (1980) for *Pelvetia fastigiata*, both in southern California. For such demographic study *P. fastigiata* has several advantages. Its life cycle is direct (haplobiontic) and it is the only fucoid in the very high intertidal, so that all juvenile fucoids in the zone can reasonably be assumed to be *P. fastigiata*. In Gunnill's study, the number of zygotes that settled could not be determined; recruitment referred to plants about 10 mm long and already at least several weeks old. Recruitment took place throughout the year, but most recruits appeared 3 to 6 months after peak release of eggs and sperm from conceptacles. Most losses were of recruits rather than established plants. Plants that

Figure 9.10. Theoretical population growth: a net process of logistic growth (dashed line) with overshoots of juvenile recruitment and subsequent mortality. K = carrying capacity of the environment. (From J.S. Levinton, *Marine Ecology*, © 1982. Reprinted by permission of Prentice-Hall, Inc., Englewood Cliffs, N.J.)

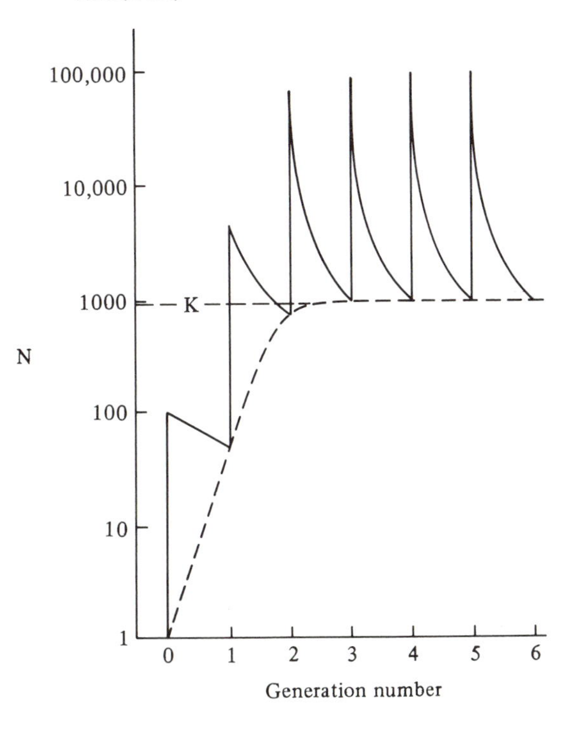

Figure 9.11. Changes with time in percentage of survivorship of 314 individuals in a population of *Pelvetia fastigiata* in California. (From Gunnill 1980, with permission of Springer-Verlag)

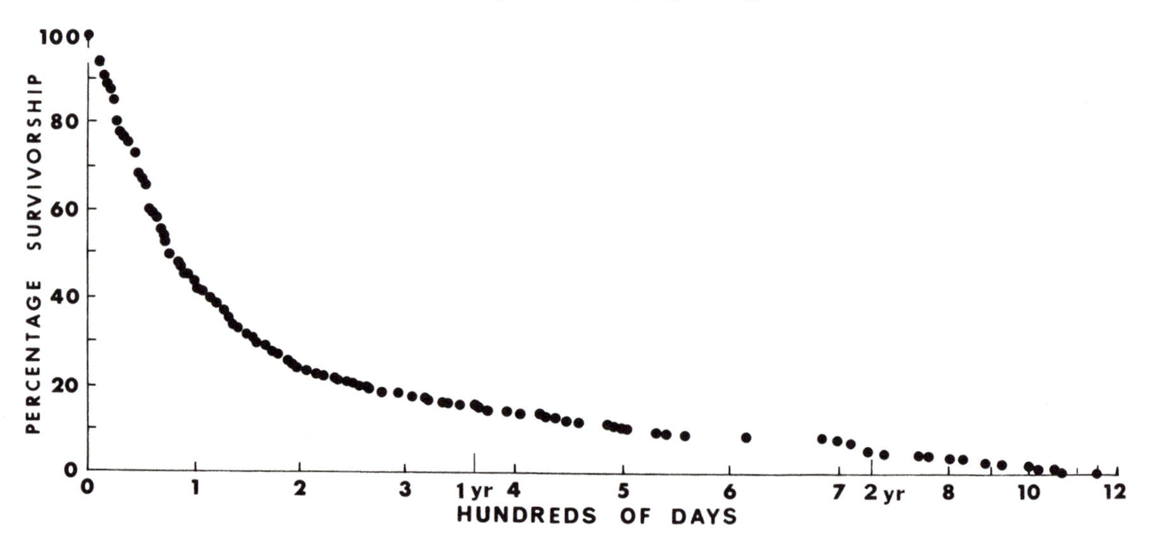

survived to 1.5 years, the time of first reproduction, had long life expectancies. Nine percent of the recruits survived this long (Fig. 9.11). The life-table for this species is shown in Table 9.1.

In environments populated to the carrying capacity, competition is high and demand for resources approximates supply. Natural selection here tends to favor competitive ability (including grazer resistance) with slow growth rate and delayed reproduction as the price; this is called *K*-selection. (The letters *K*- and *r*- refer to terms in an equation for growth rate.) In unstable areas where there is low competition because resources exceed demand, there is *r*-selection, favoring rapid growth, early reproduction, and short life spans. In an unpredictable environment, where a population or community may be suddenly removed, species with high growth rate (*r*-selected) will have the advantage and so become primary settlers. However, populations with very high growth rates tend to overshoot the carrying capacity – there is a delay in the feedback that controls growth – and this typically results in a population crash (Fig. 9.10). This crash in turn allows slower-growing species (*K*-selected) to take over. *K*-selected species fare best in stable environments. There is, however, a continuum of characteristics between these extremes (Pianka 1970).

The attributes that should hypothetically increase the fitness of *r*- and *K*-selected macroalgae have been more precisely formulated by Littler & Littler (1980) (Table 9.2), and the costs and benefits of each attribute compared (Table 9.3). By stating the hypotheses in this form, Littler & Littler were able to test some of the predictions (see also Littler et al. 1983). As expected, they found that thin, rapidly growing, short-lived algae were characteristic of unstable (temporally fluctuating) environments, whereas coarse, slower growing, long-lived algae were characteristic of stable environments. Some species, however, through dissimilar alternate phases, have attributes of both extremes: examples are *Gigartina papillata*, *Scytosiphon*, and *Petalonia*, all of which have a crustose stage (*K*-selected) and an erect phase (*r*-selected). In terrestrial environments stress-resistant plants tend to be *r*-selected, but in the case of marine algae the more stress-resistant species, such as blue-greens, *Ulva*, and *Enteromorpha*, are *r*-selected (Littler & Littler 1980).

9.3 Desiccation and zonation

9.3.1 Effects of desiccation on seaweed physiology

Except under cool, very humid conditions, loss of water from marine plants begins as soon as they are emersed (exposed to the atmosphere). Removal of seaweeds from seawater also deprives them of their source of nutrients, including most of the inorganic carbon, although postdesiccation enhancement of limited nutrient uptake has been reported for *Fucus distichus* (Thomas & Turpin 1980), which would partly compensate for the intermittent availability of nutrients. In a narrow sense the term "desiccation" is equivalent to "dehydration," but it is used to encompass the other changes as well, since they normally occur together. However, on humid days these other stresses can occur in the absence of dehydration.

The amount of water lost depends on the duration of exposure, the atmospheric conditions during exposure, and the evaporating surface-to-volume ratio of the plant (Dromgoole 1980). A brief exposure to cool, humid air may hardly affect a seaweed, whereas prolonged ex-

Table 9.1. *Life-table for* Pelvetia fastigiata *in California*

Age (days)	l_x	d_x	q_x	e_x	q_x 15 d^{-1}
0–15	1000	96	0.96	75	
16–30	904	127	0.141	78	
31–45	777	51	0.066	90	0.123
46–60	726	134	0.184	84	
61–75	592	95	0.161	99	
76–90	497	45	0.090	120	
91–105	452	38	0.085	126	
106–120	414	24	0.062	141	
121–135	389	51	0.131	144	0.082
136–150	338	23	0.066	183	
151–165	315	19	0.061	195	
166–180	296	25	0.086	201	
181–195	271	29	0.106	243	
196–210	242	10	0.039	252	
211–225	232	12	0.055	263	0.059
226–240	220	10	0.043	265	
241–255	210	6	0.030	258	
256–270	204	16	0.078	249	
271–285	188	6	0.034	266	
286–300	182	4	0.018	253	0.030
301–330	178	12	0.071	258	
331–360	166	10	0.058	354	
361–390	156	13	0.082	333	
391–420	143	3	0.022	316+	
421–450	140	19	0.136	291+	0.041
451–480	121	3	0.026	276+	
481–510	118	16	0.135	246+	
511–540	102	10	0.094	231+	
541–570	92	3	0.034	217+	
571–600	89	0	0	204+	
601–630	89	3	0.036	174+	0.015
631–660	86	0	0	149+	
661–690	86	6	0.074	119+	
691–720	80	3 + 6?	0.040 + .080?	102+	
721–800	71	10 + 19?	0.143 + .286?	140+	
801–900	41	0 + 16?	0 + .385?	190+	
901–1000	25	0 + 6?	0 + .250?	150+	
1001–1200	19	0 + 19?	0 + 1.0?	65+	
possible 690–1200	80	13	0.159	—	0.005

Of 314 individuals recruiting after November 1973, those marked ? were still alive in July 1977. Standard life-table abbreviations are used for the column headings: l_x = number living at start of each time interval (the 314 actual plants scaled to 1000 at age 0–15); d_x = number dying during the interval; q_x = proportionate mortality; e_x = mean life expectancy in days of the individuals alive during the interval; q_x 15 d^{-1} = mortality rate, the mean proportionate mortality per 15 days, averaged over 90-day periods.

Source: Gunnill (1980), with permission of Springer-Verlag.

Table 9.2. *Attributes that would seem, a priori, to improve the fitness of opportunistic macroalgae (representative of young or temporally fluctuating communities) versus late-successional macroalgae (characteristic of mature or temporally constant communities)*

Opportunistic forms	Late-successional forms
Rapid colonizers on newly cleared surfaces	Not rapid colonizers (present mostly in late seral stages), invade pioneer communities on a predictable seasonal basis
Ephemerals, annuals, or perennials with vegetative short-cuts to life history	More complex and longer life histories, reproduction optimally timed seasonally
Thallus form relatively simple (undifferentiated), small with little biomass per thallus; high thallus area:volume ratio	Thallus form differentiated structurally and functionally with much structural tissue (large thalli high in biomass); low thallus area:volume ratio
Rapid growth potential and high net primary productivity per entire thallus, nearly all tissue photosynthetic	Slow growth and low net productivity per entire thallus unit due to respiration of nonphotosynthetic tissue and reduced protoplasm per algal unit
High total reproductive capacity with nearly all cells potentially reproductive, many reproductive bodies with little energy invested in each propagule; released throughout the year	Low total reproductive capacity, specialized reproductive tissue with relatively high energy contained in individual propagules
Calorific value high and uniform throughout the thallus	Calorific value low in some structural components and distributed differentially in thallus parts. May store high-energy compounds for predictably harsh seasons
Different parts of life history have similar opportunistic strategies, isomorphic alternation, young thalli just smaller versions of old.	Different parts of life history may have evolved markedly different strategies, heteromorphic alternation, young thalli may possess strategies paralleling opportunistic forms
Escape predation by nature of their temporal and spatial unpredictability or by rapid growth (satiating herbivores)	Reduce palatability to predators by complex structural and chemical defenses

Source: Reprinted with permission from Littler & Littler (1980), *Amer. Nat.* vol. 116, pp. 25–44, © 1980 The University of Chicago Press.

posure, particularly during hot summer days, may cause severe effects. The higher up on the shore a species grows, the longer it is exposed to desiccation. The small-scale habitat has to be considered since seaweeds may be partially protected by clumping and by the slope, direction, and porosity of the shore. Some intertidal seaweeds appear to avoid desiccation by growing under overhangs or other algae, in crevices, or in other wet areas, although the distribution reflects survival in certain places rather than selection of favorable sites. For example, *Corallina vancouveriensis* in parts of southern California receives protection from desiccation where it grows among *Anthopleura elegantissima*, a colonial sea anemone that forms water-retaining carpets (Taylor & Littler 1982).

Intertidal fucoids tolerate desiccation rather than having means to avoid the stress (i.e., to maintain high tissue water potential). Moreover, some have an ability to harden to drought conditions (Schonbeck & Norton 1979a, b). The mechanism of hardening is not known but two factors are potentially involved. First, dry matter increase (hence better water retention) probably causes the cells to collapse less severely and sustain less mechanical stress, especially to the plasmalemma. (This membrane, according to Levitt 1972, is the primary site of drought injury.) Second, there may be changes in the degree of saturation of membrane lipids (i.e., number of C=C double bonds). A saturated lipid layer offers best protection against disruption by desiccation at high temperatures, whereas an unsaturated layer is most advantageous during dehydration caused by freezing (Levitt 1972: work on higher plants). The degree of saturation of fucoid lipids has been shown to be maximum in summer (Pham Quang & Laur 1976). Hardening and dehardening can take place rapidly, the few days of brief daily exposures being adequate to prepare upper shore plants for the following prolonged exposure.

Loss of water from the surface of a seaweed results in the flow of water out of the cells, with the possibility of plasmolysis if the cell wall is rigid and unable to collapse along with the vacuole and cytoplasm. The increase in salt concentration of the extracellular water

Table 9.3. *Hypothetical costs and benefits of the attributes listed in Table 9.2 for opportunistic and late-successional species of macroalgae*

Opportunistic forms	Late-successional forms
Costs	
Reproductive bodies have high mortality	Slow growth, low net productivity per entire thallus unit results in long establishment times
Small and simple thalli are easily outcompeted for light by tall canopy formers	Low and infrequent output of reproductive bodies
Delicate thalli are more easily crowded out and damaged by less delicate forms	Low surface-to-volume ratios relatively ineffective for uptake of low nutrient concentrations
Thallus relatively accessible and susceptible to grazing	Overall mortality effects more disastrous because of slow replacement times and overall lower densities
Delicate thalli are easily torn away by shearing forces of waves and abraded by sedimentary particles	Must commit a relatively large amount of energy and materials to protect long-lived structures (energy that is thereby unavailable for growth and reproduction)
High surface:volume ratio results in greater desiccation when exposed to air	Specialized physiologically and thus tend to have a narrow range of morphology
Limited survival options due to less heterogeneity of life history phases	Respiration costs high due to maintenance of structural tissues (especially during unfavorable growth conditions)
Benefits	
High productivity and rapid growth permits rapid invasion of primary substrates	High quality of reproductive bodies (more energy per propagule) reduces mortality
High and continuous output of reproductive bodies	Differentiated structure and large size increase competitive ability for light
High surface:volume ratio favors rapid uptake of nutrients	Structural specialization increases toughness and competitive ability for space
Rapid replacement of tissues can minimize predation and overcome mortality effects	Photosynthetic and reproductive structures relatively inaccessible and resistant to grazing by epilithic herbivores
Escape from predation by nature of their temporal and spatial unpredictability	Resistance to physical stresses such as shearing and abrasion
Not physiologically specialized and tend to have a broader range of morphology	Low surface:volume ratio decreases water loss during exposure to air
	More available survival options due to complex (heteromorphic) life history strategies
	Mechanisms for storing nutritive compounds, dropping costly parts, or shifting physiological patterns permit survival during unfavorable but predictable seasons

Source: Reprinted with permission from Littler & Littler (1980), *Amer. Nat.*, vol. 116, pp. 25–44, © 1980 The Univeristy of Chicago Press.

adds to the drying effect of the water loss itself. Since the effects of water potential changes on cells were already examined in Chapter 4, the following deals with other effects of desiccation.

Marine algae, which are essentially aquatic plants, are generally assumed to have severely restricted metabolism when they are exposed to air. Yet upper littoral seaweeds spend half or more of their time out of water. Have any adapted to make use of the atmospheric environment? Are there even any to which *submergence* is a stress? The answer to the first question seems to be yes, to the second no. Many experiments have tested the recovery of seaweeds from emersion, often by measuring rates of photosynthesis or respiration (see review by Gessner & Schramm 1971). Dring & Brown (1982) tried to correlate recovery rate with the normal position of the plant on the shore. Characteristically the experiments consist of drying out a selection of intertidal algae to

Figure 9.12. Recovery of photosynthesis in two intertidal fucoids, *Fucus vesiculosus* (a) and *Pelvetia canaliculata* (b), following desiccation for several days. Upper curve in (a) is rate in a thallus resubmerged immediately after reaching 10 to 12% of the original water content. Photosynthesis rate, as O_2 output, is expressed as percentage of the rate in undehydrated control plants. (From Gessner & Schramm 1971; a modified from Kaltwasser, reused with permission of Springer-Verlag; b modified from Reid, reused with permission of Gustav Fischer Verlag)

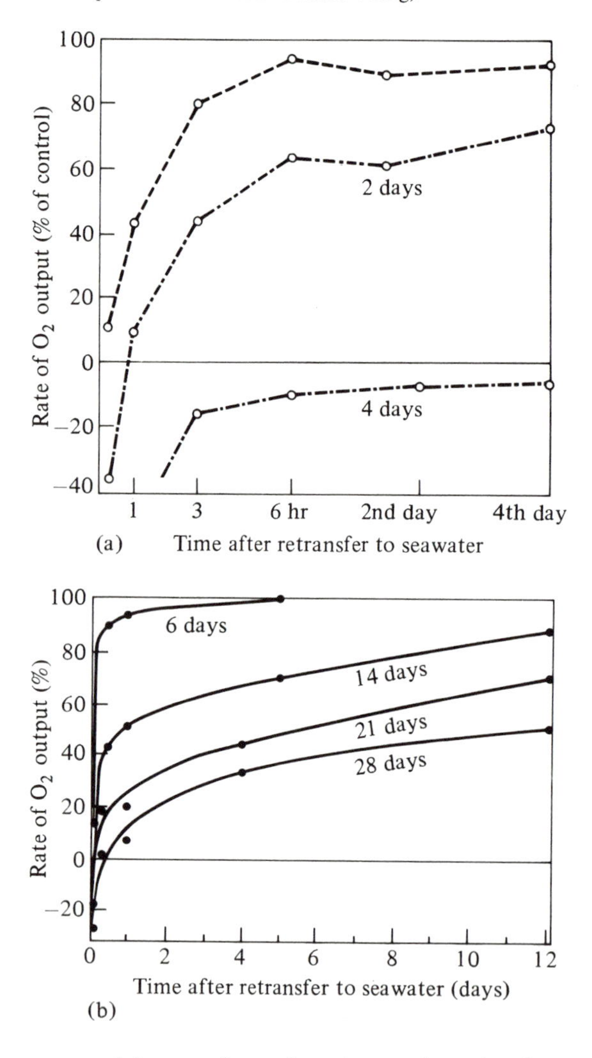

measured degrees of water loss, then resubmerging them in water and measuring rate of gas exchange at various times. Some representative results are shown in Figure 9.12, where recoveries of *Fucus vesiculosus* (mid-intertidal) and *Pelvetia canaliculata* (high intertidal) are compared. Not surprisingly, *Pelvetia* is able to withstand longer periods of desiccation. The variation in the extent of desiccation that can be withstood by different species is illustrated in Figure 9.13. *Fucus* was shown to tolerate desiccation to 25% of its original water content with virtually complete recovery after some hours, whereas

Figure 9.13. Recovery of photosynthesis (O_2 output) by *Fucus vesiculosus* and *Ulva lactuca* following dehydration to the indicated percentages of maximum water content. (From Gessner & Schramm 1971; modified from Kaltwasser, reused with permission of Springer-Verlag)

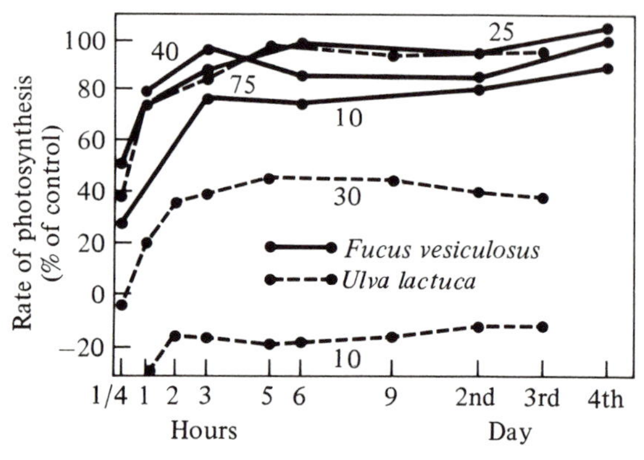

Ulva was greatly impaired by such extreme desiccation. These experiments demonstrate that emersion can be stressful. But what takes place during emersion?

As soon as a seaweed is removed from water its photosynthetic rate drops sharply, even before any desiccation takes place (e.g., Chapman 1966). This is because the inorganic carbon supply is greatly restricted. A small amount of bicarbonate in the surface film of water on the plant is available for photosynthesis, but it is not quickly replenished. CO_2 must diffuse from the air into the water film and dissolve, but the concentration of CO_2 in air is low compared with the bicarbonate content of seawater. As desiccation begins, photosynthesis often declines even further, owing perhaps to the stress on the cells or to the evaporation of the surface water film. In some species, photosynthesis actually increases again (Johnson et al. 1974, Brinkhuis et al. 1976), although continued desiccation leads to a reduction. The reason for the increase seems to be that when the water film has evaporated, CO_2 from the air can penetrate more quickly into the cells. (This explanation was given by Stocker & Holdheide 1937 and has yet to be confirmed or improved upon – Dring & Brown 1982.)

If relative humidity is experimentally maintained high enough to prevent desiccation, the photosynthetic rate may remain the same over long periods, as found in *Fucus serratus* by Dring & Brown (1982). This is evidence that emersion itself is not detrimental to photosynthesis. A species of *Ulva* studied in Israel by Beer & Eshel (1983) maintained constant photosynthetic rate from 0 to 20% water loss and positive photosynthesis to about 35% water loss. The plant was predicted to maintain positive photosynthesis for some 90 min after exposure in the morning but for only 30 min at midday. Only plants in the highest part of the population were unable to maintain positive photosynthesis during low

Figure 9.14. Carbon dioxide fixation in the light (a) and in darkness (b) by intertidal algae when submerged (black bars) and when in air (white bars). The algae, arranged in order from highest to lowest intertidal position, were not desiccated during emersion. (From Kremer & Schmitz 1973, with permission of Gustav Fischer Verlag)

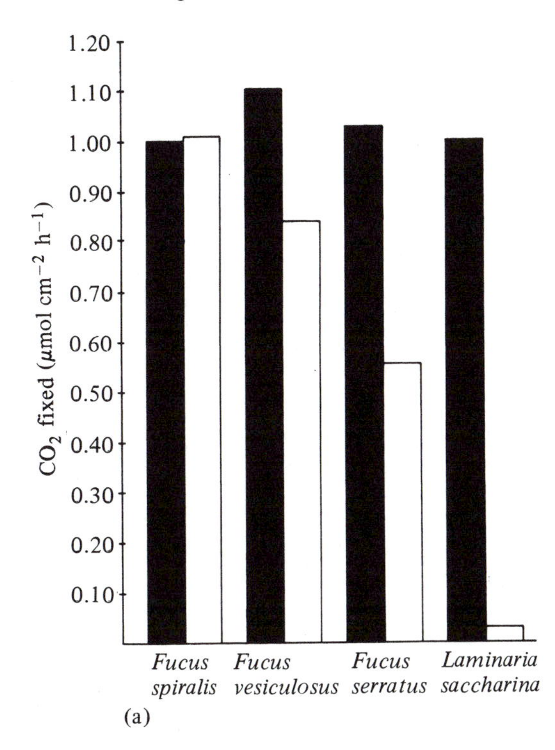

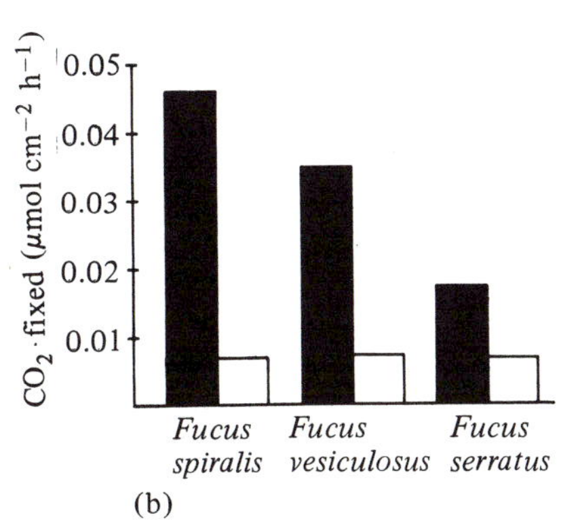

tide. The relative rates of gas exchange by plants during submersion and emersion generally correlate with the position of the alga on the shore; high intertidal species do as well or even better in air than in water, whereas lower species and sublittoral species do not do well in air (Fig. 9.14) (Kremer & Schmitz 1973, Johnson et al. 1974, Quadir et al. 1979). While some algae take up HCO_3^- as a carbon source, others incorporate only unhydrated CO_2 (Kremer 1981a). Too few species have been studied yet for any generalizations to be made. What is clear is that many intertidal algae are able to take some advantage of the atmospheric environment.

The photosynthetic apparatus of high-shore species does not appear to be more resistant to water loss than that of low-shore species (Fig. 9.15). Nor is the rate of water loss significantly lower in higher intertidal species. What does differ is the ability of the photosynthetic apparatus (and the cells in general) to *recover* from desiccation stress when resubmerged (Fig. 9.12). On the basis of their own data and previous studies, Dring & Brown (1982) assessed three hypotheses that might explain the effects of desiccation on intertidal plants and on zonation: (1) species from the upper shore are able to maintain active photosynthesis at lower tissue water contents than are species from lower on the shore (this is refuted by the data in Fig. 9.15); (2) the rate of recovery of photosynthesis after a period of emersion is more rapid in species from the upper shore (this hypothesis is also refuted by available data); (3) the recovery of photosynthesis after a period of emersion is more complete in species from the upper shore (this hypothesis was supported by Dring & Brown's data).

There is little evidence yet on how higher shore algae are able to recover more completely, but some other work suggests that the explanation may lie in biophysical properties of the photosynthetic apparatus. Red algae, such as *Porphyra*, have a mechanism to control the transfer of light energy from PSII, which is connected to the phycobilisomes, to PSI, which contains most of the chlorophyll *a*. Wiltens et al. (1978) found evidence that the parts of the photosynthetic sequence sensitive to desiccation are the transfer of electrons from PSII to PSI and the splitting of water. In a desiccation-tolerant plant, *Porphyra sanjuanensis*, rehydration led to recovery first of the intersystem electron transfer process, then of the water-splitting process. In *P. perforata* under high light and desiccation, PSII, which is the more sensitive to photo-oxidation, is protected by a cycling of electrons from the PSII acceptor (I: see Fig. 2.16) back to oxidants produced on the water-splitting side of PSII. Light energy is thus dissipated as heat (Satoh & Fork 1983).

A further consequence of drying out may be an increased rate of exudation of organic matter (Sieburth 1969). The amount of carbon released in 10 min by *Fucus vesiculosus* after resubmersion increased in relation to the duration of exposure (and hence the amount

Figure 9.15. Photosynthesis during exposure to air as a function of water content in three fucoids from four heights in the intertidal zone. Linear regression lines for the four populations are not significantly different. (From Dring & Brown 1982, with permission of Inter-Research)

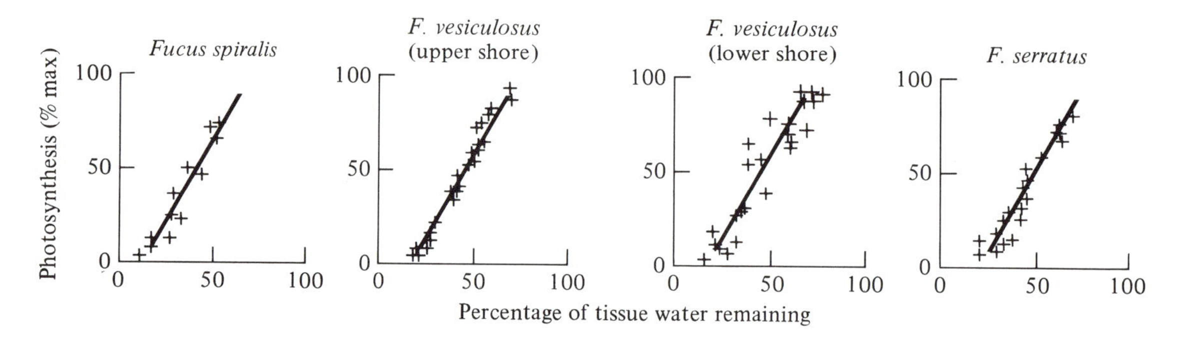

of water lost). Specimens from higher on the shore lost more water but released less carbon. While these observations need to be extended and verified, one may conclude that photosynthesis probably is not the only metabolic process on which desiccation has different effects in different species.

Are there any species for which a periodic emersion is beneficial or even essential? All species tested in a tide simulator by Edwards (1977b) grew fastest when continuously submerged, but some high intertidal species, such as *Pelvetia canaliculata* or *Fucus spiralis*, which Edwards did not test, may require periodic emersion. *P. canaliculata* showed necrosis when submerged for 11 h out of every 12 h in a tide simulator, suggesting that prolonged submergence is harmful to it (Schonbeck & Norton 1978). Strömgren (1983) argued that much of the growth in length of *Ascophyllum nodosum* in summer takes place while the plants are exposed because (1) warming of the plants by air stimulates elongation (see Sec. 3.3.3), (2) this effect is greater at higher irradiance, as found when the plants are not covered by water, and (3) photosynthesis is stimulated during brief periods of desiccation (just discussed). Some algae release their reproductive cells when rewetted and may be expected to reproduce poorly if continuously submerged. The effects of desiccation must be considered for the entire life cycle of a plant, not just for growth rates of adults.

9.3.2 Critical tide levels and zonation

Correlations between zone boundaries and critical tide levels (CTLs) may be expected because the stress of exposure to the atmosphere increases with the duration of exposure and seaweeds have differing abilities to recover from the stress. Colman (1933) made the first attempt at such correlations. His CTLs were based on average percentage exposures, which did not take into account duration and time of exposure. Moreover, Underwood (1978) subsequently showed that Colman's exposure curve was inaccurate. Doty (1946) defined CTLs differently, using maximum durations of exposure or submergence; this is the scheme that Swinbanks (1982) amplified (Fig. 9.7). When all the four possible orders of CTLs are considered, there is such a profusion of them that the probability of one of them coinciding with a zone boundary is high. On the other hand, several factors complicate possible correlations, most notably wave action and the ability of organisms to become acclimated during periods of subcritical conditions. Further, the critical period may apply to reproduction, settlement, or germling survival rather than to adults, so that one cannot simply take a correlation as evidence for a relationship. Some element of chance is eliminated because organisms are cued to seasons. Late spring, for example, is a time of reproduction for some species and also a time of some annual extreme CTLs and the first warm, sunny weather of the year.

Seasons, tides, and the life cycles of many organisms are governed by the same astronomical cycles. Sheltered habitats and even gently shelving shores in wave-exposed areas may be exposed close to predicted times (Fig. 9.8a). However, exposure times on steep-sided, wave-beaten promontories must be measured, and indeed in such places the concept of CTLs becomes meaningless owing to the unpredictability of wave height from day to day. Druehl & Green (1982) found that upper limits of seaweeds correlated most closely with accumulated time submersed, whereas lower limits correlated best with duration of the longest single exposure. They did not directly test CTLs as causative factors but nevertheless argued in favor of physical factors controlling species limits in the intertidal.

An appropriate direct test, suggested by Swinbanks (1982), is a series of experiments testing growth, survival, and reproduction just above and just below a CTL that is likely to be important. So far there are no such data for seaweeds. A correlation that Swinbanks pointed out as a good candidate for testing is the upper limit of *Pelvetia canaliculata*. This coincides with the lowest spring higher high water (see Fig. 9.7), where maximum exposure in May–June jumps from approximately 12 to 24 days (data of Schonbeck & Norton 1978 reanalyzed by Swinbanks 1982). In early spring *P. can-*

aliculata germlings develop higher on the shore than the normal population limit and are pruned back in summer.

The physical factors and their physiological effects are only part of the explanation of zonation patterns. There are also various biological controls that in some cases determine zone limits: grazing, competition, and other interspecies and intraspecies interactions.

9.4 Grazing

9.4.1 Seaweed–herbivore interactions

Every community that has vegetation also has herbivores. While some animals have means to maximize their foraging abilities, some plants have characteristics that minimize the impact of grazing. These characteristics include rapid growth followed by reproduction, production of deterrent chemicals, and production of calcareous or otherwise tough cell walls. Furthermore, some growth types are less susceptible to grazing than others. Expected herbivore damage has three components: (1) the probability that an individual plant will be encountered by a herbivore; (2) the probability of the plant being eaten, at least partially, if it is encountered; and (3) the cost, in terms of fitness, of the grazing damage (Lubchenco & Gaines 1981). Minimum damage will occur to plants that are out of contact with grazers, unpalatable, or able to sustain damage without significant cost. These three aspects will be examined in turn.

The likelihood of a particular plant being encountered is greater if the plant is large (for visual herbivores). The longer the plant lives, the more likely it is to be encountered. Lubchenco & Cubit (1980) and Slocum (1980) have advanced the hypothesis that heteromorphic algal life histories may allow a response to grazing. Crustose or boring stages (or crustose parts of heterotrichous plants) are more resistant to attack by grazers, being either out of reach of the grazers or not readily removed by them. (These stages are also more resistant to removal by abiotic agents such as logs and ice. For instance, the crustose base of *Corallina* is important in its survival in physically disturbed environments–Littler & Kauker 1984.) Shell-boring algae, such as gomontia and conchocelis stages, are especially well protected against grazing. Upright stages, in contrast, may be totally removed through damage to the relatively narrow stipes. The disadvantage to the crustose or boring stage is the increased susceptibility to being overgrown, but this problem may be mitigated by grazers removing the epiphytes. Lubchenco & Cubit (1980) worked with several winter ephemerals (*Ulothrix, Urospora, Petalonia, Scytosiphon, Bangia*, and *Porphyra*) in areas where grazing intensity varied seasonally. When grazers were removed experimentally, upright forms of the algae were found at times of the year when normally only the small phase would occur. Slocum (1980) found that the relative amounts of crust and blade of *Gigartina papillata* depended on grazing intensity.

When variations in grazing intensity are predictable (e.g., seasonal), the algae might respond to cues in the physical environment, such as temperature or daylength. Dethier (1982), however, showed that seasonal changes in abundance may be largely or entirely due to herbivore abundance and feeding rates. If variations are unpredictable, population survival will be highest if there is continuous production of both morphs, although only one would survive at a time (Lubchenco & Cubit 1980). "Bet-hedging" (as Slocum calls it) maximizes the total progeny of each such generation. Clearly, some conflicts are yet to be resolved, since the reproduction of some species of *Porphyra* and *Scytosiphon* is known to be cued by light or temperature and one would not expect to find erect forms developing out of season. In some species the upright form is produced directly, not as a result of reproduction, while in other cases the upright form has a perennating base or holdfast that is capable of regeneration if older shoots are lost due to grazing (or other factors such as ice scouring).

The chances of plant–grazer encounter are also affected by characteristics of the herbivores, including mobility, sensitivity, and abundance, plus characters of the environment, which may stimulate or suppress grazer activity. Many herbivores that have been tested have shown certain preferences among food species and, in some cases the ability to detect the preferred species. Sea urchins, for example, will move upstream in a current of water that has passed over a kelp plant, but their feeding behavior is influenced by both preference and availability (Vadas 1977). Tests of the abilities of algal compounds to attract herbivores have usually involved offering crude algal extract, or simply passing water over a food alga, but Carefoot (1982) tested the response of sea hares to specific algal compounds. Starch, glutamic acid, and aspartic acid, which might leach from algae or be released through damage, attracted sea hares and induced them to feed on agar wafers.

Feeding preference studies involve giving an animal a choice between two (sometimes more) algal species and measuring the amounts of each eaten. There are several pitfalls to this kind of study. First, the age of the alga may affect its palatability. *Littorina* readily eat *Chondrus* sporelings less than a few weeks old but have low preference for older plants. The vulnerability of the youngest stages is obviously very important in setting the distribution pattern of a species (Sousa et al. 1981, Cheney 1982). Germlings of *Fucus vesiculosus*, which are more susceptible to periwinkle grazing than are older plants, escape predation when they are in crevices (Lubchenco 1983). Second, there may be intraspecific resource partitioning such that preferences differ among individuals, for example of different ages. Finally, the amount of alga eaten may not give the best indication of the importance of the species to the grazer. *Littorina littorea* in New England preferred *Enteromorpha* in terms of biomass consumed but they grew best on a mixed diet of *Enteromorpha, Fucus*, and *Chondrus* (Cheney 1982).

Figure 9.16. Algal functional groups as related to molluscan grazer activity. (a) The functional groups; grazing difficulty refers only to structural toughness of the algae and does not take into account the additional difficulty of grazing algal groups 3, 4, and 6 due to their size refuge. (b) Relative importance of each algal group in the diets of 106 species of herbivorous mollusk. (From Steneck & Watling 1982, with permission of Springer-Verlag)

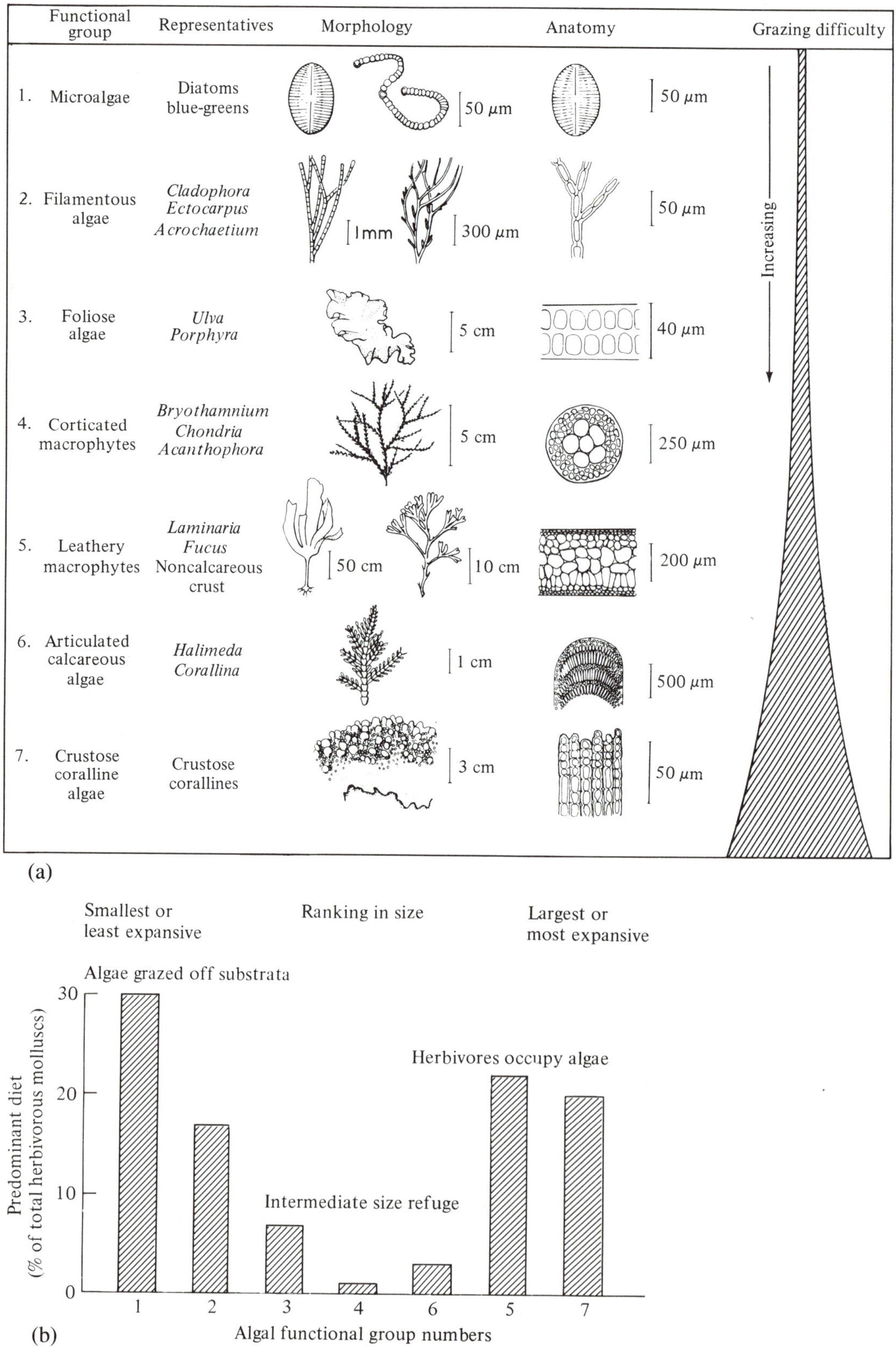

Functional group	Representatives	Morphology	Anatomy	Grazing difficulty
1. Microalgae	Diatoms blue-greens	50 μm	50 μm	Increasing
2. Filamentous algae	*Cladophora* *Ectocarpus* *Acrochaetium*	1mm; 300 μm	50 μm	
3. Foliose algae	*Ulva* *Porphyra*	5 cm	40 μm	
4. Corticated macrophytes	*Bryothamnium* *Chondria* *Acanthophora*	5 cm	250 μm	
5. Leathery macrophytes	*Laminaria* *Fucus* Noncalcareous crust	50 cm; 10 cm	200 μm	
6. Articulated calcareous algae	*Halimeda* *Corallina*	1 cm	500 μm	
7. Crustose coralline algae	Crustose corallines	3 cm	50 μm	

(a)

Smallest or least expansive — Ranking in size — Largest or most expansive

Algae grazed off substrata

Herbivores occupy algae

Intermediate size refuge

Predominant diet (% of total herbivorous molluscs)

0 10 20 30

1 2 3 4 6 5 7

(b) Algal functional group numbers

The probability of the grazer feeding on the seaweed once the encounter has taken place depends on the form and palatability of the seaweed. Steneck & Watling (1982) have assessed susceptibility of seaweeds to molluscan grazers (not including opisthobranchs) by arranging both into functional groups. Algal groups (Fig. 9.16a) were based on morphology and toughness. Some algae, of course, pass through two or more groups as they grow, while species with heteromorphic life histories occupy two groups. Grazers were grouped according to the type of radula and grazing action they have: "brooms" (e.g., keyhole limpets), "rakes" (e.g., *Littorina littorea*), "shovels" (e.g., true limpets), and "multipurposed tools" (e.g., chitons). Algal groups 1 and 2 are readily scraped off the substrata by grazers with broomlike or rakelike radulae, while the largest or most expansive algae (groups 5 and 7) are preferred by the other two groups of herbivores, which can gouge out tissue (Fig. 9.16b). Intermediate-size algae, of moderate toughness (groups 3, 4, and 6), appear to have a size refuge from these molluscan grazers, being too big for the sweepers and too small for the gougers. (Note that Steneck & Watling did not include sea urchins in their study.) Specialization for single food sources by small, slow-moving grazers such as these mollusks is usually found in those groups able to gouge the larger algae, which are, from the small grazer's vantage, abundant and long-lived food supplies. Steneck & Watling's study did not include the sea hares, which have a different method of feeding. *Elysia hedgpethi* is small and slow moving, yet selective in choosing only siphonous green algae such as *Codium* and *Bryopsis* (Greene 1970).

Chemical deterrents are produced by some algae, especially among the larger tropical Rhodophyceae and Chlorophyceae, although some of these are taken up and put to use by grazers. The best-known temperate example is the high sulfuric acid concentration in cells of some species of the brown alga *Desmarestia*: some species have a vacuolar pH less than 1 (McClintock et al. 1982), which seems to deter sea urchins unless there is no other food source available. Udoteal is a lipid in *Udotea flabellum* (Caulerpales) that deters herbivorous fish from feeding on the plant. The compound is a diterpenoid (cf. the phytol tail of chlorophyll *a*) (Fig. 9.17) (Paul et al. 1982). Sun et al. (1983) found a brominated phenolic playing the same role in the related genus *Avrainvillea*. Sea urchins have been found to have a slight avoidance of *Agarum* in preference trials, and a distinct avoidance of the red alga *Opuntiella*, but the compounds involved have not been isolated. On the other hand, sea hares, *Aplysia* spp., are not repelled by the halogenated compounds of the red algae they eat. The compounds are stored in the animal's cells and serve to deter predators (Fenical 1975). Indeed, if the larval stages of *Aplysia californica* fail to settle on *Laurencia pacifica*, they cannot undergo metamorphosis and they die (Kriegstein et al. 1974). *Oxynoe panamensis*, another opisthobranch mollusk, can also retain toxic substances, in this case from *Caulerpa* spp. (Doty & Aguilar-Santos 1970, Lewin 1970).

The effect of grazing on the alga depends on the amount and kind of tissue lost and on the timing, particularly with respect to reproduction. One aspect that has received very little attention is the survival of algal spores during digestion by herbivores. Santelices et al. (1983) showed that opportunistic species survive much better than do late successional species. Great damage to *Macrocystis* can be done by urchins, which chew through the bases of the stipes, eating little of the plant but causing the loss of entire fronds. At the other extreme, crustose coralline algae are susceptible to grazing by limpets but do not suffer much harm from it. The weakly calcified epithallial layer is removed, but it is replaced by the meristem beneath, which survives. Indeed, in the absence of limpet grazing the epithallial layer becomes too thick and overgrown by epiphytes, and the plant dies (Fig. 9.18) (Adey 1973, Steneck 1982). Erect portions of articulated corallines cannot be climbed by limpets and are relatively free of predators (Fig. 9.16b).

Although there is some new information about the interactions between seaweeds and herbivores, often from studies of either the algae or the herbivores, a better

Figure 9.17. Structure of the fish-deterrent udoteal. (Reprinted with permission from Paul et al. 1982, Udoteal, a linear diterpenoid feeding deterrent from the tropical green alga *Udotea flabellum*, *Phytochemistry*, vol. 21, pp. 468–469, © 1982 Pergamon Press Ltd.)

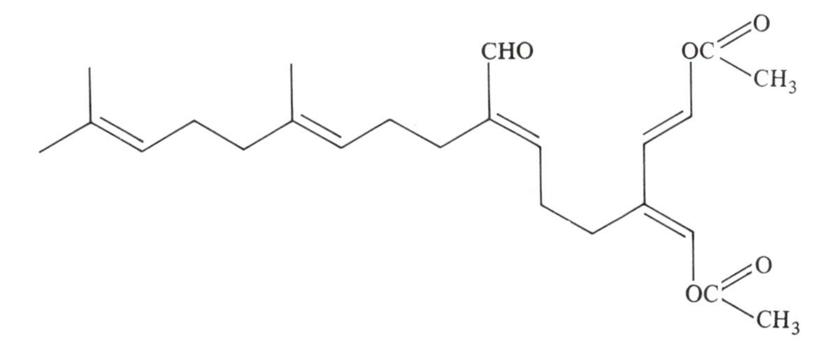

Figure 9.18. Structure of *Clathromorphum circumscriptum* in cross-section, showing (a) field specimen that has had moderate limpet grazing; (b) maintenance of the meristem even under heavy limpet grazing; and (c) death of cells due to overgrowth by diatoms in the absence of grazing. (From "A limpet-coralline alga association: adaptation and defenses between a selective herbivore and its prey," by R.S. Steneck, *Ecology*, 1982, *63*, 507–522. Copyright © 1982 The Ecological Society of America. Reprinted by permission)

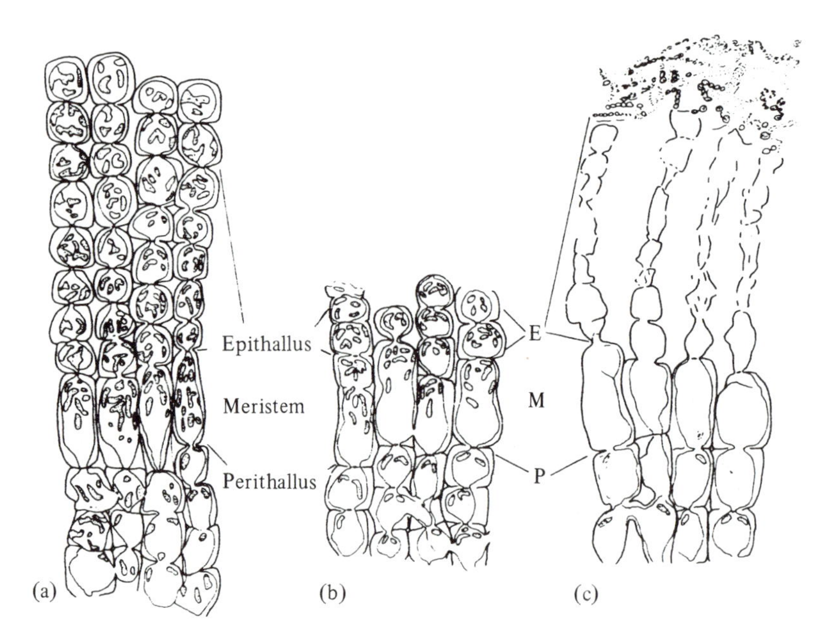

understanding of the complex interactions between seaweeds, grazers, and the predators of the herbivores requires much more work at the community level. Lubchenco & Gaines (1981) gave a good summary of the topics needing attention.

9.4.2 Impact of grazing

The impact of herbivores on seaweed populations can be devastating, as in the destruction of kelp beds by urchins in California and Nova Scotia. In many communities, however, vegetation and herbivores live more or less in balance. Tropical reef algae are kept to a turf partly by grazing but the turf form is also better able to withstand desiccation during low tides than are large plants (Hay 1981). Limpets and snails are frequently cited as preventing the upward extension of ephemeral algae in the high intertidal (Lewis 1964). When these gastropods are removed, blooms of ephemeral species such as *Enteromorpha* and *Porphyra* often appear. Frequently overlooked are arthropod grazers, such as crabs and the larvae of beach flies (Robles & Cubit 1981).

The severe attacks of urchins on kelp beds have received considerable attention because of the commercial value of the plants and some associated animals, such as abalone and lobster. The purple urchins, *Strongylocentrotus purpuratus*, and the large red urchin, *S. franciscanus*, have been responsible for the devastation of large areas of kelp forest in southern California (Leighton et al. 1966). The reason for the tremendous population explosion of these urchins is often attributed to the removal, by human harvesting, of the top predator, the sea otter. However, the last sea otters were harvested about 150 years ago, whereas the urchin explosion is a recent phenomenon. What triggered the increase in urchin numbers, according to Tegner (1980), was the harvesting of the various species of abalone, which began on a large scale in the 1930s. Abalones do not eat urchins but are competitive herbivores. In stable ecosystems both abalones and urchins collect drift seaweed and rarely forage on attached plants. Removal of the abalones resulted in a vastly increased food supply for urchins; as a result urchin numbers exceeded the level at which natural predators (fish, starfish, spiny lobster) controlled them. When the drift kelp supply was outstripped, the urchins moved onto attached plants, and the populations did not crash. Sportfishing of the sheephead, an urchin predator, has also contributed to the problem, as has the ability of urchins to use dissolved amino acids from sewage. In Nova Scotia the cause of the urchin population explosion and concomitant decrease in kelp beds has been attributed to overfishing of lobster, an urchin predator. There are other urchin predators, however, and not enough is known about their impact, nor about interactions of juvenile lobsters and urchins, to determine the cause of the increase in urchins (Pringle et al. 1982).

In the Strait of Georgia, British Columbia, the green urchin *Strongylocentrotus droebachiensis* is at its southern limit and frequently fails to recruit because of too high water temperatures. Foreman (1977) found that populations of this species undergo periodic, environmentally controlled outbreaks, which are responsible for localized perturbations of the seaweed communities. Such short-term grazing, in which urchins did not remain in the area, resulted in a temporary reduction of both biomass and species diversity (Fig. 9.19). Mean number of species per quadrat, mean dry wt per m^2, total dry wt, and species diversity fell, respectively, by 50%, 65%, 60%, and 60%. The urchins removed the foliose macrophytes first, then filamentous forms, and finally, to a small extent, crustose algae and corallines. The period of time required for the community parameters to be restored to their preurchin levels (not necessarily the exact preurchin flora) was 4 to 6 years in the mid-subtidal regions dominated by *Agarum* or *Laminaria*, but a little less, 3 to 5 years, for shallower subtidal communities.

Between the extremes of total destruction of the vegetation and light grazing in which individual plants survive, there are effects of grazers on the communities in which vegetation patterns are modified. Connell's (1978) intermediate-disturbance hypothesis predicts that local species diversity is greatest with moderate disturbance. One example of this effect is in the diversity of tropical reef algae in territories of the damselfish (Hixon & Brostoff 1983, Sammarco 1983). Beyond the territories, where fish grazing was severe, and in fish exclusion cages after natural succession had taken place, algal diversity was lower. In tidepools and the subtidal zone on the coast of Washington state, the presence of sea urchins maintains a population of small, ephemeral algae. When grazers were removed experimentally, a succession of algae ended in dominance of the pools by the kelp *Hedophyllum sessile* and of the subtidal rocks by *Laminaria* spp. (Paine & Vadas 1969, Duggins 1980). In the outer coast subtidal of British Columbia and Alaska, sea urchins limit the downward extent of kelp growth (except for *Agarum*, to which, as previously noted, urchins have an aversion). There is a seasonal shift in the minimum depth of the urchin zone owing to the animal's tolerance to wave action (Pace 1975). In winter the animals move deeper and kelp plants are able to establish temporarily below the kelp zone before the urchins move up again in late spring. In Alaska, this picture is complicated by the presence of the sea otter, which preys on the urchins in the upper part of their zone, thus allowing algal growth to extend deeper than in areas from which the otter is absent (Dayton 1975, Duggins 1980).

In the intertidal zone the major grazers are limpets and littorinids. Among the examples of zone control attributed to these grazers is the upper limit of a red *Laurencia–Gigartina* belt in parts of Britain by midshore *Patella* populations (Lewis 1964). Remarkably sparse grazers are capable of preventing the establishment of foliose algae (Underwood 1980). Removal of the grazers from plots in New South Wales, Australia, allowed foliose algae to establish, but many did not survive as

Figure 9.19. Recovery of an urchin-grazed community in British Columbia, after the urchins had moved on. Grazing took place in 1973 and greatly reduced species diversity (H′) in the 3- to 9-m depth range. By 1974 H′ was nearly restored to its 1972 (pregrazing) value, except for a patch in mid-depth dominated by *Nereocystis*. By 1975 recovery was virtually complete (but the species present were not exactly the same then as before grazing). (From Foreman 1977, with permission of *Helgoländer Meeresuntersuchungen*)

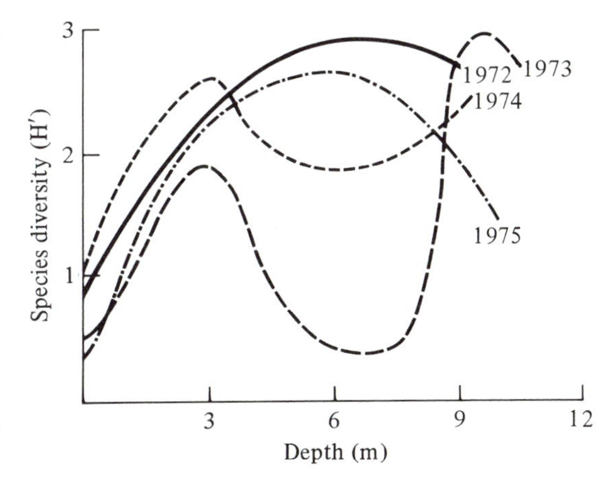

adults owing to physical conditions of the intertidal (Underwood 1980). In New England, intertidal rock pools tend to be dominated by either *Chondrus crispus* or *Enteromorpha*. *Chondrus* is successful when high densities of *Littorina* keep the rapidly growing *Enteromorpha* in check. Removal of the snails from a *Chondrus*-dominated pool resulted in a takeover by *Enteromorpha*. The snails were controlled in turn by primary and secondary carnivores: shore crabs and seagulls (Lubchenco 1978). Limpets and snails have a major role in controlling the seasonal abundance of *Collinsiella*, *Rhodomela*, and diatoms in Washington tide pools (Dethier 1982). Competition between *Ulva* and *Chondrus* in an aquaculture system is discussed in the next section. From several examples presented, it is apparent that grazing is not always a negative phenomenon for seaweeds.

9.4.3 Grazing and geographic distribution

Seaweed distributions are determined by several physical and biological factors. The roles of temperature and salinity have been discussed in Section 3.4 and 4.4. Mutualism may limit the distribution of species that depend on a particular host or basiphyte, such as *Polysiphonia lanosa* on *Ascophyllum nodosum*. Dispersal is important, particularly in regard to dispersal barriers and longevity of propagules. Finally, competition and herbivory play major, often interacting, roles, and are themselves influenced by physical conditions.

A striking pattern in algal distribution is the preponderance of smaller and more calcified plants in the tropics compared with temperate waters. Suggested reasons for this pattern include more severe desiccation, warmer water temperatures, lower nutrient availability, and changes in herbivory. Gaines & Lubchenco (1982) argued that of these, only the changes in herbivory can account for the latitudinal change in seaweed form. The pertinent characteristics of the tropical herbivore fauna include high diversity and biomass of urchins and herbivorous fishes, year-round foraging, and the well-developed visual capabilities and mobility of the fishes.

Interactions among grazing, competition, and physical factors are illustrated in the following examples. In the New England sheltered intertidal zone, *Chondrus crispus* dominates the sublittoral fringe, outcompeting fucoids. Further north, in the Gulf of St. Lawrence, ice scouring alters the interaction in favor of fucoids, and in Scotland heavy limpet grazing has the same result (Lubchenco 1980). Wave action and predation affect the competition between *Chondrus* and competitively dominant mussels. In wave-exposed areas, starfish are absent and mussels outcompete *Chondrus* (Lubchenco & Menge 1978).

9.5 Competition

Several field studies recently have indicated that the lower limits of intertidal zones can be set by interspecific competition (e.g., Foster 1982). Competition for space, an example of interference competition, takes place among algae and between algae and sessile animals. Carnivores can play a role in algae–animal competition by consuming sessile animals. Exploitation competition for nutrients and light takes place among algae. There is also intraspecific competition for space, nutrients, and light among individuals of one species.

9.5.1 Interference competition

Turf-forming algae are faster growing than crustose species and spread over the substratum, occupying large amounts of space. In two cases the competitive interactions of these algae with other species have been studied. In central Chile (Santelices et al. 1981), *Codium dimorphum* forms a thick, spongy crust and is able to overgrow and exclude most other lower intertidal algae. During spring and summer, with irradiance and temperatures increasing and the time of low tides shifting to the daytime, the *Codium* crust is bleached in the lower intertidal and killed in the mid-intertidal, creating new borders along which grazers can attack. During fall, winter, and early spring, *Codium* reinvades the mid-intertidal.

In southern California, a red algal turf comprising *Gigartina canaliculata, Laurencia pacifica,* and *Gastroclonium coulteri*, outcompetes the brown alga *Egregia laevigata*. The kelp recruits only from spores and only at certain times of year, whereas the red algae expand vegetatively at all seasons via prostrate axes, encroaching on any space that becomes available. A 100-cm^2 clearing in the middle of a *G. canaliculata* bed was completely filled in 2 years. The turf traps sediment, which fills the spaces between the axes and prevents the settlement of other algal spores on the rock (Sousa 1979, Sousa et al. 1981). *E. laevigata* may settle if clearings become available at the right time, but by the time it is adult and reproductive (it lives only 8–15 months) the turf has encroached all around and prevents the kelp from replacing itself in that area. Periodic space clearing is necessary also for the maintenance of *Postelsia palmaeformis* populations further north; in this case mussels are the space-competitive dominant (Dayton 1973). Larger algae do not form turfs but some, such as the fucoid *Halidrys dioica*, are nevertheless able to preempt space from other species by vegetatively spreading (Kastendiek 1982).

Competition for space also takes place between algae and sessile animals. According to Foster (1975), the outcome of this competition depends on three things: irradiance, presence or absence of grazers, and presence or absence of predators on sessile animals. Under very low irradiance (less than 5% surface irradiance) sessile animals will dominate whether there are predators or not, but under moderate irradiance the algae may predominate over sessile animals if predators reduce the numbers of animals. Foster speculated that the lower limits of sublittoral algal growth might be partly regulated by the inability of the algae to compete against sessile animals, rather than the inadequacy to the algae of the light flux per se. An interesting aside to this is that in California the ocean goldfish, or garibaldi, establishes nests of filamentous red algae on poorly lit vertical walls dominated by sessile animals. It does this by systematically removing the animals from the area, with the result that the algae can successfully compete for space (Foster 1972).

9.5.2 Exploitation competition

Intraspecific competition is manifest in effects of density on plant size and survival. Populations of newly recruited juveniles may be very dense, whereas by the time the plants have matured there are far fewer of them (Fig. 9.11); part of the decline is due to intraspecific competition (i.e., self-thinning). Plants in dense stands tend to be small, whereas those in more dispersed populations may be larger. On logarithmic plots of mean plant weight versus plant density, data points lie on or below a line of slope $-3/2$ (Fig. 9.20). This line seems to represent a boundary for weight (w) $\times$ density (d). The equation for the line is $w = Kd^{-3/2}$, where K is a constant equal to 4.3 for perennial herbs. The equation can be rewritten $\log w = 4.3 - 1.5 \log d$. Although data for seaweeds are scarce, Cousens & Hutchings (1983) have shown that these data also generally fit the $-3/2$ power law (Fig. 9.20). Self-thinning is often attributed to light attenuation by the plant canopy. However, there is no evidence as yet that thinning via competition drives the $-3/2$ power law (Cousens, personal communication).

Figure 9.20. Relationship between mean frond weight and frond density for various marine macrophytic algae growing in unialgal stands. Data for *Saccorhiza polyschides* and *Chordaria flagelliformis* are for time-courses of growth of stands. Both lines have a slope of −3/2. The upper line has the equation log $w = 4.3 - 1.5 \log d$. (From Cousens & Hutchings 1983, reprinted by permission from *Nature*, vol. 301, pp. 240–241, © 1983 Macmillan Journals Limited)

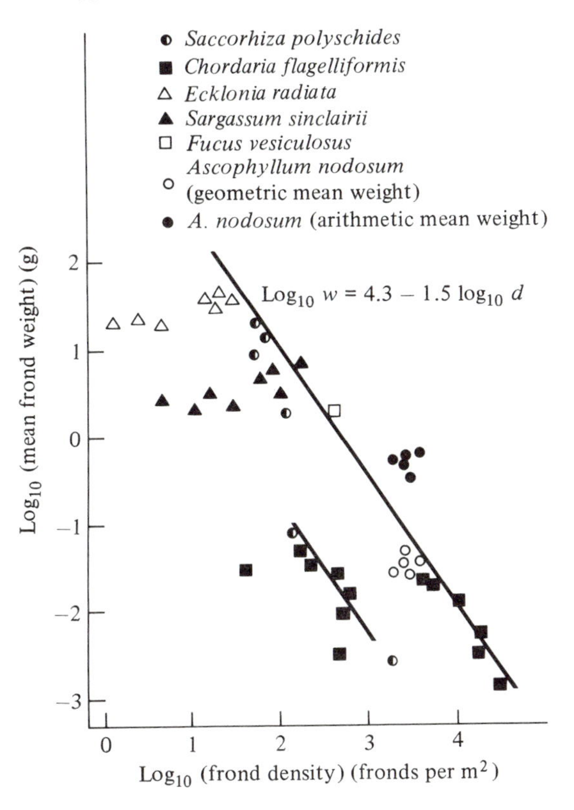

Interspecific competition between seaweeds has been studied in the laboratory only twice (by Russell & Fielding 1974 and by Enright 1979), although there have been several studies on phytoplankton (e.g., Kayser 1979, Maestrini & Bonin 1981, Tilman et al. 1982). Russell & Fielding devised a "triangular" method, in which three species were grown in pairs (AB, BC, AC). In this way any differences in yield of a species when in culture with the other two, respectively, are likely to be due to competition from the other species. Competitive ability was judged from the sizes of the differences. The limitation of the method is that if no difference is seen it may mean either that there was no competition with either of the other two species or that the species was equally affected by the other two. The three species used were *Ectocarpus siliculosus*, *Erythrotrichia carnea*, and *Ulothrix flacca*. Russell & Fielding found that *Ectocarpus* was involved in many competitive interactions. For example, as shown in Table 9.4, the yield of *Ulothrix* at 10 C was lower in the presence of *Ectocarpus* than in the presence of *Erythrotrichia*, except in three of the nine combinations (high irradiance with U:Co inoculum ratio 3:1; and low irradiance with U:Co::1:3 and 2:2). Under some conditions every species became noncompetitive: *Ulothrix* tended to be noncompetitive in dim light whatever the temperature, while *Ectocarpus* and *Erythrotrichia* in dim light became noncompetitive only at high temperatures. All species lost competitive ability in reduced salinity, and the loss was greater than anticipated from results with monocultures. *Ulothrix* at 15 C and approximately 38 $\mu E\ m^{-2}\ s^{-1}$ grew well against *Erythrotrichia* but was nearly extinguished by *Ectocarpus* (Table 9.4). Russell & Fielding concluded that "the presence of a competitor . . . seems to sharpen the sensitivity of a species to its environmental conditions and to reduce the amplitude of those conditions in which the species is vigorous."

Viewed against this study, results of experiments in which the competitive abilities of different species have been assessed from growth in monoculture may be misleading. For example, although Kain (1969) showed that *Saccorhiza polyschides* grew faster at 17 C than at 10 C, whereas *Laminaria hyperborea* and *L. digitata* grew equally at both temperatures (Fig. 3.8), this does not mean that, growing together, all would survive in accordance with the predictions from monoculture.

Enright (1979) studied competition between *Chondrus crispus* and one of its epiphytes, *Ulva lactuca*, in a flow-through seawater system. The two species were planted in several ratios as well as alone. The possible outcomes of the experiment, called a de Wit replacement series, are shown in Figure 9.21, where the two species are called A and B. If there is no interaction (Fig. 9.21a), the yield of each species is directly proportional to the amount sown. Interaction between species may increase the proportion of A (Fig. 9.21b) or B (Fig. 9.21c), or may increase or decrease the yield of both species (Fig. 9.21d, e). Enright tested the growth of the two species under several light and temperature combinations (Fig. 9.22). Under all conditions *Ulva* grew faster than *Chondrus* in monoculture, and grew better still (average 1.6 ×) in mixed culture. *Chondrus* in mixed culture showed only 60% of its monoculture growth rate. Dominance by *Ulva* was greater at higher irradiance and temperature. Enright concluded that in all of these experiments the eventual outcome would have been monocultures of *Ulva*, since her results did not show any evidence of competitive equilibrium. This conclusion seems to conflict with field observations on these two species and on other epiphyte–basiphyte associations, where basiphytes do in fact survive. The answer to the paradox may lie in the role of grazers in nature, as indicated in the example of the effect of snails on *Chondrus* and *Enteromorpha* in tide pools, given previously. However, the outcome of de Wit replacement series experiments can depend on the initial density of plants; furthermore, such ratio diagrams give no information about competitive interaction

Table 9.4. *Yields of* Ulothrix *in co-culture with its competitors* Ectocarpus *(Ec) and* Erythrotrichia *(Er)*

Temp. (C)	Approximate irradiance (μE m^{-2} s^{-1})	Competitor	Ratio: *Ulothrix* to Competitor 1:3		2:2		3:1	
10	38	Ec	12	13	26	24	37	36
		Er	25	26	31	34	35	36
	19	Ec	4	4	15	14	25	23
		Er	15	15	31	27	32	34
	6	Ec	4	6	9	8	10	10
		Er	5	5	9	9	14	15
15	38	Ec	0	0	0	0	8	8
		Er	17	20	25	23	27	23
	19	Ec	2	1	5	6	8	9
		Er	8	8	10	12	16	17
	6	Ec	2	2	7	8	10	10
		Er	4	5	8	9	8	10

Yield is given in relative units related to packed cell volume.
Salinity in all experiments was 34‰. Boxed-in parts of the table are those that show significant differences in yield of *Ulothrix* in culture against each of the two competitors.
Source: Russell & Fielding (1974), with permission of Blackwell Scientific Publications.

Figure 9.21. Possible outcomes of de Wit replacement series experiments. The initial ratio of species A to species B goes from 1 to 0 in each experiment. In the top row, the yield of both species in monoculture is identical, in the bottom row species B is lower yielding than species A. The vertical axis is any measure of species response (e.g., fresh weight). Cases (a) to (e) apply to both rows. (From Bannister 1976, with permission of Blackwell Scientific Publications)

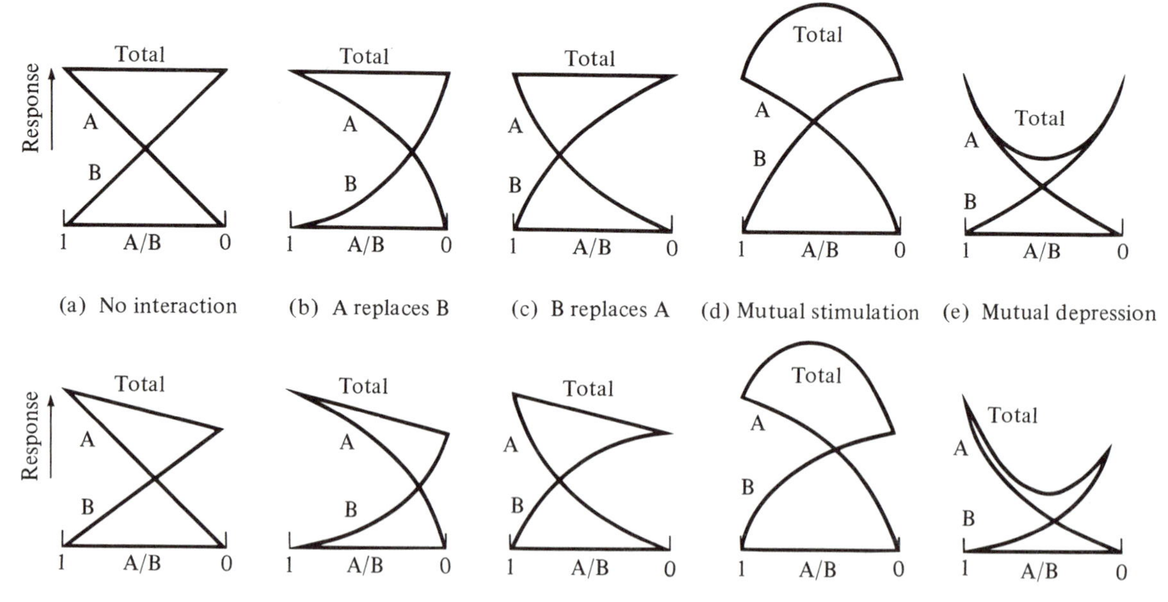

in the field, where initial densities are not experimentally constrained (Inouye & Schaffer 1981).

Continued experiments with mixed cultures will help to determine the nature of competitive interactions, including the extent to which different resources are important, and perhaps whether extracellular products are important in any positive or negative ways.

9.6 **Antibiosis**

Antibiosis is a step beyond competition: it is the deleterious effects of one organism on another via some chemical (antibiotic). Antibiosis by seaweeds is generally directed toward epiphytes or grazers. Since grazer deterrence was already dealt with in Section 9.4.1, this section will deal only with epiphyte deterrence (also

Figure 9.22. Replacement series diagrams (see Fig. 9.21) showing changes in fresh weight of *Chondrus crispus* (c) and *Ulva lactuca* (u) when grown at a range of initial densities and at several temperatures and irradiances. Means and standard deviations of measurements are shown on the lines for each species; summed yields are also given. (Biomass of *Ulva* in the 10 C, 100 μE m^{-2} s^{-1} experiment declined because the thalli sporulated and then disintegrated.) (From Enright 1979, with permission of Science Press)

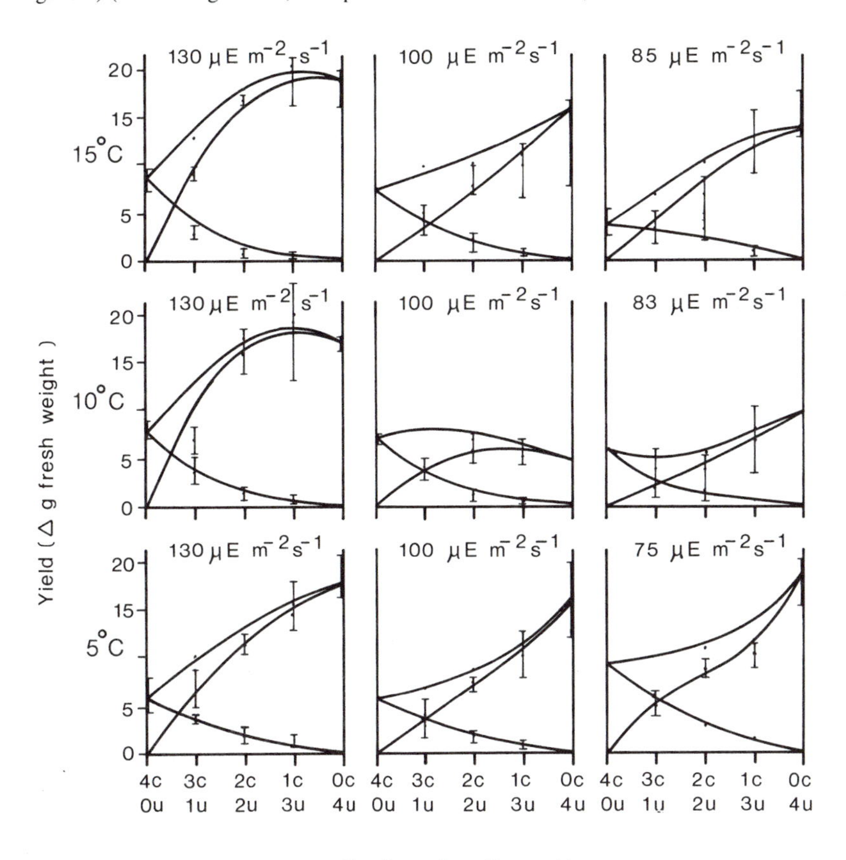

called antifouling). Microorganisms, larvae of sessile animals, and spores of algae settle everywhere, including on other algae (beautifully illustrated by Sieburth & Tootle 1981). If the larvae and spores that settle on algae are able to develop, they shade the "host" – more appropriately called the basiphyte – and interfere with water motion (hence nutrient exchange) and light. In some cases the epiphytes can become large enough or numerous enough to significantly increase drag on the basiphyte. Thus the basiphyte benefits from restricting or preventing the growth of epiphytes, or sloughing them off. Both means are used and are found particularly in growing tissues.

Some algae, including *Ulva*, *Enteromorpha*, and *Cladophora*, prevent epiphytism simply by very rapid growth and the changes in pH at the thallus surface caused by rapid metabolic rate (den Hartog 1972a). When growth of these species slows, epiphytes soon cover them. Several cases of epiphyte inhibition via secondary metabolites have now been documented, although the compounds involved have rarely been identified. Crustose germlings of *Chondrus crispus* can inhibit diatom growth (Khfaji & Boney 1979). Although a great many halogenated compounds are found in red algae and are suspected of providing protection against epiphytes and grazers, the possible roles are so far largely speculative. Antibiotic properties of seaweeds or seaweed extracts have been tested several times but frequently with application to pharmacology rather than to plant ecology. For example, Fenical et al. (1979) tested a variety of volatile halogenated red algal lipids for toxicity to various microorganisms (Table 9.5) and found them to be extremely toxic. Lanosol (Fig. 7.17c), which has been isolated from *Polysiphonia lanosa* and several other Florideophycidae, has been shown to be toxic to several (nonepiphytic) microalgae. Fenical (1975) pointed out that although these results suggest the involvement of lanosol as an antifouling agent, bromophenols are thought to occur largely as their sulfates, which are not toxic, and may even be stimulatory to microalgae. Possibly the sulfate is cleaved off after the compounds are released, but this has not been established. The effectiveness in

Table 9.5. *Antibiotic activities of halogenated 1-octen-3-ones from the red alga* Bonnemaisonia asparagoides *tested on three bacteria and a yeast*

Compounds	*Staphylococcus aureus*	*Escherichia coli*	*Candida albicans*	*Vibrio anguillarium*
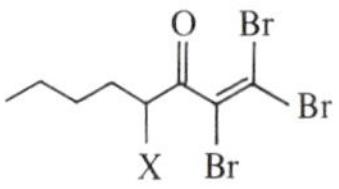 O, Br; Br; X, Br; X = H, Br (2:1)	(1) 12 (0.25) (2) 12	(1) N (1.0) (2) —	(1) 10 (0.25) (2) 2.2	(1) N (1.0) (2) —
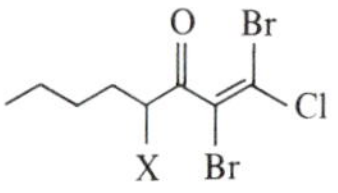 O, Br; Cl; X, Br; X = H, Br (1:3)	(1) 13 (0.25) (2) 1.5	(1) 2 (1.0) (2) —	(1) 15 (0.25) (2) 3.0	(1) N (1.0) (2) —
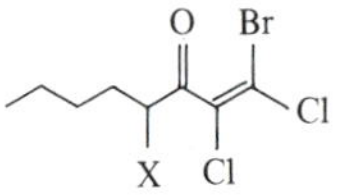 O, Br; Cl; X, Cl; X = H, Cl (1:3)	(1) 6 (0.25)	(1) N (1.0)	(1) 8 (0.25)	(1) 1 (1.0)
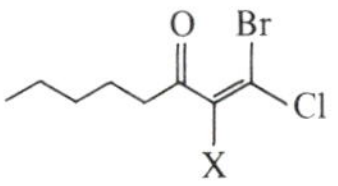 O, Br; Cl; X; X = Cl, Br (3:2)	(1) 2 (0.25)	(1) N (2.0)	(1) 8 (0.25)	(1) 0–1 (1.0)

(1) = width of inhibitory zone, in millimeters, around an agar disk impregnated with the compound (quantity of compound, in milligrams, given in parentheses). (2) = minimum inhibitory concentration of the compound in $\mu g\ mL^{-1}$. N = no activity; — = not tested. (For comparison, examples of minimum inhibitory concentrations of standard antibiotics are: penicillin G, 0.02 $\mu g\ mL^{-1}$ against *S. aureus* and more than 20 $\mu g\ mL^{-1}$ against *E. coli*; tetracycline, 1.17 $\mu g\ mL^{-1}$ against *E. coli*, 100 $\mu g\ mL^{-1}$ against *C. albicans*.) Of the microbes tested, *S. aureus* and *E. coli* are human pathogenic bacteria; *C. albicans* is a yeast; *V. anguillarium* is a marine bacterium.
Source: Fenical et al. (1979), with permission of Science Press.

nature of the lipids tested by Fenical et al. is unknown at present. Their role, if they are active, is still speculative, since the organisms against which the compounds were tested are unlikely to be a problem for the algae.

Brown algal cells characteristically contain numerous refractive bodies called physodes (Fig. 9.23; see also Fig. 2.9d), which appear to be a kind of vacuole, but may not be membrane-bound (Ragan 1976, Pellegrini 1980). These bodies originate in the plastids and contain tannins – phenolic and polyphenolic compounds – which are active as antifouling agents (Craigie & McLachlan 1964). Phenolics are toxic because they bind and inactivate proteins. The production of tannins by *Sargassum* species in the Sargasso Sea correlates with the number of physodes in the cells. Activity is concentrated in the growing tips of plants near the center of the Sargasso Sea, and only these tips are free of epiphytes (Conover & Sieburth 1964, Sieburth & Conover 1965). Barnacles (*Balanus balanoides*) and mussels (*Mytilus edulis*) are suppressed in *Ralfsia*-dominated tide pools (Conover & Sieburth 1965) but not in *Hildenbrandia*-dominated pools; again, phenolics are implicated. Phenolics could conceivably be the cause of inhibition of epiphytic animals on distal tissue of *Laminaria* (Al-ogily & Knight-Jones 1977), and there may also be a correlation in Rawlence's (1972) observation that physodes were absent from tissues of *Ascophyllum* near invading rhizoids of *Polysiphonia lanosa*, but these casual observations need experimental support.

The antibacterial activity of several seaweeds was studied by Hornsey & Hide (1976a, b), who found three activity distribution patterns in the thalli. Highest activity was not correlated with growing tissue. In *Chondrus*, antibiotic activity was higher in the youngest tissue, at the tips of the fronds, but in *Laminaria* the pattern was quite the reverse. Activity in *L. saccharina* and *L. digitata* increased toward the distal (older) part of the thallus, except that the very oldest tissue of *L. digitata* had reduced activity and was heavily epiphytized. The third distribution, exhibited by *Ulva*, was uniform. There were also seasonal changes in activities. The antibacterial substances were not identified. Unfortunately, in this kind of study the tissues are undoubtedly dying under experimental conditions, causing substances to leach from the cells; hence the results may not reflect the natural situation (Fenical 1975).

Figure 9.23. Physodes in the fucoid *Cystoseira stricta*. (a) Fresh section stained with caffeine, which precipitates the physodes as white globules. (b) Section of thallus seen in electron microscope; the cells at the top of the section are meristodermal, those at the bottom promeristematic cells. c = cuticle; ci = irridescent body; d = dictyosome (Golgi body); mi = mitochondrion; n = nucleus; p = plastid; ph = physode. (From Pellegrini 1980, with permission of The Company of Biologists)

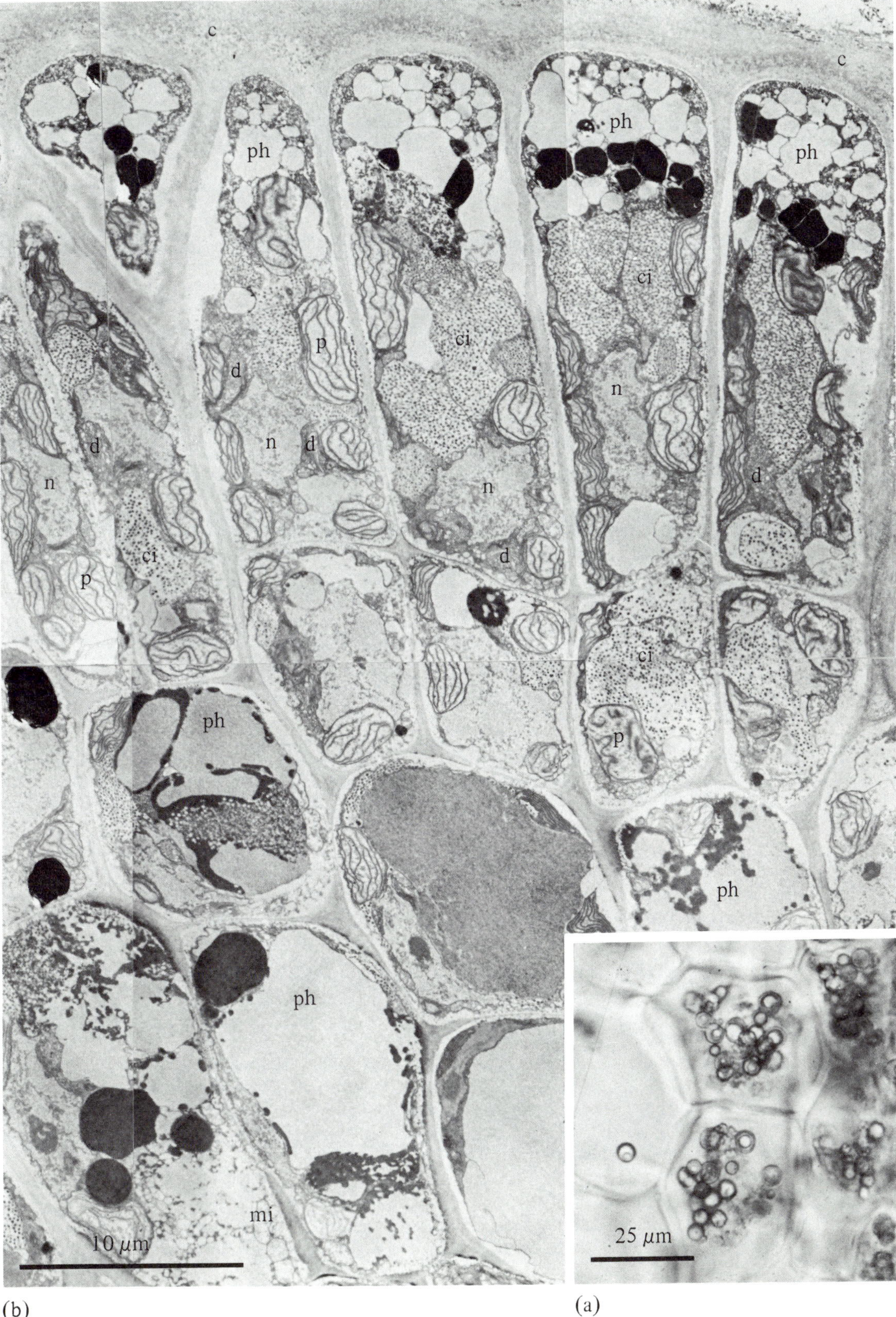

Figure 9.24. Sloughing of outer cell wall layers with removal of epiphytes. Scanning EM view of the surface of *Chondrus crispus* showing the cuticle *c* with bacteria *b* sloughing away, leaving a clean algal surface *a*. Scale: 10μm. (From Sieburth & Tootle 1981, with permission of *Journal of Phycology*)

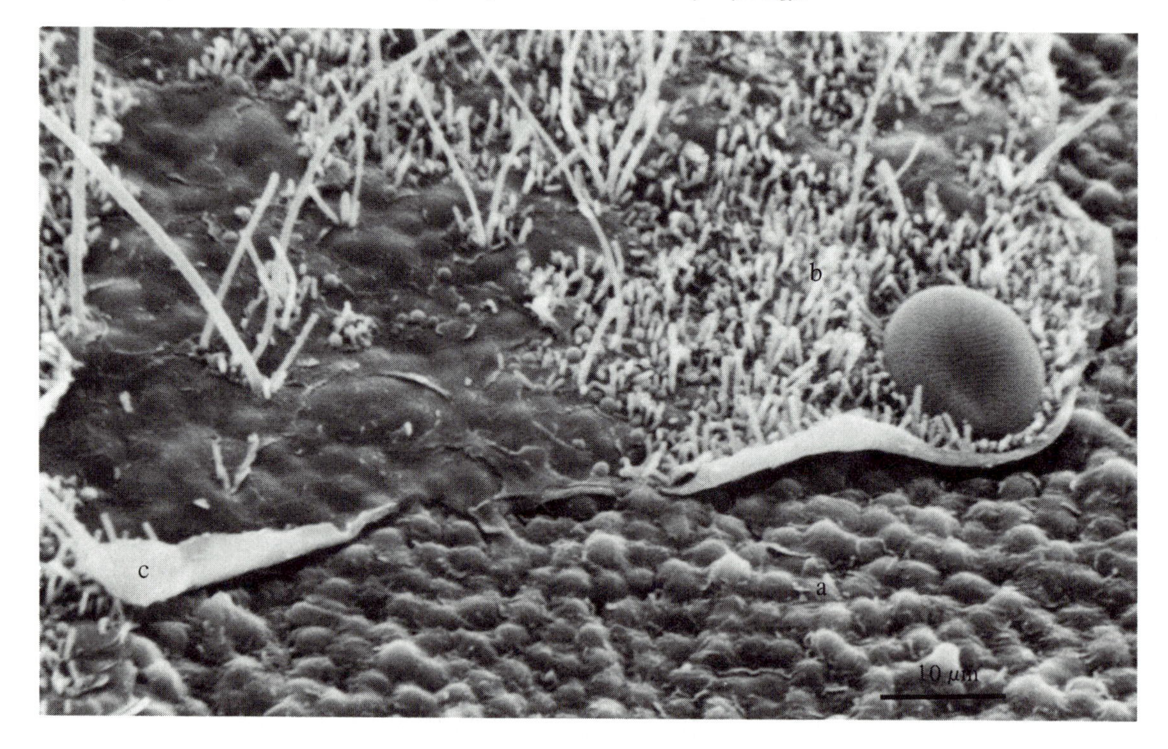

Several algae have recently been found to slough their outer layer, which has the effect of ridding the thallus of epiphytes (Fig. 9.24). *Enteromorpha intestinalis* continuously produces new wall layers and sloughs off the outer layers of glycoprotein (McArthur & Moss 1977). This species has an unusual wall structure, and its ability to remain free of epiphytes is probably essential in minimizing drag on the thallus and allowing the alga to colonize ship hulls, where hydrodynamic forces are great. In *Halidrys siliquosa* (Moss 1982) and *Himanthalia* (Russell & Veltkamp 1982), the old walls of the epidermal cells are shed as "skins" following production of a new wall underneath. The whole outer epidermal layer may be shed by *Ascophyllum nodosum* (Filion-Myklebust & Norton 1981), although Moss (1982) questions this. The epiphyte *Polysiphonia lanosa* is able to retain a hold on *Ascophyllum* by having rhizoids that penetrate below the sloughing layer.

9.7 Factor interactions

The environment of an organism comprises many factors, each almost constantly varying and some interacting with others. In this last section of environmental factors, a brief recapitulation of the examples of factor interactions is followed by some philosophical comments on experimental phycology. Factor interactions can be grouped into four categories: (1) multifaceted factors; (2) interaction between environmental variables; (3) interaction between environmental variables and biological factors; and (4) sequential effects.

Many of the environmental variables that have been discussed have many components that do not necessarily change together. Light quality and quantity both change with depth but the changes depend on turbidity and the nature of the particles. In submarine caves, light quantity diminishes with little change in quality. Natural light has the further important component of daylength. Salinity is another complex factor, of which the two chief components are the osmotic potential of the water and the ionic composition, in particular the concentrations of Ca^{2+} and HCO_3^-. The hydrodynamic aspects of water motion are critical to thallus survival and to spore settling, and water motion has important effects on the boundary layers over plant surfaces and thus on nutrient uptake and gas exchange. Each nutrient must be considered not simply in its absolute concentration but also in the amounts present in biologically available forms; concentrations of trace metals may create toxicity problems, particularly in polluted areas. Pollution as a factor may include not only the toxic effects of component chemicals but also an increase in turbidity, hence a reduction in irradiance. Emersion often involves desiccation, heating or chilling, removal of most nutrients (carbon can be an exception), and, frequently, changes in salinity of the water in the surface film on the plants and in the free space between cells.

Interactions among environmental variables are the rule rather than the exception. Bright light is often associated with increased heating, particularly of plants exposed at low tide. Temperature and salinity affect the density of seawater hence the mixing of nutrient-rich bottom water with nutrient-depleted surface water. Temperature also affects cellular pH and thus some enzyme activities. The carbonate equilibrium, and especially the concentration of free CO_2, is greatly affected by pH, salinity, and temperature, while the availability of NH_4^+ is pH-dependent because at high pH the ion escapes as free ammonia. Water motion can affect turbidity and siltation as well as nutrient availability. These are examples of one environmental variable affecting another; there are also examples of two environmental variables acting synergistically on plants. The mitigating effect of normal salinity on the effects of warm temperatures was summarized in temperature–salinity diagrams for survival limits. There are several cases of combined effects of temperature and photoperiod on development and reproduction of seaweeds (these have been brought up in Sec. 3.3.4 but will be treated more fully in Chap. 10). Once photosynthesis is light-saturated, temperature shows a large effect on photosynthetic rate.

Interactions between environmental variables and biological factors include both the way the biological parameters such as age, phenotype, and genotype affect a plant's response to an environmental variable and also the effects that organisms have on the environment. Moreover, the environment of a given plant includes other organisms, with which it interacts through intraspecific and interspecific competition, prey–predator relationships, and basiphyte–epiphyte relationships. The chief biological parameters that condition a given plant's response to its environment are age, reproductive condition, nutrient status (including stores of N, P, and C), and past history. By past history is meant the effects of past environment on plant development, which is the topic of Chapter 10. The season can also affect certain physiological responses aside from those involved in life history changes; these responses include acclimation of temperature optima and tolerance limits.

Responses of organisms may be environmentally determined, genetically determined, or partly both; the differences in rhizoid production by *Enteromorpha* populations (Sec. 4.3.4) and in morphological response of *Laminaria* to wave action (Sec. 5.3.3) are examples. Other organisms may greatly modify the environment of a given individual. Protection from strong irradiance and desiccation by canopy seaweeds is important to the survival of understory algae, including germlings of the larger species. Organisms shade each other (and sometimes themselves) and have large effects on nutrient concentrations and water flow. They also alter the environment by their excretions, including antibiotic compounds. Competition perhaps sharpens an organism's sensitivity to environmental variables. Grazing damages or may destroy plants, yet some species depend on grazers for their own survival – the importance of the removal of surface layers and epiphytes of *Clathromorphum* is one example; the dominance of *Chondrus* in areas where grazers keep down ephemeral species is another.

Finally, there are factor interactions through sequential effects. Nitrogen limitation may cause red algae to catabolize some of their phycobiliproteins, which will in turn reduce the light-harvesting ability. In general, any factor that alters the growth, form, reproductive, or physiological condition is apt to change the response of the plant to other factors both at the same time and in the future.

For the interpretation of experiments to be clear, usually one variable is tested at a time. In such experiments all other factors must be held constant, or at least equal in all treatments. Variations in extra factors confound the results. For example, Underwood (1980) criticized some field experiments designed to determine effects of grazer exclusion because the fences and cages used to keep out grazers also affected water motion over the rock surface and provided some shade. The statistical design of experiments is considerably more complicated when more than one variable is being assessed at a time; practical reviews in this area include Box et al. (1978), Green (1979), and Underwood (1981).

Laboratory studies provide much more controlled conditions than nature, but are limited in some important ways and contain some implicit assumptions, such as (1) high nutrient levels do not alter the plants' response to the factor under study; (2) the reaction of the plant to uniform conditions (including the factor under study) is not different from its response to the factor(s) under fluctuating conditions. To a certain extent these assumptions are valid. Culture media are very rich in nutrients to compensate for lack of water movement and exchange, but whether the substitution gives precisely the same results with all parameters is doubtful. (An example of interference by high nutrient levels is given in Sec. 10.4.1) Other culture conditions are also generally optimal, except for the variable under study, and the results may not elucidate behavior of plants in the field, which are subject to competition and often suboptimal conditions (Neushul 1981). Another important difference between culture and nature is that in culture, species are usually tested in isolation, away from interspecific competition and grazing. Furthermore, culture conditions are uniform (at least on a large scale), whereas in nature there are often large and unpredictable fluctuations in the environment (Fréchette & Legendre 1978, Turpin et al. 1981). Microscale heterogeneity in culture conditions should not be overlooked (Allen 1977, Norton & Fetter 1981). In the culture flask one cell may shade another, and cells form nutrient-depleted zones around them, creating a mosaic of nutrient concentrations through which cells pass.

The study of individual factors is only the beginning. Eventually, the isolated threads have to be woven together into models of the fabric of nature. This can be done in part by experimentally assessing factor combinations, in part through mathematical modeling. Marine zoologists have recently come to the stage in their studies at which multivariable analysis is possible (Newell 1979), and some of their broad conclusions are pertinent to discussion of the ecology of seaweeds.

Alderdice (1972) and Newell (1979) consider that an organism has a multidimensional "zone of tolerance," the boundaries of which are defined by its tolerance to all environmental variables. These boundaries depend not only on the species and genotype of the organism, but also on its size, age, stage of life history, and previous environmental experience; the boundaries change as these change. Within the overall zone of tolerance there are smaller multidimensional zones, which are defined by the local conditions under which the organism is operating; acclimation to other conditions, such as during seasonal changes, involves changes in the boundaries of these smaller zones. These zones can be visualized on paper as far as three axes, but computers are necessary for analysis along further axes. One may

Figure 9.25. Three-dimensional zones of photosynthetic ratio, PS/R, as a function of light, temperature, and salinity. (a) young and (b) old thalli of the red alga *Delesseria sanguinea* from the western Baltic Sea. Ratios calculated from O_2 exchange. The contour lines are drawn for PS/R ratios of 20 (inner), 15, and 10 (outer). (From Lehnberg 1978, with permission of Walter de Gruyter & Co.)

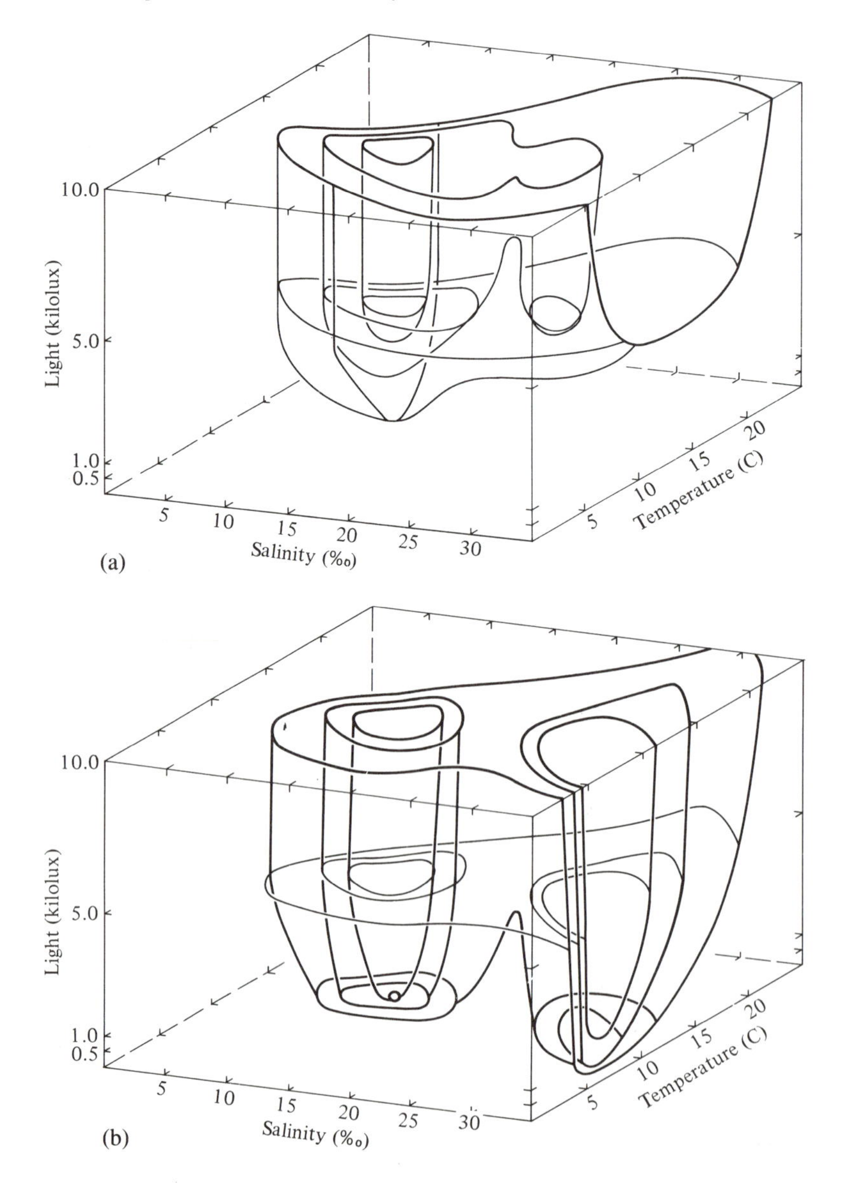

imagine, for example, constructing temperature–salinity (T–S) diagrams for a seaweed at various ages, or at various irradiances. If these T–S diagrams were stood up at the appropriate places along the third axis, the lines on the T–S diagrams could be regarded as contours, and the diagrams themselves as cross-sections through a three-dimensional space. Lehnberg (1978) has constructed such three-dimensional diagrams for the red seaweed *Delesseria sanguinea*. Two examples shown in Figure 9.25 combine salinity, temperature, and light, and show the responses of photosynthetic ratio at two different ages (compare with Fig. 4.7). In a computer one could begin to stack such diagrams along a fourth axis: age. Lehnberg found that there was increased tolerance to salinity toward optimal light and temperature, and that this tolerance also developed with the age of the specimen. There were two optimum salinities: one at about 15‰, the salinity of the culture conditions and the Baltic Sea, the other at about 30‰, the salinity of open coast seawater. However, the description of tolerances in terms of response surfaces is still only a refined description of the physical environment, and does not take account of competition.

9.8 **Synopsis**

Seashore communities, especially in the intertidal zone, form characteristic horizontal bands. Small-scale variations in the environment cause patchiness within these broad zonation patterns. Among the complicating factors are the slope, aspect, and texture of the surface, with concomitant differences in drainage and evaporation, sedimentation, and shade. Biological and physical schemes have been used to describe seashore zonation.

In the intertidal zone, duration of exposure to the atmosphere is a major gradient, and the effects of desiccation on seaweeds can be severe. However, high intertidal seaweeds are able to carry out some photosynthesis using CO_2 from the air, and are able to withstand considerable desiccation. Sometimes there may be correlations of zone limits with critical tide levels (CTLs), levels at which there are abrupt, large changes in duration of the longest emersion or submersion. Convincing evidence is still wanting for the hypothesis that there is a causal relationship between CTLs and zone limits. However, there is no doubt that in some cases emersion stresses, largely dehydration and extreme temperature, control the upper limit of intertidal populations.

Other factors also limit populations. Wave action is sometimes critical, and, in the subtidal region, light forms the major environmental gradient.

An environment has a carrying capacity, and populations are maintained generally at or below that capacity. Interspecific and intraspecific competition play major roles in reducing population size from the initial, often high settlement densities. Species with rapid growth and short life spans are said to be *r*-selected. Species with slower growth, longer life spans, and delayed reproduction are *K*-selected; other species have intermediate attributes. Interference competition takes place among algae and between algae and sessile animals for space; the outcome can be influenced by both herbivores and carnivores. Exploitation competition takes place among algae for light and nutrients. Some algae produce chemicals or slough their outer layers, which inhibit growth of epiphytes.

Complex relationships exist between herbivores and seaweeds. Damage to seaweeds by grazers depends on encounters between the two, how much is eaten or broken off, what parts are lost and when, especially as it affects reproduction and hence fitness of the individual. Some grazers are attracted to their prey by chemicals leaking out of the algae, but algae also produce toxic or acidic compounds that deter some grazers, or have tough cell walls as protection against grazer damage. Heteromorphic life histories may also be beneficial to seaweeds in grazer avoidance. In a few instances grazing is beneficial to seaweeds. Grazers may cause only modest damage to individuals; they may alter the community composition by selectively grazing some species; or, in rare cases, they may virtually destroy the vegetation. Predators play key but poorly understood roles in the balance between seaweeds and herbivores.

Competition, grazing, and physical factors occur together in the environment and interact. A major task of physiological ecologists will be to assemble data from experiments on isolated factors into multidimensional models of nature.

10 Morphogenesis

10.1 Introduction

A seaweed begins life as a single undifferentiated cell, with the potential to produce the whole organism through the expression of its genetic information. The genotype interacts with the environment to produce the phenotype. The environment of a cell consists of the physical and chemical influences of the other cells in the plant, plus the environment of the plant itself. The environmental history of a plant, because it affects growth and form, becomes in a sense recorded in the plant body. Thus individual plants of the same genotype, planted in exactly the same place on the same day, but in different years, will grow into phenotypically distinct individuals. As an example, a period of low light may produce a plant with long internodes; stronger light may later cause internodes to be shorter, but branches will remain sparse at the base of the plant.

Successful growth and reproduction of plants are possible over a very wide range of form, size, and relative proportions of the parts (Evans 1972). Under exceptional circumstances, such as in moderate tidal rapids, seaweeds may become unusually large, but whether they have an intrinsic size limit is not known. Size is normally determined by environmental constraints. However, among the Desmarestiales are examples, such as *Himantothallus grandifolius*, with closed growth. In these the number and position of blades are determined when the plant is only a few millimeters long, even though this species eventually reaches 10 m in length (Moe & Silva 1981). The successful competitors in a population will not necessarily be the largest, nor even the fastest growing. The switch from growth (i.e., vegetative growth) to reproduction (which in most seaweeds involves very little growth) depends on environmental factors, such as temperature and light. Kelp gametophytes, for example, may reproduce when only a few cells in size, or they may grow vegetatively almost indefinitely, depending on light quality and quantity.

There are four basic types of growth. In diffuse growth, all (or many) of the cells undergo division, as in *Ulva* and *Porphyra*. Localized growth restricts cell division to one of three regions of the plant. Apical division is the most common type, occurring throughout the Florideophycidae and in such classic examples as *Fucus* and *Dictyota* (although there is now some evidence that the apical cell of *Fucus* is not very active – Moss 1974a). Basal growth is very rare among the algae but it is found in *Urospora*. Division of cells in a localized region in the middle of the plant is called intercalary, and it is found in the kelps and in *Desmarestia* among other seaweeds (Bold & Wynne 1978). Each of these types of growth may involve cell divisions in one plane (elongation of filaments), a second plane (formation of sheets; branching), or a third plane (thickening).

This chapter treats cell growth and differentiation, then thallus morphogenesis and reproduction, finally a number of related phenomena such as wound healing, regeneration, and apical dominance. In part, this chapter recapitulates material from the foregoing chapters on environmental factors, but presents it in a different perspective.

10.2 Cell morphogenesis

10.2.1 Totipotency

The environment of a cell (or its nucleus) is critical in determining when particular genes will be expressed. Cells isolated from the parent individual may behave as zygotes or spores and form a whole new organism; the ability of a cell to do this is called totipotency. The process of cloning depends on totipotency. Single cells from simple thalli such as *Prasiola* (Bingham & Schiff 1979) readily regenerate the whole thallus, but cells from complex algae, such as cortical cells from *Laminaria* (Saga et al. 1978), can also regenerate the thallus under appropriate culture conditions. There are probably few cells in algal thalli that are irreversibly differentiated;

Figure 10.1. The multinet concept of cell wall growth showing (left to right) reorientation of microfibrils at successive stages of wall extension. (With permission from P.A. Roelofsen 1965, *Advances in Botanical Research*, vol. 2, © Academic Press Inc. [London] Ltd.)

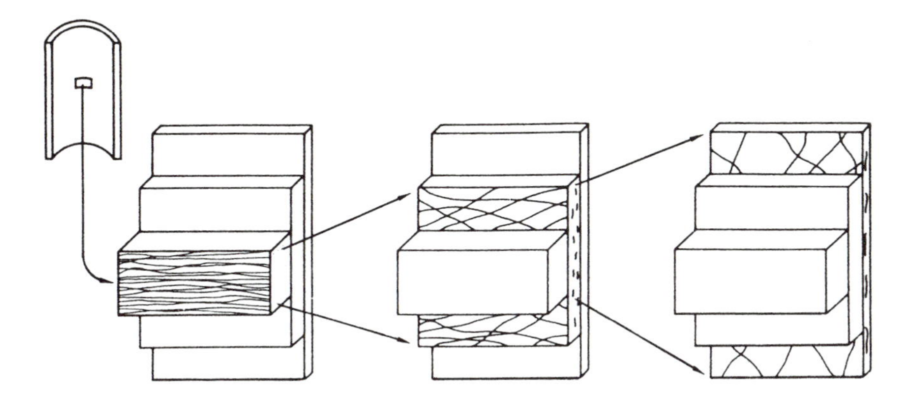

anucleate sieve elements of *Macrocystis* would obviously be an exception. Yet, cells released from constraints imposed by neighboring cells do not always grow into a plant of the same generation. Examples are seen in the phenomenon of apospory: diploid sporophytes of three kelp species raised in stagnant culture for 3 to 4 months bleached, leaving only scattered epidermal cells alive. A few of these epidermal cells germinated on the thallus, giving rise to gametophytes, which were shown to be diploid (Nakahara & Nakamura 1973). Similar results have been obtained by isolating cells of moss sporophytes (Bold et al. 1980). These findings emphasize that morphological phase and ploidy level are not consistently correlated, but they raise the question, Why do these cells not stay in the sporophyte phase?

10.2.2 Cell growth

The growth of seaweed cells is most often constrained in one or more directions. In parenchymatous thalli the packing of cells frequently leads to more or less cuboidal shapes, whereas in filaments the cells are often cylindrical as a result of their own cell wall properties. Cells that are unconstrained are spherical.

Cylindrical cells in filaments grow in length and divide transversely. The cytoplasm in apical cells may be polarized (as in *Sphacelaria* – Katsaros et al. 1983). The partitioning of cytoplasm and, in multinucleate species the nuclei, is not always equal, especially in apical cell division (e.g., *Acrosiphonia* – Kornmann 1970). The reason for the growth in length rather than girth is, at least in some green algae and in higher plants, that the polysaccharide microfibrils in the primary wall are laid down preferentially transverse to the filament axis. One may imagine by analogy the way that hoops around a barrel prevent outward expansion. In green algal walls, the transverse fibrils are not actually in a hooplike arrangement but in gradual helices (Mackie & Preston 1974). As the cell grows these fibrils become stretched along the axis of the cell, hence slightly reoriented, while new (secondary) wall material is laid down next to the plasmalemma (Fig. 10.1). The secondary wall fibrils may again be transverse or, once cell expansion and differentiation have been completed, may be oriented along the axis. This so-called multinet growth hypothesis has been recently discussed by Preston (1979, 1982).

Although many green algae are thought to have such wall structure, others, including *Valonia* and *Chaetomorpha*, have multilamellar walls (Fig. 10.2). The cellulose fibrils in these walls are laid down in alternating helices, more or less at right angles to each other, one helix running almost transversely around the cell, the other much steeper (Fig. 10.3) (Mackie & Preston 1974). During cell growth the microfibrils in the steeper helices become almost parallel (Fig. 10.1), while the transverse fibrils tend to become more randomly oriented. Yet, the steeper helices themselves (as opposed to the fibrils in them) become flatter, owing to the cell twisting as it elongates. The multinet growth hypothesis (or passive reorientation hypothesis – Preston 1982) can also explain the reorientation in these crossed microfibril cases. Until 1972 the process just described was assumed to be common throughout the plant kingdom, but a different means of cell elongation is now known to occur in at least some red algal intercalary cells (Waaland et al. 1972) (discussed subsequently). The mechanism of wall loosening is not known in algae. The established dogma for higher plants, that auxin stimulates a proton flux, which in turn breaks the bonds between microfibrils, has recently been critiqued by Hanson & Trewavas (1982).

Cellulose and xylans form microfibrils but mannans form short rods embedded in amorphous mucilage (McCandless 1981). The relation between wall structure and cell shape in algae such as *Acetabularia* (sporophytes) and *Porphyra*, which have mannan as the structural component of their walls, has not yet been investigated. The biochemical alternation of generations is puzzling: Why do the sporophytes of some green algae produce mannans whereas their gametophytes produce

Figure 10.2. A hypothetical reconstruction of the wall of *Chaetomorpha melagonium* and *Cladophora rupestris*, showing several amorphous layers (PAL and CAL) alternating with microfibrillar layers (PML and CML). An outer protein-rich cuticular layer (PL), with a possible additional spongy layer (SL) in *Cladophora*, merges through a transition zone (TZ) into a carbohydrate-rich region (CL). The microfibrillar layers consist of sublamellae (SB) of parallel oriented microfibrils, with microfibrils of one layer crossing those of the next at somewhat less than 90°. (From Hanic & Craigie 1969, with permission of *Journal of Phycology*)

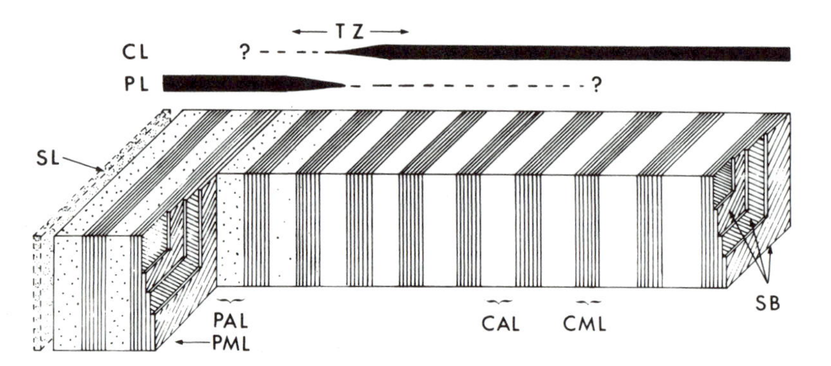

Figure 10.3. Schematic representation of the run of the microfibrils in the wall of a vesicle of *Valonia ventricosa*. The vesicle is considered to be spherical and the helical lines represent microfibril directions. (From Mackie & Preston 1974, with permission of Blackwell Scientific Publications)

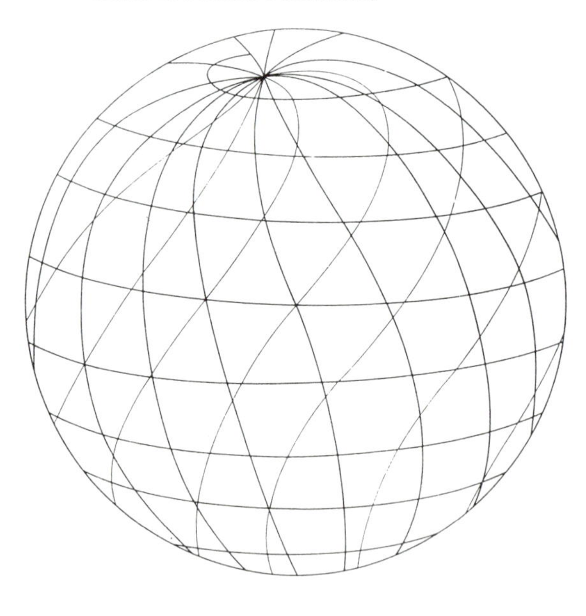

xylans or cellulose, and why do the n and 2n phases of *Chondrus* and *Gigartina* produce different carrageenans?

The location of cell growth can be followed by labeling existing cell wall polysaccharides with a fluorescent stain, Calcofluor White (ST or M2R) (a brightener at one time used in laundry detergents). The dye does not harm the cells if they are dipped in it for only 15 to 30 min. If cell growth occurs by extension of existing wall material, the dye will be uniformly diluted. If, on the other hand, cell growth occurs by localized synthesis of new wall, dark bands will appear on the cells when seen under UV light, because the new wall will not be stained. Intercalary cell extension in some Ceramiales, studied by Waaland & Waaland (1975), takes place through localized additions of wall material at each end of the cell (Fig. 10.4). The number and location of the bands are characteristic of a species. The type illustrated in Figure 10.4 is called a bipolar band growth. This is in contrast to green algal cells discussed above, in which extension of preexisting wall takes place. (Growth of the apical cells of the Ceramiales may be nonlocalized; this has not been studied.)

The importance of structural control by the cell wall is illustrated by a mutant, *lumpy*, of *Ulva mutabilis*, which grows as an aggregate of undifferentiated cells rather than first forming a filament and later a holdfast plus blade. The large, round cells of this mutant contain 1.3 times as much polysaccharide as wild type cells, but 80% of it is water soluble, compared with only 50% for the wild type (Bryhni 1978). The result may be that there is less cross-linking between the insoluble cellulose fibrils. This would increase plasticity of the cell walls, causing the cells to swell and preventing local differences in wall expansion that are required for normal, oriented morphogenesis.

A special case of cell growth is lateral branch initiation. This may begin by local wall loosening and the development of a new region of growth (Green 1980). The cytology of this process has been studied in only one seaweed, the filamentous brown *Sphacelaria furcigera* (Burns et al. 1982a, b). The longitudinal walls of this regularly branched alga consist of alternating layers of fucans and alginates (Fig. 10.5a, b). Branch initiation begins as a protrusion of the cell wall, below which is a thickened layer of alginate, and a noticeable concentration of protein (which is possibly involved in wall loosening and/or deposition). Changes also take

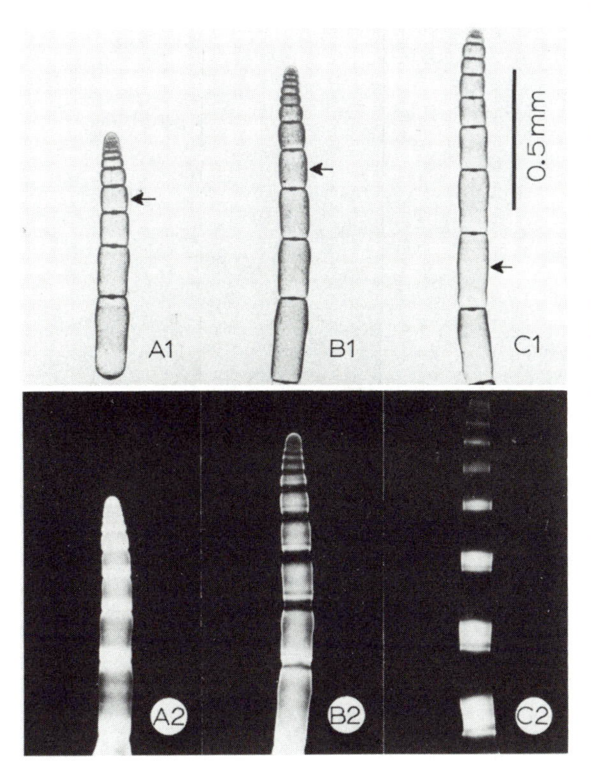

Figure 10.4. Cell elongation in *Griffithsia tenuis* showing bipolar band growth. In the top row of photographs a filament stained with Calcofluor White is shown in white light immediately after staining and after 24 h and 48 h. In the lower row the filament is shown at the same times, as seen in UV light, which causes the stain to fluoresce. The newly deposited wall is unstained and appears as black bands at the top and bottom of each cell. (From Waaland & Waaland 1975, with permission of Springer-Verlag)

place in the cytoplasm, the most evident being the loss of vacuoles (Fig. 10.5c). As the bud bulges outward, chloroplasts and mitochondria migrate, then the nucleus moves into the base of the bulge. At this time vacuoles begin to reappear in the axial part of the cell but not in the bulge (Fig. 10.5 d, e). After nuclear division, a cell wall cuts off the meristematic, nonvacuolated bud cell from the now highly vacuolated axial cell. Although high turgor pressure may be necessary for bud initiation, the subsequent disappearance of vacuoles suggests that continued protrusion of the bud does not require turgor pressure.

10.2.3 Nucleocytoplasmic interactions

The formation of most materials in the cytoplasm is directly or indirectly under the control of the nucleus (more specifically the genetic information in the chromosomes). Yet the cytoplasm is also the environment of the nucleus; through or from the cytoplasm come stimuli that turn on or off parts of the genetic code. Before the first division of a settled zygote the components of the cytoplasm become unequally distributed, so that after division the nuclei of the two cells are in different environments; this leads eventually to their forming the holdfast or frond (Quatrano 1978). In the division of the apical cell of *Acrosiphonia* more of the nuclei are partitioned to the apical cell than to the subapical cell; the apical cell remains meristematic whereas the other cell rarely divides again (Kornmann 1970). (Nuclei are equally distributed during intercalary divisions in this genus – Hudson & Waaland 1974.) But these observations only show that there are interactions between the nucleus and cytoplasm; an investigation of the nature of the interactions requires a cell–nucleus system that can be experimentally manipulated.

Giant-celled uninucleate algae, especially *Acetabularia* spp., have provided such a system since Hämmerling's classic studies in the 1930s (reviewed by Puiseux-Dao 1970). Accumulating evidence has led recently to the conclusion that the large primary nucleus is diploid, with meiosis occurring when this nucleus breaks up prior to cap formation (see Koop 1979). This is important to a consideration of nucleocytoplasmic interactions, and a point to be kept in mind when reading older literature, where meiosis is taken to occur in the cysts and the secondary nuclei are assumed to be diploid. A further advantage of *Acetabularia* for these studies is that interspecific grafts can be made; both nuclei and cytoplasm can be transferred between species. (A second system with potential for study of nucleocytoplasmic interactions is the red algal alloparasite–host connection. As recently discovered by Goff & Coleman (1983), nuclei from the parasite *Choreocolax polysiphoniae* are transferred into host cells with which secondary pit connections are formed and may direct the host's response to parasitism.)

Hämmerling concluded, long before messenger RNA was known, that "morphogenetic substances" are released from the nucleus into the cytoplasm, where they could be stored for some time but were gradually used up. *Acetabularia* cells can still form a cap if, after reaching about one third of their final length, the nucleus is removed. There is an apicobasal gradient of morphogenetic substances, shown by apical portions of the stalk forming a cap more readily than basal sections. The type of cap formed by an enucleated stalk is characteristic of the species, but if a nucleus from another species is inserted the caps formed are first intermediate and then have the characteristics of the nucleus donor species. This, and the fact that enucleate cells can form a cap only once, is evidence that the morphogenetic substances are used up in cap formation. A nonspecies-specific substance triggers cap formation, but a cap will not form in darkness.

Although Hämmerling's experiments are now interpreted in terms of long-lived mRNA, there is as yet no proof of the identity of the morphogenetic substances

Figure 10.5. Diagrams of bud development in *Sphacelaria furcigera*. (a) Morphology of the *Sphacelaria* apex; (b) and (c) show the structure of the cell wall during development, with four alternating layers of fucan (CW1, CW3) and alginate (CW2, CW4) changing to two (CWO and CWI). (d–g) Changes in vacuoles (V) and nucleus (N) during bud development. CrW = cross wall. (a from Bold & Wynne, *Introduction to the Algae: Structure and Reproduction*, © 1978. Reprinted with permission of Prentice-Hall Inc., Englewood Cliffs, N.J.; b, c redrawn from Burns et al. 1982a; d–g redrawn from Burns et al. 1982b, with permission of *The New Phytologist*)

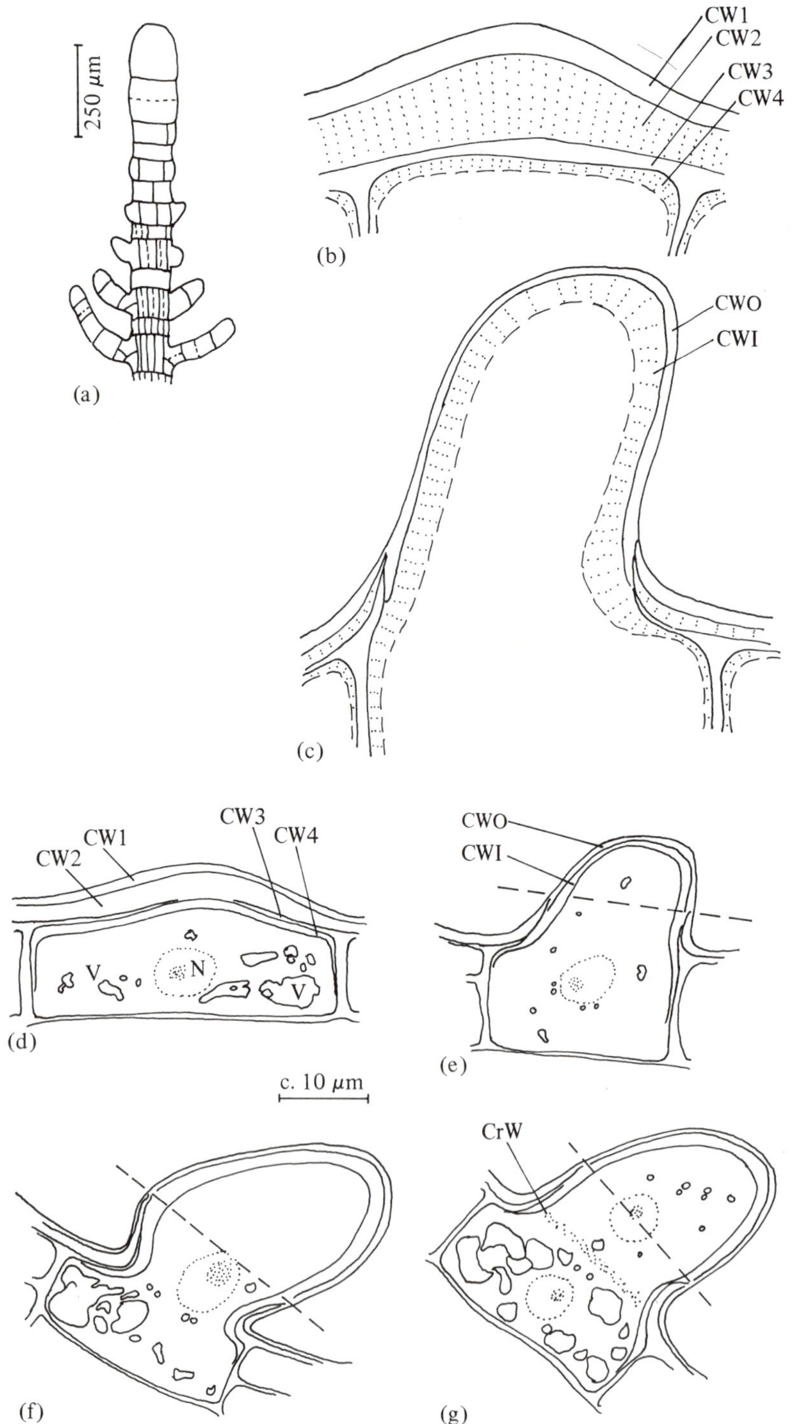

Figure 10.6. Ultrastructure of swimming (a) and newly settled (b) zoospores of *Enteromorpha intestinalis*. In the anterior of the swimming cell can be seen numerous vesicles filled with adhesive (arrows). Also visible are part of the nucleus (N), chloroplast, and flagella bases, Golgi body (g), vacuole (v), and mitochondrion (m). In (b) a mass of secreted adhesive lies in the triangle between the two cells, and there are virtually no vesicles remaining in the cell. (The attachment surface is parallel to the bottom of the photograph.) c = chloroplast; p = pyrenoid. Scales: (a) × 9,000; (b) × 10,000. (From Evans & Christie 1970, with permission of the Annals of Botany Company)

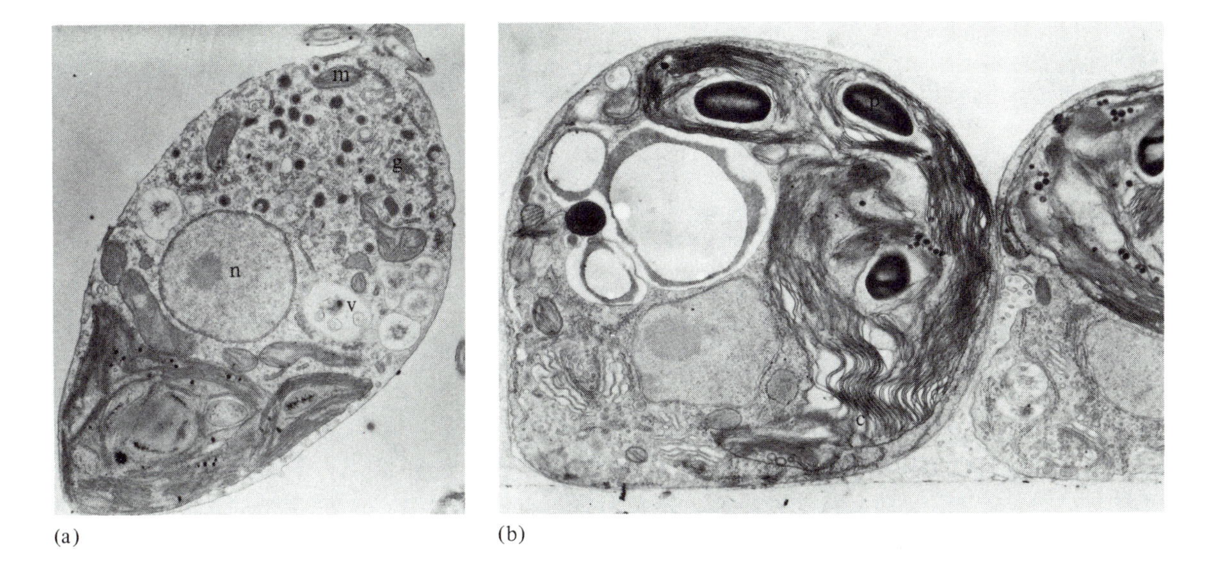

(Green 1976). Some evidence for their identity has been adduced, however. RNA absorbs and can be inactivated by UV light, with a peak at about 260 nm (see Lüning 1981a); when nucleated bases of *A. mediterranea* were grafted to irradiated stalks of *A. crenulata* there was a decrease in the percentage of *A. crenulata*-type caps as the irradiation dosage increased, suggesting destruction of *A. crenulata* morphogenetic substances (Werz 1965). Further work with various inhibitors tends to implicate mRNA's as the morphogenetic substances, but there are some problems in interpreting the results (Green 1976). One problem is that much RNA synthesis takes place in the chloroplasts. The extent to which chloroplast DNA may be involved in cell morphogenesis is unknown, but an interesting hint is that there is more DNA in apical chloroplasts than in basal or middle chloroplasts (Mazzo et al. 1977). Also interesting is a finding that at least some higher plant chloroplasts have some mobility of their own, in addition to passive movement in cytoplasmic streaming (DeRobertis & DeRobertis 1980). Whatever the nature of the messages sent out by the nucleus, the cytoplasm has some control over when they are read.

10.3 Thallus morphogenesis

A seaweed begins life as a spore or a zygote (or occasionally as some other one- or few-celled propagule). Once this has settled to the substratum, through chiefly passive forces (Sec. 5.3.1), active processes of attachment, germination, growth, and differentiation begin, leading to an adult thallus capable of reproduction. This section traces the ecological and physiological aspects of these processes.

10.3.1 Attachment, polarity, and germination

Once a spore or propagule has been deposited, it attaches to the substratum either by a general mucilage layer or, in the case of motile cells, first by the flagella tips. Subsequently, the attachment is consolidated by hardening of the mucilage, then by the development of rhizoids.

Zoospore ultrastructure and settling have been studied in a few species, including *Enteromorpha intestinalis* (Evans & Christie 1970, Callow & Evans 1974) and a variety of Laminariales (Henry & Cole 1982). Zoospores lack a cell wall and have among their organelles numerous cytoplasmic vesicles containing adhesive material (Fig. 10.6a). Attachment initially takes place by the tip of the anterior flagellum in kelps, and presumably all four flagella in *Enteromorpha* swarmers. Within a short time the flagella are withdrawn into the cell, and the cytoplasmic vesicles are released, adhering the cell to the substratum (Fig. 10.6b). Cell wall secretion begins once the spore has attached.

Attachment of nonmotile reproductive bodies has been studied with fucoid zygotes. Fucoid eggs are initially covered by a layer of alginates and fucoidan, which attaches them to the oogonium wall. The eggs are expelled from the conceptacle still enclosed in this layer, which is called the mesochiton. In *Fucus* and *Himan-*

thalia the mesochiton soon breaks down, and the zygotes attach by the zygote wall, which is formed soon after fertilization. In *Pelvetia canaliculata*, however, the mesochiton persists, probably to protect zygotes from drying out in the very high shore habitat of this species (Moss 1974b, Hardy & Moss 1979). The mesochiton, rather than the zygote walls, attaches the pairs of *Pelvetia* zygotes to the substratum. Within 24 h of settling, the zygote develops a firm alginate wall inside the mesochiton. Each zygote divides once or twice, then pushes out a group of up to four rhizoids, each from a single cell. These rhizoids grow down into the substratum, entering minute crevices if they are available. At this stage the mesochiton splits in a ring between the two zygotes (Fig. 10.7). The time between fertilization and the formation of rhizoids in this species is about one week. Other rhizoids quickly form and spread out over the substratum, and the germling begins to enlarge. Germination of *Fucus evanescens* zygotes at 3 C has been found to proceed slowly only as far as the formation of a protuberance; complete rhizoid formation required higher temperatures (Abe 1969).

Hardening of the attachment mucilage in various seaweeds apparently involves the formation of cross-links between polymer molecules, such as Ca^{2+} bridges between sulfate ester groups of fucoidan or between polyguluronic acid chains of alginate. *Fucus* embryos grown in sulfate-free seawater form normal rhizoids, but their fucan is not sulfated and the rhizoids cannot adhere to the substratum (Crayton et al. 1974). Moreover, sulfation is necessary for intracellular transport of fucan (discussed subsequently). Attachment of single cells mechanically released from *Prasiola stipitata* thalli also required sulfation of a cell wall polysaccharide: inhibitors of sulfation, such as molybdate, and of protein synthesis prevented attachment (Bingham & Schiff 1979). The protein may be complexed with the polysaccharide or may be an enzyme involved in the sulfation process.

Morphogenesis is an oriented process. Cells and plants have polarities, especially apicobasal polarities, which distinguish holdfast from frond. A few hours after fertilization in *Fucus*, asymmetrical jelly secretion begins and fucoidan first appears in the wall (Schröter 1978; see also Brawley & Wetherbee 1981). The new material is normally secreted on the side in contact with the substratum, although in the laboratory, fucoidan secretion and rhizoid formation, which follows, can be induced to take place on other parts of the cell by altering the polarity of the cell. Various natural and artificial gradients, including light, pH, Ca^{2+}, and K^{+}, and the proximity of other zygotes can serve as orientation cues in *Fucus*.

The research to date has been very well reviewed by Quatrano (1978) and Evans et al. (1982). When a *Fucus* zygote is placed in an orienting gradient, a polar axis begins to form parallel to the gradient. Division of the cell into rhizoid- and frond-forming cells will later

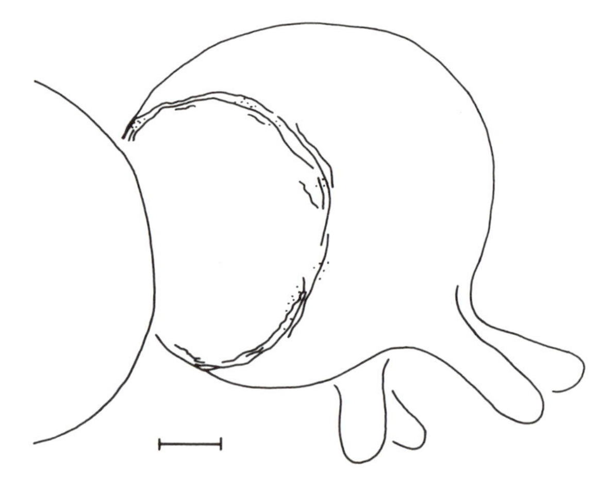

Figure 10.7. *Pelvetia* germling approximately one week old, with a cluster of four rhizoids, and showing the mesochiton split in a ring between the pair of zygotes (only one zygote visible). Scale: 20 μm. (Drawn from a photo in Moss 1974b)

take place perpendicular to this axis. For the first 8 to 14 h, the polar axis can be reoriented if the gradient is moved, but after that it is fixed. During the labile period, changes take place in membrane patches on the side where the rhizoid will form (e.g., the shady side in a light gradient). These membrane patches become fixed to the underlying cytoplasm by microfilaments; cytochalasin B, which disrupts microfilaments, inhibits axis fixation. The membrane patches then generate an influx of Ca^{2+} ions, which remain near this pole and may initiate other localized events. The cytoskeleton, a system of contractile protein microtubules and microfibrils throughout the soluble cytoplasm, may become oriented along the polar axis and provide a track along which particles and molecules could move. The movement could be caused by the contractile proteins or could be driven by an electrical gradient set up by the localized accumulation of Ca^{2+} ions. The latter mechanism is essentially electrophoresis.

The biochemical and biophysical properties of cell membranes and the cytoskeleton are not yet well enough known to detail precisely how these events take place (Quatrano 1978). After axis fixation, several cellular changes can be seen. Numerous extensions of the nuclear membrane project toward the rhizoidal pole, where there is an accumulation of vesicles, apparent in the light microscope as a "cortical clearing." The vesicles, derived from the Golgi apparatus, are filled with sulfated fucoidan, which is deposited in the cell wall at the pole and anchors the cell to the substratum. Negative charges on the sulfate ester groups or perhaps on the vesicle surfaces may be needed to draw the material toward the positively charged pole. Sulfation of the fucan requires new enzyme synthesis; if synthesis is prevented by cycloheximide, or if SO_4^{2-} is lacking, there is no movement of fucan to the rhizoid pole.

Are all nonmotile reproductive cells completely unpolarized at first? (Flagellated reproductive cells have a very obvious polarity and tend to settle with the flagella in contact with the substratum.) And are the first two cells always differentiated? Kelp zygotes might conceivably receive some stimulus from the female gametophytes to which they remain attached. Yet, because the lower cell need not form a rhizoid immediately, the first two cells may be identical; electron micrographs by Bisalputra et al. (1971) show no obvious differences. Very young sporophytes of *Macrocystis* are sticky and can reattach to the substratum if broken loose (North 1976); what happens if they are reoriented during reattachment?

Not all germinating spores or zygotes first divide parallel to the substratum (or perpendicular to the light gradient). Horizontal germination is common, in which a single filament or germ tube is formed. In some species, such as *Coelocladia arctica* (Dictyosiphonales), the protoplast migrates into the germ tube, leaving the spore wall empty (Pedersen 1981). Other brown algal zoospores push out several lobes, which are then cut off as cells; this stellate kind of germination leads to a monostromatic crust. Some red algal spores are known to coalesce to give a single germling (e.g., *Smithora naiadum* – McBride & Cole 1972). Dixon (1973) describes five types of germination among the Rhodophyta. Germination of carpospores of *Bangia fuscopurpurea* may be unipolar, leading to the conchocelis stage, or bipolar, leading directly back to the erect stage. The type of germination in this case is regulated by photoperiod – unipolar germination occurs in more than 12 h daylength, bipolar in less than 12 h (Dixon & Richardson 1970).

Thallus morphogenesis requires not only cell division but also cell adhesion. Many seaweeds can grow normally in axenic culture but in a few known cases cell adhesion requires external morphogenetic factors. Such cases, which include the green algae *Ulva lactuca*, *Enteromorpha* spp., and a variety of plants formerly grouped in *Monostroma*, provide useful clues to the morphogenetic process (Provasoli & Pintner 1977, Provasoli et al. 1977, Tatewaki & Provasoli 1977). In the normal course of development of *Monostroma oxyspermum*, the biflagellate swarmer produces a filament that divides in three planes to give a little sac, which subsequently ruptures, yielding a flat, monostromatic sheet (Tatewaki 1970). However, if placed in axenic culture the germinating swarmer forms only a two-cell thallus, consisting of an apical cell, which sloughs off cells during subsequent divisions, and a basal, rhizoidal cell. Normal morphology can be restored by the addition of exudates of axenically cultured brown and red seaweeds, by growing *Monostroma* in bialgal axenic culture with a red or brown seaweed, and by extracts of seven marine bacteria (out of over 200 isolates tested) in the genera *Caulobacter*, *Cytophaga*, *Flavobacterium*, and *Pseudomonas* (Tatewaki et al. 1983).

In nature, *Monostroma* does not bear a diverse microflora but is colonized only by a *Pseudomonas* sp. If the factor, which is apparently a small polypeptide residing in the cement between the cells, is added to rhizoidal cells, they coalesce and form a complete plant, whereas apical cells coalesce to form sheets without rhizoids (Provasoli et al. 1977). The critical cell division in this species, in terms of cell potential, therefore seems to be the very first one. A continued supply of the factor is needed to maintain thallus integrity. *Ulva* apparently induces its epiphytic bacteria to produce the morphogenetic substance, since the bacteria cultured in isolation stop producing it (Provasoli & Pintner 1980).

10.3.2 Cell differentiation

The organs of higher plants (roots, stems, leaves, and reproductive organs) each consist of several tissue types: epidermis, phloem, and so forth. Seaweeds have fewer organs, tissues, and cell types than vascular plants. Even the most complex seaweeds have photosynthetic and absorptive capacities combined into one organ, the frond. Nevertheless, all seaweeds show some cellular differentiation, if only between rhizoid and frond cells. In addition, differentiation must usually occur when vegetative cells become or form reproductive cells, even if no special branches or other reproductive structures are formed. The larger seaweeds, especially Laminariales and Fucales, have several different tissue and cell types, including photosynthetic epidermis, cortex, sieve tubes, gland cells, and medulla.There is also some evidence, and a priori reason to suppose, that there is a certain amount of specialization over a thallus, at least in the larger browns. Some of the evidence comes from the study by Fagerberg et al. (1979) of morphological and physiological differences between stipes and blades of *Sargassum filipendula*. Blades of this species have relatively more epidermis and cortex, while stipes have more thick-walled medulla. There are also greater areas of thylakoid and mitochondrial cristae per unit volume of blades than of stipes, and correspondingly greater photosynthesis and respiration.

The cytological processes of differentiation have been studied in only a few cases, mostly gametogenesis and sporogenesis. There, changes include formation of vesicles to store reserve carbohydrates and adhesive polysaccharides, plus flagella, eyespots, and so forth. Shih et al. (1983) have studied the cytology of development of *Macrocystis* sieve elements, but while some aspects of physiology of phloem differentiation in higher plants are known (e.g., the roles of hormones and sugars – Bidwell 1979), no such information is yet available for algal tissues. A detailed consideration of the cytology of differentiation is beyond the scope of this book; further information can be found in Brawley & Wetherbee's

(1981) review and the original literature they cite. Some environmental morphogenetic cues have been discovered (such as effects of daylength on reproduction or of temperature on the switch from crustose to erect growth), but nothing is known about how these cues are processed. (This aspect remains poorly understood even for such well-studied higher plant processes as flowering.)

During normal development of a germling of an erect thallus, the basal cell forms the rhizoids while the apical cell forms the frond. The cells resulting from further divisions are locked into their developmental pattern if only because the thallus is part of their regulatory environment. Abnormalities occur, however, and isolated cells may also behave differently. Both of these circumstances potentially can be exploited to yield information on morphogenesis. Among the interesting abnormalities recorded are those in *Callithamnion hookeri* carposporelings (Edwards 1977a). Some spores germinated into two axial filaments or two rhizoidal filaments instead of one each; cells in axial filaments sometimes gave rise to rhizoidal branches, and vice versa.

Cells of many simple algae, if isolated, will regenerate into a whole plant, but there are useful exceptions. Blade cells of *Ulva mutabilis* are not able to form rhizoidal cells but form vesicular thalli one cell thick (Fjeld & Løvlie 1976). Isolated rhizoidal cells of this species can form the whole plant; thus there is presumably some repressor present in the thallus cells. This possibility was investigated in a mutant form, *bubble* (*bu*), by Fjeld (1972; and see Fjeld & Løvlie 1976), which behaves like isolated blade cells. The mutant gene is recessive and chromosomal. Curiously, *bu* spores from meiotic sporangia on heterozygous plants (bu^+/bu) develop a partly or competely wild-type phenotype in the first generation. If these are propagated asexually, subsequent generations are completely the mutant type. The explanation appears to be that there is a repressor of rhizoid-forming genes in normal blade cells and *bubble* mutant cells and that this repressor is removed during sporogenesis so that spores can form rhizoids when they settle. The bu^+ wild-type gene is thus responsible for removal of the repressor and its transcription takes place before meiosis, so that a derepressor is present in the cytoplasm of *bu* spores as well as in wild-type spores. (This substance is not diluted through many cell generations since the number of rhizoidal cells is small.) The rhizoid-forming genes of both types are re-repressed early in development, but the mutant gene cannot derepress them when it forms spores.

A less well-understood developmental mutant has been found in *Enteromorpha lingulata* (Baca & Cox 1979).The mutant is spherical but has rhizoids; only haploid females exhibit the mutation. The change in developmental pattern seems to be the result of oblique cell walls forming earlier than usual. Baca & Cox speculated that this in turn might be due to changes in the ability of organelles to reorient prior to cell division. Unlike the *bu* mutation of *Ulva mutabilis*, inheritance of the spherical mutation of *E. lingulata* is non-Mendelian.

10.3.3 Development of adult form

Many seaweeds are filamentous or have a thallus made of filaments – simple filaments, coalesced filaments, or corticated axes (e.g., *Ulothrix, Corallina, Desmarestia*, respectively). Linear filamentous growth requires cell division in one plane; branch formation requires division in a second plane. Other seaweeds (e.g., *Laminaria*) have parenchymatous thalli, formed by cell division in three planes.

Filamentous thalli can be prostrate, erect, or heterotrichous (both). The proportion of prostrate to erect filaments has been used as a taxonomic criterion in groups such as Ectocarpales, but recent work has begun to show that the proportions are environmentally variable (Russell 1978). Plants with reproductive structures have been assumed to be full-grown, but plants can reproduce over a very wide range of form, size, and relative proportions of parts. Since species of Ectocarpales are opportunistic, the timing of their reproduction, and therefore their size at maturity, is liable to be very flexible; indeed, their life cycles as a whole are very flexible (Wynne & Loiseaux 1976). Young plants or microscopic stages of larger genera can be easily mistaken for full-grown specimens of the smaller genera (as the genera are presently conceived).

The extent of cortication, likewise a taxonomic feature in some groups, may also be environmentally variable. In *Callithamnion* (Price 1978), robust, heavily corticated forms with usually short axial cells tend to be found in wave-exposed areas, whereas more delicate forms with longer axial cells and little or no cortication are characteristic of sheltered habitats. The means by which wave exposure and other environmental factors bring about such morphological variation are not yet understood.

The formation of erect uniaxial or multiaxial thalli from crustose germlings or microthalli involves the conversion of the tips of one or a number of erect filaments into meristems (Fig. 10.8) (Dixon 1973, Rietema & Klein 1981). Formation of the meristems (macrothallus initials) and the outgrowth of the erect thalli are separate events and, in *Dumontia contorta* (*D. incrassata*) these events are controlled by different environmental cues (Rietema 1982). Production of the initials in this species depends solely on photoperiod: short days (long nights) are required. Outgrowth of the initials also requires short days, but in addition the temperature must be 16 C or lower. The formation of erect multiaxial thalli in *Codium*, which has a feltlike filamentous juvenile stage, involves the coalescence of the filaments, a process that Ramus (1972) found requires water motion. *Codium fragile* filaments twist together into knots that develop polarity and thus become primordia (or macrothallus initials). If primordia are kept in shaken cultures they develop the characteristic

Figure 10.8. Development of erect thalli from crustose bases. (a) Formation of multiaxial macrothallus in *Platoma bairdii*; (b) formation of uniaxial frond in *Gloiosiphonia capillaris*. Scales: 25 μm. (From Dixon 1973, with permission of the author)

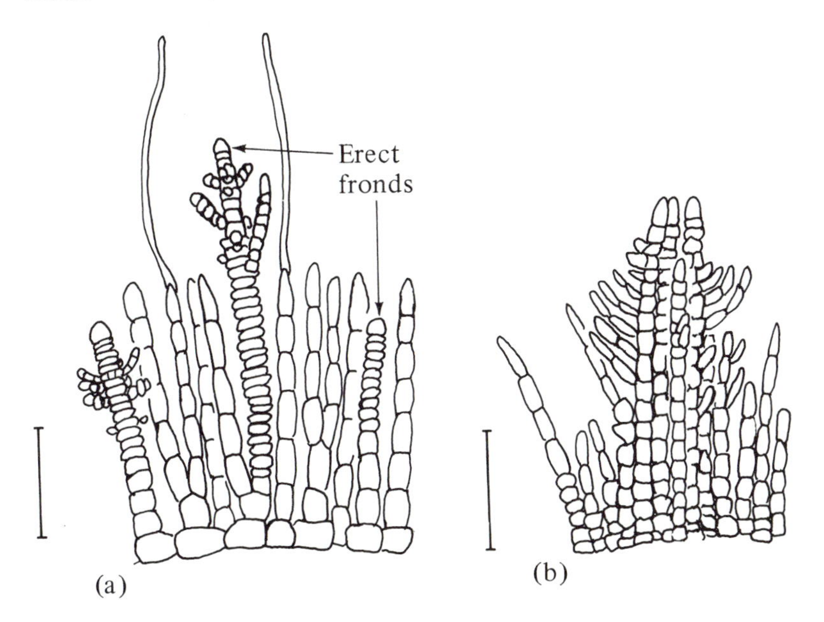

adult thallus structure, but if placed into calm water they revert to nonoriented filamentous growth.

A few photomorphogenetic effects are known in which aspects of growth are regulated by light quality. Preliminary evidence has been adduced for red or far-red light effects on growth of kelp stipes (*Laminaria saccharina* and *Nereocystis luetkeana* – Lüning 1981b, Duncan & Foreman 1980). Red light causes specimens of the red alga *Calosiphonia vermicularis* to grow shorter and bushier than white or blue light (Mayhoub et al. 1976). Several photomorphogenetic effects are due to blue light: for instance, *Acetabularia* spp. require blue light (or white light, which includes blue) for normal growth and cap formation. *Dictyota* and *Scytosiphon* produce normal growth form, including abundant hairs, only in blue or white light, not in green or red light (Lüning 1981b).These blue light effects are not necessarily direct; indeed, that there seems to be general improvement of growth suggests that the effects are complex. For example, hair production in these cases may be a response to increased nutrient demands of vigorous growth. Nutrient shortage is known to cause hair formation in several species, including *Acetabularia acetabulum* (*A. mediterranea*), *Ceramium rubrum*, *Fucus spiralis*, and *Codium fragile* (DeBoer 1981, Norton et al. 1981, Benson et al. 1983). The hairs are a means of increasing the absorptive surface area of the thallus. Nevertheless, in some unicellular green algae and a variety of fungi and higher plants, blue light effects on individual enzymes have been demonstrated (Senger 1980).

A role of nutrition in morphogenesis has been shown in *Petalonia fascia* (Hsiao 1969) and *Scytosiphon lomentaria* (Roberts & Ring 1972). In the case of *Petalonia* from Newfoundland, Hsiao found that zoospores from plurilocular sporangia on the blade could form protonemata (sparsely branched uniseriate filaments), plethysmothalli (profusely branched filaments), or *Ralfsia*-like thalli, any one of which could reproduce itself via zoospores or give rise directly to the blade. Protonemata and plethysmothalli survived in iodine-free medium, but formation of *Ralfsia*-like thalli or blades required iodine (5.1 mg I L^{-1} and 508 μg I L^{-1}, respectively). Plethysmothalli formed blades in progressively shorter times as the iodine concentration increased. Roberts & Ring (1972) found changes in proportions of filamentous and crustose microthalli correlated with nitrogen and phosphorus levels.

Growth of kelp haptera is oriented by negative phototropism, not geotropism, with blue light being the most strongly orienting part of the spectrum (Buggeln 1974). Thigmotropism takes over when the elongating hapteron touches the substratum (Lobban 1978a). Several other examples of phototropism are reviewed by Buggeln (1981). Orientation of unicellular rhizoids is more rapid and easier to interpret than orientation of multicellular haptera. Unilateral irradiance is detected by some as yet unidentified pigment. The information can be stored for several hours, with the response exhibited in subsequent darkness.

Gravity may be the stimulus for rhizoid orientation in *Caulerpa prolifera*. When rhizomes were inverted by Matilsky & Jacobs (1983), rhizoid initiation was preceded by a movement (sinking) of amyloplasts toward the lower side and rhizoid initials contained numerous amyloplasts. Amyloplasts play a role in root-tip orientation in angiosperms and may interact with movement of growth substances (Bidwell 1979).

Thallus orientation is related to attachment, if only because this gives consistency in the direction of environmental cues. A few seaweeds are known to exist in free-living populations, morphologically distinct from detached specimens in the drift whose floating existence is temporary. A characteristic of free-floating plants is their lack of polarity; growth may continue in all directions to form a ball (Norton et al. 1981). *Sargassum natans* and *S. fluitans* in the Sargasso Sea are the most famous populations of free-floating seaweeds, and also the most distinct. These species are entirely pelagic and are not recognizably derived from any present-day benthic populations. Other cases include forms of *Ascophyllum nodosum* (South & Hill 1970, Brinkhuis 1976), *Cladophora* aff. *albida* (Gordon et al. 1980), *Pilayella littoralis* (Wilce et al. 1982), and the rhodolith forms of coralline algae (Johansen 1981). Species such as *Chondrus crispus* in floating cultivation also form balls. The free-floating plants, like driftweed, probably came originally from attached plants, but under peculiar current regimes have been able to form persistent, self-perpetuating populations, rather than being washed up on shore. A further characteristic of free-floating populations is reproduction almost solely by fragmentation (vegetative propagation). Reproductive structures typical of attached plants are rare and also have not formed in culture conditions under which detached specimens became fertile. Fragmentation may occur simply by breakage of large balls or following senescence near the center. In the case of the *Pilayella* population studied by Wilce et al. (1982), fungal infection by the chytid *Eurychasma dicksonii* plays a key positive role.

Branching is a characteristic developmental step in many algae. Branching patterns, like cortication and the proportions of erect and creeping filaments, are constant enough in many species, for instance, among Ceramiales, to be a taxonomic criterion. Yet in many other plants branching is irregular and controlled as much by environment as by genotype. *Enteromorpha* spp. are particularly variable in response to salinity (Norton et al. 1981, Pringle 1982). In *Ascophyllum nodosum*, branch initials are formed, some of which grow out into vegetative or reproductive lateral shoots, while others remain dormant. This pattern, reminiscent of buds in flowering plants, suggests internal control. In other large seaweeds, such as *Egregia* and *Macrocystis*, branches (new fronds) arise when lateral blades resume indeterminate growth; again there is the suggestion of internal control in timing of these outgrowths, but no direct evidence.

Among the intriguing morphogenetic questions posed by *Macrocystis* is, What determines whether a lateral blade will form a new frond (with unilateral branching and indeterminate stipe growth), or a sporophyll (with dichotomous branching and short stipes), or will remain a single, undivided lamina? The apical meristem normally produces two frond initials, followed by several sporophylls (not all of which branch), then determinate, undivided vegetative blades (Lobban 1978a). Frond initials and even determinate blades may bear sori (although whether these form viable spores or not is unknown), while rare cases are known in which determinate blades well long a frond grew out into fronds as do frond initials. Thus, although sporophylls are the usual reproductive organ in this genus, there is not a precise correlation between reproduction and branch type.

A developmental characteristic of a number of the large brown algae is the formation of gas vesicles (pneumatocysts). They occur in various positions on the frond, but all are essentially hollows in the medulla of the stipe or stipe/lamina. Some plants produce one large pneumatocyst (e.g., *Nereocystis*); others produce numerous small vesicles, sometimes on special lateral branches (e.g., *Sargassum*). In the austral fucoid *Durvillaea*, air chambers form a honeycomb within the thallus. As the outer cell layers of a pneumatocyst grow, the medulla tears apart and the space is filled with gas. The air chambers of *Durvillaea* are initiated by tight interweaving of medullary hyphae to form diaphragms (Lindauer et al. 1961). Oxygen and nitrogen, in roughly the proportions in air, form the bulk of the gas but there are also small and variable amounts of CO_2 and, in those kelps with a single large pneumatocyst, carbon monoxide (CO) (Foreman 1976, Dromgoole 1981a, b). The O_2 and CO_2 derive partly from the metabolic activities of the cells in the pneumatocyst wall, and diurnal changes in the composition and pressure of pneumatocyst gases have been shown in *Carpophyllum* (Dromgoole 1981a). However, equilibration takes place between the gases in the pneumatocyst and in the surrounding water (or air); this is the source of the nitrogen in the vesicles and the major source of O_2 and CO_2 (Hurka 1971). CO is postulated to be a by-product of degradative metabolism involved in the formation of the pneumatocyst, and its concentration quickly diminishes once pneumatocyst growth ceases. The reason that plants with many small pneumatocysts lack CO is probably that pneumatocyst growth is soon over and the CO is lost through equilibration with the surroundings, although this hypothesis has not been tested (Foreman 1976). The shape and wall thickness of a pneumatocyst can be related to the pressures it must withstand, hence to the depth of water (Dromgoole 1981b). The environmental influences on the abundance of vesicles on a plant are obscure, even though in *Fucus vesiculosus* there appears to be a correlation with wave exposure (Sec. 5.3.3; Fig. 5.15).

Although coenocytes such as *Caulerpa* and *Bryopsis* are technically single cells, there are differences between regions of cytoplasm, allowing the same kinds of differentiation that occur in multicellular thalli. Not all nuclei look alike (e.g., in *Cymopolia barbata* – Liddle

Figure 10.9. The prosthetic, light-absorbing group of phytochrome in the red absorbing form. In vivo this is attached to a protein. (Reprinted with permission from Goodwin & Mercer, *Introduction to Plant Biochemistry*, © 1983 Pergamon Press Ltd.)

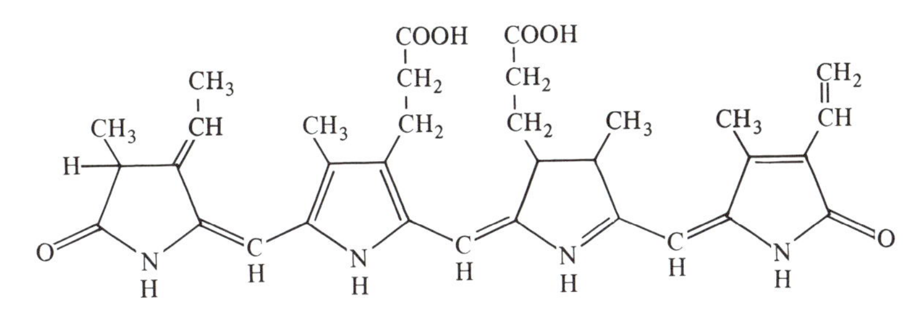

et al. 1982), and there are undoubtedly many further differences that are not ultrastructurally obvious. Further, cytoplasm and organelle movement is under cellular control. In *Caulerpa prolifera* there is a concentrated "meristemplasm" at the growing tips (Dawes & Lohr 1978) and there are diurnal movements of cytoplasm between shoot and rhizome (Dawes & Barilotti 1969). In this species gravity also affects the distribution of organelles, important to polarity (Jacobs & Olson 1980). The mobility of chloroplasts has already been mentioned.

10.4 **Reproduction**

The culmination of life for seaweeds, as for other organisms, is reproduction. The change from the vegetative to the reproductive condition in simpler algae involves virtually no further growth. In *Ulva* or *Cladophora*, for example, vegetative cells simply cleave into gametes or spores, which are released from the thallus. In other plants, sporangia or gametangia are produced on special branches (e.g., *Ectocarpus*, *Alaria*). In the Florideophycidae, sexual reproduction involves complex development of the female gametophyte both before and after fertilization. Reproduction involves several stages: (1) onset of reproduction – conversion of vegetative tissue into reproductive tissue, or formation of reproductive branches, including in some cases determination of the type of reproductive body; (2) release of reproductive products; and (3), in the case of sexual reproduction, the approach and union of gametes. Some aspects of the physiology of each of these stages are known.

10.4.1 Onset of reproduction

Plants can reproduce over a wide range of thallus size, but as yet there has been no systematic study of the effect of plant size alone on timing of reproduction. So far, investigators have concentrated on species that show a seasonal change in phase.

Several life history phase changes have now been shown to be cued by photoperiod (Lüning 1980, 1981b). These include sporangium formation in the filamentous stages of *Porphyra tenera* and *Bonnemaisonia hamifera*; formation of erect thalli of *Scytosiphon lomentaria* and *Petalonia zosterifolia* from their crustose stages; and the sporulation of the microscopic stages of *Monostroma grevillei* and *Protomonostroma undulatum*. In all these cases the events take place in short days (8 h daylight, 16 h darkness). Photoperiod effects are due to the length of uninterrupted darkness rather than to the length of the light period. This has been demonstrated by giving the plants a short "night-break" of light in the middle of a long night (e.g., 8 h light, 7.5 h darkness, 1 h light, 7.5 h darkness in a 24-h cycle). Such a night-break inhibits sporulation, even though the *day* length has not changed.

These events are similar to those reported for flowering of angiosperms, and there is preliminary evidence that photoperiod is detected by the pigment phytochrome (Fig. 10.9) in algae as it is in angiosperms. This pigment responds to relative amounts of red and far-red light (Noggle & Fritz 1983, Goodwin & Mercer 1983). The evidence for this pigment being active in seaweed photoperiodic responses comes from detection of the pigment in red algae (Dring 1967, van der Velde & Hemrika-Wagner 1978; and see Nultsch 1974), and the demonstration that red light – but not far-red light – is effective in the night-break (Dring 1974, Rietema & Breeman 1982). Evidence is accumulating for the involvement of a variety of pigments in algal photomorphogenesis. The photoperiod response of *Scytosiphon* is mediated by *blue* light (Dring & Lüning 1975). Tetrasporangium formation in *Rhodochorton purpureum* takes place in short days and is inhibited by a night-break of red but not far-red light, yet the red light inhibition is not reversed by a subsequent exposure to far-red light. This is in contrast to the reversibility of phytochrome in angiosperms. Moreover, a night-break of narrow-band blue light (448 nm) also inhibited tetrasporangium formation in this species (Dring & West 1983). Blue light could be detected by phytochrome, or it might be detected by a xanthophyll or a flavin (e.g., riboflavin) (see Leopold & Kriedemann 1975). Although the phenomena cited above are all "short-day" (i.e., long night) responses, seaweeds also show long-day responses; the formation of crusts by *Scytosiphon* is an example (Clayton 1976, Lüning 1981b).

At more northerly latitudes spring nights become progressively shorter and the critical daylength for the

Figure 10.10. Effect of daylength on erect thallus formation by different geographic isolates of *Scytosiphon lomentaria* at 10 C (open circles) and 15 C (filled circles). Each value is based on a count of 250 plants. (From Lüning 1980, with permission of The Systematics Association)

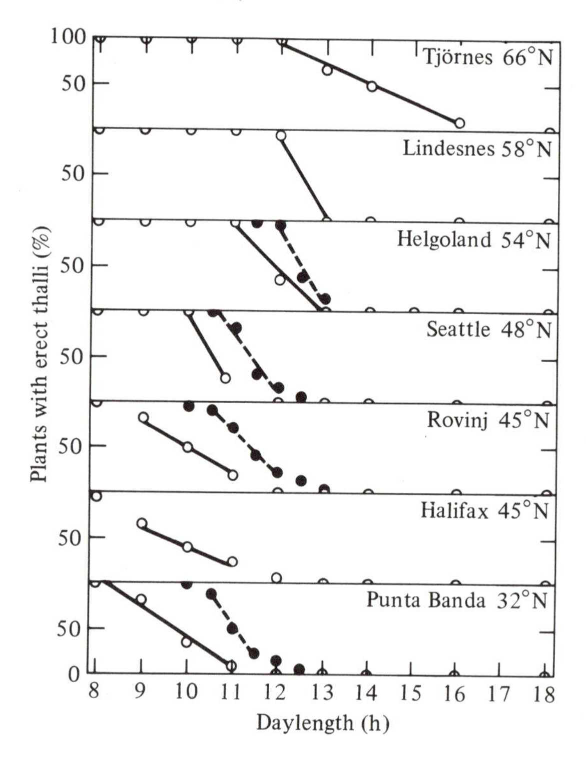

short-day *Scytosiphon lomentaria* becomes longer (Fig. 10.10); that is, the critical night length becomes shorter. Unfortunately, intraspecific differences in critical daylength do not always correlate with latitude, as Rietema & Breeman (1982) found in *Dumontia contorta*. However, photoperiod responses are sometimes altered by temperature; one such effect is seen in Figure 10.10. Data on tetraspore formation by the trailliella stage of *Bonnemaisonia hamifera* (Table 10.1) further illustrate the narrow limits of conditions under which reproduction can occur. In the northern part of its range, the trailliella stage reproduces asexually, presumably by apomeiotic spores. This example is even further complicated because the photoperiod response is not exhibited under high nitrogen levels (such as are created in standard culture media).

Photoperiod is not the only aspect of light that cues reproduction. Although examples of photoperiod-sensitive seaweeds have been recorded, many seaweeds are expected to be day neutral. So far gametophytes of two species of *Laminaria, L. saccharina* and *L. digitata*, have been shown to be insensitive to photoperiod. Some seaweeds, including kelp and *Desmarestia* gametophytes, have minimum requirements for accumulated total daily irradiance or a certain irradiance intensity in order to reproduce (Chapman & Burrows 1970, Cosson 1977, Lüning & Neushul 1978). Other factors also trigger reproduction in various cases. Gamete production in *Derbesia* and *Bryopsis* and egg production in *Dictyota* have been shown to be controlled by endogenous rhythms of 4 to 5 days and 16 to 17 days, respectively (the latter a semilunar cycle) (Round 1981, Tanner 1981). Sudden changes in medium can induce reproduction in some simple seaweeds, a method exploited in culture work (Chapman 1973a). In one case, *Ulva mutabilis*, this may be because healthy vegetative thalli release substances that inhibit sporulation. Nilsen & Nordby (1975) showed that one of the substances is heat labile, but they were not able to isolate and identify the compounds. Temperature and salinity shocks can also induce reproduction. The mechanisms are unknown but might involve nutrient depletion or osmotic effects. Nitrogen availability has been shown to influence reproduction in a few instances, notably *Ulva* spp.; high nitrogen levels favor vegetative growth and asexual reproduction, low nitrogen stimulates gametogenesis (DeBoer 1981).

The kind of sporangium formed by *Ectocarpus siliculosus* depends on temperature (see Fig. 3.10); individuals of this species are capable of producing either unilocular or plurilocular sporangia, or even both (Fig. 10.11). Sporangium type in *Ectocarpus* does not indicate whether meiosis or mitosis has taken place. Mixed phases have also been recorded in several red algae (West & Hommersand 1981): both gametophytes with tetrasporangia and tetrasporophytes with gametangia have been found. In most cases the ploidy levels of the spores or gametes have not been determined. An exception is *Gracilaria tikvahiae* (van der Meer & Todd 1977). Here the appearance of male and female tissue (evidenced by mutant color patches as well as gametangia) was explained by mitotic recombination (Fig. 10.12). Crossing over of chromosomes normally occurs during meiosis but can also take place during mitosis, with the result that one cell gets both copies of one gene (wild type, +, in the example illustrated in Fig. 10.12), while the other cell gets both copies of the mutant gene (grn). Any gametes produced are therefore diploid. This explanation does not apply to cases where tetraspores appear on gametophytes. Some other examples of reproductive anomalies might also have a genetic explanation. For instance, *Macrocystis* canopy blades sometimes become fertile, even though the sporophylls are normally a few blades close to the base of the frond. Sori of *Alaria* are sometimes seen on the blade as well as on the sporophylls.

10.4.2 Reproductive cell release

Several intertidal algae, including *Postelsia* and *Pelvetia*, release their spores or gametes when wetted after a period of desiccation. Storing kelp sori overnight in air followed by rewetting is sometimes used as a means of inducing zoospore release (e.g., Lüning & Neushul 1978). Gamete release in *Ulva* also occurs when the thalli

Table 10.1. *The effects of photoperiod and temperature on tetrasporangium formation in the trailliella phase of* Bonnemaisonia hamifera

Response to daylength (at 15 C)											
Hours of light per day	8	9	10	10.5	11	12	12.5	13	14	15	16
Percent fertile	93	92	48	16	6	0	0	0	0	0	0
Response to water temperature (at 8 h light per day)											
Temperature (C)	10	12	15	17	20	23					
Percent fertile	0	0	97	73	0	0					

In each experiment 150 plants were grown in slightly enriched seawater (containing less than 20 μM NO_3^-).
Source: Lüning (1981b), with permission of Gustav Fischer Verlag.

Figure 10.11. *Ectocarus siliculosus* with unilocular (u) and plurilocular (p) sporangia on the same filament. (From Kornmann & Sahling 1977, with permission of *Helgoländer Meeresuntersuchungen*)

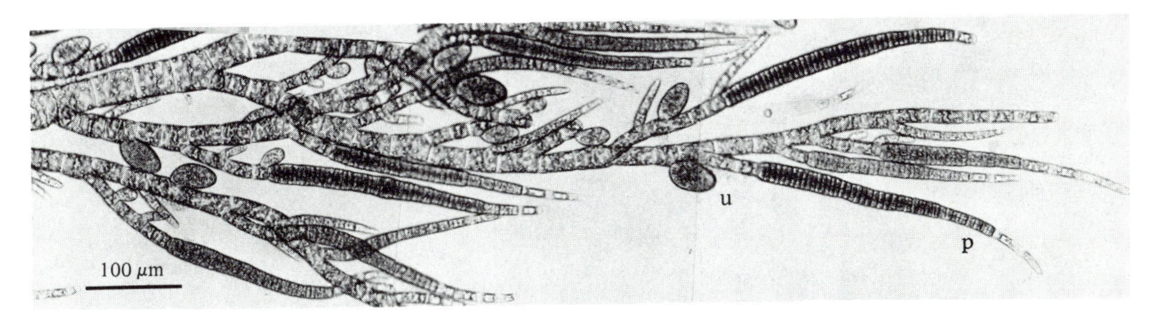

Figure 10.12. Mitotic recombination (right) compared with normal mitosis in heterozygous, diploid tetrasporophytes of *Gracilaria tikvahiae*. Diploid pairs of chromosomes are shown, each with two bivalents; the chromatids are numbered 1 to 4. + = wild-type color gene; grn = green mutant gene. (From van der Meer & Todd 1977, reproduced by permission of the National Research Council of Canada from the *Canadian Journal of Botany*, vol. 55)

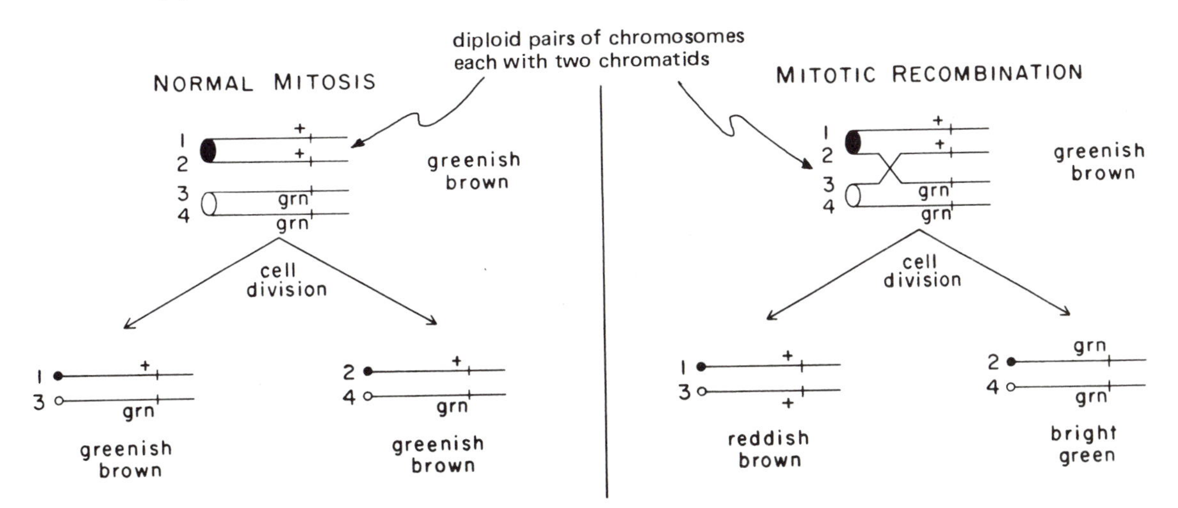

are rewetted by the tide. In some populations there is a periodicity of release (which implies periodicity in gamete formation). On the Pacific coast of the United States *Ulva* spp. release gametes at the beginning of the spring tide series, spores 2 to 5 days later (Smith 1947), while *U. pertusa* in Japan releases gametes on the neap tides (Sawada 1972). Some *Monostroma* and *Enteromorpha* spp. also show periodicity of gamete or spore release. However, *Ulva* on the Atlantic coast of the United States has no periodicity.

Gamete release by *Derbesia tenuissima* in culture takes place at the beginning of the photoperiod. It is

Table 10.2. *Male gamete attractants in the Phaeophyta*

Species	Type of reproduction	Attractant
Ectocarpus siliculosus	Isogamous	Ectocarpene
Cutleria multifida	Anisogamous	Multifidene
Dictyota dichotoma	Oogamous	Dictyopterene C′
Desmarestia viridis	Oogamous	Desmarestene
Laminaria digitata	Oogamous	Unidentified attractant and antheridium releaser
Fucus vesiculosus and *F. serratus*	Oogamous	Fucoserratene
Ascophyllum nodosum	Oogamous	Finavarrene

triggered by an instantaneous light-induced turgor pressure increase that ruptures a weak area of the wall, forming a pore (Wheeler & Page 1974). The antheridia of Laminariales do not release their sperm until they receive a messenger substance from mature female gametophytes (Lüning & Müller 1978). Such a mechanism is important in plants with obligatory alternation of generations and with gamete production limited by the small size of the gametophytes.

10.4.3 Gamete attraction and syngamy

In dioicous (heterothallic) species with unisexual gametophytes, coordination of gamete release and attraction of one gamete to the other increases the chance of successful syngamy. Several brown algae have so far been discovered in which one gamete releases a volatile attractant for the other (Table 10.2) (Müller 1981, Müller et al. 1982a, b). There may be other brown algae and possibly some green algae in which this mechanism will also be found. (No attractant is expected in the red algae since the spermatia are nonmotile. However, Buggeln [1981] has speculated that ooblast filaments might be guided to auxiliary cells by some chemical attractant.) The diverse taxa and reproductive types (Table 10.2) are reasons to expect further examples to be found. All genera except the fucoids produce a compound ectocarpene (Fig. 10.13a), but this compound is active as an attractant only in *Ectocarpus*. The other genera each produce their own attractants (Table 10.2, Fig. 10.13b–f). All the compounds are simple, volatile hydrocarbons; those of the fucoids are straight-chain, whereas the others are cyclic compounds. Their insolubility and volatility prevent their concentrations building up in the water and enable the female gametes to maintain steep concentration gradients. In *Fucus* the same attractant works for two species, but syngamy is prevented because surface phenomena do not permit egg and sperm to unite. Indeed, two races of *Ectocarpus siliculosus*, one from Naples, one from Cape Cod, are unable to fuse (Müller 1976).

The behavior of *Ectocarpus* male gametes is complex (Fig. 10.14). In "open" water they swim in straight lines, periodically changing direction abruptly. When they encounter a surface they change to a wide, looping path along the surface. In the presence of attractant from the female gamete, the male changes to a circular path, the diameter of which decreases as the hormone concentration increases.

Gamete recognition and fusion is a critical stage in sexual reproduction. Whereas attractants are generalized, recognition has to be species-specific. The processes of recognition and fusion have been studied in some Fucales (Evans et al. 1982). The eggs initially lack a wall and the membrane appears lumpy due to protrusion of cytoplasmic vesicles (Fig. 10.15a) (Callow et al. 1978). The spermatozoid probes the surface of the egg with the tip of its anterior flagellum, apparently seeking specific binding sites (Friedmann 1961, Callow et al. 1978). Attachment takes place first by the flagellum tip, later also by the body of the cell (Fig. 10.15b). After fertilization the zygote rapidly secretes a smooth membrane (Fig. 10.15a), which prevents further sperm attachment. Cell recognition and binding apparently takes place through fucose- and mannose-containing glycoprotein ligands on the egg membrane and matching carbohydrate-binding protein on the sperm, analogous to a lock and key mechanism (Bolwell et al. 1979, 1980). The evidence, which is somewhat equivocal, has been critically reviewed by Evans et al. (1982).

10.5 Wound healing, regeneration, and apical dominance

10.5.1 Wound healing

Different events are necessary for wound healing in multicellular and coenocytic seaweeds. In cellular seaweeds there is no need for the cut cells to recover, and the sealing of the wound involves changes in the underlying cells. In coenocytic algae the formation of a plug, or septum, is necessary to prevent loss of cytoplasm.

Wound healing in multicellular algae has been most thoroughly studied in *Fucus vesiculosus* (Moss 1964, Fulcher & McCully 1969, 1971). The thin, perforated cross-walls of the medullary filaments are plugged after about 6 h with newly synthesized sulfated polysaccharide (presumably fucoidan). Later there is general accumu-

Figure 10.13. Sex attractants of the brown algae. (a) Ectocarpene; (b) dictyopterene C′; (c) desmarestene; (d) multifidene; (e) fucoserratene; (f) finavarrene. (From Müller et al. *Science*, vol. 218, pp. 1119–1129, © 1982 by the American Association for the Advancement of Science)

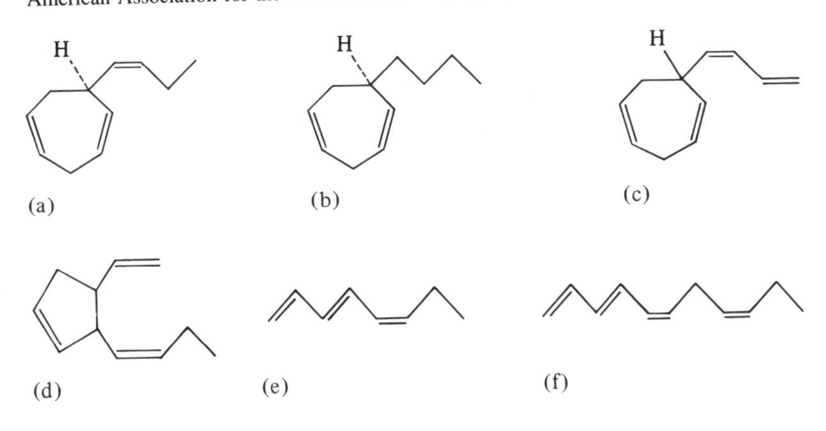

Figure 10.14. Diagram of approach of a male gamete to female of *Ectocarpus siliculosus*. See text for details. (Based on description in Müller 1981)

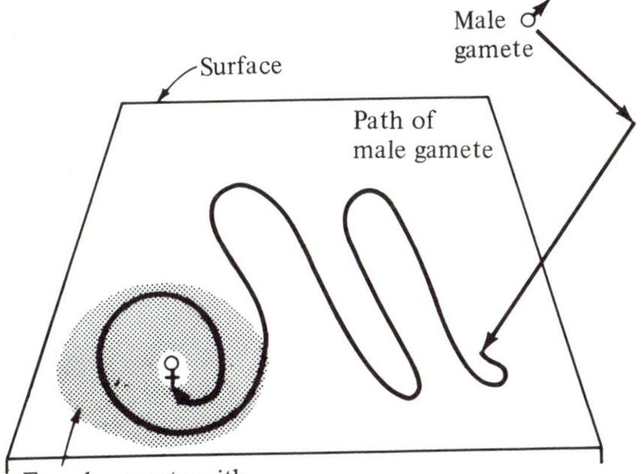

lation of polysaccharide at the wound surface. Medullary cells adjacent to the damaged cells round off and become pigmented. After about a week they give rise to lateral filaments that elongate and push through to the wound surface, where they branch repeatedly to form a protective layer. According to Fulcher & McCully (1969) these filaments are short-lived and full of antibiotic polyphenolics. Cortical cells undergo longitudinal division (parallel to the wound surface), and the outer cells assume the cytological and functional characteristics of epidermal cells (e.g., they become pigmented). Cells of the medulla may also contribute to the formation of new epidermis. There is no formation of undifferentiated callus tissue. A more detailed cytological study of wound regeneration and regrowth, using *Sargassum filipendula*, was carried out by Fagerberg & Dawes (1977). The cytological changes indicated that regenerating medullary cells became more active in respiration and secretion.

A rapid response to wounding is essential in coenocytes. One may induce cross-wall formation in *Caulerpa prolifera* merely by 30 s local pressure followed by 60 s recovery time (Jacobs 1970). Squeezing *Bryopsis* filaments similarly causes rapid cross-wall formation (Burr & West 1971). The mechanism by which the stimulus is perceived and transmitted from the cell wall to the vacuoles, where the fibers for wound repair are stored, is unknown. The repair material in *Bryopsis* is protein that is stored as fractured protein bodies in the vacuole. The protein is subsequently altered to a fibrillar-granular form, in which it participates in plug formation (Burr & West 1971).

The material and wound repair process are different in *Caulerpa* (Dreher et al. 1978). Here, the first response to a cut is the immediate formation of a gelatinous (glucan) plug. The material appears to be formed from vacuolar contents that change from a sol to a gel as they are extruded when the turgor pressure is released. The cytoplasm retracts from the wound and vesicles fill the space. Then, beginning within 1 min but continuing over the next 11 h, a second plug is deposited inside the first. This second plug contains protein and sulfated polysaccharide. Gradually the cytoplasm migrates back to the internal plug and new wall formation begins. Dreher et al. likened this complex process to blood clotting in mammals.

Insoluble polyphenolics formed by the action of peroxidase on coumarins may also form part of the plug in Dasycladales (Menzel et al. 1983). Behavior of the hemisiphonous Siphonocladales is different again (La Claire 1982). Here, as found in *Ernodesmis verticellata*, the response to wounding is a rapid contraction of the protoplasm both away from the wound and centripetally to close the wound. Healing is complete within 30 min and there is no plug. La Claire found evidence for Ca^{2+}-activated, ATP-driven contractile proteins, which would be consistent with what is known of contractile proteins in other plants.

Figure 10.15. Scanning electron microscope views of eggs, sperm, and zygotes of *Fucus serratus*. (a) Group of cells 10 min after mixing eggs and sperm. Smooth cells have been fertilized and have formed a fertilization membrane; the rough cell in the foreground is an unfertilized egg. (× 450) (b) Detail of fertilized egg with three sperm (arrows); the tip of the anterior flagellum of the middle sperm is embedded in secreted cell wall material. (× 1,600) (From Callow et al. 1978, with permission of the Company of Biologists)

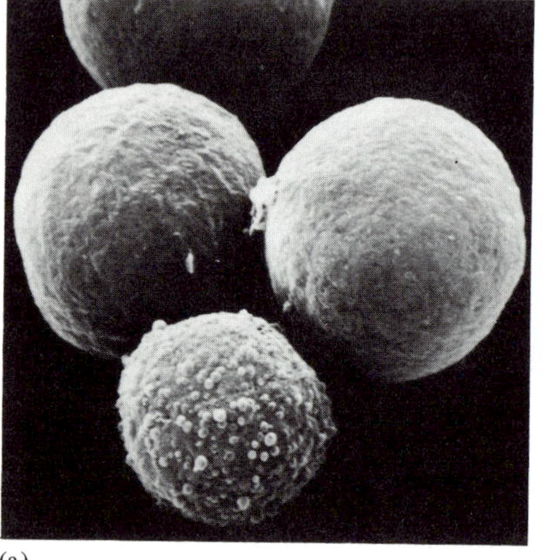

(a)

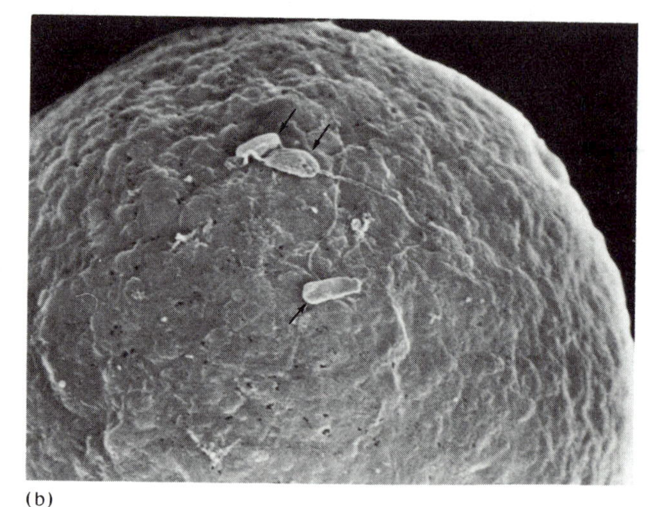

(b)

10.5.2 Regeneration and thallus polarity

Wound healing is commonly followed by either regeneration, the replacement of lost tissue by redifferentiation of medullary cells (as in *Fucus*, previously described, and in *Schottera* and *Gigartina*), or proliferation, the production of lateral outgrowths by renewed growth of the filaments forming the cortex, as in *Schottera* and *Gigartina* (Perrone & Felicini 1972, 1976). The type of tissue produced – rhizoidal or bladelike – typically depends on the position of the wound with respect to the apex or base of the thallus. In other words, there is a correlation with an internal thallus polarity. Proliferations in *Schottera nicaeensis* clearly show these correlation effects (Fig. 10.16) (Perrone & Felicini 1972). In general, leafy outgrowths arise from the apical sides of cut surfaces and rhizoidal outgrowths arise from basal sides. When the apex of the thallus is intact, horizontal cuts near the base give leafy proliferations, near the apex, rhizoids (Fig. 10.16a). Regeneration of thallus segments (Fig. 10.16b–d) results in leafy outgrowths from the apical end, rhizoids from the basal end. The number of rhizoids formed by 2-mm excised fragments of *Dictyota dichotoma* increased with distance of the fragment from the apical cell (Russell 1970).

In *Caulerpa*, excised "leaves" regenerate rhizomes and rhizoids from the basal end and new leafy shoots from the apical end. Rhizomes regenerate first rhizoids from the apical end, later rhizoids from the basal end plus rhizome and leafy shoots from the apical end (Fig. 10.17) (Jacobs 1970). "Leaf" segments 30 mm long formed only rhizoids; if 40 mm long, half the specimens also formed a rhizome and a new leaf; if 50 mm long, all regenerated completely. However, leafy shoot production and rhizoid production from the rhizome of *Caulerpa* also respond to gravity, as shown by Jacobs & Olson's (1980) experiments in which uninjured thalli were turned upside down. Rhizoids were produced from the new lower side, leafy shoots from the new upper side (the rhizome did not twist, so polarity had been reoriented).

The simplest kinds of regeneration involve uniseriate (branched or unbranched) filaments having apical growth. Two examples illustrate the events: *Acrosiphonia arcta*, a hemisiphonous green alga with multinucleate cells (Kornmann 1970), and *Rhodochorton purpureum*, a red alga with uninucleate cells (Pearlmutter & Vadas 1978). Wound healing is different in the two species. In *A. arcta*, the cut cells survive, presumably by forming a plug in a manner analogous to the siphonous algae previously described. *R. purpureum* does not have this ability, and the cell next to the cut grows through the old cell wall. Growth of the apical cell of *Acrosiphonia* is rapid and not impaired by excision. The cut cell regenerates into a rhizoidal filament, while the uninjured apical cell continues to divide as usual (Kornmann 1970). Kornmann did not follow the fate of filaments left without the apical cell.

In *Rhodochorton*, single cells or fragments of two or three cells usually regenerated a rhizoid first, and later, from the opposite end, began to extend new photosynthetic cells. However, fragments sometimes produced a rhizoid or, less frequently, photosynthetic cells from both ends. Pearlmutter & Vadas (1978) stated that "intact apical cells on fragments did not appear to inhibit re-

Figure 10.16. Regeneration (R) and proliferation (P) in *Schottera nicaeensis* wounded thalli or thallus pieces. Initial condition is shown on the left in each pair; discarded tissue is shaded. Holdfast was removed in all cases. (From Perrone & Felicini 1972, with permission of Blackwell Scientific Publications)

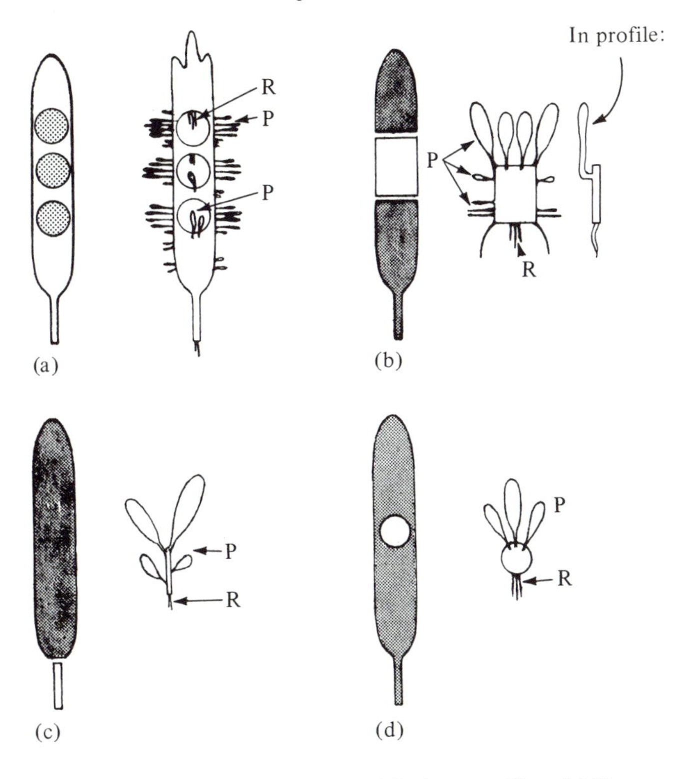

Figure 10.17. Regeneration of pieces of *Caulerpa prolifera*. (a) Diagram of the time course of regeneration of a portion of the "leaf." (b) Two stages in regeneration of a rhizome fragment. (From Jacobs 1970, with permission of the New York Academy of Sciences)

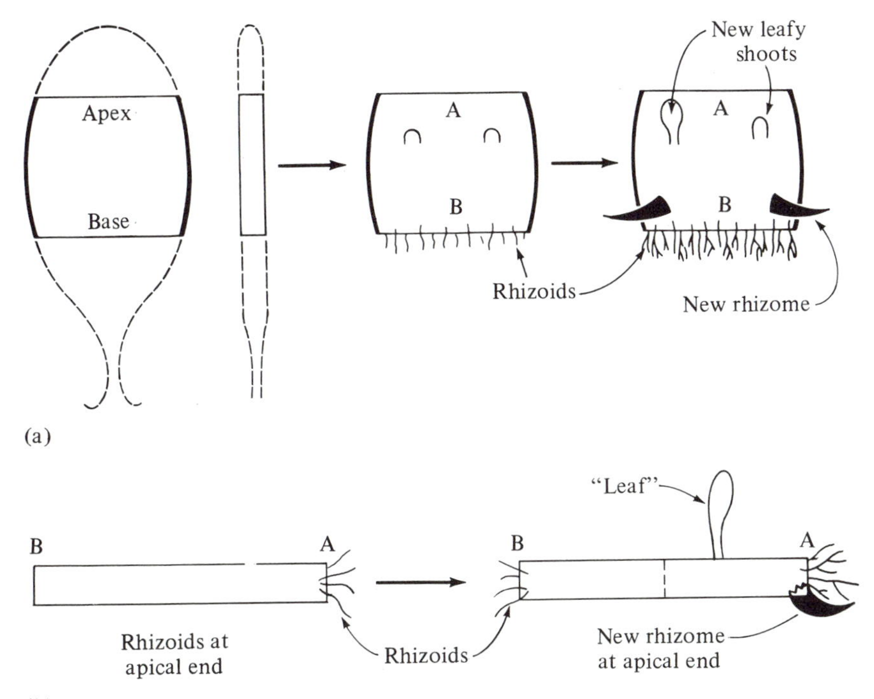

generation,'' but they did not observe whether there was a correlation between the position, relative to the apical cell, from which the fragment came, and the type of outgrowth produced. Nor could they tell whether rhizoids normally grew from the basal end of cells or fragments, since they had 40 or more fragments in each culture dish.

The physiologically best-known cases of wound healing and regeneration are in *Anotrichium tenue* (*Griffithsia tenuis*) and *Griffithsia pacifica* (Waaland & Cleland 1974, Waaland 1975). If filaments are cut and allowed to regenerate in isolation, a rhizoid is produced from the base of the apical portion and a new apical cell is regenerated on the basal portion (Fig. 10.18b), as in the examples just cited. However, the two fragments can be slipped into a cytoplasm-free *Nitella* cell wall cylinder so that the cut surfaces are apposed. The *Nitella* wall provides constraint but, unlike a glass capillary, is permeable. In this case a regenerating rhizoid is produced by the apical fragment and a special repair-shoot cell, not an apical cell, is produced by the basal fragment (Fig. 10.18a). This repair-shoot cell is induced by a hormone called rhodomorphin, which diffuses out of the regenerating rhizoid. The repair-shoot cell grows toward and fuses with the regenerating rhizoid. If apical and basal fragments of different species are apposed, no repair occurs. Rhodomorphins are thus species-specific. Their precise structure has yet to be elucidated, but the rhodomorphin from *G. pacifica* has recently been shown to be a glycoprotein of molecular weight ca. 14,000 (Watson & Waaland 1983).

The process of regeneration in *Fucus* is unusual in that distinct embryos, rather than lateral branches, are formed. During the process of wound healing in this genus, epidermal cells in certain regions of the wound begin to divide perpendicular to the wound surface, forming groups of branch initials (visible macroscopically after 4 to 6 weeks in culture), which develop directly into adventive embryos (Fulcher & McCully 1969, 1971). (Adventive signifies that the embryos arise from the frond rather than from zygotes.) The midrib region of the thallus regenerates very much more rapidly than the wings (Moss 1964), correlating with the abundance in the midrib of medullary filaments, which are primarily responsible for formation of new epidermis. Regeneration from vegetative branches always gives rise to vegetative shoots. Regeneration of strips cut from the discolored frond beneath spent receptacles of the dioicous species *F. vesiculosus*, although extremely slow, resulted in branches with small receptacles at their tips. Branches regenerated from strips cut from male thalli bore male receptacles, those from female thalli bore female receptacles (Moss 1964). (The nature of sex determination in this diploid genus has not been established.)

Polarity is also expressed in apical dominance. This phenomenon is well known in seed plants, where the apical meristem often inhibits the growth of the

Figure 10.18. (a) Formation of rhizoid (pr) and repair-shoot cell (rs) when apical and basal fragments of *Griffithsia* are apposed within a cytoplasm-free *Nitella* cell wall (nw). The apical fragment is at the bottom of the photograph. (b) Formation of normal apical cell (ns) and rhizoid (gr) rather than repair-shoot cell when fragments from different species are apposed. (From Waaland 1975, with permission of Springer-Verlag)

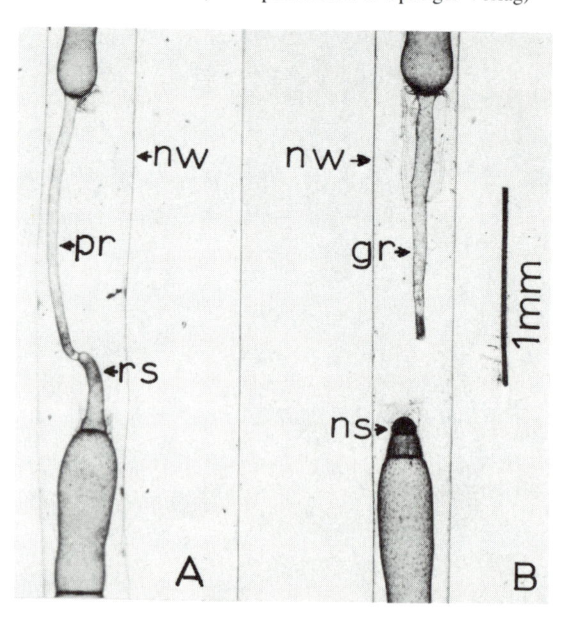

branch meristems in the axils of the leaves. Among the seaweeds are a few known examples of lateral branch inhibition by the apical cell or region (reviewed by Buggeln 1981). Among these is *Schottera nicaeensis*, which was discussed earlier. Differences in regeneration and proliferation with and without the apex in this species are shown in Figure 10.19. With the apex present rhizoidal regeneration occurs from the upper (i.e., downward facing) side of the cuts near the base, and there are no proliferations near the apex. If the apex is cut off, no rhizoidal regeneration occurs in the slits but bladelike proliferation takes place from the lower sides of all cuts and at the tip of the blade. Other examples of apical dominance include the inhibition of development of lateral fronds in *Sargassum* (Chamberlain et al. 1979) and probably also in *Macrocystis* (Lobban 1978a).

10.5.3 Evidence for internal growth regulators

Discussions of apicobasal polarity and of apical dominance lead naturally to a consideration of growth regulator compounds in seaweeds. (The word ''hormone,'' originally coined for animal morphogenetic substances, implies substances produced in specific sites – glands – and having specific targets. In plants the term is best restricted to compounds like rhodomorphin and the sex attractants. Plant growth regulators are produced

Figure 10.19. Effect of apex removal on regeneration (R) and proliferation (P) in *Schottera nicaeensis* with slits cut across the middle of the thallus. (a) Apex present; (b) apex removed. (From Perrone & Felicini 1972, with permission of Blackwell Scientific Publications)

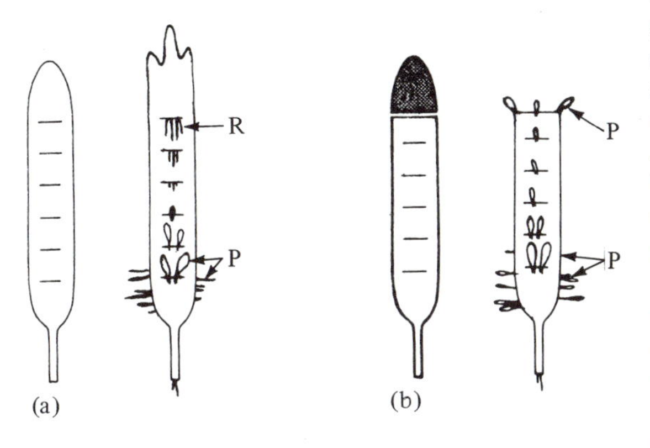

in unspecialized tissues and work together on many aspects of morphogenesis.) To begin with, growth and development of seaweeds clearly are coordinated processes, implying control. Some of these controls are external, such as photoperiod, whereas others appear to be internal, mediated by chemical growth regulators. External chemicals required for growth, such as vitamins and the compounds required for cell adhesion in *Ulva* and *Monostroma*, are not considered to be growth regulators, although like nutrients and light they certainly affect growth. Many studies have looked at the effects on seaweeds of applying higher plant growth regulators such as auxins and cytokinins. Many other studies have attempted to extract and characterize seaweed compounds with growth regulatory effects. A comprehensive review of such studies was published in a series of papers by Augier (most recently Augier 1978). Provasoli & Carlucci (1974) pointed out some of the problems with such studies and Buggeln (1981) critically assessed the current evidence for higher plant growth regulators in seaweeds. The evidence for the presence and nature of growth regulators in seaweeds remains equivocal.

Among the examples of effects of added growth regulators are some of relevance to earlier discussion. The auxin indole-3-acetic acid (IAA) at 1 ppm inhibited growth of excised lateral fronds of *Sargassum muticum*, suggesting that this compound might be the agent of apical dominance in this alga, as it is in higher plants (Chamberlain et al. 1979). IAA has also been shown to promote rhizoid formation in several seaweeds in culture, among them *Bryopsis*, *Ulva*, and *Caulerpa* (Moss 1974a). Interestingly, *Caulerpa* growing tips contain endosymbiotic bacteria (Dawes & Lohr 1978), and a substratum of sand rich in microorganisms was just as effective as added IAA in promoting rhizoids. The cytokinin kinetin (6-furfuryl amino purine) induced *Ectocarpus fasciculatus* in axenic culture to produce erect axes; without the cytokinin only prostrate axes were produced (Pedersén 1973). Specificity for particular chemicals was not established in any of these cases.

One means of demonstrating specificity is the use of compounds having similar structures to the putative growth regulator, but having, in higher plants, no growth regulatory activity or producing opposite effects. The specificity for IAA can be tested with a number of antiauxins, including 2,4-D, 3,5-D, and naphthoxyacetic acids (NOAA). Growth inhibition of *Alaria esculenta* blades by these compounds could not be distinguished from the response to IAA, so that this response in this plant is probably not auxin-specific (Buggeln 1976).

Among the major problems in demonstrating the presence of growth regulators is that they are present in exceedingly small quantities. To get enough material for satisfactory identification, huge quantities of seaweed must be extracted, yet such quantities cannot be obtained from axenic culture. Thus, included in the extraction are all the microbiota epiphytic on natural plants. Some pathogenic terrestrial bacteria and fungi are known producers of higher plant growth regulators (Bidwell 1979) (although this ability may have been acquired in these special cases by gene transfer from the hosts – Chapman & Ragan 1980). In principle, as Buggeln points out, the origin of the substance is unimportant provided it has a specific action on seaweed growth. Identification is also problematical. Standard tests such as the oat coleoptile bending test for auxins (see Bidwell 1979) are unreliable for seaweeds whose chemical composition is different from angiosperms. These tests must be complemented by chromatographic comparison of seaweed substances and authentic compounds, gas chromatography, and mass spectrometry (see Brenner 1981). Abe et al. (1972), using a fairly reliable combination of these methods, reported IAA and indole-3-carboxylic acid in young *Undaria pinnatifida* (Laminariales). The yield was 10 μg from 1 kg of fresh kelp from nature.

There are also many difficulties in interpreting the results of adding higher plant growth regulators to seaweeds in nature or in culture. Are the substances ever taken up by the seaweeds or are they removed by microorganisms? If they are taken up by microbes, do these organisms release other substances, such as vitamins, which affect seaweed growth? If the growth regulator is taken up by the seaweed, a negative response to it may mean that there was already an optimal concentration of it in the alga; on the other hand, a positive response does not prove that the specific compound is the native regulator in the seaweed.

The presence of growth regulators in seaweeds seems certain. The evidence suggests that at least some of these compounds may be similar, perhaps even identical, to some higher plant growth regulators, and that they are probably produced by epiphytic microbiota as well as by the seaweeds themselves.

10.6 Synopsis

The life of a seaweed is a complex sequence of interactions between the genetic information and external stimuli and constraints. Development, from the initial polarization of the spore or zygote to the production and release of reproductive cells, is a highly coordinated process. Light (quality and quantity), photoperiod (actually the length of uninterrupted darkness), and temperature are the principal environmental cues, with gravity and water motion sometimes playing roles. Cells in a thallus also receive influences, probably both physical and chemical, from other cells. Cells released from these constraints may exhibit totipotency and regenerate an entire thallus, or some genes may remain repressed so that certain parts (e.g., rhizoids) cannot be regenerated. Erect thalli characteristically have an apicobasal polarity, which is expressed in the position and kind of regenerative outgrowths on wounded thalli and sometimes in apical dominance. Growth regulating substances similar or identical to higher plant growth regulators are almost certainly present, although most evidence is still circumstantial and some of these compounds may in fact be produced by epiphytic microorganisms. Vitamins and cell-binding factors are also produced by the microbiota. Other substances, such as rhodomorphin, are probably unique to the algae.

11 Mariculture

11.1 Introduction

Mariculture is the cultivation of the sea: the large-scale culture of commercially useful organisms, including seaweeds, as economically successful crops. It is distinct from simple harvesting of wild stocks. In Japan, China, and other Asian countries, where seaweeds are an important part of the human diet, seaweed farming is a major business. In other regions of the world, where seaweeds are used chiefly as animal fodder, fertilizer, or for phycocolloid extraction, harvesting is usually of wild stocks (Hoppe & Schmid 1969). Occasionally some habitat improvement is practiced. In recent years seaweeds have also been considered as potential solar energy converters, providing biomass as a source of food for methane-producing bacteria. Mariculture depends on improving conditions found in the sea, improving the plant material, or creating artificial environments that to a greater or lesser extent mimic the habitat of the plant. Thus, just as agriculture depends on vascular plant physiology for basic understanding of the crops, mariculture depends on a knowledge of what factors are most important to the plants, the extent to which these factors can be manipulated to improve yield, and the consequences of unavoidable differences from the original habitat.

Ancient records establish that people collected seaweeds for food as long ago as 2500 B.P. in China and 1500 B.P. in Europe (Levring 1977, Tseng 1981). Yet only in the last 250 years has the practice grown, first in Japan and then in China, from simply harvesting wild stands to the development of methods of selection, breeding, and cultivation of certain species. As part of the human diet, seaweeds provide protein, vitamins, and minerals, especially iodine. Four major crop plant genera in Asia are the reds *Eucheuma* and *Porphyra* and the browns *Laminaria* and *Undaria*.

The commercially important phycocolloids are agars and carrageenans from red algae, alginates from browns. Agars, obtained from such genera as *Gelidium* and *Gracilaria*, are used extensively in microbiology and tissue culture for solidifying growth media. Carrageenans, chiefly from *Eucheuma* and *Chondrus*, are widely used as a thickener in dairy products. Alginates, from *Macrocystis* and *Laminaria*, are also used as thickeners in a multitude of products from salad dressings to oil drilling fluids, and as coatings for papers (Waaland 1981, Chapman & Chapman 1980). Numerous products are refined from each raw material for specific applications. For example, Kelco Company produces sodium, potassium, ammonium, and calcium salts and the propylene glycol ester of alginic acid. Each salt is prepared in several formulations, including granular or fibrous particles of various mesh sizes, and different viscosities (molecular weights), dispersion characteristics, and pH (Kelco, no date).

Some 400 to 500 species of seaweed are collected for food, fodder, or chemicals, but fewer than 20 species in 11 genera are commercially cultivated (Michanek 1978, Tseng 1981, van der Meer 1983). In the following section, the major aspects of *Porphyra* mariculture are considered as an example of the application of the ecological and physiological principles and parameters described in the preceding chapters.

11.2 Present-day seaweed mariculture

11.2.1 *Porphyra* cultivation

The farming practices developed in Japan and China for *Porphyra* illustrate the basic principles of seaweed mariculture for food. *Porphyra* is known in Japan as *nori*, in China as *zicai*, and in the west as purple laver. Its cultivation began in Tokyo Bay some 300 years ago and remained there until the early nineteenth century, when the practice gradually spread to other areas of Japan (Okazaki 1971, Tseng 1981). Cultivation was originally done by pushing tree branches or bamboo shoots into the mud of the bay bottom to catch drifting *Porphyra* spores. Later, horizontal nets were slung between poles. The nets were more readily transported from the collecting grounds to the fisherman's cultivation area. Until 1949, when Drew showed that the genus *Conchocelis* is a stage in the *Porphyra* life cycle, the fishermen did not know where the spores came from, nor that the habitat of the unknown plant was quite different from that of the crop. Drew's revelation transformed the nori industry, allowing indoor mass cultivation of the filaments in disinfected oyster shells and the seeding of conchospores directly onto nets for outplanting in the sea.

Mass culture of conchocelis takes place in tanks in greenhouses (Miura 1975, Tseng 1981). In February or March *Porphyra* thalli are induced to release carpospores by being dried overnight and then being reimmersed in seawater for 4 to 5 h. Between 15 and 150 g of *Porphyra* blades, depending on the species, are sufficient for coverage of about 3 m^2 of shells. Sterile oyster shells or artificial substrate treated with calcite granules are placed in seawater tanks with the fertile *Porphyra* blades or sprinkled with a suspension of carpospores. The best conditions for conchocelis to bore into the shells are a temperature of 10 to 15 C and bright but not direct sunlight. Good growth of conchocelis requires bright daylight and abundant nutrients. Nitrogen, phosphorus, and potassium are added and the water in the culture tanks is stirred to improve exchange and uptake. During the early summer the water temperature increases from <15 C to >25 C. Midday irradiance is kept to about 3000 lux (ca. 55 $\mu E\ m^{-2}\ s^{-1}$) by using screens.

From early July to late August–September the water temperature rises from about 22 C to 28 – 30 C and then gradually decreases. This is a critical time for formation of conchosporangia and maximum conchospore production, which are dependent on temperature and photoperiod. Light is manipulated to promote accumulation of reserves and to delay sporulation (Tseng 1981). Irradiance is first reduced to about one quarter (ca. 15 $\mu E\ m^{-2}\ s^{-1}$) in early to mid-July and held there as temperatures continue to rise. In September, when temperatures have fallen to about 23 C, sporulation is encouraged by artificially reducing the photoperiod to 8 to 10 h per day. Conchospores are collected on the growth nets either by placing the nets over the shells in the indoor tanks or by transferring the shells into large floating containers and placing these under nets in the field. Since the spores are nonmotile, water motion is essential to transfer the spores to the nets. Dense settlement of spores (3–5 mm^{-2}) is an important method of weed control. Conchospore adherence and germination require brighter light so irradiance is increased back up to 50 $\mu E\ m^{-2}\ s^{-1}$ or more. Conchospore germination is usually carried out in the sea.

Several methods are used for suspending the nets, depending on the depth of water and the tidal amplitude. In very shallow areas the nets can be suspended from fixed poles, so that the plants are regularly exposed to the atmosphere. If the tide range is greater than about 2 m, the nets are attached to poles that rest on the bottom at low tide but float up on the rising tide. This avoids too much shading by the water column. In deep water, *Porphyra* is grown on nets that can float near the surface, but the nets are placed in intertidal cultivation until the germlings are 10 to 20 mm long. Periodic exposure of the proper duration reduces diseases and growth of competitive (weed) species (Tseng 1981). As discussed in Section 3.3.3, the temperature growth optimum decreases as the thallus ages. Thus timing of outplanting is important. Delay slows the growth of the germlings and results in a later harvest. Seawater is considered infertile for nori growth if $NH_4^+ + NO_3^-$ concentration is <50 $\mu g\ L^{-1}$. Best-quality nori is obtained when nitrogen is > 200 $\mu g\ L^{-1}$. Fertilizer such as ammonium sulfate is applied as a spray over the beds or allowed to diffuse from bottles hung on the support poles. The best-quality plants are harvested from October to December under normal growth conditions. At the end of the season, in January and February, mature plants are of lesser quality (Okazaki 1971). The harvested nori is thoroughly washed in seawater and all epiphytes and dead tissues are removed. The thalli are chopped and mixed with freshwater, spread over screens, and dried. Finished nori sheets are approximately 200 × 180 mm. One net, 18 × 1.2 m, with a stretched mesh size of 300 mm, produces between 300 and 2000 sheets of dried nori (Miura 1975).

The discovery that *Porphyra* germlings could survive deep freezing added a new dimension to nori farming. If the thalli are allowed to dry to between 20% and 30% of their moisture content, frozen, and stored at −20 C, they can resume normal growth as much as a year later. This practice can be used to extend the useful harvest period into March or April, and also serves as insurance against failure of the early crop.

Each commercially cultivated seaweed has some reported diseases, some caused by pathogens, others by adverse physical conditions (Andrews 1976, Tseng 1981). For example, *Porphyra* is susceptible to "red wasting disease" caused by a chytrid fungus, and to "green spot disease" caused by pathogenic species of the bacteria *Vibrio* and *Pseudomonas*. In Japan, fog has been known to seriously damage nori crops, which are exposed to the atmosphere part of each day. Sulfites in air pollution are part of the cause. Plants in greenhouses may be affected by H_2S coming from sulfate-reducing bacteria in the water pipes. Disease need not kill the plants to destroy or reduce their value as a crop.

11.2.2 Phycocolloid production

The great bulk of the phycocolloids is extracted from wild plants at present, although there is much interest in the development of mariculture techniques. Perhaps the major hurdle is economics. Whereas edible seaweeds fetch good prices, seaweed gums are of relatively low value and must also compete with cheap gums from terrestrial plants, such as tragacanth gum, gum arabic, and fruit pectin, and from bacterial fermentation, such as xanthan gum. Moreover, much of the value of the product is added during processing and refining. The various seaweed colloids have unique properties that suit them to particular applications. For example, carrageenan is especially suitable for interaction with milk protein, hence its widespread use in the dairy industry. Successful seaweed mariculture in the Orient is labor-intensive in a region where labor is inexpensive. The only successful mariculture for a phycocolloid is production of *Eucheuma* for carrageenan, which takes place in the Phillipines (Doty & Alvarez 1975). In the West, where labor is expensive, capital-intensive, low-labor methods must be sought. Such schemes are financially risky given the low value of the product and the availability of wild stocks.

The demand for phycocolloids continues to increase, however, while wild stocks decrease. The 1970s were years of much experimentation in the design of semienclosed seawater systems to grow phycocolloid seaweeds on a large scale. In the 1980s, when economic conditions became tighter, some of the optimism dissipated in the face of the technical problems and the realization of the economic feasibility of the schemes. Nevertheless, work continues on solving the problems of scaling up from laboratory cultures to commercial production. One may expect that successful mariculture will occur first with species producing special application, higher-value polysaccharides or other organic molecules.

11.3 Applications of laboratory cultures

Successful mariculture depends on knowledge of the basic biology of the crop species. What is its life history, and how do environmental variables impinge on this? What conditions give the best growth or the best yield of the valuable extract? How do growth and chemical composition change from time to time? Many of the answers to such questions must be sought through laboratory experimentation, as was discussed in Section 9.7. Here we summarize a few studies that have been or will be important to mariculture.

The importance of knowledge of the life history is evident in the payoff from Drew's discovery of the *Conchocelis* stage in the life history of *Porphyra umbilicalis* and Iwasaki's (1961) subsequent corroboration with the commercial *P. tenera*. The Japanese erected a shrine to Kathleen Drew (Baker 1965). Often the laboratory culture of a species reveals stages that had not been found in nature owing to their small size. A recent example is van der Meer and Todd's (1980) discovery of the extremely small female gametophyte of the commercial species *Palmaria palmata* (dulse). Field phycologists had found only males and tetrasporophytes, so the plant was believed to have no sexual reproduction (see Bold & Wynne 1978). Another fairly recent example is the connection of some species of *Gigartina* (a carrageenophyte) and the crustose *Petrocelis* (Polanshek & West 1977).

Genetic studies are at the heart of breeding programs. A lengthy series of studies on the agarophyte *Gracilaria tikvahiae* has been carried out by van der Meer and co-workers in Halifax. These studies began in 1977 (van der Meer & Bird) with an investigation of inheritance of two spontaneous green mutants (recall Sec. 10.4.1). In the most recent study, Patwary & van der Meer (1983) compared frond composition and agar quality between morphological mutants and wild-type plants. Among seven mutant strains there were at least three different agars, whereas wild-type clones were relatively uniform. One mutant, MP-40, yielded not only exceptionally strongly gelling agar, it also had rapid growth and excellent epiphyte resistance. These results have immediate practical application. The Chinese have carried out extensive selection and hybridization programs and developed strains of *Laminaria japonica* that can be farmed in the relatively warm waters of China (Tseng 1981). They have produced several successful strains, including one with high iodine content.

Tissue culture studies offer potential in breeding programs. While tissue culture is becoming common in work with vascular crop plants, few attempts have been made with seaweeds. Saga et al. (1978) made some progress in cloning *Laminaria* from isolated single cells of the sporophyte. Chen & Taylor (1978) developed techniques for establishing axenic tissue cultures from medullary tissue of *Chondrus crispus*. They obtained fronds with apparently normal morphology of the parent plant, starting from a 2-mm cube of tissue in medium enriched with growth regulators.

Perhaps the most common application of laboratory cultures is in screening wild populations for superior individuals. Domesticated *Porphyra* in Japan is greatly different from the original wild plants (Miura et al. 1979). The Chinese are considered world leaders in special breeding of commercial seaweeds, notably *Laminaria japonica* (van der Meer 1983). Laboratory cultures have been the basis for selection of fast-growing strains of *Eucheuma* spp. (Doty & Alvarez 1975), *Gigartina exasperata* (Waaland 1978), and *Gracilaria tikvahiae* (Ryther et al. 1979).

11.4 Synopsis

Seaweeds are used by humans as sources of food, fodder, fertilizer, and chemicals. A few of the more valuable food plants are cultivated, whereas the others are harvested from wild stocks. Cultivation is restricted to the Orient at present because of the high value of the crops there and the low labor costs. In the West, attempts at initiating commercially viable seaweed mariculture have not gone beyond the pilot-project stage.

Porphyra spp. and kelps, chiefly *Laminaria* and

Undaria, are cultivated. Spores from the macroscopic stages are collected and the microscopic stages raised indoors. Young sporophytes on nets or ropes are hung out in the sea to mature. Fertilizer application is frequently necessary and the farmer must also be alert for problems from too high or too low irradiance and temperatures, epiphytes and competition from weed species, and diseases.

Breeding programs are a key element in mariculture. Just as our terrestrial crops have been improved over the original wild plants, so the strains of *Porphyra* and *Laminaria* under cultivation are already different from wild plants.

Laboratory experiments provide the understanding of crop plant life histories, ecology, and physiology that is essential for successful mariculture.

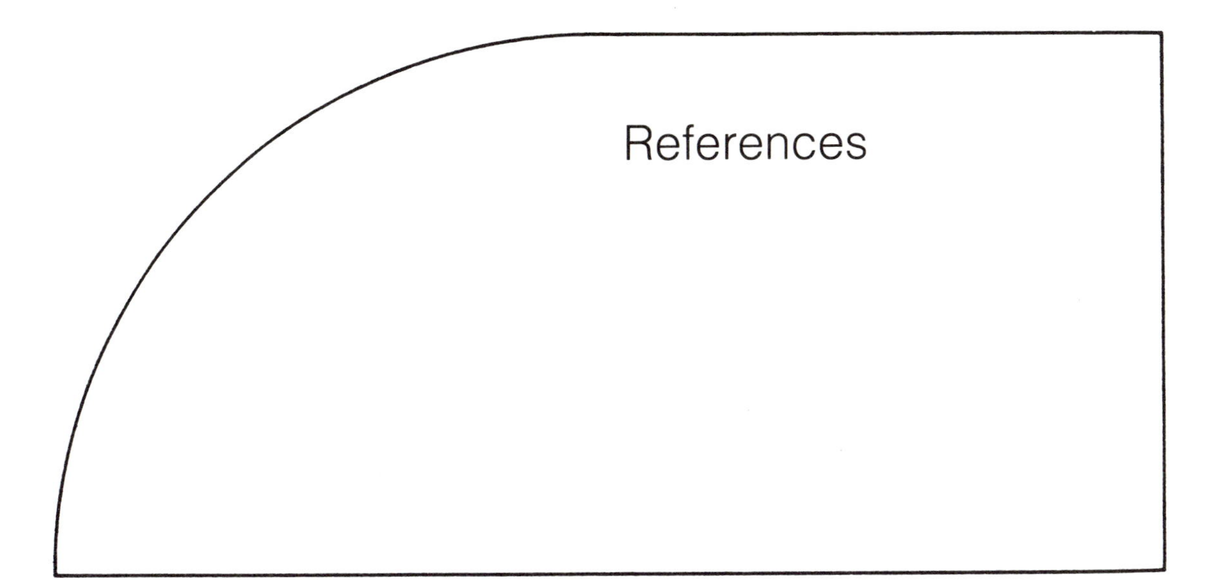

References

Aaronson, S. 1973. Effect of incubation temperature on the macromolecular and lipid content of the phytoflagellate *Ochromonas danica*. *J. Phycol.* 9:111–113.

Abe, H., M. Uchiyama, & R. Sato. 1972. Isolation and identification of native auxins in marine algae. *Agric. Biol. Chem.* 36:2259–2260.

Abe, M. 1969. Rhizoid differentiation in *Fucus* eggs in low temperatures. *Bot. Mag. Tokyo* 82:53–55.

Adamich, M., A. Gibor, & B.M. Sweeney. 1975. Effects of low nitrogen levels and various nitrogen sources on growth and whorl development in *Acetabularia* (Chlorophyta). *J. Phycol.* 11:364–367.

Adey, W.H. 1970. The effects of light and temperature on growth rates in boreal-subarctic crustose corallines. *J. Phycol.* 6:269–276.

– 1973. Temperature control of reproduction and productivity in a subarctic coralline alga. *Phycologia* 12:111–118.

Aghajanian, J.G., & M.H. Hommersand. 1978. The fine structure of the pit connections of *Batrachospermum sirodotii* Skuja. *Protoplasma* 96:247–265

Akazawa, T., & C.B. Osmond. 19[illegible] Structural properties and ribulosebisphosphate carboxylase and oxygenase activity of fraction-1 protein from the marine alga *Halimeda cylindracea* (Chlorophyta). *Austr. J. Plant Physiol.* 3:93–103.

Albright, L.J., J. Chocair, K. Masuda, & M. Valdés. 1980. *In situ* degradation of the kelps *Macrocystis integrifolia* and *Nereocystis luetkeana* in B[illegible]h Columbia coastal waters. *Naturaliste Can.* 107:3–10.

Alderdice, D.F. 1972. Factor combinations. Responses of marine poikilotherms to environmental factors acting in concert. In O. Kinne (ed.), *Marine Ecology*, vol. I, pt. 3, pp. 1659–1722. Wiley, New York.

Allen, T.F.H. 1977. Scale in microscopic algal ecology: a neglected dimension. *Phycologia* 16:253–257.

Al-ogily, S.M., & E.W. Knight-Jones. 1977. Anti-fouling role of antibiotics produced by marine algae and bryozoans. *Nature* 265:728–729.

Anderson, F.E., & J. Green. 1980. Estuaries. *The New Encyclopaedia Brittanica, Macropaedia*, vol. 6, pp. 968–976.

Anderson, J.M., & J. Barrett. 1979. Chl *a*–protein complexes of brown algae: P700 reaction center and LH complexes. In *Chlorophyll Organization and Energy Transfer in Photosynthesis*, CIGA Foundation Symposium 61, pp. 81–96. Excerpta Medica, Amsterdam.

Anderson, M.A., & F.M.M. Morel. 1980. Uptake of Fe(II) by a diatom in oxic culture medium. *Mar. Biol. Lett.* 1:263–268.

– 1982. The influence of aqueous iron chemistry on the uptake of iron by the coastal diatom *Thalassiosira weissflogii*. *Limnol. Oceanogr.* 27:789–813.

Anderson, S.M., & A.C. Charters. 1982. A fluid dynamics study of seawater flow through *Gelidium nudifrons*. *Limnol. Oceanogr.* 27:399–412.

Andrews, J.H. 1976. The pathology of marine algae. *Biol. Rev.* 51:211–253.

Archambault, D., & E. Bourget. 1983. Importance du régime de dénudation sur la structure et la succession des communautés intertidales de substrat rocheux en milieu subarctique. *Can. J. Fish. Aquat. Sci.* 40:1278–1292.

Arnon, D.I., & P.R. Stout. 1939. The essentiality of certain elements in minute quantity for plants, with special reference to copper. *Plant Physiol.* 14:371–375.

Asare, S.O., & M.M. Harlin. 1983. Seasonal fluctuations in tissue nitrogen for five species of perennial macroalgae in Rhode Island Sound. *J. Phycol.* 19:254–257.

Asensi, A.O., R. Delépine, & G. Gugliemi. 1977. Nouvelles observations sur l'ultrastructure du plastidome des Phéophycées. *Bull. Soc. Phycol. Fr.* 22:192–205.

Atkins, G.L., & I.A. Nimmo. 1980. Current trends in the estimation of Michaelis-Menten parameters. *Anal. Biochem.* 104:1–9.

Atkinson, M.J., & S.V. Smith. 1983. C:N:P ratios of benthic marine plants. *Limnol. Oceanogr.* 28:568–574.

Augier, H. 1978. Les hormones des algues. Etat actuel des connaissances. VII. Applications, conclusion, bibliographie. *Bot. Mar.* 21:175–197.

Avers, C.J. 1981. *Cell Biology*, 2nd ed. Van Nostrand, New York. 566 pp.

Avron, M. 1981. Photosynthetic electron transport and photophosphorylation. In M.D. Hatch & N.K. Boardman (eds.), *The Biochemistry of Plants*, vol. 8, *Photosynthesis*, pp. 163–191. Academic Press, New York.

Baardseth, E. 1970. A square-scanning, two-stage sampling method of estimating seaweed quantities. *Norwegian Inst. Seaweed Res. Rep.* 33, 41 pp.

Baca, B.J., & E.R. Cox. 1979. A developmental mutant in *Enteromorpha lingulata* J.Ag. (Chlorophyta, Ulvales). *Phycologia* 18:369–377.

Baker, H.W. 1965. The story of nori in Japan. *Brit. Phycol. Bull.* 2:497–500.

Baker, J.M. 1970. The effects of oils on plants. *Environ. Pollut.* 1:27–44.

Bannister, P. 1976. *Introduction to Physiological Plant Ecology*. Blackwell Scientific Publications, Oxford. 273 pp.

Barber, R.T., & J.H. Ryther. 1969. Organic chelators: factors affecting primary production in the Cromwell Current upwelling. *J. Exp. Mar. Biol. Ecol.* 3:191–199.

Barber, R.T., R.C. Dugdale, J.J. MacIsaac, & R.L. Smith. 1971. Variations in phytoplankton growth associated with the source and conditioning of upwelling water. *Invest. Pesq.* 35:171–199.

Barnes, H., & J.A. Topinka. 1969. Effect of the nature of the substratum on the force required to detach a common littoral alga. *Amer. Zool.* 9:753–758.

Bauernfeind, J.C., & E. DeRitter. 1970. *Handbook of Biochemistry*, 2nd ed. CRC Press, Boca Raton, Fla.

Beardall, J., H. Griffiths, & J.A. Raven. 1982. Carbon isotope discrimination and the CO_2 accumulating mechanism in *Chlorella emersonii*. *J. Exp. Bot.* 33:729–737.

Beer, S., & A. Eshel. 1983. Photosynthesis of *Ulva* sp. I. Effects of desiccation when exposed to air. *J. Exp. Mar. Biol. Ecol.* 70:91–97.

Beevers, L. 1976. *Nitrogen Metabolism in Plants*. Arnold, London. 339 pp.

Beijer, K., & A. Jerenlöv. 1979. Sources, transport and transformation of metals in the environment. In L. Friberg, G.F. Nordberg, & V.B. Vouk (eds.), *Handbook on the Toxicology of Metals*, pp. 47–63. Elsevier, Amsterdam.

Bell, E.A. 1980. The possible significance of secondary compounds in plants. In E.A. Bell & B.V. Charlwood (eds.), *Secondary Plant Products. Encycl. Plant Physiol.*, New Ser. 8:11–21.

Bender, M.E., G.P. Klinkhammer, & D.W. Spencer, 1977. Manganese in seawater and the marine manganese balance. *Deep Sea Res.* 24:799–812.

Benson, E.E., J.C. Rutter, & A.H. Cobb. 1983. Seasonal variation in frond morphology and chloroplast physiology of the intertidal alga *Codium fragile* (Suringar) Hariot. *New Phytol.* 95:569–580.

Bidwell, R.G.S. 1979. *Plant Physiology*, 2nd ed. Macmillan, New York. 726 pp.

– 1983. Carbon nutrition in plants. In F.C. Steward & R.G.S. Bidwell (eds.), *Plant Physiology: A Treatise*, vol. VII, pp. 287–457. Academic Press, New York.

Biebl, R. 1962. Seaweeds. In R.A. Lewin (ed.), *Physiology and Biochemistry of the Algae*, pp. 799–815. Academic Press, New York.

– 1970. Vergleichende Untersuchungen zur Temperaturresistenz von Meeresalgen entlang der pazifischen Küste Nordamerikas. *Protoplasma* 69:61–83.

Bingham, S., & J.A. Schiff. 1979. Conditions for attachment and development of single cells released from mechanically-disrupted thalli of *Prasiola stipitata* Suhr. *Biol. Bull.* 156: 257–271.

Birch, P.B., J.O. Gabrielson, & K.S. Hamel. 1983. Decomposition of *Cladophora*. I. Field studies in the Peel-Harvey Estuary System, Western Australia. *Bot. Mar.* 26:165–171.

Bird, K.T. 1976. Simultaneous assimilation of ammonium and nitrate by *Gelidium nudifrons* (Gelidiales: Rhodophyta). *J. Phycol.* 12:238–241.

Bird, K.T., C.J. Dawes, & J.T. Romeo. 1980. Patterns of non-photosynthetic carbon fixation in dark held, respiring thalli of *Gracilaria verrucosa*. *Z. Pflanzenphysiol.* 98:359–364.

Bird, K.T., C. Habig, & T. DeBusk. 1982. Nitrogen allocation and storage patterns in *Gracilaria tikvahiae* (Rhodophyta). *J. Phycol.* 18:344–348.

Bisalputra, T. 1974. Plastids. In W.D.P. Stewart (ed.), *Algal Physiology and Biochemistry*, pp. 124–160. Blackwell Scientific Publications, Oxford.

Bisalputra, T., C.M. Shields, & J.W. Markham. 1971. *In situ* observations of the fine structure of *Laminaria* gametophytes and embryos in culture. *J. Microscopie* 10:83–98.

Bishop, P.L. 1983. *Marine Pollution and Its Control*. McGraw-Hill, New York. 357 pp.

Bisson, M.A., & G.O. Kirst. 1979. Osmotic adaptation in the marine alga *Griffithsia monilis* (Rhodophyceae): the role of ions and organic compounds. *Austr. J. Plant Physiol.* 6:523–538.

Black, W.A.P. 1949. Seasonal variation in chemical composition of some of the littoral seaweeds common to Scotland. Part II. *Fucus serratus, Fucus vesiculosus, Fucus spiralis* and *Pelvetia canaliculata*. *J. Soc. Chem. Ind.* 68:183–189.

– 1950. The seasonal variation in weight and chemical composition of the common British Laminariaceae. *J. Mar. Biol. Ass. U.K.* 29:45–72.

Bogorad, L. 1975. Phycobiliproteins and complementary chromatic adaptation. *Ann. Rev. Plant Physiol.* 26:369–401.

Bold, H.C., & M.J. Wynne. 1978. *Introduction to the Algae: Structure and Reproduction*. Prentice-Hall, Englewood Cliffs, N.J. 706 pp.

Bold, H.C., C.J. Alexopoulos, & T. Delevoryas. 1980. *Morphology of Plants and Fungi*, 4th ed. Harper & Row, New York. 819 pp.

Bolton, J.J. 1979. Estuarine adaptation in populations of *Pilayella littoralis* (L.) Kjellm. (Phaeophyta, Ectocarpales). *Estu. Cstl. Mar. Sci.* 9:273–280.

– 1981. Community analysis of vertical zonation patterns on a Newfoundland rocky shore. *Aquat. Bot.* 10:299–316.

– 1983. Ecoclinal variation in *Ectocarpus siliculosus* (Phaeophyceae) with respect to temperature growth optima and survival limits. *Mar. Biol.* 73:131–138.

Bolwell, G.P., J.A. Callow, M.E. Callow, & L.V. Evans. 1979. Fertilization in brown algae. II. Evidence for lectin-sensitive complementary receptors involved in gamete recognition in *Fucus serratus*. *J. Cell Sci.* 36:19–30.

Bolwell, G.P., J.A. Callow, & L.V. Evans. 1980. Fertilization in brown algae. III. Preliminary characterization of putative gamete receptors from eggs and sperm of *Fucus serratus*. *J. Cell Sci.* 43:209–224.

Boney, A.D. 1963. The effects of 3 amino-1,2,4-triazole on the growth of sporelings of marine red algae. *J. Mar. Biol. Ass. U.K.* 43:643–652.

– 1970. Toxicity studies with an oil-spill emulsifier and the green alga *Prasinocladus marinus*. *J. Mar. Biol. Ass. U.K.* 50:461–473.

– 1975. Mucilage sheaths of spores of red algae. *J. Mar. Biol. Ass. U.K.* 55:511–518.

Borowitzka, M.A. 1972. Intertidal algal species diversity and

the effect of pollution. *Austr. J. Mar. Freshwater Res.* 23:73–84.
– 1977. Algal calcification. *Oceanogr. Mar. Biol. Ann. Rev.* 15:189–223.
– 1979. Calcium exchange and the measurement of calcification rates in the calcareous coralline red alga *Amphiroa foliacea*. *Mar. Biol.* 50:339–347.
– 1982. Mechanisms in algal calcification. In F.E. Round & D.J. Chapman (eds.), *Progress in Phycological Research*, vol. 1, pp. 137–177. Elsevier, Amsterdam.
Borowitzka, M.A., A.W.D. Larkum, & C.E. Nockolds. 1974. A scanning electron microscope study of the structure and organization of the calcium carbonate deposits of algae. *Phycologia* 13:195–203.
Borstlap, A.C. 1981. Invalidity of the multiphasic concept of ion absorption in plants. *Plant Cell Environ.* 4:189–195.
Box, G.E.P., W.G. Hunter, & J.S. Hunter. 1978. *Statistics for Experimenters: An Introduction to Design, Data Analysis and Model Building*. Wiley, New York. 653 pp.
Boyle, E.A. 1979. Copper in natural waters. In J.O. Nriagu (ed.), *Copper in the Environment*, pt. 1, pp. 77–88. Wiley, New York.
Brawley, S.H., & J.R. Sears. 1982. Septal plugs in a green alga. *Amer. J. Bot.* 69:455–463.
Brawley, S.H., & R. Wetherbee. 1981. Cytology and ultrastructure. In C.S. Lobban & M.J. Wynne (eds.), *The Biology of Seaweeds*, pp. 248–299. Blackwell Scientific Publications, Oxford.
Brenner, M.L. 1981. Modern methods for plant growth substance analysis. *Ann. Rev. Plant Physiol.* 32:511–538.
Brinkhuis, B.H. 1976. The ecology of temperate salt-marsh fucoids. I. Occurrence and distribution of *Ascophyllum nodosum* ecads. *Mar. Biol.* 34:325–338.
– 1977. Comparisons of salt-marsh fucoid production estimated from three different indices. *J. Phycol.* 13:328–335.
Brinkhuis, B.H., N.R. Tempel, & R.F. Jones. 1976. Photosynthesis and respiration of exposed salt-marsh fucoids. *Mar. Biol.* 34:349–359.
Britz, S.J., & W.R. Briggs. 1976. Circadian rhythms of chloroplast orientation and photosynthetic capacity in *Ulva*. *Plant Physiol.* 58:22–27.
Broecker, W.S. 1974. *Chemical Oceanography*. Harcourt Brace Jovanovich, New York. 214 pp.
Brown, J.C. 1978. Mechanism of iron uptake by plants. *Plant Cell Environ.* 1:249–257.
Brown, L.M., & J. McLachlan. 1982. Atypical carotenoids for the Rhodophyceae in the genus *Gracilaria* (Gigartinales). *Phycologia* 21:9–16.
Brown, V., S.C. Ducker, & K.S. Rowan. 1977. The effect of orthophosphate concentrations on the growth of the coralline algae (Rhodophyta). *Phycologia* 16:125–131.
Bryan, G.W. 1969. The absorption of zinc and other metals by the brown seaweed *Laminaria digitata*. *J. Mar. Biol. Ass. U.K.* 49:225–243.
Bryan, G.W. & L.G. Hummerstone. 1973. Brown seaweed as an indicator of heavy metals in estuaries in south-west England. *J. Mar. Biol. Ass. U.K.* 53:705–720.
Bryhni, E. 1978. Quantitative differences between polysaccharide compositions in normal differentiated *Ulva mutabilis* and the undifferentiated mutant *lumpy*. *Phycologia* 17:119–124.
Brylinski, M. 1977. Release of dissolved organic matter by marine macrophytes. *Mar. Biol.* 39:213–220.
Buggeln, R.G. 1974. Negative phototropism of the haptera of *Alaria esculenta* (Laminariales). *J. Phycol.* 10:80–82.
– 1976. Auxin, an endogenous regulator of growth in algae? *J. Phycol.* 12:355–358.
– 1981. Morphogenesis and growth substances. In C.S. Lobban & M.J. Wynne (eds.), *The Biology of Seaweeds*, pp. 627–660. Blackwell Scientific Publications, Oxford.
– 1983. Photoassimilate translocation in brown algae. In F.E. Round & D.J. Chapman (eds.), *Progress in Phycological Research*, vol. 2, pp. 283–332.
Bunt, J.S. 1970. Uptake of cobalt and vitamin B_{12} by tropical macroalgae. *J. Phycol.* 6:339–343.
Burdon-Jones, C., G.R.W. Denton, G.B. Jones, & K.A. McPhie. 1982. Regional and seasonal variations of trace metals in tropical Phaeophyceae from North Queensland. *Mar. Environ. Res.* 7:13–30.
Burns, A.R., L. Oliveira, & T. Bisalputra. 1982a. A histochemical study of bud initiation in the brown alga *Sphacelaria furcigera*. *New Phytol.* 92:297–307.
– 1982b. A morphological study of bud initiation in the brown alga *Sphacelaria furcigera*. *New Phytol.* 92:309–325.
Burr, A.H., & M.J. Duncan. 1972. Portable spectroradiometer for underwater environments. *Limnol. Oceanogr.* 17:466–475.
Burr, F.A. & J.A. West. 1971. Protein bodies in *Bryopsis hypnoides*: their relationship to wound-healing and branch septum development. *J. Ultrastruct. Res.* 35:476–498.
Burris, J.E. 1977. Photosynthesis, photorespiration, and dark respiration in eight species of algae. *Mar. Biol.* 39:371–379.
Burrows, E.M. 1971. Assessment of pollution effects by the use of algae. *Proc. Roy. Soc. Lond.* 177:295–306.
Burton, J.D. 1979. Physico-chemical limitations in experimental investigations. *Phil. Trans. Roy. Soc. Lond.* B 286:443–456.
Butterworth, J., P. Lester, & G. Nickless. 1972. Distribution of heavy metals in the Severn estuary. *Mar. Pollut. Bull.* 3:72–74.
Byrne, R.H., & D.R. Kester. 1976. Solubility of hydrous ferric oxide and iron speciation in seawater. *Mar. Chem.* 4:255–274.
Cairns, J., Jr. 1983. Are single species toxicity tests alone adequate for estimating environmental hazard? *Hydrobiologia* 100:47–57.
Cairns, J., Jr., A.G. Heath, & B.C. Parker. 1975. Temperature influence on chemical toxicity to aquatic organisms. *J. Water Poll. Control Fed.* 47:267–280.
Callow, M.E., & L.V. Evans. 1974. Studies on the ship-fouling alga *Enteromorpha*. III. Cytochemistry and autoradiography of adhesive production. *Protoplasma* 80:15–27
Callow, M.E., L.V. Evans, G.P. Bolwell, & J.A. Callow. 1978. Fertilization in brown algae. I. SEM and other observations on *Fucus serratus*. *J. Cell Sci.* 32:45–54.
Callow, J.A., M.E. Callow, & L.V. Evans. 1979. Nutritional studies on the parasitic red alga *Choreocolax polysiphoniae*. *New Phytol.* 83:451–462.
Calvin, M. 1962. The path of carbon in photosynthesis. *Science* 135:879–889.
Campbell, G.S. 1977. *An Introduction to Environmental Biophysics*. Springer-Verlag, New York. 159 pp.
Carefoot, T.H. 1977. *Pacific Seashores: A Guide to Intertidal Ecology*. Douglas, Vancouver. 208 pp.
– 1982. Phagostimulatory properties of various chemical compounds to sea hares (*Aplysia kurodai* and *A. dactylomela*). *Mar. Biol.* 68:207–215.

Carlberg, S.R. 1980. Oil pollution of the marine environment – with emphasis on estuarine studies. In E. Olausson & I. Cato (eds.), *Chemistry and Biogeochemistry of Estuaries*, pp. 367–402. Wiley, New York.

Carpenter, E.J., & J.S. Lively. 1980. Review of estimates of algal growth using ^{14}C tracer techniques. In P.G. Falkowski (ed.), *Primary Production in the Sea*, pp. 161–178. Plenum Press, New York.

Cembella, A., N.J. Antia, & P.J. Harrison. 1983. The utilization of inorganic and organic phosphorus compounds as nutrients by eukaryotic microalgae: a multidisciplinary perspective. *Crit. Rev. Microbiol.* 10:317–391.

Chamberlain, A.H.L., J. Gorham, D.F. Kane, & S.A. Lewey. 1979. Laboratory growth studies on *Sargassum muticum* (Yendo) Fensholt. II. Apical dominance. *Bot. Mar.* 22:11–19.

Chan, G. 1975. A study of the effects of the San Francisco oil spill on marine life. Part II: recruitment. In *Proc. 1975 Conference on Prevention and Control of Oil Pollution*, pp. 457–461. Amer. Petroleum Inst., Washington, D.C.

Chan, K., P.K. Wong, & S.L. Ng. 1982. Growth of *Enteromorpha linza* in sewage effluent and sewage effluent-seawater mixtures. *Hydrobiologia* 97:9–13.

Chapman, A.R.O. 1973a. Methods for macroscopic algae. In J.R. Stein (ed.), *Handbook of Phycological Methods: Culture Methods and Growth Measurements*, pp. 87–104. Cambridge University Press, New York.

– 1973b. Phenetic variability of stipe morphology in relation to season, exposure and depth in the non-digitate complex of *Laminaria* Lamour. (Phaeophyta, Laminariales) in Nova Scotia. *Phycologia* 12:53–57.

– 1973c. A critique of prevailing attitudes towards the control of seaweed zonation on the sea shore. *Bot. Mar.* 16:80–82.

– 1974. The genetic basis of morphological differentiation in some *Laminaria* populations. *Mar. Biol.* 24:85–91.

– 1975. Inheritance of mucilage canals in *Laminaria* (section Simplices) in eastern Canada. *Brit. Phycol. J.* 10:219–223.

Chapman, A.R.O., & E.M. Burrows. 1970. Experimental investigations into the controlling effects of light conditions on the development and growth of *Desmarestia aculeata* (L.) Lamour. *Phycologia* 9:103–108.

Chapman, A.R.O., & J.S. Craigie. 1977. Seasonal growth in *Laminaria longicruris*: relations with dissolved inorganic nutrients and internal reserves of nitrogen. *Mar. Biol.* 40:197–205.

Chapman, A.R.O., & J.E. Lindley. 1980. Seasonal growth of *Laminaria solidungula* in the Canadian High Arctic in relation to irradiance and dissolved nutrient concentrations. *Mar. Biol.* 57:1–5.

Chapman, A.R.O., J.W. Markham, & K. Lüning. 1978. Effects of nitrate concentration on the growth and physiology of *Laminaria saccharina* (Phaeophyta) in culture. *J. Phycol.* 14:195–198.

Chapman, D.J., & M.A. Ragan. 1980. Evolution of biochemical pathways: evidence from comparative biochemistry. *Ann. Rev. Plant Physiol.* 31:639–678.

Chapman, V.J. 1966. The physiological ecology of some New Zealand seaweeds. *Proc. Intl. Seaweed Symp.* 5:29–54.

Chapman, V.J., & D.J. Chapman. 1980. *Seaweeds and Their Uses*, 3rd ed. Chapman & Hall, London. 334 pp.

Charters, A.C., & M. Neushul. 1979. A hydrodynamically defined culture system for benthic seaweeds. *Aquat. Bot.* 6:67–78.

Charters, A.C., M. Neushul, & C. Barilotti. 1969. The functional morphology of *Eisenia arborea*. *Proc. Intl. Seaweed Symp.* 6:89–105.

Charters, A.C., M. Neushul, & D. Coon. 1973. The effect of water motion on algal spore adhesion. *Limnol. Oceanogr.* 18:884–896.

Chen, L.C.-M., & A.R.A. Taylor. 1978. Medullary tissue culture of the red alga *Chondrus crispus*. *Can. J. Bot.* 56:883–886.

Chen, L.C.-M., T. Edelstein, E. Ogata, & J. McLachlan. 1970. The life history of *Porphyra miniata*. *Can J. Bot.* 48:385–389.

Cheney, D.P. 1977: R&C/P – a new and improved ratio for comparing seaweed floras. *J. Phycol.* 13 (Suppl.):12.

– 1982. The determining effects of snail herbivore density on intertidal algal recruitment and composition. *Abstr. 1st Intl. Phycol. Congr.*, p. a8.

Christie, A.O., & M. Shaw. 1968. Settlement experiments with zoospores of *Enteromorpha intestinalis* (L.) Link. *Brit. Phycol. Bull.* 3:529–534.

Christie, A.O., L.V. Evans, & M. Shaw. 1970. Studies on the ship-fouling alga *Enteromorpha*. II. The effect of certain enzymes on the adhesion of zoospores. *Ann. Bot.* 34:467–482.

Clayton, M.N. 1976. The morphology, anatomy and life history of a complanate form of *Scytosiphon lomentaria* (Scytosiphonales, Phaeophyta) from southern Australia. *Mar. Biol.* 38:201–208.

Clayton, R.K. 1980. *Photosynthesis – Physical Mechanisms and Chemical Patterns*. Cambridge University Press, Cambridge, 281 pp.

Clendenning, K.A., & W.J. North. 1960. Effects of wastes on the giant kelp, *Macrocystis pyrifera*. *Proc. Intl. Conf. Waste Disposal Mar. Environ.* 1:82–91.

Cloutier-Mantha, L., & P.J. Harrison. 1980. Effects of sublethal concentrations of mercuric chloride on ammonium limited *Skeletonema costatum*. *Mar. Biol.* 56:219–231.

Cole, H.A. 1979a. Pollution of the sea and its effects. *Proc. Roy. Soc. Lond.* B 205:17–30.

– 1979b. Summing-up: deficiencies and future needs. *Phil. Trans. Roy. Soc. Lond.* B 286:625–633.

Colijn, F., & C. van den Hoek. 1971. The life-history of *Sphacelaria furcigera* Kütz. (Phaeophyceae). II. The influence of daylength and temperature on sexual and vegetative reproduction. *Nova Hedwigia* 21:901–922.

Colman, J. 1933. The nature of the intertidal zonation of plants and animals. *J. Mar. Biol. Ass. U.K.* 18:435–476.

Connell, J.H. 1978. Diversity in tropical forests and coral reefs. *Science* 199:1302–1310.

Conover, J.T., & J. McN. Sieburth. 1964. Effect of *Sargassum* distribution on its epibiota and antibacterial activity. *Bot. Mar.* 6:147–157.

– 1965. Effect of tannins excreted from Phaeophyta on planktonic animal survival in tide pools. *Proc. Intl. Seaweed Symp.* 5:99–100.

Conover, R.J. 1978. Transformation of organic matter. In O. Kinne (ed.), *Marine Ecology*, vol. 4, pp. 221–499. Wiley, New York.

Conway, H.L., P.J. Harrison, & C.O. Davis. 1976. Marine diatoms grown in chemostats under silicate or ammonium limitation. II. Transient response of *Skeletonema costatum* to a single addition of the limiting nutrient. *Mar. Biol.* 35:187–189.

Coon, D.A., M. Neushul, & A.C. Charters. 1972. The settling behavior of marine algal spores. *Proc. Intl. Seaweed Symp.* 7:237–242.

Cornish-Bowden, A. 1979. *Fundamentals of Enzyme Kinetics.* Butterworth, London. 203 pp.

Cosgrove, D.J. 1981. Analysis of the dynamic and steady-state responses of growth rate and turgor pressure to changes in cell parameters. *Plant Physiol.* 68:1439–1446.

Cosson, J. 1977. Action de l'éclairement sur la morphogénèse des gamétophytes de *Laminaria digitata* (L.) Lam. (Phaeophycée, Laminariale). *Bull. Soc. Phycol. Fr.* 22:19–26.

Coughlan, S. 1977. Sulphate uptake in *Fucus serratus. J. Exp. Bot.* 28:1207–1215.

Coughlan, S., & L.V. Evans. 1978. Isolation and characterization of Golgi bodies from vegetative tissue of the brown alga *Fucus serratus. J. Exp. Bot.* 29:55–68.

Court, G.J. 1980. Photosynthesis and translocation studies of *Laurencia spectabilis* and its symbiont *Janczewskia gardneri* (Rhodophyceae). *J. Phycol.* 16:270–279.

Cousens, R., & M.J. Hutchings. 1983. The relationship between density and mean frond weight in monospecific seaweed stands. *Nature* 301:240–241.

Craigie, J.S. 1974. Storage products. In W.D.P. Stewart (ed.), *Algal Physiology and Biochemistry*, pp. 206–235. Blackwell Scientific Publications, Oxford.

Craigie, J.S., & J. McLachlan. 1964. Excretion of coloured ultraviolet-absorbing substances by marine algae. *Can. J. Bot.* 42:23–33.

Craik, G.J.S. 1980. Simple method for measuring relative scouring of intertidal areas. *Mar. Biol.* 59:257–260.

Crayton, M.A., E. Wilson, & R.S. Quatrano. 1974. Sulfation of fucoidan in *Fucus* embryos. II. Separation from initiation of polar growth. *Devel. Biol.* 39:134–137.

Cross, F.A., & W.G. Sunda. 1977. Relationship between bioavailability of trace metals and geochemical processes in estuaries. In M.L. Wiley (ed.), *Estuarine Interactions*, pp. 429–442. Academic Press, New York.

Crothers, J.H. 1983. Field experiments on the effects of crude oil and dispersant on the common animals and plants of rocky seashores. *Mar. Environ. Res.* 8:215–239.

Dalby, D.H. 1980. Monitoring and exposure scales. In J.H. Price, D.E.G. Irvine, & W.F. Farnham (eds.), *The Shore Environment*, vol. 1, *Methods*, pp. 117–136. Academic Press, New York.

Daly, M.A., & A.C. Mathieson. 1977. The effects of sand movement on intertidal seaweeds and selected invertebrates at Bound Rock, New Hampshire, USA. *Mar. Biol.* 43:45–55.

Darley, W.M. 1974. Silicification and calcification. In W.D.P. Stewart (ed.), *Algal Physiology and Biochemistry*, pp. 665–675. Blackwell Scientific Publications, Oxford.

Davavin, I.A., O.G. Mironov, & I.M. Tsimbal. 1975. Influence of oil on nucleic acids of algae. *Mar. Pollut. Bull.* 6:13–15.

Davies, A.G. 1978. Pollution studies with marine plankton. Part II. Heavy metals. *Adv. Mar. Biol.* 15:381–508.

Davison, I.R., & W.D.P. Stewart. 1983. Occurrence and significance of nitrogen transport in the brown alga *Laminaria digitata. Mar. Biol.* 77:107–112.

– 1984. Studies on nitrate reductase activity in *Laminaria digitata* (Huds.) Lamour. I. Longitudinal and transverse profiles of nitrate reductase activity within the thallus. *J. Exp. Mar. Biol. Ecol.* 74:201–210.

Dawes, C.J. 1979. Physiological and biochemical comparisons of *Eucheuma* spp. (Florideophyceae) yielding *iota*-carrageenan. *Proc. Intl. Seaweed Symp.* 9:188–207.

– 1981. *Marine Botany*. Wiley, New York. 628 pp.

Dawes, C.J., & C. Barilotti. 1969. Cytoplasmic organization and rhythmic streaming in growing blades of *Caulerpa prolifera. Amer. J. Bot.* 56:8–15.

Dawes, C.J., & C.A. Lohr. 1978. Cytoplasmic organization and endosymbiotic bacteria in the growing points of *Caulerpa prolifera. Rev. Algol. N.S.* 13:309–314.

Dawes, C.J., & R.P. McIntosh. 1981. The effect of organic material and inorganic ions on the photosynthetic rate of the red alga *Bostrychia binderi* from a Florida estuary. *Mar. Biol.* 64:213–218.

Dawes, C.J., J.M. Lawrence, D.P. Cheney, & A.C. Mathieson. 1974. Ecological studies of Floridian *Eucheuma* (Rhodophyta, Gigartinales). III. Seasonal variation of carrageenan, total carbohydrate, protein and lipid. *Bull. Mar. Sci.* 24:286–299.

Dawes, C.J., N.F. Stanley, & D.J. Stanicoff. 1977. Seasonal and reproductive aspects of plant chemistry, and *iota*-carrageenan from Floridian *Eucheuma* (Rhodophyta, Gigartinales). *Bot. Mar.* 20:137–147.

Dawson, E.Y. 1951. A further study of upwelling and associated vegetation along Pacific Baja California, Mexico, *J. Mar. Res.* 10:39–58.

– 1959. A preliminary report on the benthic marine flora of southern California. In *Oceanographic Survey of the Continental Shelf Area of Southern California*, pp. 169–264. Publ. Calif. State Water Qual. Contr. Bd. No. 20, Sacramento.

Dayton, P.K. 1973. Dispersion, dispersal, and persistence of the annual intertidal alga, *Postelsia palmaeformis* Ruprecht. *Ecology* 54:433–438.

– 1975. Experimental studies of algal canopy interactions in a sea otter dominated kelp community at Amchitka Island, Alaska. *Natl. Oceanogr. Atmos. Admin. Fish. Bull.* 73:230–237.

DeBoer, J.A. 1979. Effects of nitrogen enrichment on growth rate and phycocolloid content in *Gracilaria foliifera* and *Neoagardhiella baileyi* (Florideophyceae). *Proc. Intl. Seaweed Symp.* 9:263–271.

– 1981. Nutrients. In C.S. Lobban & M.J. Wynne (eds.), *The Biology of Seaweeds*, pp. 356–391. Blackwell Scientific Publications, Oxford.

DeBoer, J.A., & J.H. Ryther. 1977. Potential yields from a waste-recycling algal mariculture system. In R. Krauss (ed.), *The Marine Plant Biomass of the Pacific Northwest Coast*, pp. 231–249. Oregon State University Press, Corvallis.

DeBoer, J.A., H.J. Guigli, T.L. Israel, & C.F. D'Elia. 1978. Nutritional studies of two red algae. I. Growth rate as a function of nitrogen source and concentration. *J. Phycol.* 14:261–266.

de Lestang-Brémond, G., & M. Quillet. 1981. The turnover of sulphates on the *lambda*-carrageenan of the cell-walls of the red seaweed Gigartinale: *Catenella opuntia* (Grev.). *Proc. Intl. Seaweed Symp.* 10:449–454.

D'Elia, C.F., & J.A. DeBoer. 1978. Nutritional studies of two red algae. II. Kinetics of ammonium and nitrate uptake. *J. Phycol.* 14:266–272.

DeManche, J.M., H.C. Curl, D.W. Lundy, & P.L. Donaghay. 1979. The rapid response of the marine diatom *Skeletonema costatum* to changes in external and internal nutrient concentration. *Mar. Biol.* 53:323–333.

den Hartog, C. 1972a. Substratum. Multicellular plants. In O. Kinne (ed.), *Marine Ecology*, vol. I, pt. 3, pp. 1277–1289. Wiley, New York.

– 1972b. The effect of the salinity tolerance of algae on their distribution, as exemplified by *Bangia*. *Proc. Intl. Seaweed Symp.* 7:274–276.

Derenbach, J.B., & M.V. Gereck. 1980. Interference of petroleum hydrocarbons with the sex pheromone reaction of *Fucus vesiculosus* (L.). *J. Exp. Mar. Biol. Ecol.* 44:61–65.

DeRobertis, E.D.P., & E.M.F. DeRobertis, Jr. 1980. *Cell and Molecular Biology*, 7th ed. Saunders, Philadelphia. 673 pp.

Dethier, M.N. 1981. Heteromorphic algal life histories: the seasonal pattern and response to herbivory of the brown crust, *Ralfsia californica*. *Oecologia* 49:333–339.

– 1982. Pattern and process in tidepool algae: factors influencing seasonality and distribution. *Bot. Mar.* 25:55–66.

– 1984. Disturbance and recovery in intertidal pools: maintenance of mosaic patterns. *Ecol. Monogr.* 54:99–118.

Devinny, J.S., & L.A. Volse. 1978. Effects of sediments on the development of *Macrocystis pyrifera* gametophytes. *Mar. Biol.* 48:343–348.

Deysher, L., & T.A. Norton. 1982. Dispersal and colonization in *Sargassum muticum* (Yendo) Fensholt. *J. Exp. Mar. Biol. Ecol.* 56:179–195.

Dickson, D.M., R.G. Wyn-Jones, & J. Davenport. 1980. Steady state osmotic adaptation in *Ulva lactuca*. *Planta* 150:158–165.

– 1982. Osmotic adaptation in *Ulva lactuca* under fluctuating salinity regimes. *Planta* 155:409–415.

Diouris, M., & J.-Y. Floc'h. 1984. Long-distance transport of ^{14}C-labelled assimilates in the Fucales: directionality, pathway and velocity. *Mar. Biol.* 78:199–204.

Dixon, P.S. 1973. *Biology of the Rhodophyta*. Oliver & Boyd, Edinburgh. 285 pp.

Dixon, P.S., & W.N. Richardson. 1970. Growth and reproduction in red algae in relation to light and dark cycles. *Ann. N.Y. Acad. Sci.* 175:764–777.

Dortch, Q. 1982. Effect of growth conditions on accumulation of internal nitrate, ammonium, and protein in three marine diatoms. *J. Exp. Mar. Biol. Ecol.* 61:243–264.

Doty, M.S. 1946. Critical tide factors that are correlated with the vertical distribution of marine algae and other organisms along the Pacific Coast. *Ecology* 27:315–328.

– 1971. Measurement of water movement in reference to benthic algal growth. *Bot. Mar.* 14:32–35.

Doty, M.S., & G. Aguilar-Santos. 1970. Transfer of toxic algal substances in marine food chains. *Pacific Sci.* 24:351–355.

Doty, M.S., & V.B. Alvarez. 1975. Status, problems, advances and economics of *Eucheuma* farms. *Mar. Technol. Soc. J.*. 9:30–35.

Dowd, J.E., & D.S. Riggs. 1965. A comparison of estimates of Michaelis-Menten kinetic constants from various linear transformations. *J. Biol. Chem.* 240:863–869.

Drebes, G. 1977. Sexuality. In D. Werner (ed.), *The Biology of Diatoms*, pp. 250–283. Blackwell Scientific Publications, Oxford.

Dreher, T.W., B.R. Grant, & R. Wetherbee. 1978. The wound response in the siphonous alga *Caulerpa simpliciuscula* C. Ag.: fine structure and cytology. *Protoplasma* 96:189–203.

Drew, E.A. 1977. The physiology of photosynthesis and respiration in some antarctic marine algae. *Br. Antarct. Survey Bull.* 46:59–76.

Drew, K.M. 1949. Conchocelis phase in the life history of *Porphyra umbilicalis* (L.) Kütz. *Nature* 164:748.

Dring, M.J. 1967. Phytochrome in red alga, *Porphyra tenera*. *Nature* 215: 1411–1412.

– 1974. Reproduction. In W.D.P. Stewart (ed.), *Algal Physiology and Biochemistry*, pp. 814–837. Blackwell Scientific Publications, Oxford.

– 1981. Photosynthesis and development of marine macrophytes in natural light spectra. In H. Smith (ed.), *Plants and the Daylight Spectrum*, pp. 297–314. Academic Press, London.

Dring, M.J., & F.A. Brown. 1982. Photosynthesis of intertidal brown algae during and after periods of emersion: a renewed search for physiological causes of zonation. *Mar. Ecol. Progr. Ser.* 8:301–308.

Dring, M.J., & K. Lüning. 1975. A photoperiodic response mediated by blue light in the brown alga *Scytosiphon lomentaria*. *Planta* 125:25–32.

Dring, M.J., & J.A. West. 1983. Photoperiodic control of tetrasporangium formation in the red alga *Rhodochorton purpureum*. *Planta* 159:143–150.

Dromgoole, F.I. 1978. The effects of oxygen on dark respiration and apparent photosynthesis of marine macroalgae. *Aquat. Bot.* 4:281–297.

– 1980. Desiccation resistance of intertidal and subtidal algae. *Bot. Mar.* 23:149–159.

– 1981a. Form and function of the pneumatocysts of marine algae. I. Variations in the pressure and composition of internal gases. *Bot. Mar.* 24:257–266.

– 1981b. Form and function of the pneumatocysts of marine algae. II. Variations in morphology and resistance to hydrostatic pressure. *Bot. Mar.* 24:299–310.

– 1982. The buoyant properties of *Codium*. *Bot. Mar.* 25:391–397.

Droop, M.R. 1968. Vitamin B_{12} and marine ecology. IV. The kinetics of uptake, growth and inhibition in *Monochrysis lutheri*. *J. Mar. Biol. Ass. U.K.* 48:689–733.

– 1973. Some thoughts on nutrient limitation in algae. *J. Phycol.* 9:264–272.

– 1974. The nutrient status of algal cells in continuous culture. *J. Mar. Biol. Ass. U.K.* 54:825–855.

– 1983. 25 years of algal growth kinetics. A personal view. *Bot. Mar.* 26:99–112.

Druehl, L.D. 1967. Distribution of two species of *Laminaria* as related to some environmental factors. *J. Phycol.* 3:103–108.

– 1978. The distribution of *Macrocystis integrifolia* in British Columbia as related to environmental parameters. *Can. J. Bot.* 56:69–79.

– 1981. Geographic distribution. In C.S. Lobban & M.J. Wynne (eds.), *The Biology of Seaweeds*, pp. 306–325. Blackwell Scientific Publications, Oxford.

Druehl, L.D., & J.M. Green. 1970. A submersion-emersion sensor, for intertidal biological studies. *J. Fish. Res. Bd. Can.* 27:401–403.

– 1982. Vertical distribution of intertidal seaweeds as related to patterns of submersion and emersion. *Mar. Ecol. Progr. Ser.* 9:163–170.

Duggins, D.O. 1980. Kelp beds and sea otters: an experimental approach. *Ecology* 61:447–453.

Duncan, M.J., & R.E. Foreman. 1980. Phytochrome-mediated stipe elongation in the kelp *Nereocystis* (Phaeophyceae). *J. Phycol.* 16:138–142.

Duncan, M.J., & P.J. Harrison. 1982. Comparison of solvents for extracting chlorophylls from marine macrophytes. *Bot. Mar.* 25:445-447.

Durako, M.J., & C.J. Dawes. 1980. A comparative seasonal study of two populations of *Hypnea musciformis* from the east and west coasts of Florida, U.S.A. II. Photosynthetic and respiratory rates. *Mar. Biol.* 59:157–162.

Duursma, E.K., & M. Marchand. 1974. Aspects of organic marine pollution. *Oceanogr. Mar. Biol. Ann. Rev.* 12:315–431.

Edelstein, T., & J. McLachlan. 1975. Autecology of *Fucus distichus* ssp. *distichus* (Phaeophyceae: Fucales) in Nova Scotia, Canada. *Mar. Biol.* 30:305–324.

Edwards, P. 1977a. An analysis of the pattern and rate of cell division and morphogenesis of sporelings of *Callithamnion hookeri* (Dillw.) S.F. Gray (Rhodophyta, Ceramiales). *Phycologia* 16:189–196.

– 1977b. An investigation of the vertical distribution of selected benthic marine algae with a tide-simulating apparatus. *J. Phycol.* 13:62–68.

Eide, I., S. Myklestad, & S. Melson. 1980. Long-term uptake and release of heavy metals by *Ascophyllum nodosum* (L). Le Jol. (Phaeophyceae) *in situ*. *Environ. Pollut.* 23:19–28.

Emerson, C.J., R.G. Buggeln, & A.K. Bal. 1982. Translocation in *Saccorhiza dermatodea* (Laminariales, Phaeophyceae): anatomy and physiology. *Can. J. Bot.* 60:2164–2184.

Emerson, R., & W. Arnold. 1932. A separation of the reactions in photosynthesis by means of artificial light. *J. Gen. Physiol.* 15:391–420.

Emerson, R., R. Chalmers, & C. Cedarstrand. 1957. Some factors influencing the long wave limit of photosynthesis. *Proc. Natl. Acad. Sci.* 43:133–143.

Engelmann, T.W. 1883. Farbe und Assimilation. *Bot. Zeit.* 41:1–13.

– 1884. Untersuchungen über die quantitativen Beziehungen zwischen Absorption des Lichtes und Assimilation in Pflanzenzellen. *Bot. Zeit.* 42:81–93.

Enright, C.T. 1979. Competitive interaction between *Chondrus crispus* (Florideophyceae) and *Ulva lactuca* (Chlorophyceae) in *Chondrus* aquaculture. *Proc. Intl. Seaweed Symp.* 9:209–218.

Eppley, R.W. 1958. Sodium exclusion and potassium retention by the marine red alga *Porphyra perforata*. *J. Gen. Physiol.* 41:901–911.

– 1978. Nitrate reductase in marine phytoplankton. In J.A. Hellebust & J.S. Craigie (eds.), *Handbook of Phycological Methods. Physiological and Biochemical Methods*, pp. 217–223. Cambridge University Press, New York.

Eppley, R.W., & L.R. Blinks. 1957. Cell space and apparent free space in the red alga, *Porphyra perforata*. *Plant Physiol.* 32:63–64.

Eppley, R.W., & B.S. Cyrus. 1960. Cation regulation and survival of the red alga *Porphyra perforata* in diluted and concentrated seawater. *Biol. Bull.* 118:55–65.

Eppley, R.W., E.H. Renger, & P.M. Williams. 1976. Chlorine reactions with seawater constituents and the inhibition of photosynthesis of natural marine phytoplankton. *Estu. Cstl. Mar. Sci.* 4:147–161.

Epstein, E. 1972. *Mineral Nutrition of Plants: Principles and Perspectives*. Wiley, New York.

Estep, M.F., F.R. Tabita, & C. van Baalen. 1978. Purification of ribulose 1,5-bisphosphate carboxylase and carbon isotope fractionation by whole cells and carboxylase from *Cylindrotheca* sp. (Bacillariophyceae). *J. Phycol.* 14:183–188.

Evans, G.C. 1972. *The Quantitative Analysis of Plant Growth*. Blackwell Scientific Publications, Oxford. 734 pp.

Evans, L.V., & A.O. Christie. 1970. Studies on the ship-fouling alga *Enteromorpha*. I. Aspects of the fine-structure and biochemistry of swimming and newly settled zoospores. *Ann. Bot.* 34:451–466.

Evans, L.V., J.A. Callow, & M.E. Callow. 1973. Structural and physiological studies on the parasitic red alga *Holmsella*. *New Phytol.* 72:393–402.

– 1978. Parasitic red algae: an appraisal. In D.E.G. Irvine & J.H. Price (eds.), *Modern Approaches to the Taxonomy of Red and Brown Algae*, pp. 87–109. Academic Press, New York.

– 1982. The biology and biochemistry of reproduction and early development in *Fucus*. In F.E. Round & D.J. Chapman (eds.), *Progress in Phycological Research*, vol. 1, pp. 67–110. Elsevier, New York.

Fagerberg, W.R., & C.J. Dawes. 1977. Studies on *Sargassum*. II. Quantitative ultrastructural changes in differentiated stipe cells during wound regeneration and regrowth. *Protoplasma* 92:211–227.

Fagerberg, W.R., R. Moon, & E. Truby. 1979. Studies on *Sargassum*. III. A quantitative ultrastructural and correlated physiological study of the blade and stipe organs of *S. filipendula*. *Protoplasma* 99:247–261.

Falkowski, P.G., & R.B. Rivkin. 1976. The role of glutamine synthetase in the incorporation of ammonium in *Skeletonema costatum* (Bacillariophyceae). *J. Phycol.* 12:448–450.

Fenical, W. 1975. Halogenation in the Rhodophyta. A review. *J. Phycol.* 11:245–259.

Fenical, W., O.J. McConnell, & A. Stone. 1979. Antibiotics and antiseptic compounds from the family Bonnemaisoniaceae (Florideophyceae). *Proc. Intl. Seaweed Symp.* 9:387–400.

Ferguson, R.L., G.W. Thayer, & T.R. Rice. 1980. Marine primary producers. In F.J. Vernberg & W.B. Vernberg (eds.), *Functional Adaptations of Marine Organisms*, pp. 9–69. Academic Press, New York.

Fielding, A.H., & G. Russell. 1976. The effect of copper on competition between marine algae. *J. Appl. Ecol.* 13:871–876.

Filion-Myklebust, C., & T.A. Norton. 1981. Epidermis shedding in the brown seaweed *Ascophyllum nodosum* (L.) Le Jolis, and its ecological significance. *Mar. Biol. Lett.* 2:45–51.

Fisher, N.S., & C.F. Wurster. 1973. Individual and combined effects of temperature and polychlorinated biphenyls on the growth of three species of phytoplankton. *Environ. Pollut.* 5:205–212.

Fisher, N.S., E.J. Carpenter, C.C. Remsen, & C.F. Wurster. 1974. Effects of PCB on interspecific competition in natural and gnotobiotic phytoplankton communities in continuous and batch cultures. *Microbial Ecol.* 1:39–50.

Fitter, A.H., & R.K.M. Hay. 1981. *Environmental Physiology of Plants*. Academic Press, New York. 355 pp.

Fjeld, A. 1972. Genetic control of cellular differentiation in *Ulva mutabilis*. Gene effects in early development. *Devel. Biol.* 28:326–343.

Fjeld, A., & A. Løvlie. 1976. Genetics of multicellular marine algae. In R.A. Lewin (ed.), *The Genetics of Algae*, pp. 219–235. Blackwell Scientific Publications, Oxford.

Floc'h, J.-Y. 1982. Uptake of inorganic ions and their long distance transport in Fucales and Laminariales. In L.M. Srivastava (ed.), *Synthetic and Degradative Processes in Marine Macrophytes*, pp. 139–165. Walter de Gruyter, Berlin.

Floc'h, J.-Y., & M. Penot. 1978. Changes in ^{32}P-phosphorus compounds during translocation in *Laminaria digitata* (L.) Lamouroux. *Planta* 143:101–107.

Foreman, R.E. 1976. Physiological aspects of carbon monoxide production by the brown alga *Nereocystis luetkeana*. *Can. J. Bot.* 54:352–360.

– 1977. Benthic community modification and recovery following intensive grazing by *Strongylocentrotus droebachiensis*. *Helgol. wiss. Meeresunters.* 30:468–484.

Fork, D.C. 1963. Observations on the function of chlorophyll *a* and accessory pigments in photosynthesis. In *Photosynthetic Mechanisms in Green Plants*, pp. 352–361. Publ. No. 1145, NAS-NRC, Washington, D.C.

Forstner, M., & K. Rützler. 1970. Measurements of the microclimate in littoral marine habitats. *Oceanogr. Mar. Biol. Ann. Rev.* 8:225–249.

Förstner, U. 1980. Inorganic pollutants, particularly heavy metals in estuaries. In E. Olausson & I. Cato (eds.), *Chemistry and Biogeochemistry of Estuaries*, pp. 307–348. Wiley, New York.

Förstner, U., & G. Wittman. 1979. *Metal Pollution in Aquatic Environments*. Springer-Verlag, New York. 486 pp.

Fortes, M.D., & K. Lüning. 1980. Growth rates of North Sea macroalgae in relation to temperature, irradiance and photoperiod. *Helgol. Meeresunters.* 34:15–29.

Foster, M.S. 1972. The algal turf community in the nest of the ocean goldfish (*Hypsypops rubicunda*). *Proc. Intl. Seaweed Symp.* 7:55–60.

– 1975. Regulation of algal community development in a *Macrocystis pyrifera* forest. *Mar. Biol.* 32:331–342.

– 1982. Factors controlling the intertidal zonation of *Iridaea flaccida* (Rhodophyta). *J. Phycol.* 18:285–294.

Foster, P. 1976. Concentrations and concentration factors of heavy metals in brown algae. *Environ. Pollut.* 10:45–54.

Foy, C.D., R.L. Chaney, & M.C. White. 1978. The physiology of metal toxicity in plants. *Ann. Rev. Plant Physiol.* 29:511–566.

Fralick, R.A., & A.C. Mathieson. 1975. Physiological ecology of four *Polysiphonia* species (Rhodophyta, Ceramiales). *Mar. Biol.* 29:29–36.

Francke, J.A., & H. Hillebrand. 1980. Effects of copper on some filamentous Chlorophyta. *Aquatic Bot.* 8:285–289.

Fréchette, M., & L. Legendre. 1978. Photosynthèse phytoplanctonique: réponse à un stimulus simple, imitant les variations rapides de la lumière engendrées par les vagues. *J. Exp. Mar. Biol. Ecol.* 32:15–25.

French, C.S., J.S. Brown, & M.C. Lawrence. 1972. Four universal forms of chlorophyll *a*. *Plant Physiol.* 49:421–429.

Friedmann, (E.) I. 1961. Cinemicrography of spermatozoids and fertilization in Fucales. *Bull. Res. Counc. Israel* 10D:73–83.

Friedmann, E.I., & W.C. Roth. 1977. Development of the siphonous green alga *Penicillus* and the *Espera* state. *J. Linn. Soc. Bot.* 74:189–214.

Friedrich, H. 1969. *Marine Biology; An Introduction to Its Problems and Results*. Sidgwick & Jackson, London. 474 pp.

Fries, L. 1966. Temperature optima of some algae in axenic culture. *Bot. Mar.* 9:12–14.

– 1975. Requirements of bromine in a red alga. *Z. Pflanzenphysiol.* 76:366–368.

– 1982a. Selenium stimulates growth of marine macroalgae in axenic culture. *J. Phycol.* 18:328–331.

– 1982b. Vanadium, an essential element for some marine macroalgae. *Planta* 154:393–396.

Fries, N. 1979. Physiological characteristics of *Mycosphaerella ascophylli*, a fungal endophyte of the marine brown alga *Ascophyllum nodosum*. *Physiol. Plant.* 45:117–121.

Fuge, R. & K.H. James. 1973. Trace metal concentrations in brown seaweeds, Cardigan Bay, Wales. *Mar. Chem.* 1:218–293.

Fulcher, R.G., & M.E. McCully. 1969. Histological studies on the genus *Fucus*. IV. Regeneration and adventive embryony. *Can. J. Bot.* 47:1643–1649.

– 1971. Histological studies on the genus *Fucus*. V. An autoradiographic and electron microscopic study of the early stages of regeneration. *Can. J. Bot.* 49:161–165.

Gadd, G.M., & A.J. Griffiths. 1978. Microorganisms and heavy metal toxicity. *Microbial Ecol.* 4:303–317.

Gagné, J.A., K.H. Mann, & A.R.O. Chapman. 1982. Seasonal patterns of growth and storage in *Laminaria longicruris* in relation to differing patterns of availability of nitrogen in the water. *Mar. Biol.* 69:91–101.

Gaines, S.D., & J. Lubchenco. 1982. A unified approach to marine plant-herbivore interactions. II. Biogeography. *Ann. Rev. Ecol. System.* 13:111–138.

Gantt, E. 1975. Phycobilisomes: light harvesting pigment complexes. *BioScience* 25:781–788.

– 1980. Structure and function of phycobilisomes: light harvesting pigment complexes in red and blue-green algae. *Int. Rev. Cytol.* 66:45–80.

– 1981. Phycobilisomes. *Ann. Rev. Plant Physiol.* 32:327–347.

Garbary, D.J., D. Grund, & J. McLachlan. 1978. The taxonomic status of *Ceramium rubrum* (Huds.) C. Ag. (Ceramiales, Rhodophyceae) based on culture experiments. *Phycologia* 17:85–94.

Gayler, K.R., & W.R. Morgan. 1976. An NADP-dependent glutamate dehydrogenase in chloroplasts from the marine green alga *Caulerpa simpliciuscula*. *Plant Physiol.* 58:283–287.

Geesink, R. 1973. Experimental investigations on marine and freshwater *Bangia* (Rhodophyta) from the Netherlands. *J. Exp. Mar. Biol. Ecol.* 11:239–247.

Gemmill, E.R., & R.A. Galloway. 1974. Photoassimilation of ^{14}C-acetate by *Ulva lactuca*. *J. Phycol.* 10:359–366.

Gerard, V.A. 1976. Some aspects of material dynamics and energy flow in a kelp forest in Monterey Bay, California. Ph.D. Thesis, University of California, Santa Cruz. 173 pp.

– 1982a. *In situ* water motion and nutrient uptake by the giant kelp *Macrocystis pyrifera*. *Mar. Biol.* 69:51–54.

– 1982b. *In situ* rates of nitrate uptake by giant kelp, *Macrocystis pyrifera* (L.) C. Agardh: tissue differences, environmental effects, and predictions of nitrogen-limited growth. *J. Exp. Mar. Biol. Ecol.* 62:211–224.

– 1982c. Growth and utilization of internal nitrogen reserves by the giant kelp *Macrocystis pyrifera* in a low-nitrogen environment. *Mar. Biol.* 66:27–35.

Gerard, V.A., & K.H. Mann. 1979. Growth and production of *Laminaria longicruris* (Phaeophyta) populations exposed to different intensities of water movement. *J. Phycol.* 15:33–41.

Gerlach, S.A. 1982. *Marine Pollution: Diagnoses and Therapy*. Springer-Verlag, New York, 218 pp.

Gerloff, G.C., & P.H. Krombholz. 1966. Tissue analysis as a measurement of nutrient availability for the growth of angiosperm aquatic plants. *Limnol. Oceanogr.* 11:529–537.

Gessner, F. 1970. Temperature: plants. In O. Kinne (ed.), *Marine Ecology*, vol. 1, pt. 1, pp. 363–406. Wiley, New York.

– 1971. Wasserpermeabilität und Photosynthese bei marinen Algen. *Bot. Mar.* 14:29–31.

Gessner, F., & L. Hammer. 1968. Exosmosis and "free space" in marine benthic algae. *Mar. Biol.* 2:88–91.

Gessner, F., & W. Schramm. 1971. Salinity: plants. In O. Kinne (ed.), *Marine Ecology*, vol. 1, pt. 2, pp. 705–820. Wiley, New York.

Geyer, R.A. (ed.) 1980. *Marine Environmental Pollution* vol. I. *Hydrocarbons*. Elsevier Scientific Publications, Amsterdam. 591 pp.

Gibor, A. 1973. *Acetabularia*. Physiological role of their deciduous organelles. *Protoplasma* 78:461–465.

Gieskes, J.M. 1982. The practical salinity scale 1978: a reply to comments by T.R. Parsons. *Limnol. Oceanogr.* 27:387–389.

Glynn, P.W. 1965. Community composition, structure, and interrelationships in the marine intertidal *Endocladia muricata–Balanus glandula* association in Monterey Bay, California. *Beaufortia* 12 (148):1–198.

Goff, L.J. 1979a. The biology of *Harveyella mirabilis* (Cryptonemiales, Rhodophyceae). VI. Translocation of photoassimilated C. *J. Phycol.* 15:82–87.

– 1979b. The biology of *Harveyella mirabilis* (Cryptonemiales, Rhodophyceae). VII. Structure and proposed function of host-penetrating cells. *J. Phycol.* 15:87–100.

– 1982. Biology of parasitic red algae. In F.E. Round & D.J. Chapman (eds.), *Progress in Phycological Research*, vol. 1, pp. 289–369. Elsevier, New York.

Goff, L.J., & A.W. Coleman. 1983. Intergeneric nuclear transfer in a parasitic red algal association. *J. Phycol.* 19 (Suppl.):7.

Goldman, J.C., & R. Mann. 1980. Temperature-influenced variations in speciation and chemical composition of marine phytoplankton in outdoor mass culture. *J. Exp. Mar. Biol. Ecol.* 46:29–39.

Goldman, J.C., & J.J. McCarthy. 1978. Steady state growth and ammonium uptake of a fast-growing marine diatom. *Limnol. Oceanogr.* 23:695–703.

Goldman, J.C., J.J. McCarthy, & D.G. Peavey. 1979. Growth rate influence on the chemical composition of phytoplankton in oceanic waters. *Nature* 279:210–215.

Goodwin, T.W., & E.I. Mercer. 1983. *Introduction to Plant Biochemistry*, 2nd ed. Pergamon Press, New York. 677 pp.

Gordon, D.M., P.B. Birch, & A.J. McComb. 1980. The effects of light, temperature and salinity on photosynthetic rates of an estuarine *Cladophora*. *Bot. Mar.* 29:749–755.

– 1981. Effects of inorganic phosphorus and nitrogen on the growth of an estuarine *Cladophora* in culture. *Bot. Mar.* 24:93–106.

Gould, J.M. 1976. Inhibition by triphenyltin chloride of a tightly bound membrane component involved in photophosphorylation. *European J. Biochem.* 62:567–575.

Govinjee, & B.Z. Braun. 1974. Light absorption, emission and photosynthesis. In W.D.P. Stewart (ed.), *Algal Physiology and Biochemistry*, pp. 346–390. Blackwell Scientific Publications, Oxford.

Graham, D., & B.D. Patterson. 1982. Responses of plants to low, non-freezing temperatures: proteins, metabolism, and acclimation. *Ann. Rev. Plant Physiol.* 33:347–372.

Green, B.R. 1976. Approaches to the genetics of *Acetabularia*. In R.A. Lewin (ed.), *The Genetics of Algae*, pp. 236–256. Blackwell Scientific Publications, Oxford.

Green, P.B. 1980. Organogenesis – a biophysical view. *Ann. Rev. Plant Physiol.* 31:51–82.

Green, R.H. 1979. *Sampling Design and Statistical Methods for Environmental Biologists*. Wiley, New York. 257 pp.

Greene, R.W. 1970. Symbiosis in sacoglossan opisthobranchs: functional capacity of symbiotic chloroplasts. *Mar. Biol.* 72:138–142.

Grice, G.D., & M.R. Reeve (eds.). 1982. *Marine Microcosms: Biological and Chemical Research in Experimental Ecosystems*. Springer-Verlag, New York. 430 pp.

Groen, P. 1980. Oceans and seas. I. Physical and chemical properties. In *The New Encyclopaedia Britannica, Macropaedia*, vol. 13, pp. 484–497.

Gross, M.G. 1982. *Oceanography: A View of the Earth*, 3rd ed. Prentice-Hall, Englewood Cliffs, N.J. 498 pp.

Guillard, R.R.L., & S. Myklestad. 1970. Osmotic and ionic requirements of the marine centric diatom *Cyclotella nana*. *Helgol. wiss. Meeresunters.* 20:104–110.

Gundlach, E., & M. Hayes. 1978. Vulnerability of coastal environments to oil spill impacts. *Mar. Technol. Soc. J.* 12:18–27.

Gundlach, E.R., P.D. Boehm, M. Marchand, R.M. Atlas, D.M. Ward, & D.A. Wolfe. 1983. The fate of *Amoco Cadiz* oil. *Science* 221:122–129.

Gunnill, F.C. 1980. Demography of the intertidal brown alga *Pelvetia fastigiata* in southern California, U.S.A. *Mar. Biol.* 59:169–179.

Gutknecht, J. 1963. Zinc-65 uptake by benthic algae. *Limnol. Oceanogr.* 8:31–38.

– 1965. Uptake and retention of cesium 137 and zinc 65 by seaweeds. *Limnol. Oceanogr.* 10:58–66.

Gutknecht, J., & J. Dainty. 1968. Ionic relations of marine algae. *Oceanogr. Mar. Biol. Ann. Rev.* 6:163–200.

Hagmeier, E. 1971. Turbidity – plants. In O. Kinne (ed.), *Marine Ecology*, vol. I, pt. 2, pp. 1177–1180. Wiley, New York.

Haines, K.C., & P.A. Wheeler. 1978. Ammonium and nitrate uptake by the marine macrophyte *Hypnea musciformis* (Rhodophyta) and *Macrocystis pyrifera* (Phaeophyta). *J. Phycol.* 14:319–324.

Hall, A. 1980. Heavy metal co-tolerance in a copper-tolerant population of the marine fouling alga, *Ectocarpus siliculosus* (Dillw.) Lyngbye. *New Phytol.* 85:73–78.

Hall, A., A.H. Fielding, & M. Butler. 1979. Mechanisms of copper tolerance in the marine fouling alga *Ectocarpus siliculosus*: evidence for an exclusion mechanism. *Mar. Biol.* 54:195–199.

Halldal, P. 1964. Ultraviolet action spectra of photosynthesis and photosynthetic inhibition in a green alga and a red alga. *Physiol. Plant.* 17:414–421.

Hamelink, J.L., R.C. Waybrant, & R.C. Ball. 1971. A proposal: exchange equilibria control the degree chlorinated hydrocarbons are biologically magnified in lentic environments. *Trans. Amer. Fish. Soc.* 100:207–214.

Hammer, L. 1968. Salzgehalt und Photosynthese bei marinen Pflanzen. *Mar. Biol.* 1:185–190.

– 1969. "Free space-photosynthesis" in the algae *Fucus virsoides* and *Laminaria saccharina*. *Mar. Biol.* 4:136–138.

Hanic, L.A., & J.S. Craigie. 1969. Studies on the algal cuticle. *J. Phycol.* 5:89–109.

Hanisak, M.D. 1979. Nitrogen limitation of *Codium fragile* ssp. *tomentosoides* as determined by tissue analysis. *Mar. Biol.* 50:333–337.

– 1983. The nitrogen relationships of marine macroalgae. In E.J. Carpenter & D.G. Capone (eds.), *Nitrogen in the Marine Environment*, pp. 699–730. Academic Press, New York.

Hanisak, M.D., & M.M. Harlin. 1978. Uptake of inorganic nitrogen by *Codium fragile* subsp. *tomentosoides* (Chlorophyta). *J. Phycol.* 14:450–454.

Hanson, J.B., & A.J. Trewavas. 1982. Regulation of plant cell growth: the changing perspective. *New Phytol.* 90:1–18.

Harborne, J.B. 1980. Plant phenolics. In E.A. Bell & B.V. Charlwood (eds.), *Secondary Plant Products, Encycl. Plant Physiol.*, New Ser. 8:329–402.

Harding, L.W., Jr., & J.H. Phillips, Jr. 1978. Polychlorinated biphenyls: transfer from microparticulates to marine phytoplankton and the effects on photosynthesis. *Science* 202:1189–1192.

Hardy, F.G., & B.L. Moss. 1979. Attachment and development of the zygotes of *Pelvetia canaliculata* (L.) Dcne. et Thur. (Phaeophyceae, Fucales). *Phycologia* 18:203–212.

Harlin, M.M., & J.S. Craigie. 1975. The distribution of photosynthate in *Ascophyllum nodosum* as it relates to epiphytic *Polysiphonia lanosa*. *J. Phycol.* 11:109–113.

– 1978. Nitrate uptake by *Laminaria longicruris* (Phaeophyceae). *J. Phycol.* 14:464–467.

Harlin, M.M., & J.M. Lindbergh. 1977. Selection of substrata by seaweeds: optimal surface relief. *Mar. Biol.* 40:33–40.

Harper, J.L. 1977. *Population Biology of Plants*. Academic Press, New York. 892 pp.

– 1982. After description. In E.I. Newmann (ed.), *The Plant Community as a Working Mechanism*, pp. 11–25. Blackwell Scientific Publications, Oxford.

Harris, G.P. 1980. The measurement of photosynthesis in natural populations of phytoplankton. In I. Morris (ed.), *The Physiological Ecology of Phytoplankton*, pp. 129–187. Blackwell Scientific Publications, Oxford.

Harrison, P.J., & L.D. Druehl. 1982. Nutrient uptake and growth in the Laminariales and other macrophytes: a consideration of methods. In L.M. Srivastava (ed.), *Synthetic and Degradative Processes in Marine Macrophytes*, pp. 99–120. Walter de Gruyter, Berlin.

Harrison, W.G. 1983. Nitrogen in the marine environment: use of isotopes. In E.J. Carpenter & D.G. Capone (eds.), *Nitrogen in the Marine Environment*, pp. 763–807. Academic Press, New York.

Hartmann, T., & W. Eschrich. 1969. Stofftransport in Rotalgen. *Planta* 85:303–312.

Hatcher, B.G. 1977. An apparatus for measuring photosynthesis and respiration of intact large marine algae and comparison of results with those from experiments with tissue segments. *Mar. Biol.* 43:381–385.

Hatcher, B.G., A.R.O. Chapman, & K.H. Mann. 1977. An annual carbon budget for the kelp *Laminaria longicruris*. *Mar. Biol.* 44:85–96.

Haug, A. 1961. The affinity of some divalent metals to different types of alginates. Acta Chem. Scand. 15:1794–1795.

– 1976. The influence of borate and calcium on the gel formation of a sulfated polysaccharide from *Ulva lactuca*. *Acta Chem. Scand.* B 30:562–566.

Haug, A., & B. Larsen. 1974. Biosynthesis of algal polysaccharides. In J.B. Pridham (ed.), *Plant Carbohydrate Biochemistry*, pp. 207–218. Academic Press, New York.

Haug, A., & O. Smidsrød. 1967. Strontium, calcium and magnesium in brown algae. *Nature* 215:1167–1168.

Hawkes, M.J. 1983. Anatomy of *Apophlaea sinclairii* – an enigmatic red alga endemic to New Zealand. *Jap. J. Phycol. (Sôrui)* 31:55–64.

Hawkins, S.J., & R.G. Hartnoll. 1983. Changes in a rocky shore community: an evaluation of monitoring. *Mar. Environ. Res.* 9:131–181.

Haxen, P.G., & O.A.M. Lewis. 1981. Nitrate assimilation in the marine kelp, *Macrocystis angustifolia* (Phaeophyceae). *Bot. Mar.* 24:631–635.

Haxo, F.T., & L.R. Blinks. 1950. Photosynthetic action spectra of marine algae. *J. Gen. Physiol.* 33:389–422.

Hay, M.E. 1981. The functional morphology of turf-forming seaweeds: persistence in stressful marine habitats. *Ecology* 62:739–750.

Hayden, R.E., L. Dionne, & D.S. Fensom. 1972. Electrical impedance studies of stem tissue of *Solanum* clones during cooling. *Can. J. Bot.* 50:1547–1554.

Healey, F.P. 1973. Inorganic nutrient uptake and deficiency in algae. *Crit. Rev. Microbiol.* 3:69–113.

– 1980. Slope of the Monod equation as an indicator of advantage in nutrient competition. *Microbial Ecol.* 5:281–286.

Helfferich, C., & C.P. McRoy. 1980. Introduction: the spaces in the pattern. In R.C. Phillips & C.P. McRoy (eds.), *Handbook of Seagrass Biology. An Ecosystem Perspective*, pp. 1–5. Garland STPM Press, New York.

Hellebust, J.A. 1976. Osmoregulation. *Ann. Rev. Plant Physiol.* 27:485–505.

Hellebust, J.A., & J.S. Craigie (eds.). 1978. *Handbook of Phycological Methods, II. Physiological and Biochemical Methods*. Cambridge University Press, Cambridge. 512 pp.

Hellenbrand, K. 1978. Effect of pulp mill effluent on productivity of seaweeds. *Proc. Intl. Seaweed Symp.* 9:161–171.

Henry, E.C., & K.M. Cole. 1982. Ultrastructure of swarmers in the Laminariales. I. Zoospores. *J. Phycol.* 18:550–569.

Hershelman, G.P., H.A. Schafer, T.K. Jan, & D.R. Young. 1981. Metals in marine sediments near a large California municipal outfall. *Mar. Pollut. Bull.* 12:131–134.

Herth, W. 1980. Calcofluor white and Congo red inhibit chitin microfibril assembly of *Poterioochromonas*: evidence for a gap between polymerization and microfibril formation. *J. Cell Biol.* 87:442–450.

Hetherington, J.A. 1976. Radioactivity in surface and coastal waters of the British Isles, 1974. *Fish. Radiobiol. Labor. Tech. Rep.* 11:1–35.

Hiller, R.G., & D.J. Goodchild. 1981. Thylakoid membranes and pigment organization. In M.D. Hatch & N.K. Boardman (eds.), *The Biochemistry of Plants*, vol. 8, *Photosynthesis*, pp. 1–49. Academic Press, New York.

Hillman, W.S. 1976. Biological rhythms and physiological timing. *Ann. Rev. Plant Physiol.* 27:159–179.

Hixon, M.A., & W.N. Brostoff. 1983. Damselfish as keystone species in reverse: intermediate disturbance and diversity of reef algae. *Science* 220:511–513.

Hodgson, L.M. 1981. Photosynthesis of the red alga *Gastroclonium coulteri* (Rhodophyta) in response to changes in temperature, light intensity, and desiccation. *J. Phycol.* 17:37–42.

Hodson, P.V., U. Borgmann, & H. Shear. 1979. Toxicity of copper to aquatic biota. In J.O. Nriagu (ed.), *Copper in the Environment. Part II. Health Effects*, pp. 308–365. Wiley, New York.

Hoffmann, A.J., & B. Santelices. 1982. Effect of light intensity and nutrients on gametophytes and gametogenesis of *Lessonia nigrescens* Bory (Phaeophyta). *J. Exp. Mar. Biol. Ecol.* 60:77–89.

Holmes, R.W. 1957. Solar radiation, submarine daylight, and photosynthesis. In J.W. Hedgpeth (ed.), *Treatise on Marine Ecology and Paleoecology*, vol. 1, pp. 109–128. Geol. Soc. Amer. Mem. 67.

Hopkin, R., & J.M. Kain. 1978. The effects of some pollutants on the survival, growth and respiration of *Laminaria hyperborea*. *Estu. Cstl. Mar. Sci.* 7:531–553.

Hoppe, H.A., & O.J. Schmid. 1969. Commercial products. In T. Levring, H.A. Hoppe, & O.J. Schmid (eds.), *Marine*

Algae: A Survey of Research and Utilization, pp. 288–368. Cram, de Gruyter, Hamburg.

Horne, R.A. 1978. *The Chemistry of Our Environment*. Wiley, New York, 869 pp.

Hornsey, I.S., & D. Hide. 1976a. The production of antimicrobial compounds by British marine algae. II. Seasonal variation in production of antibiotics. *Brit. Phycol. J.* 11:63–67.

– 1976b. The production of antimicrobial compounds by British marine algae. III. Distribution of antimicrobial activity within the algal thallus. *Brit. Phycol. J.* 11:175–181.

Howard, R.J., K.R. Gayler, & B.R. Grant. 1975. Products of photosynthesis in *Caulerpa simpliciuscula*. *J. Phycol.* 11:463–471.

Hsiao, S.I.-C. 1969. Life history and iodine nutrition of the marine brown alga *Petalonia fascia* (O.F. Müll.) Kuntze. *Can. J. Bot.* 47:1611–1616.

– 1972. Nutritional requirements for gametogenesis in *Laminaria saccharina* (L.) Lamouroux. Ph.D. Thesis, Simon Fraser University, Burnaby, B.C.

Hsiao, S.I.-C., D.W. Kittle, & M.G. Foy. 1978. Effects of crude oils and the oil dispersant Corexit on primary production of Arctic marine phytoplankton and seaweed. *Environ. Pollut.* 15:209–221.

Hudson, P.R., & J.R. Waaland. 1974. Ultrastructure of mitosis and cytokinesis in the multinucleate green alga *Acrosiphonia*. *J. Cell Biol.* 62:274–294.

Humm, H.J. 1969. Distribution of marine algae along the Atlantic coast of North America. *Phycologia* 7:43–53.

Huntsman, S.A., & W.G. Sunda. 1980. The role of trace metals in regulating phytoplankton growth. In I. Morris (ed.), *The Physiological Ecology of Phytoplankton*, pp. 285–328. Blackwell Scientific Publications, Oxford.

Hurka, H. 1971. Factors influencing the gas composition in the vesicles of *Sargassum*. *Mar. Biol.* 11:82–89.

– 1974. A new hypothesis concerning the adaptive value of buoyancy increasing devices of algae. *Nova Hedwigia* 25:429–432.

Hurlbert, S.H. 1975. Secondary effects of pesticides on aquatic ecosystems. *Residue Rev.* 57:81–148.

Hutchins, L.W. 1947. The basis for temperature zonation in geographical distribution. *Ecol. Monogr.* 17:325–335.

Ikawa, M., V.M. Thomas, L.J. Buckley, & J.J. Uebel. 1973. Sulfur and the toxicity of the red alga *Ceramium rubrum* to *Bacillus subtilis*. *J. Phycol.* 9:302–304.

Ikawa, T., T. Watanabe, & K. Nisizawa. 1972. Enzymes involved in the last steps of the biosynthesis of mannitol in brown algae. *Plant Cell Physiol.* 13:1017–1029.

Inouye, R.S., & W.M. Schaffer. 1981. On the ecological meaning of ratio (de Wit) diagrams in plant ecology. *Ecology* 62:1679–1681.

Iseki, K., M. Takahashi, E. Bauernfiend, & C.S. Wong. 1981. Effects of polychlorinated biphenyls (PCBs) on a marine plankton population and sedimentation in controlled ecosystem enclosures. *Mar. Ecol. Prog. Ser.* 5:207–214.

Iverson, W.P., & F.E. Brinckman. 1978. Microbial metabolism of heavy metals. In R.M. Mitchell (ed.), *Water Pollution Microbiology*, pp. 201–231. Wiley, New York.

Iwasaki, H. 1961. The life cycle of *Porphyra tenera* in vitro. *Biol. Bull.* 121:173–187.

Jackson, G.A. 1977. Nutrients and production of the giant kelp, *Macrocystis pyrifera*, off southern California. *Limnol. Oceanogr.* 22:979–995.

Jackson, S.G., & E.L. McCandless. 1982. The effect of sulphate concentration on the uptake and incorporation of [^{35}S]sulphate in *Chondrus crispus*. *Can. J. Bot.* 60:162–165.

Jacobs, W.P. 1970. Development and regeneration of the algal giant coenocyte *Caulerpa*. *Ann. N.Y. Acad. Sci.* 175:732–748.

Jacobs, W.P., & J. Olson. 1980. Developmental changes in the algal coenocyte *Caulerpa prolifera* (Siphonales) after inversion with respect to gravity. *Amer. J. Bot.* 67:141–146.

Jensen, A. 1978. Chlorophylls and carotenoids. In J.A. Hellebust & J.S. Craigie (eds.), *Handbook of Phycological Methods. II. Physiological and Biochemical Methods*, pp. 59–70. Cambridge University Press, Cambridge.

Jensen, A., & A. Haug. 1956. Geographical and seasonal variation in the chemical composition of *Laminaria hyperborea* and *Laminaria digitata* from the Norwegian coast. *Norwegian Inst. Seaweed Res. Rep.* 14.

Jensen, R.G., & J.T. Bahr. 1977. Ribulose 1,5-bisphosphate carboxylase-oxygenase. *Ann. Rev. Plant Physiol.* 28:379–400.

Jerlov, N.G. 1968. *Optical Oceanography*. Elsevier, Amsterdam. 194 pp.

– 1970. Light: general introduction. In O. Kinne (ed.), *Marine Ecology*, vol. I, pt. 1, pp. 95–102. Wiley, New York.

– 1976. *Marine Optics*. Elsevier, Amsterdam. 231 pp.

Johannes, R.E. 1980. The ecological significance of the submarine discharge of groundwater. *Mar. Ecol. Progr. Ser.* 3:365–373.

Johansen, H.W. 1981. *Coralline Algae, A First Synthesis*. CRC Press, Boca Raton, Fla. 233 pp.

John, D.M., D. Lieberman, M. Lieberman, & M.D. Swaine. 1980. Strategies of data collection and analysis of subtidal vegetation. In J.H. Price, D.E.G. Irvine, & W.F. Farnham (eds.), *The Shore Environment*, vol. 1, *Methods*, pp. 265–284. Academic Press, New York.

Johnson, W.S., A. Gigon, S.L. Gulmon, & H.A. Mooney. 1974. Comparative photosynthetic capacities of intertidal algae under exposed and submerged conditions. *Ecology*. 55:450–453.

Johnstone, I.M. 1977. *Draparnaldiopsis*: a filamentous green alga (Chlorophyta, Chaetophoraceae) requiring vitamin B_{12}. *Phycologia* 16:183–187.

Joly, A.B., & E.C. de Oliveira, Fil. 1967. Two Brazilian *Laminaria*. *Publ. Inst. Pesq. Mar.* 4:1–13.

Jones, W.E., & M.S. Babb. 1968. The motile period of swarmers of *Enteromorpha intestinalis* (L.) Link. *Brit. Phycol. Bull.* 3:525–528.

Jones, W.E., & A. Demetropoulos. 1968. Exposure to wave action: measurements of an important ecological parameter on rocky shores on Anglesey. *J. Exp. Mar. Biol. Ecol.* 2:46–63.

Josselyn, M.N., & A.C. Mathieson. 1978. Contribution of receptacles from the fucoid *Ascophyllum nodosum* to the detrital pool of a north temperate estuary. *Estuaries* 1:258–261.

Kageyama, A., Y. Yokohama, & K. Nisizawa. 1979. Diurnal rhythm of apparent photosynthesis of a brown alga, *Spatoglossum pacificum*. *Bot. Mar.* 22:199–201.

Kageyama, A., Y. Yokohama, S. Shirmura, & T. Ikawa. 1977. An efficient excitation energy transfer from a carotenoid, siphonaxanthin, to chlorophyll *a* observed in a deep-water species of Chlorophycean seaweed. *Plant Cell Physiol.* 18:477–480.

Kain, J.M. 1964. Aspects of the biology of *Laminaria hyperborea*. III. Survival and growth of gametophytes. *J. Mar. Biol. Ass. U.K.* 44:415–433.

– 1969. The biology of *Laminaria hyperborea*. V. Comparison with early stages of competitors. *J. Mar. Biol. Ass. U.K.* 49:455–473.

Kalle, K. 1971. Salinity: general introduction. In O. Kinne (ed.), *Marine Ecology*, vol. I, pt. 2, pp. 683–688. Wiley, New York.

– 1972. Dissolved gases: general introduction. In O. Kinne (ed.), *Marine Ecology*, vol. I, pt. 3, pp. 1451–1457. Wiley, New York.

Kastendiek, J. 1982. Competitor-mediated coexistence: interactions among three species of benthic macroalgae. *J. Exp. Mar. Biol. Ecol.* 62:201–210.

Katsaros, C., B. Galatis, & K. Mitrakos. 1983. Fine structural studies on the interphase and dividing apical cells of *Sphacelaria tribuloides* (Phaeophyta). *J. Phycol.* 19:16–30.

Kauss, H. 1973. Turnover of galactosylglycerol and osmotic balance in *Ochromonas*. *Plant Physiol.* 52:613–615.

– 1978. Osmotic regulation in algae. In L. Reinhold, J.B. Harborne, & T. Swain (eds.), *Progress in Phytochemistry*, vol. 5, pp. 1–27. Pergamon Press, New York.

Kauss, H., & K.-S. Thomson. 1982. Biochemistry of volume control in *Poterioochromonas*. In D. Marmé, E. Marrè, & R. Hertel (eds.), *Plasmalemma and Tonoplast: Their Functions in the Plant Cell*, pp. 255–262. Elsevier, New York.

Kauss, H., K.-S. Thomson, M. Tetour, & W. Jeblick. 1978. Proteolytic activation of a galactosyl transferase involved in osmotic regulation. *Plant Physiol.* 61:35–37.

Kautsky, L. 1982. Primary production and uptake kinetics of ammonium and phosphate by *Enteromorpha compressa* in an ammonium sulfate industry outlet area. *Aquat. Bot.* 12:23–40.

Kayser, M. 1979. Growth interactions between marine dinoflagellates in multispecies culture experiments. *Mar. Biol.* 52:357–370.

Kelco. No date. *Kelco Algin. Hydrophilic Derivatives of Alginic Acid for Scientific Water Control*, 2nd ed. Available from Kelco Division of Merck & Co. Inc., 20 N. Wacker Drive, Chicago, Ill. 60606.

Kerby, N.W., & L.V. Evans. 1978. Isolation and partial characterization of pyrenoids from the brown alga *Pilayella littoralis* (L.) Kjellm. *Planta* 142:91–95.

– 1983. Phosphoenolpyruvate carboxykinase activity in *Ascophyllum nodosum* (Phaeophyceae). *J. Phycol.* 19:1–3.

Khailov, K.M., & Z.P. Burlakova. 1969. Release of dissolved organic matter by marine seaweeds and distribution of their total organic production to inshore communities. *Limnol. Oceanogr.* 14:521–527.

Khailov, K.M., V.I. Kholodov, Yu.K. Firsov, & A.V. Prazukin. 1978. Thalli of *Fucus vesiculosus* in ontogenesis: changes in morpho-physiological parameters. *Bot. Mar.* 21:289–311.

Khfaji, A.H., & A.D. Boney. 1979. Antibiotic effects of crustose germlings of the red alga *Chondrus crispus* Stackh. on benthic diatoms. *Ann. Bot.* 43:231–232.

Kimpel, D.L., R.K. Togasaki, & S. Miyachi. 1983. Carbonic anhydrase in *Chlamydomonas reinhardtii*. I. Localization. *Plant Cell Physiol.* 24:255–259.

Kindig, A.C., & M.M. Littler. 1980. Growth and primary productivity of marine macrophytes exposed to domestic sewage effluents. *Mar. Environ. Res.* 3:81–100.

King, R.J., & W. Schramm. 1982. Calcification in the maerl coralline alga *Phymatolithon calcareum*: effects of salinity and temperature. *Mar. Biol.* 70:197–204.

Kingham, D.L., & L.V. Evans. 1977. The *Pelvetia/Ascophyllum-Mycosphaerella* inter-relationship. *Brit. Phycol. J.* 12:120.

Kinne, O. 1970. Temperature – general introduction. In O. Kinne (ed.), *Marine Ecology*, vol. I, pt. 1, pp. 321–346. Wiley, New York.

– 1971. Salinity: invertebrates. In O. Kinne. (ed.), *Marine Ecology*, vol. I, pt. 2, pp. 821–995. Wiley, New York.

Kirkman, H. 1981. The first year in the life history and the survival of the juvenile marine macrophyte, *Ecklonia radiata* (Turn.) J. Agardh. *J. Exp. Mar. Biol. Ecol.* 55:243–254.

Kirst, G.O. 1977. Coordination of ionic relations and mannitol concentrations in the euryhaline unicellular alga *Platymonas subcordiformis* (Hazen) after osmotic shock. *Planta* 135:69–75.

Kirst, G.O., & M.A. Bisson. 1979. Regulation of turgor pressure in marine algae: ions and low-molecular-weight organic compounds. *Austr. J. Plant Physiol.* 6:539–556.

Kitching, J.A., & F.J. Ebling. 1967. Ecological studies at Lough Ine. *Adv. Ecol. Res.* 4:197–291.

Kitoh, S., & S. Hori. 1977. Metabolism of urea in *Chlorella ellipsoidea*. *Plant Cell Physiol.* 18:513–519.

Klumpp, D.W. 1980. Characteristics of arsenic accumulation by the seaweeds *Fucus spiralis* and *Ascophyllum nodosum*. *Mar. Biol.* 58:257–264.

Knauer, G.A., & J.H. Martin. 1972. Mercury in a marine pelagic food chain. *Limnol. Oceanogr.* 17:868–876.

Knight, M., & M. Parke. 1950. A biological study of *Fucus vesiculosus* L. and *F. serratus* L. *J. Mar. Biol. Ass. U.K.* 29:439–514.

Koehl, M.A.R. 1982. The interaction of moving water and sessile organisms. *Sci. Amer.* 247(6):124–135.

Koehl, M.A.R., & S.A. Wainright. 1977. Mechanical adaptations of a giant kelp. *Limnol. Oceanogr.* 22:1067–1071.

Kohler, A., & B.C. Labus. 1983. Eutrophication processes and pollution of freshwater ecosystems including waste heat. In O. Lange, P.S. Nobel, C.B. Osmond, & H. Ziegler (eds.), *Physiological Plant Ecology IV, Encycl. Plant Physiol*, vol. 12D, pp. 413–464. Springer-Verlag, New York.

Kohlmeyer, J., & E. Kohlmeyer. 1968. *Icones Fungorum Maris*. Cramer, Lehre.

– 1972. Is *Ascophyllum* lichenized? *Bot. Mar.* 15:109–112.

Kok, B., B. Forbush, & M. McGloin. 1970. Cooperation of charges in photosynthetic O_2 evolution. I. A linear 4-step mechanism. *Photochem. Photobiol.* 11:457–475.

Koop, H.-U. 1979. The life cycle of *Acetabularia* (Dasycladales, Chlorophyceae): a compilation of evidence for meiosis in the primary nucleus. *Protoplasma* 100:353–366.

Koop, K., R.C. Newell, & M.I. Lucas. 1982. Microbial regeneration of nutrients from the decomposition of macrophyte debris on the shore. *Mar. Ecol. Progr. Ser.* 9:91–96.

Kornmann, P. 1970. Advances in marine phycology on the basis of cultivation. *Helgol. wiss. Meeresunters.* 20:39–61.

Kornmann, P., & P.-H. Sahling. 1977. Meeresalgen von Helgoland. Benthische Grün-, Braun- und Rotalgen. *Helgol. wiss. Meeresunters.* 29:1–289.

Kremer, B.P. 1977. Biosynthesis of polyols in *Pelvetia canaliculata*. *Z. Pflanzenphysiol.* 81:68–73.

– 1979a. Biosynthesis and metabolism of polyhydroxy alcohols in marine benthic algae. *Proc. Intl. Seaweed Symp.* 9:421–428.

– 1979b. Photoassimilatory products and osmoregulation in marine Rhodophyceae. *Z. Pflanzenphysiol.* 93:139–148.

– 1981a. Carbon metabolism. In C.S. Lobban & M.J. Wynne (eds.), *The Biology of Seaweeds*, pp. 493–533. Blackwell Scientific Publications, Oxford.

– 1981b. Metabolic implications of non-photosynthetic carbon fixation in brown macroalgae. *Phycologia* 20:242–250.

– 1983. Carbon economy and nutrition of the alloparasitic red alga *Harveyella mirabilis*. *Mar. Biol.* 76:231–239.

Kremer, B.P., & J.W. Markham. 1979. Carbon assimilation by different developmental stages of *Laminaria saccharina*. *Planta* 144:497–501.

– 1982. Primary metabolic effects of cadmium in the brown alga, *Laminaria saccharina*. *Z. Pflanzenphysiol.* 108:125–130.

Kremer, B.P., & K. Schmitz. 1973. CO_2-Fixierung und Stofftransport in benthischen marinen algen. IV. Zur ^{14}C-Assimilation einiger litoraler Braunalgen in submersen und emersen Zustand. *Z. Pflanzenphysiol.* 68:357–363.

Kriegstein, A.R., V. Castellucci, & E.R. Kandel. 1974. Metamorphosis of *Aplysia californica* in laboratory culture. *Proc. Natl. Acad. Sci.* 71:3654–3658.

Kulh, A. 1962. Inorganic phosphorus uptake and metabolism. In R.A. Lewin (ed.), *Physiology and Biochemistry of Algae*, pp. 211–229. Academic Press, New York.

– 1974. Phosphorus. In W.D.P. Stewart (ed.), *Algal Physiology and Biochemistry*, pp. 636–654. Blackwell Scientific Publications, Oxford.

Küppers, U., & B.P. Kremer. 1978. Longitudinal profiles of carbon dioxide fixation capacities in marine macroalgae. *Plant Physiol.* 62:49–53.

Küppers, U., & M. Weidner. 1980. Seasonal variation of enzyme activities in *Laminaria hyperborea*. *Planta* 148:222–230.

Kurogi, M., & K. Hirano. 1956. Influences of water temperature on the growth, formation of monosporangia and monospore-liberation in the *Conchocelis*-phase of *Porphyra tenera* Kjellm. *Bull. Tohoku Reg. Fish. Res. Lab.* 8:45–61. (Japanese with extensive English summary.)

Kuwabara, J.S. 1982. Micronutrients and kelp cultures: evidence for cobalt and manganese deficiency in southern California deep seawater. *Science* 216:1219–1221.

La Claire, J.W., II. 1982. Wound-healing motility in the green alga *Ernodesmis*: calcium ions and metabolic energy are required. *Planta* 156:466–474.

Lang, J.C. 1974. Biological zonation at the base of a reef. *Amer. Sci.* 62:271–281.

Larkum, A.W.D., E.A. Drew, & R.N. Crossett. 1967. The vertical distribution of attached marine algae in Malta. *J. Ecol.* 55:361–371.

Larsen, B., A. Haug, & T.J. Painter. 1970. Sulphated polysaccharides in brown algae. III. The native state of fucoidan in *Ascophyllum nodosum* and *Fucus vesiculosus*. *Acta Chem. Scand.* 24:3339–3352.

Laue, E.G., & A.J. Drummond. 1968. Solar constant. First direct measurements. *Science* 161:888.

Laws, E.A. 1981. *Aquatic Pollution*. Wiley, New York. 482 pp.

Laws, E.A., & D.G. Redalje. 1982. Sewage diversion effects on the water column of a subtropical estuary. *Mar. Environ. Res.* 6:265–279.

Laycock, M.V., & J.S. Craigie. 1977. The occurrence and seasonal variation of gigartinine and L-citrullinyl-L-arginine in *Chondrus crispus* Stackh. *Can. J. Biochem.* 55:27–30.

Laycock, M.V., K.C. Morgan, & J.S. Craigie. 1981. Physiological factors affecting the accumulation of L-citrullinyl-L-arginine in *Chondrus crispus*. *Can. J. Bot.* 59:522–527.

Lean, D.R.S., & F.R. Pick. 1981. Photosynthetic response of lake plankton to nutrient enrichment: a test of nutrient limitation. *Limnol. Oceanogr.* 26:1001–1019.

Lee, R.B. 1980. Sources of reductant for nitrate assimilation in non-photosynthetic tissue: a review. *Plant Cell Environ.* 3:65–90.

Lee, R.F. 1980. Processes affecting the role of oil in the sea. In R.A. Gayer (ed.), *Marine Environmental Pollution*, vol. 1, *Hydrocarbons*, pp. 338–352. Elsevier Scientific Publishing Company, Amsterdam.

Leedale, G.F. 1968. Editorial note: the adoption of SI. *Brit. Phycol. Bull.* 3:589–591.

Lehnberg, W. 1978. Die Wirkung eines Licht-Temperatur-Salzgehalt Komplexes auf den Gaswechsel von *Delesseria sanguinea* (Rhodophyta) aus der westlichen Ostsee. *Bot. Mar.* 21:485–497.

Lehninger, A.L. 1975. *Biochemistry*, 2nd ed. Worth, New York. 1104 pp.

Leighton, D.L., L.G. Jones, & W.J. North. 1966. Ecological relationships between giant kelp and sea urchins in southern California. *Proc. Intl. Seaweed Symp.* 5:141–153.

Leopold, A.C., & P.E. Kriedemann. 1975. *Plant Growth and Development*, 2nd ed. McGraw-Hill, New York. 545 pp.

Levinton, J.S. 1982. *Marine Ecology*. Prentice-Hall, Englewood Cliffs, N.J. 526 pp.

Levitt, J. 1969. *Introduction to Plant Physiology*. Mosby, Saint Louis. 304 pp.

– 1972. *Responses of Plants to Environmental Stresses*. Academic Press, New York. 687 pp.

Levring, T. 1977. Potential yields of European marine algae. In R.W. Krauss (ed.), *Marine Plant Biomass of the Pacific Northwest Coast*, pp. 251–270. Oregon State University Press, Corvallis.

Lewin, J., & C. Chen. 1971. Available iron: a limiting factor for marine phytoplankton. *Limnol. Oceanogr.* 16:670–675.

Lewin, R.A. 1970. Toxin secretion and tail autotomy by irritated *Oxynoe panamensis* (Opisthobranchiata: Sacoglossa). *Pac. Sci.* 24:356–358.

– 1974. Biochemical taxonomy. In W.D.P. Stewart (ed.), *Algal Physiology and Biochemistry*, pp. 1–39. Blackwell Scientific Publications, Oxford.

Lewis, A.G., & W.R. Cave. 1982. The biological importance of copper in oceans and estuaries. *Oceanogr. Mar. Biol. Ann. Rev.* 20:471–695.

Lewis, J.R. 1964. *The Ecology of Rocky Shores*. English University Press, London. 323 pp.

– 1980. Objectives in littoral ecology – a personal viewpoint. In J.H. Price, D.E.G. Irvine, & W.F. Farnham (eds.), *The Shore Environment*, vol. 1, *Methods*, pp. 1–18. Academic Press, New York.

Li, W.K.W. 1978. Kinetic analysis of interactive effects of cadmium and nitrate on growth of *Thalassiosira fluviatalis* (Bacillariophyceae). *J. Phycol.* 14:454–460.

Liaaen-Jensen, S. 1978. Algal carotenoids and chemosystematics. In D.J. Faulkner & W.H. Fenical (eds.), *Marine Natural Products Chemistry*, pp. 239–259. Plenum, New York.

Liddle, L.B., J.P. Thomas, & J. Scott. 1982. Morphology and distribution of nuclei during development in *Cymopolia barbata* (Chlorophycophyta, Dasycladales). *J. Phycol.* 18:257–264.

Lin, C.K. 1977. Accumulation of water soluble phosphorus and hydrolysis of polyphosphates by *Cladophora glomerata* (Chlorophyceae). *J. Phycol.* 13:46–51.

Lin, T.-Y., & W.Z. Hassid. 1966. Pathway of alginic acid synthesis in the marine brown alga, *Fucus gardneri* Silva. *J. Biol. Chem.* 241:5284–5297.

Lindauer, V.W., V.J. Chapman, & M. Aiken. 1961. The marine algae of New Zealand. III. Phaeophyceae. *Nova Hedwigia* 3:129–350.

Lindstrom, S.C., & R.E. Foreman. 1978. Seaweed associations of the Flat Top Islands, British Columbia: a comparison of community methods. *Syesis* 11:171–185.

Linley, E.A.S., R.C. Newell, & S.A. Bosma. 1981. Heterotrophic utilization of mucilage released during fragmentation of kelp (*Ecklonia maxima* and *Laminaria pallida*). I. Development of microbial communities associated with the degradation of kelp mucilage. *Mar. Ecol. Progr. Ser.* 4:31–41.

Lipschultz, C.A., & E. Gantt. 1981. Association of phycoerythrin and phycocyanin: *in vitro* formation of a functional energy transferring phycobilisome complex of *Porphyridium soridium*. *Biochemistry* 20:3371–3376.

Littler, M.M. 1976. Calcification and its role among the macroalgae. *Micronesica* 12:27–41.

– 1979. The effects of bottle volume, thallus weight, oxygen saturation levels, and water movement on apparent photosynthetic rates in marine algae. *Aquat. Bot.* 7:21–34.

Littler, M.M., & K.E. Arnold. 1980. Sources of variability in macroalgal primary productivity: sampling and interpretative problems. *Aquat. Bot.* 8:141–156.

– 1982. Primary productivity of marine macroalgal functional-form groups from southwestern North America. *J. Phycol.* 18:307–311.

Littler, M.M., & B.J. Kauker. 1984. Heterotrichy and survival strategies in the red alga *Corallina officinalis* L. *Bot. Mar.* 27:37–44.

Littler, M.M., & D.S. Littler. 1980. The evolution of thallus form and survival strategies in benthic marine macroalgae: field and laboratory tests of a functional form model. *Amer. Nat.* 116:25–44.

Littler, M.M., & S.N. Murray. 1975. Impact of sewage on the distribution, abundance and community structure of rocky intertidal macro-organisms. *Mar. Biol.* 30:277–291.

Littler, M.M., D.S. Littler, & P.R. Taylor. 1983. Evolutionary strategies in a tropical barrier reef system: functional-form groups of marine macroalgae. *J. Phycol.* 19:229–237.

Lloyd, N.D.H., J.L. McLachlan, & R.G.S. Bidwell. 1981. A rapid infra-red carbon dioxide analysis screening technique for predicting growth and productivity of marine algae. *Proc. Intl. Seaweed Symp.* 10:461–466.

Lobban, C.S. 1978a. The growth and death of the *Macrocystis* sporophyte (Phaeophyceae, Laminariales). *Phycologia* 17:196–212.

– 1978b. Translocation of ^{14}C in *Macrocystis pyrifera* (giant kelp). *Plant Physiol.* 61:585–589.

– 1978c. Translocation of ^{14}C in *Macrocystis integrifolia* (Phaeophyceae). *J. Phycol.* 14:178–182.

– 1981. Physiology and biochemistry: introduction. In C.S. Lobban & M.J. Wynne (eds.), *The Biology of Seaweeds*, pp. 455–457. Blackwell Scientific Publications, Oxford.

Lockhart, J.C. 1979. Factors affecting various forms in *Cladosiphon zosterae* (Phaeophyceae). *Amer. J. Bot.* 66:836–844.

Loya, Y., & B. Rinkevich. 1980. Effects of oil pollution on coral reef communities. *Mar. Ecol. Prog. Ser.* 3:167–180.

Lubchenco, J. 1978. Plant species diversity in a marine intertidal community: importance of herbivore food preference and algal competitive abilities. *Amer. Nat.* 112:23–39.

– 1980. Algal zonation in the New England rocky intertidal community: an experimental analysis. *Ecology* 61:333–344.

– 1983. *Littorina* and *Fucus*: effects of herbivores, substratum heterogeneity, and plant escapes during succession. *Ecology* 64:1116–1123.

Lubchenco, J., & J. Cubit. 1980. Heteromorphic life histories of certain marine algae as adaptations to variations in herbivory. *Ecology* 61:676–687.

Lubchenco, J., & S.D. Gaines. 1981. A unified approach to marine plant-herbivore interactions. I. Populations and communities. *Ann. Rev. Ecol. Syst.* 12:405–437.

Lubchenco, J., & B.A. Menge. 1978. Community development and persistence in a low rocky intertidal zone. *Ecol. Monogr.* 48:67–94.

Lubimenko, V., & Q. Tichovskaya. 1928. *Recherches sur la Photosynthèse et l'Adaptation Chromatique chez les Algues Marines*. Acad. Sci. U.S.S.R., Moscow.

Lüning, K. 1969. Growth of amputated and dark-exposed individuals of the brown alga *Laminaria hyperborea*. *Mar. Biol.* 2:218–223.

– 1975. Kreuzungsexperimente an *Laminaria saccharina* von Helgoland und von der Isle of Man. *Helgol. wiss. Meeresunters.* 27:108–114.

– 1980. Control of algal life-history by daylength and temperature. In J.H. Price, D.E.G. Irvine, & W.F. Farnham (eds.), *The Shore Environment*, vol. 2, *Ecosystems*, pp. 915–945. Academic Press, New York.

– 1981a. Light. In C.S. Lobban & M.J. Wynne (eds.), *The Biology of Seaweeds*, pp. 326–355. Blackwell Scientific Publications, Oxford.

– 1981b. Photomorphogenesis of reproduction in marine macroalgae. *Ber. Deutsch. Bot. Ges.* 94:401–417.

Lüning, K., & M.J. Dring. 1979. Continuous underwater light measurements near Helgoland (North Sea) and its significance for characteristic light limits in the sublittoral region. *Helgol. wiss. Meeresunters.* 32:403–424.

Lüning, K., & D.G. Müller. 1978. Chemical interactions in sexual reproduction of several Laminariales (Phaeophyceae): release and attraction of spermatozoids. *Z. Pflanzenphysiol.* 89:333–341.

Lüning, K., & M. Neushul. 1978. Light and temperature demands for growth and reproduction of laminarian gametophytes in southern and central California. *Mar. Biol.* 45:297–309.

Lüning, K., A.R.O. Chapman, & K.H. Mann. 1978. Crossing experiments in the non-digitate complex of *Laminaria* from both sides of the Atlantic. *Phycologia* 17:293–298.

Lüning, K., K. Schmitz, & J. Willenbrink. 1973. CO_2 fixation and translocation in benthic marine algae. III. Rates and ecological significance of translocation in *Laminaria hyperborea* and *L. saccharina*. *Mar. Biol.* 23:275–281.

Lüttge, U., & N. Higinbotham. 1979. *Transport in Plants*. Springer-Verlag, New York. 468 pp.

Lüttge, U., & M.G. Pitman (eds.) 1976. *Transport in Plants. II, Part B. Tissues and Organs*. Encyclopedia of Plant Physiology. Springer-Verlag, New York, 456 pp.

Luoma, S.N., G.W. Bryan, & W.T. Langston. 1982. Scavenging of heavy metals from particulates by brown seaweed. *Mar. Pollut. Bull.* 13: 394–396.

Lyons, J.M. 1973. Chilling injury in plants. *Ann. Rev. Plant Physiol.* 24:445–466.

Lyons, J.M., D. Graham, & J.K. Raison. (eds.) 1979. *Low Temperature Stress in Crop Plants: The Role of the Membrane*. Epilogue, pp. 543–548. Academic Press, New York.

MacDonald, M.A., D.S. Fensom, & A.R.A. Taylor. 1974. Electrical impedance in *Ascophyllum nodosum* and *Fucus vesiculosus* in relation to cooling, freezing, and desiccation. *J. Phycol.* 10:462–469.

MacIsaac, J.J., R.C. Dugdale, S.A. Huntsman, & H.L. Conway. 1979. The effect of sewage on uptake of inorganic nitrogen and carbon by natural populations of marine phytoplankton. *J. Mar. Res.* 37:51–66.

Mackie, W., & R.D. Preston. 1974. Cell wall and intercellular region polysaccharides. In W.D.P. Stewart (ed.), *Algal Physiology and Biochemistry*, pp. 40–85. Blackwell Scientific Publications, Oxford.

MacRobbie, E.A.C. 1974. Ion uptake. In W.D.P. Stewart (ed.), *Algal Physiology and Biochemistry*, pp. 714–740. Blackwell Scientific Publications, Oxford.

Maestrini, S.Y., & D.J. Bonin. 1981. Competition among phytoplankton based on inorganic nutrients. In T. Platt (ed.), *Physiological Bases of Phytoplankton Ecology*, *Can. Bull. Fish. Aquat. Sci.* 210:264–278.

Malkin, R. 1982. Photosystem I. *Ann. Rev. Plant Physiol.* 33:455–479.

Manley, S.L. 1981. Iron uptake and translocation by *Macrocystis pyrifera*. *Plant Physiol.* 68:914–918.

– 1983. Composition of sieve tube sap from *Macrocystis pyrifera* (Phaeophyta) with emphasis on the inorganic constituents. *J. Phycol.* 19:118–121.

Mann, K.H. 1973. Seaweeds: their productivity and strategy for growth. *Science* 182:975–981.

Markham, J.W. 1973. Observations on the ecology of *Laminaria sinclairii* on three northern Oregon beaches. *J. Phycol.* 9:336–341.

Markham, J.W., & P.R. Newroth. 1972. Observations on the ecology of *Gymnogongrus linearis* and related species. *Proc. Intl. Seaweed Symp.* 7:127–130.

Markham, J.W., B.P. Kremer, & K.R. Sperling. 1980. Effect of cadmium on *Laminaria saccharina* in culture. *Mar. Ecol. Prog. Ser.* 3:31–39.

Martens, C.S., R.A. Berner, & J.K. Rosenfeld. 1978. Interstitial water chemistry of anoxic Long Island Sound sediments. 2. Nutrient regeneration and phosphate removal. *Limnol. Oceanogr.* 23:605–617.

Martin, J.H., & G.A. Knauer. 1973. The elemental composition of plankton. *Geochim. Cosmochim. Acta.* 37:1639–1653.

Mathieson, A.C., & T.L. Norall. 1975. Photosynthetic studies of *Chondrus crispus*. *Mar. Biol.* 33:207–213.

Mathieson, A.C., N.B. Reynolds, & E.J. Hehre. 1981. Investigations of New England marine algae. II: The species composition, distribution and zonation of seaweeds in the Great Bay Estuary System and the adjacent open coast of New Hampshire. *Bot. Mar.* 24:533–545.

Mathieson, A.C., E. Tveter, M. Daly, & J. Howard. 1977. Marine algal ecology in a New Hampshire tidal rapid. *Bot. Mar.* 20:277–290.

Matilsky, M.B., & W.P. Jacobs. 1983. Accumulation of amyloplasts on the bottom of normal and inverted rhizome tips of *Caulerpa prolifera* (Forsskål) Lamouroux. *Planta* 159:189–192.

Matsumoto, F. 1959. Studies on the effects of environmental factors on the growth of "nori" (*Porphyra tenera* Kjellm.), with special reference to the water current. *J. Fac. Fish. Anim. Husband. Hiroshima U.* 2:249–333. (In Japanese with summary, tables, and figure legends in English.)

Mayhoub, H., P. Gayral, & R. Jacques. 1976. Action de la composition spectrale de la lumière sur la croissance et la reproduction de *Calosiphonia vermicularis* (J. Agardh) Schmitz (Rhodophycées, Gigartinales). *C.R. Acad. Sci. Paris* 283(D):1041–1044.

Mazzo, A., S. Bonotto, & B. Felluga. 1977. Ultrastructure of DNA of basal, middle and apical chloroplasts of individual *Acetabularia mediterranea* cells. In C.L.F. Woodcock (ed.), *Progress in Acetabularia Research*, pp. 123–136. Academic Press, New York.

McArthur, D.M., & B.L. Moss. 1977. The ultrastructure of cell walls in *Enteromorpha intestinalis* (L.) Link. *Brit. Phycol. J.* 12:359–368.

McBride, D.L., & K. Cole. 1972. Ultrastructural observations on germinating monospores in *Smithora naiadum* (Rhodophyceae, Bangiophycidae). *Phycologia* 11:181–191.

McBride, D.L., P. Kugrens, & J.A. West. 1974. Light and electron microscope observations on red algal galls. *Protoplasma* 79:249–264.

McCandless, E.L. 1981. Polysaccharides of the seaweeds. In C.S. Lobban & M.J. Wynne (eds.), *The Biology of Seaweeds*, pp. 559–588. Blackwell Scientific Publications, Oxford.

McCandless, E.L., & J.S. Craigie. 1979. Sulfated polysaccharides in red and brown algae. *Ann. Rev. Plant Physiol.* 30:41–53.

McCandless, E.L., J.S. Craigie, & J.A. Walter. 1973. Carrageenans in the gametophytic and sporophytic stages of *Chondrus crispus*. *Planta* 112:201–212.

McClintock, M., N. Higinbotham, E.G. Uribe, & R.E. Cleland. 1982. Active, irreversible accumulation of extreme levels of H_2SO_4 in the brown alga, *Desmarestia*. *Plant Physiol.* 70:771–774.

McCracken, D.A., & J.R. Cain. 1981. Amylose in floridean starch. *New Phytol.* 88:67–71.

McKenzie, G.H., A.L. Ch'ng, & K.R. Gayler. 1978. Glutamine synthetase/glutamine: α-ketoglutarate amino transferase in chloroplasts from the marine alga *Caulerpa simpliciuscula*. *Plant Physiol.* 63:578–582.

McLachlan, J. 1977. The effects of nutrients on growth and development of embryos of *Fucus edentatus* Pyl. (Phaeophyceae, Fucales). *Phycologia* 16:329–338.

– 1982. Inorganic nutrition of marine macro-algae in culture. In L.M. Srivastava (ed.), *Synthetic and Degradative Processes in Marine Macrophytes*, pp. 71–98. Walter de Gruyter, Berlin.

McLachlan, J., & R.G.S. Bidwell. 1978. Photosynthesis of eggs, sperm, zygotes, and embryos of *Fucus serratus*. *Can. J. Bot.* 56:371–373.

– 1983. Effects of colored light on the growth and metabolism of *Fucus* embryos and apices in culture. *Can. J. Bot.* 61:1993–2003.

McLachlan, J., L. C.-M. Chen, & T. Edelstein. 1971. The culture of four species of *Fucus* under laboratory conditions. *Can. J. Bot.* 49:1463–1469.

McLean, M.W., & F.B. Williamson. 1977. Cadmium accumulation by the marine red alga *Porphyra umbilicalis*. *Physiol. Plant.* 41:268–272.

Menzel, D., R. Kazlauskas, & J. Reichelt. 1983. Coumarins in the siphonalean green algal family Dasycladaceae Kützing (Chlorophyceae). *Bot. Mar.* 26:23–29.

Michanek, G. 1978. Trends in applied phycology, with a literature review: seaweed farming on an industrial scale. *Bot. Mar.* 21:469–475.

– 1979. Phytogeographic provinces and seaweed distribution. *Bot. Mar.* 22:375–391.

Miflin, B.J., & P.J. Lea. 1977. Amino acid metabolism. *Ann. Rev. Plant Physiol.* 28:299–329.

Millard, P., & L.V. Evans. 1982. Sulphate uptake in the unicellular marine red alga *Rhodella maculata*. *Arch. Microbiol.* 131:165–169.

Millner, P.A., & L.V. Evans. 1980. The effects of triphenyltin chloride on respiration and photosynthesis in the green algae *Enteromorpha intestinalis* and *Ulothrix pseudoflacca*. *Plant Cell Environ.* 3:339–348.

– 1981. Uptake of triphenyltin chloride by *Enteromorpha intestinalis* and *Ulothrix pseudoflacca*. *Plant Cell Environ.* 4:383–389.

Mishkind, M., D. Mauzerall, & S.I. Beale. 1979. Diurnal variation *in situ* of photosynthetic capacity in *Ulva* caused by a dark reaction. *Plant Physiol.* 64:896–899.

Mitchell, C.T., E.K. Anderson, L.C. Jones, & W.J. North. 1970. What oil does to ecology. *J. Water Pollut. Contr. Fed.* 42:812–818.

Mitchell, P. 1974. A chemiosmotic molecular mechanism for proton translocating adenosine triphosphatases. *FEBS Lett.* 43:189–194.

Miura, A. 1975. *Porphyra* cultivation in Japan. In J. Tokida & H. Hirose (eds.), *Advance of Phycology in Japan*, pp. 273–304. W. Junk, The Hague.

Miura, A., Y. Fujio, & S. Suto. 1979. Genetic differentiation between the wild and cultured populations of *Porphyra yezoensis*. *Tohoku J. Agric. Res.* 30:114–125.

Moe, R.L., & P.C. Silva. 1981. Morphology and taxonomy of *Himantothallus* (including *Phaeoglossum* and *Phyllogigas*), an Antarctic member of the Desmarestiales (Phaeophyceae). *J. Phycol.* 17:15–29.

Moebus, K., & K.M. Johnson. 1974. Exudation of dissolved organic carbon by brown algae. *Mar. Biol.* 26:117–125.

Moebus, K., K.M. Johnson, & J. McN. Sieburth. 1974. Rehydration of desiccated intertidal brown algae: release of dissolved organic carbon and water uptake. *Mar. Biol.* 26:127–134.

Mohsen, A.F., A.F. Khaleafa, M.A. Hashem, & A. Metwalli. 1974. Effect of different nitrogen sources on growth, reproduction, amino acid, fat and sugar contents in *Ulva fasciata* Petite. *Bot. Mar.* 17:218–222.

Monod, J. 1942. *Recherches sur la Croissance des Cultures Bacteriennes*. Herman, Paris. 210 pp.

Morel, A., & R.C. Smith. 1974. Relation between total quanta and total energy for aquatic photosynthesis. *Limnol. Oceanogr.* 19:591–600.

Morris, A.W., & A.J. Bale. 1975. The concentration of cadmium, copper, manganese and zinc by *Fucus vesiculosus* in the Bristol Channel. *Estu. Cstl. Mar. Sci.* 3:153–163.

Morris, I., & M. Darley. 1982. Physiology and biochemistry of algae: introduction and bibliography. In J.R. Rosowski & B.C. Parker (eds.), *Selected Papers in Phycology, II*, pp. 278–287. Phycol. Soc. Amer., Lawrence, Kan.

Morris, I., & H.E. Glover. 1974. Questions on the mechanism of temperature adaptation in marine phytoplankton. *Mar. Biol.* 24:147–154.

Moss, B. 1964. Wound healing and regeneration in *Fucus vesiculosus* L. *Proc. Intl. Seaweed Symp.* 4:117–122.

– 1974a. Morphogenesis. In W.D.P. Stewart (ed.), *Algal Physiology and Biochemistry*, pp. 788–813. Blackwell Scientific Publications, Oxford.

– 1974b. Attachment and germination of the zygotes of *Pelvetia canaliculata* (L.) Dcne. et Thur. (Phaeophyceae, Fucales). *Phycologia* 13:317–322.

– 1982. The control of epiphytes by *Halidrys siliquosa* (L.) Lyngb. (Phaeophyta, Cystoseiraceae). *Phycologia* 21:185–191.

– 1983. Sieve elements in the Fucales. *New Phytol.* 93:433–437.

Moss, B., & P. Woodhead. 1975. The effect of two commercial herbicides on the settlement, germination and growth of *Enteromorpha*. *Mar. Pollut. Bull.* 6:189–192.

Müller, D.G. 1963. Die Temperaturabhängigkeit der Sporangienbildung bei *Ectocarpus siliculosus* von verschiedenen Standorten. *Publ. Staz. Zool. Napoli* 33:310–314.

– 1976. Sexual isolation between a European and an American population of *Ectocarpus siliculosus* (Phaeophyta). *J. Phycol.* 12:252–254.

– 1981. Sexuality and sex attraction. In C.S. Lobban & M.J. Wynne (eds.), *The Biology of Seaweeds*, pp. 661–674. Blackwell Scientific Publications, Oxford.

Müller, D.G., A. Peters, G. Gassmann, W. Boland, F.-J. Marner, & L. Jaenicke. 1982a. Identification of a sexual hormone and related substances in the marine brown alga *Desmarestia*. *Naturwiss.* 69:290–291.

Müller, D.G., G. Gassmann, F.-J. Marner, W. Boland, & L. Jaenicke. 1982b. The sperm attractant of the marine brown alga *Ascophyllum nodosum* (Phaeophyceae). *Science* 218:1119–1120.

Munda, I.M. 1967. Der Einfluss der Salinität auf die chemische Zusammensetzung, das Wachstum und die Fruktifikation einiger Fucaceen. *Nova Hedwigia* 13:471–508.

– 1974. Changes and succession in the benthic algal associations of slightly polluted habitats. *Rev. Int. Oceanogr. Med.* 34:37–52.

– 1978. Salinity dependent distribution of benthic algae in estuarine areas of Icelandic fjords. *Bot. Mar.* 21:451–468.

Munda, I.M., & B.P. Kremer. 1977. Chemical composition and physiological properties of fucoids under conditions of reduced salinity. *Mar. Biol.* 42:9–16.

Murray, S.N., & M.M. Littler. 1978. Patterns of algal succession in a perturbated marine intertidal community. *J. Phycol.* 14:506–512.

Myklestad, S., I. Eide, & S. Melson. 1978. Exchange of heavy metals in *Ascophyllum nodosum* (L.) Le Jol. *in situ* by means of transplanting experiments. *Environ. Pollut.* 16:277–284.

Nagashima, H., S. Nakamura, K. Nisizawa, & T. Hori. 1971. Enzymic synthesis of floridean starch in a red alga, *Serraticardia maxima*. *Plant Cell Physiol.* 12:243–253.

Nakahara, H., & Y. Nakamura. 1973. Parthenogenesis, apogamy and apospory in *Alaria crassifolia* (Laminariales). *Mar. Biol.* 18:327–332.

Nalewajko, C., & D.R.S. Lean. 1980. Phosphorus. In I. Morris (ed.), *The Physiological Ecology of Phytoplankton*, pp. 235–258. Blackwell Scientific Publications, Oxford.

Nasr, A.H., & I.A. Bekheet. 1970. Effects of certain trace elements and soil extract on some marine algae. *Hydrobiologia* 36:53–60.

Nasr, A.H., I.A. Bekheet, & R.K. Ibrahim. 1968. The effects of different nitrogen and carbon sources on amino acid synthesis in *Ulva, Dictyota* and *Pterocladia*. *Hydrobiologia* 31:7–16.

Natl. Ocean. Atmos. Admin. 1983. *Tide Tables 1984: East Coast of North and South America Including Greenland*. U.S. Govt. Printing Office, Washington, D.C. 285 pp.

Neilands, J.B. 1973. Microbial iron transport compounds (siderochromes). In G.L. Eichhorn (ed.), *Inorganic Biochemistry*, pp. 167–202. Elsevier, New York.

– 1981. Iron absorption and transport in microorganisms. *Ann. Rev. Nutr.* 1:27–46.

Neilson, A.H., & R.A. Lewin. 1974. The uptake and utilization of organic carbon by algae: an essay in comparative biochemistry. *Phycologia* 13:227–264.

Neish, A.C., & P.F. Shacklock. 1971. Greenhouse experiments (1971) on the propagation of strain T4 of Irish Moss. Nat. Res. Counc., Canada, Atl. Res. Lab. Tech. Rep. No. 14.

Neish, A.C., P.F. Shacklock, C.H. Fox & F.J. Simpson. 1977. The cultivation of *Chondrus crispus*. Factors affecting growth under greenhouse conditions. *Can. J. Bot.* 55:2263–2271.

Nelson, W.G. 1982. Experimental studies of oil pollution on the rocky intertidal community of a Norwegian fjord. *J. Exp. Mar. Biol. Ecol.* 65:121–138.

Nelson-Smith, A. 1972. *Oil Pollution and Marine Ecology*. Elek Science Press, London. 260 pp.

Neushul, M. 1963. Studies on the giant kelp, *Macrocystis*. II. Reproduction. *Amer. J. Bot.* 50:354–359.

– 1965. SCUBA diving studies of the vertical distribution of benthic marine plants. In T. Levring (ed.), *Proc. 5th Europ. Mar. Biol. Symp.*, Botan. Gothoburg. 3:161–176.

– 1972. Functional interpretation of benthic marine algal morphology. In I.A. Abbott & M. Kurogi (eds.), *Contributions to the Systematics of Benthic Marine Algae of the North Pacific*, pp. 47–73. Jap. Soc. Phycol., Tokyo.

– 1981. The ocean as a culture dish: experimental studies of marine algal ecology. *Proc. Intl. Seaweed Symp.* 8:19–35.

Newell, R.C. 1979. *Biology of Intertidal Animals*, 3rd ed. Marine Ecological Surveys, Faversham, U.K. 781 pp.

Newell, R.C., & J.G. Field. 1983. Relative flux of carbon and nitrogen in a kelp-dominated system. *Mar. Biol. Lett.* 4:249–257.

Newell, R.C., & V.I. Pye. 1968. Seasonal variation in the effect of temperature on the respiration of certain intertidal algae. *J. Mar. Biol. Ass. U.K.* 48:341–348.

Niell, F.X. 1976. C:N ratio in some marine macrophytes and its possible ecological significance. *Bot. Mar.* 14:347–350.

Nilsen, G., & Ø. Nordby. 1975. A sporulation-inhibiting substance from vegetative thalli of the green alga, *Ulva mutabilis* Føyn. *Planta* 125:127–139.

Nissen, P. 1974. Uptake mechanisms: inorganic and organic. *Ann. Rev. Plant. Physiol.* 25:53–79.

Nixon, W.S., C.A. Oviatt, & S.S. Hale. 1976. Nitrogen regeneration and the metabolism of coastal bottom communities. In J.M. Anderson & A. MacFadyen (eds.), *The Role of Aquatic Organisms in Decomposition Processes*, pp. 269–283. Blackwell Scientific Publications, Oxford.

Noda, H., & Y. Horiguchi. 1972. The significance of zinc as a nutrient for the red alga *Porphyra tenera*. *Proc. Intl. Seaweed Symp.* 7:368–372.

Noggle, G.R., & G.J. Fritz. 1983. *Introductory Plant Physiology*, 2nd ed. Prentice-Hall, Englewood Cliffs, N.J. 627 pp.

North, W.J. 1972. Mass-cultured *Macrocystis* as a means of increasing kelp stands in nature. *Proc. Intl. Seaweed Symp.* 7:394–399.

– 1976. Aquacultural techniques for creating and restoring beds of giant kelp, *Macrocystis* spp. *J. Fish. Res. Bd. Canada* 33:1015–1023.

– 1979. Adverse factors affecting giant kelp and associated seaweeds. *Experientia* 35:445–447.

North, W.J., M. Neushul, & K.A. Clendenning. 1965. Successive biological change observed in a marine cove exposed to a large spillage of mineral oil. In *Symposium sur les Pollutions Marines par les Microorganismes et les Produits Petroliers*, Monaco, pp. 335–354.

Norton, T.A. 1977. Ecological experiments with *Sargassum muticum*. *J. Mar. Biol. Ass. U.K.* 57:33–43.

Norton, T.A., & R. Fetter. 1981. The settlement of *Sargassum muticum* propagules in stationary and flowing water. *J. Mar. Biol. Ass. U.K.* 61:929–940.

Norton, T.A., A.C. Mathieson, & M. Neushul. 1981. Morphology and environment. In C.S. Lobban & M.J. Wynne (eds.), *The Biology of Seaweeds*, pp. 421–451. Blackwell Scientific Publications, Oxford.

– 1982. A review of some aspects of form and function in seaweeds. *Bot. Mar.* 25:501–510.

Nultsch, W. 1974. Movements. In W.D.P. Stewart (ed.), *Algal Physiology and Biochemistry*, pp. 864–893. Blackwell Scientific Publications, Oxford.

Nultsch, W., J. Pfau, & U. Rüffer. 1981. Do correlations exist between chromatophore arrangement and photosynthetic activity in seaweeds? *Mar. Biol.* 62:111–117.

Nybakken, J.W. 1969. Pre-earthquake intertidal ecology of Three Saints Bay, Kodiak Island, Alaska. *Biol. Pap. U. Alaska* 9. 117 pp.

O'Brien, P.Y., & P.S. Dixon. 1976. The effects of oils and oil components on algae: a review. *Brit. Phycol. J.* 11:115–142.

Ogata, E. 1971. Growth of conchocelis in artificial medium in relation to carbon dioxide and calcium metabolism. *J. Shimonoseki U. Fish.* 19:123–129.

O'hEocha, C. 1965. Biliproteins of algae. *Ann. Rev. Plant Physiol.* 16:415–434.

Okazaki, A. 1971. *Seaweeds and Their Uses in Japan*. Tokai University Press, Tokyo. 165 pp.

Okazaki, M. 1977. Some enzymatic properties of Ca^{2+}-dependent adenosine triphosphatase from a calcareous red alga, *Serraticardia maxima* and its distribution in marine algae. *Bot. Mar.* 20:347–354.

O'Kelley, J.C. 1974. Inorganic nutrients. In W.D.P. Stewart (ed.), *Algal Physiology and Biochemistry*, pp. 610–635. Blackwell Scientific Publications, Oxford.

Olivier, S.R., I.K. de Paternoster, & R. Bastida. 1966. Estudios biocenóticos en las costas de Chubut (Argentina). I. Zonación biocenológica de Puerto Pardelas (Golfo Nuevo). *Bol. Inst. Biol. Mar. (Mar del Plata)* 10:1–74.

Oltmanns, F. 1905. Morphologie und Biologie der Algen. *Jb. Wiss. Bot.* 37:121–142.

Oohusa, T. 1980. Diurnal rhythm in the rates of cell division, growth and photosynthesis of *Porphyra yezoensis* (Rhodophyceae) cultured in the laboratory. *Bot. Mar.* 23:1–5.

Owens, N.J.P., & W.D.P. Stewart. 1983. *Enteromorpha* and the cycling of nitrogen in a small estuary. *Estu. Cstl. Shelf Sci.* 17:287–296.

Paasche, E. 1977. Growth of three plankton diatom species in Oslofjord water in the absence of artificial chelators. *J. Exp. Mar. Biol. Ecol.* 29:91–106.

Pace, D.R. 1975. Factors governing the boundary between *Macrocystis integrifolia* and the red sea urchin. Ph.D. Thesis, Simon Fraser University, Burnaby. 102 pp.

Paine, R.T. 1979. Disaster, catastrophe, and local persistence of the sea palm *Postelsia palmaeformis*. *Science* 205:685–687.

Paine, R.T., & R.L. Vadas. 1969. The effects of grazing by sea urchins, *Strongylocentrotus* spp. on benthic algal populations. *Limnol. Oceanogr.* 14:710–719.

Pallaghy, C.K., J. Minchinton, G.T. Kraft, & R. Wetherbee.

1983. Presence and distribution of bromine in *Thysanocladia densa* (Solieriaceae, Gigartinales), a marine red alga from the Great Barrier Reef. *J. Phycol.* 19:204–208.

Parker, B.C. 1963. Translocation in the giant kelp *Macrocystis*. *Science* 140:891–892.

Parsons, T.R. 1982. The new physical definition of salinity: biologists beware. *Limnol. Oceanogr.* 27:384–385.

Parsons, T.R., & P.J. Harrison. 1983. Nutrient cycling in marine ecosystems. In A. Pirson & M.H. Zimmerman (eds.), *Physiological Plant Ecology IV, Encyclop. Plant Physiol.* vol. 12D, pp. 77–105. Springer-Verlag, New York.

Parsons, T.R., M. Takahashi, & B. Hargrave. 1977. *Biological Oceanographic Processes*, 2nd ed. Pergamon Press, New York. 332 pp.

Pasciak, W.J., & J. Gavis. 1974. Transport limitation of nutrient uptake in phytoplankton. *Limnol. Oceanogr.* 19:881–888.

Patwary, M.U., & J.P. van der Meer. 1983. Genetics of *Gracilaria tikvahiae* (Rhodophyceae). IX. Some properties of agar extracted from morphological mutants. *Bot. Mar.* 26:295–299.

Paul, V.J., H.H. Sun, & W. Fenical. 1982. Udoteal, a linear diterpenoid feeding deterrent from the tropical green alga *Udotea flabellum*. *Phytochem.* 21:468–469.

Peakall, D.B. 1975. PCB's and their environmental effects. *Crit. Rev. Environ. Control* 5:469–508.

Pearlmutter, N.L., & R.L. Vadas. 1978. Regeneration of thallus fragments of *Rhodochorton purpureum* (Rhodophyceae, Nemalionales). *Phycologia* 17:186–190.

Pearse, A.S., & G. Gunter. 1957. Salinity. In J.W. Hedgpeth (ed.), *Treatise on Marine Ecology and Paleoecology*, vol. 1, pp. 129–158. Geol. Soc. Amer. Mem. 67.

Pedersén, M. 1973. Identification of a cytokinin, 6-(3 methyl-2-butenylamino) purine, in sea water and the effect of cytokinins on brown algae. *Physiol. Plant.* 28:101–105.

Pedersen, P.M. 1981. Phaeophyta: life histories. In C.S. Lobban & M.J. Wynne (eds.), *The Biology of Seaweeds*, pp. 194–217. Blackwell Scientific Publications, Oxford.

Pellegrini, L. 1980. Cytological studies on physodes in the vegetative cells of *Cystoseira stricta* Sauvageau (Phaeophyta, Fucales). *J. Cell Sci.* 41:209–231.

Penot, M., & M. Penot. 1979. High speed translocation of ions in seaweeds. *Z. Pflanzenphysiol.* 95:265–273.

Penot, M., & C. Videau. 1975. Absorption du ^{86}Rb et du ^{99}Mo par deux algues marines: le *Laminaria digitata* et le *Fucus serratus*. *Z. Pflanzenphysiol.* 76:285–293.

Percival, E. 1979. The polysaccharides of green, red and brown seaweeds: their basic structure, biosynthesis and function. *Brit. Phycol. J.* 14:103–117.

Percival, E., & R.H. McDowell. 1967. *Chemistry and Enzymology of Marine Algal Polysaccharides*. Academic Press, New York. 219 pp.

Pérès, J.M. 1982a. Zonations. In O. Kinne (ed.), *Marine Ecology*, vol. 5, pt. 1, pp. 9–45. Wiley, New York.

– 1982b. Major benthic assemblages. In O. Kinne (ed.), *Marine Ecology*, vol. 5, pt. 1, pp. 373–522. Wiley, New York.

Perkins, E.J. 1979. The need for sublethal studies. *Phil. Trans. Roy. Soc. Lond.* B 286:425–442.

Perkins, E.J., & O.J. Abbott. 1972. Nutrient enrichment and sand flat fauna. *Mar. Pollut. Bull.* 3:70–72.

Perrone, G., & G.P. Felicini. 1972. Sur les bourgeons adventifs de *Petroglossum nicaeense* (Duby) Schotter (Rhodophycées, Gigartinales) en culture. *Phycologia* 11:87–95.

– 1976. Les bourgeons adventifs de *Gigartina acicularis* (Wulf.) Lamour. (Rhodophyta, Gigartinales) en culture. *Phycologia* 15:45–50.

Perry, M.J., M.C. Larsen, & R.S. Alberte. 1981. Photoadaptation in marine phytoplankton: response of the photosynthetic unit. *Mar. Biol.* 62:91–101.

Peterson, B.J. 1980. Aquatic primary productivity and the ^{14}C-CO_2 method: a history of the productivity problem. *Ann. Rev. Ecol. System.* 11:359–385.

Pettersen, R. 1975. Control by ammonium of intercompartmental guanine transport in *Chlorella*. *Z. Pflanzenphysiol.* 76:213–223.

Pham Quang, L., & M.H. Laur. 1976. Teneur, composition et répartition cytologique des lipides polaires sulfrés et phosphorés de *Pelvetia canaliculata* (L.) Decn. et Thur., *Fucus vesiculosus* (L.) et *Fucus serratus* (L.). *Phycologia* 15:367–375.

Phillips, D.J.H. 1977. The use of biological indicator organisms to monitor trace metal pollution in marine and estuarine environments: a review. *Environ. Pollut.* 13:281–317.

Phillips, G.L., D. Eminson, & B. Moss. 1978. A mechanism to account for macrophyte decline in progressively eutrophicated freshwaters. *Aquat. Bot.* 4:103–126.

Pianka, E.R. 1970. On *r* and *K* selection. *Amer. Nat.* 104:592–597.

Polanshek, A.R., & J.A. West. 1977. Culture and hybridization studies on *Gigartina papillata* (Rhodophyta). *J. Phycol.* 13:141–149.

Popovic, R., K. Colbow, W. Vidaver, & D. Bruce. 1983. Evolution of O_2 in brown algal chloroplasts. *Plant Physiol.* 73:889–892.

Preston, R.D. 1979. Polysaccharide conformation and cell wall function. *Ann. Rev. Plant Physiol.* 30:55–78.

– 1982. The case for multinet growth in growing walls of plant cells. *Planta* 155:356–363.

Prézelin, B.B. 1981. Light reactions in photosynthesis. In T. Platt (ed.), *Physiological Bases of Phytoplankton Ecology*, Can. Bull. Fish. Aquat. Sci. No. 210, pp. 1–43.

Price, J.H. 1978. Ecological determination of adult form in *Callithamnion*: its taxonomic implications. In D.E.G. Irvine & J.H. Price (eds.), *Modern Approaches to the Taxonomy of Red and Brown Algae*, pp. 263–300. Academic Press, New York.

Pringle, J.D. 1982. Variation in *Enteromorpha*. Abstr. 1st Intl. Phycol. Congr., St. John's, Newfoundland, p. a39.

Pringle, J.D., G.J. Sharp & J.F. Caddy. 1982. Interactions in kelp bed ecosystems in the northwest Atlantic: review of a workshop. In M.C. Mercer (ed.), *Multispecies Approaches to Fisheries Management Advice*, Can. Spec. Publ. Fish. Aquat. Sci. No. 59, pp. 108–115.

Probyn, T.A., & A.R.O. Chapman. 1982. Nitrogen uptake characteristics of *Chordaria flagelliformis* (Phaeophyta) in batch mode and continuous mode experiments. *Mar. Biol.* 71:129–133.

Provasoli, L., & A.F. Carlucci. 1974. Vitamins and growth regulators. In W.D.P. Stewart (ed.), *Algal Physiology and Biochemistry*, pp. 741–787. Blackwell Scientific Publications, Oxford.

Provasoli, L., & I.J. Pintner. 1977. Effect of media and inoculum on the morphology of *Ulva*. *J. Phycol.* 13 (Suppl.):56.

– 1980. Bacteria induced polymorphism in an axenic laboratory strain of *Ulva lactuca* (Chlorophyceae). *J. Phycol.* 16:196–201.

Provasoli, L., I.J. Pintner, & S. Sampathkumar. 1977. Morphogenetic substances for *Monostroma oxyspermum* from marine bacteria. *J. Phycol.* 13 (Suppl.):56.

Puiseux-Dao, S. 1970. *Acetabularia and Cell Biology*. Logos Press, London. 162 pp.

Quader, H. 1981. Interruption of cellulose microfibril crystallization. *Naturwissensch.* 67:428.

Quadir, A., P.J. Harrison, & R.E. DeWreede. 1979. The effects of emergence and submergence on the photosynthesis and respiration of marine macrophytes. *Phycologia* 18:83–88.

Quatrano, R.S. Development of cell polarity. *Ann. Rev. Plant Physiol.* 29:487–510.

Queen, W.H. 1974. Physiology of coastal halophytes. In R.J. Reimold & W.H. Queen (eds.), *Ecology of Halophytes*, pp. 345–353. Academic Press, New York.

Quillet, M., & G. de Lestang-Brémond. 1978. Action synergique à l'égard de l'eau de certaines associations de polysaccharides dans les parois des Fucales et des Gigartinales. *Bull. Soc. Phycol. Fr.* 23:88–93.

– 1981. The MeCDPS, a carrying sulphate's nucleotide of the red seaweed *Catenella opuntia* (Grev.). *Proc. Intl. Seaweed Symp.* 10:503–507.

Ragan, M.A. 1976. Physodes and the phenolic compounds of brown algae. Composition and significance of physodes *in vivo*. *Bot. Mar.* 19:145–154.

– 1981. Chemical constituents of seaweeds. In C.S. Lobban & M.J. Wynne (eds.), *The Biology of Seaweeds*, pp. 589–626. Blackwell Scientific Publications, Oxford.

Ragan, M.A., & J.S. Craigie. 1976. Physodes and the phenolic compounds of brown algae. Isolation and characterization of phloroglucinol polymers from *Fucus vesiculosus* (L.). *Can. J. Biochem.* 54:66–73.

Ragan, M.A., & A. Jensen. 1979. Quantitative studies on brown algal phenols. III. Light mediated exudation of polyphenols from *Ascophyllum nodosum* (L.) Le Jol. *J. Exp. Mar. Biol. Ecol.* 36:91–101.

Ragan, M.A., O. Smidsrød, & B. Larsen. 1979. Chelation of divalent metal ions by brown algal polyphenols. *Mar. Chem.* 7:265–271.

Rai, L.C., J.P. Gaur, & H.D. Kumar. 1981. Phycology and heavy-metal pollution. *Biol. Rev.* 56:99–151.

Ramus, J. 1972. Differentiation of the green alga *Codium fragile*. *Amer. J. Bot.* 59:478–482.

– 1978. Seaweed anatomy and photosynthetic performance: the ecological significance of light guides, heterogenous absorption and multiple scatter. *J. Phycol.* 14:352–362.

– 1981. The capture and transduction of light energy. In C.S. Lobban & M.J. Wynne (eds.), *The Biology of Seaweeds*, pp. 458–492. Blackwell Scientific Publications, Oxford.

– 1983. A physiological test of the theory of complementary chromatic adaptation. II. Brown, green and red seaweeds. *J. Phycol.* 19:173–178.

Ramus, J., & G. Rosenberg. 1980. Diurnal photosynthetic performance of seaweeds measured under natural conditions. *Mar. Biol.* 56:21–28.

Ramus, J., F. Lemons, & C. Zimmerman. 1977. Adaptation of light-harvesting pigments to downwelling light and to consequent photosynthetic performance of the eulittoral rockweeds *Ascophyllum nodosum* and *Fucus vesiculosus*. *Mar. Biol.* 24:293–303.

Rao, V.S., & U.K. Tipnis. 1967. Chemical composition of some marine algae from the Gujurat Coast. In V. Krisnamurthy (ed.), *Proc. Seminar on Sea, Salt and Plants*, pp. 277–288. Central Salt & Mar. Chem. Res. Inst., Bhavnagar, India.

Raven, J.A. 1974. Carbon dioxide fixation. In W.D.P. Stewart (ed.), *Algal Physiology and Biochemistry*, pp. 434–455. Blackwell Scientific Publications, Oxford.

– 1980. Nutrient transport in microalgae. *Adv. Microbiol. Physiol.* 21:47–226.

– 1982. The energetics of freshwater algae: energy requirements for biosynthesis and volume regulation. *New Phytol.* 92:1–20.

Raven, J.A., & J. Beardall. 1981. Respiration and photorespiration. In T. Platt (ed.), *Physiological Bases of Phytoplankton Ecology*. Can. Bull. Fish. Aquat. Sci. No. 210, pp. 55–82.

Rawlence, D.J. 1972. An ultrastructural study of the relationship between rhizoids of *Polysiphonia lanosa* (L.) Tandy (Rhodophyceae) and the tissue of *Ascophyllum nodosum* (L.) LeJolis (Phaeophyceae). *Phycologia* 11:279–290.

Redfield, A.C., B.H. Ketchum, & F.A. Richards. 1963. The influence of organisms on the composition of seawater. In M.N. Hill (ed.), *The Sea*, vol. 2, pp. 26–77. Wiley, New York.

Reed, R.H. 1983. Measurement and osmotic significance of β-dimethylsulfoniopropionate in macroalgae. *Mar. Biol. Lett.* 4:173–181.

Reed, R.H., & J.A. Barron. 1983. Physiological adaptation to salinity change in *Pilayella littoralis* from marine and estuarine sites. *Bot. Mar.* 26:409–416.

Reed, R.H., & J.C. Collins. 1980. The ionic relations of *Porphyra purpurea* (Roth) C. Ag. (Rhodophyta, Bangiales). *Plant Cell Environ.* 3:399–407.

– 1981. The kinetics of Rb^+ and K^+ exchange in *Porphyra purpurea*. Plant Sci. Lett. 20:281–289.

Reed, R.H., & L. Moffat. 1983. Copper toxicity and copper tolerance in *Enteromorpha compressa* (L.) Grev. *J. Exp. Mar. Biol. Ecol.* 63:85–103.

Reed, R.H., & G. Russell. 1979. Adaptation to salinity stress in populations of *Enteromorpha intestinalis* (L.) Link. *Estu. Cstl. Mar. Sci.* 8:251–258.

Reed, R.H., J.C. Collins, & G. Russell. 1980a. The effects of salinity upon cellular volume of the marine red alga *Porphyra purpurea* (Roth) C. Ag. *J. Exp. Bot.* 31:1521–1537.

– 1980b. The effects of salinity upon galactosyl-glycerol content and concentration of the marine red alga *Porphyra purpurea* (Roth) C. Ag. *J. Exp. Bot.* 31:1539–1554.

– 1980c. The influence of variations in salinity upon photosynthesis in the marine alga *Porphyra purpurea* (Roth) C. Ag. (Rhodophyta, Bangiales). *Z. Pflanzenphysiol.* 98:183–187.

Rees, D.A. 1975. Stereochemistry and binding behaviour of carbohydrate chains. In W.J. Whelan (ed.), *Biochemistry of Carbohydrates*, pp. 1–42. Butterworth, London.

– 1977. *Polysaccharide Shapes*. Chapman & Hall, London. 80 pp.

Reise, K. 1983. Sewage, green algal mats anchored by lugworms, and the effects on *Turbellaria* and small Polychaeta. *Helgol. Meeresunters.* 36:151–162.

Reutergårdh, L. 1980. Chlorinated hydrocarbons in estuaries. In E. Olausson & I. Cato (eds.), *Chemistry and Biogeochemistry of Estuaries*, pp. 349–365. Wiley, New York.

Rhee, C., & W.R. Briggs. 1977. Some responses of *Chondrus crispus* to light. I. Pigment changes in the natural habitat. *Bot. Gaz.* 138:123–128.

Rhee, G.-Y. 1974. Phosphate uptake under nitrate limitation

by *Scenedesmus* sp. and its ecological implications. *J. Phycol.* 10:470–475.
– 1978. Effects of N:P atomic ratios and nitrate limitation on algal growth, cell composition, and nitrate uptake. *Limnol. Oceanogr.* 23:10–25.
– 1980. Continuous culture in phytoplankton ecology. In M.R. Droop & H.W. Jannasch (eds.), *Advances in Aquatic Microbiology*, vol. 2, pp. 151–203. Academic Press, New York.
– 1982. Effects of environmental factors and their interactions on phytoplankton growth. *Adv. Microbial Ecol.* 6:33–74.
Rhodes, R.G. 1970. Relation of temperature to development of the macrothallus of *Desmotrichum undulatum. J. Phycol.* 6:312–314.
Rice, D.L., & B.E. LaPointe. 1981. Experimental outdoor studies with *Ulva fasciata* Delile. II. Trace metal chemistry. *J. Exp. Mar. Biol. Ecol.* 54:1–11.
Rice, H.V., D.A. Leighty, & G.C. McLeod. 1973. The effects of some trace metals on marine phytoplankton. *Crit. Rev. Microbiol.* 3:27–49.
Riedl, R. 1971. Water movement: general introduction. In O. Kinne (ed.), *Marine Ecology*, vol I, pt. 2, pp. 1085–1088. Wiley, New York.
Rietema, H. 1982. Effects of photoperiod and temperature on macrothallus initiation in *Dumontia contorta* (Rhodophyta). *Mar. Ecol. Progr. Ser.* 8:187–196.
Rietema, H., & A.M. Breeman. 1982. The regulation of the life history of *Dumontia contorta* in comparison to that of several other Dumontiaceae (Rhodophyta). *Bot. Mar.* 25:569–576.
Rietema, H., & A.W.O. Klein. 1981. Environmental control of the life cycle of *Dumontia contorta* (Rhodophyta) kept in culture. *Mar. Ecol. Progr. Ser.* 4:23–29.
Rietema, H., & C. van den Hoek. 1981. The life history of *Desmotrichum undulatum* (Phaeophyceae) and its regulation by temperature and light conditions. *Mar. Ecol. Progr. Ser.* 4:321–335.
Riley, J.P., & R. Chester. 1971. *Introduction to Marine Chemistry*. Academic Press, New York. 465 pp.
Riley, J.P., & G. Skirrow (eds.). 1965. *Chemical Oceanography*, vol. 1. Academic Press, New York. 712 pp.
Roberts, M., & F.M. Ring. 1972. Preliminary investigations into conditions affecting the growth of the microscopic phase of *Scytosiphon lomentarius* (Lyngbye) Link. *Mem. Soc. Bot. Fr.* 1972:117–128.
Robles, C.D., & J. Cubit. 1981. Influence of biotic factors in an upper intertidal community: Dipteran larvae grazing on algae. *Ecology* 62:1536–1547.
Rosenberg, D.M., et al. 1981. Recent trends in environmental impact assessment. *Can. J. Fish. Aquat. Sci.* 38:591–624.
Rosenberg, G., & H.W. Paerl. 1981. Nitrogen fixation by blue-green algae associated with the siphonous green seaweed *Codium decorticatum*: effects on ammonium uptake. *Mar. Biol.* 61:151–158.
Rosenberg, G., & J. Ramus. 1982. Ecological growth strategies in the seaweeds *Gracilaria foliifera* (Rhodophyceae) and *Ulva* sp. (Chlorophyceae): soluble nitrogen and reserve carbohydrates. *Mar. Biol.* 66:251–259.
– 1984. Uptake of inorganic nitrogen and seaweed surface area : volume ratios. *Aquat. Bot.* 19:65–72.
Rosenthal, R.J., W.D. Clarke, & P.K. Dayton. 1974. Ecology and natural history of a stand of giant kelp, *Macrocystis pyrifera*, off Del Mar, California. *Natl. Oceanogr. Atmos. Admin. (U.S.) Fish. Bull.* 72:670–684.
Round, F.E. 1981. *The Ecology of Algae*. Cambridge University Press, Cambridge. 653 pp.
Rubin, P.M., E. Zetooney, & R.E. McGowan. 1977. Uptake and utilization of sugar phosphates by *Anabaena flos-aquae. Plant Physiol.* 60:407–411.
Rudd, R.L. 1964. *Pesticides and the Living Landscape*. University of Wisconsin Press, Madison. 320 pp.
Rueness, J. 1973. Pollution effects on littoral algal communities in the inner Oslofjord, with special reference to *Ascophyllum nodosum. Helgol. wiss. Meeresunters.* 24:446–454.
Russell, G. 1970. Rhizoid production in excised *Dictyota dichotoma. Brit. Phycol. J.* 5:243–245.
– 1978. Environment and form in the discrimination of taxa in brown algae. In D.E.G. Irvine & J.H. Price (eds.), *Modern Approaches to the Taxonomy of Red and Brown Algae*, pp. 339–369. Academic Press, New York.
– 1980. Applications of simple numerical methods to the analysis of intertidal vegetation. In J.H. Price, D.E.G. Irvine, & W.F. Farnham (eds.), *The Shore Environment*, vol. 1, *Methods*, pp. 171–192. Academic Press, New York.
Russell, G., & J.J. Bolton. 1975. Euryhaline ecotypes of *Ectocarpus siliculosus* (Dillw.) Lyngb. *Estu. Cstl. Mar. Sci.* 3:91–94.
Russell, G., & A.H. Fielding. 1974. The competitive properties of marine algae in culture. *J. Ecol.* 62:689–698.
– 1981. Individuals, populations and communities. In C.S. Lobban & M.J. Wynne (eds.), *The Biology of Seaweeds*, pp. 393–420. Blackwell Scientific Publications, Oxford.
Russell, G., & C.J. Veltkamp. 1982. Epiphytes and antifouling characteristics of *Himanthalia* (brown algae). *Brit. Phycol. J.* 17:239.
Ryther, J.H., & W.M. Dunstan. 1971. Nitrogen, phosphorus and eutrophication in the coastal marine environment. *Science* 171:1008–1013.
Ryther, J.H., W.M. Dunstan, K.R. Tenore, & J.E. Huguenin. 1972. Controlled eutrophication – increasing food production from the sea by recycling human wastes. *BioScience* 22:144–152.
Ryther, J.H., L.D. Williams, M.D. Hanisak, R.W. Stenberg, & T.A. DeBusk. 1979. Biomass production by marine and freshwater plants. In *3rd Ann. Biomass Energy Systems Conf. Proc.*, pp. 13–24. Solar Res. Inst., Golden, Col.
Saenko, G.N., M.D. Koryakova, V.F. Makienko, & J.G. Dobrosmyskova. 1976. Concentration of polyvalent metals by seaweeds in Vostok Bay, Sea of Japan. *Mar. Biol.* 34:169–176.
Saga, N., T. Uchida, & Y. Sakai. 1978. Clone *Laminaria* from single isolated cell. *Bull. Jap. Soc. Sci. Fish.* 44:87.
Sammarco, P.W. 1983. Effects of fish grazing and damselfish territoriality on coral reef algae. I. Algal community structure. *Mar. Ecol. Progr. Ser.* 13:1–14.
Sanders, H.L., J.F. Grassle, G.R. Hampson, L.S. Morse, S. Garner-Price, & C.C. Jones. 1980. Anatomy of an oil spill: long-term effects from the grounding of the barge *Florida* off West Falmouth, Massachussetts. *J. Mar. Res.* 38:265–380.
Sanders, J.G. 1979. The concentration and speciation of arsenic in marine macroalgae. *Estu. Cstl. Mar. Sci.* 9:95–99.
Santelices, B., J. Correa, & M. Avila. 1983. Benthic algal spores surviving digestion by sea urchins. *J. Exp. Mar. Biol. Ecol.* 70:263–269.
Santelices, B., S. Montalva, & P. Oliger. 1981. Competitive algal community organization in exposed intertidal habitats from central Chile. *Mar. Ecol. Progr. Ser.* 6:267–276.
Satoh, K., & D.C. Fork. 1983. A new mechanism for adaptation to changes in light intensity and quality in the red alga

Porphyra perforata. III. Fluorescence transients in the presence of 3-(3,4-dichlorophenyl)-1,1-dimethylurea. *Plant Physiol*. 71:673–676.

Sauer, K. 1975. Primary events and the trapping of energy. In Govindjee (ed.), *Bioenergetics of Photosynthesis*, pp. 115–118. Academic Press, New York.

Sawada, T. 1972. Periodic fruiting of *Ulva pertusa* at three localities in Japan. *Proc. Intl. Seaweed Symp*. 7:229–230.

Schatz, S. 1980. Degradation of *Laminaria saccharina* by higher fungi: a preliminary report. *Bot. Mar*. 23:617–622.

Schiff, J.A. 1980. Pathways of assimilatory sulfate reduction in plants and microorganisms. In K. Elliot & J. Whelan (eds.), *Sulfur in Biology*, Ciba Found. Symp. 72 (New Ser.), pp. 49–79. Excerpta Medica, Amsterdam.

– 1983. Reduction and other metabolic reactions of sulfate. In A. Läuchli & R.L. Bieleski (eds.), *Encyclopedia of Plant Physiology*, vol. 15, pp. 382–399. Springer-Verlag, New York.

Schiff, J.A., & R.C. Hodson. 1970. Pathways of sulfate reduction in algae. *Ann. N.Y. Acad. Sci*. 175:555–576.

Schlesinger, M.J., M. Ashburner, & A. Tissières (eds.). 1982. *Heat Shock: From Bacteria to Man*. Cold Spring Harbor Lab. Press, Cold Spring Harbor, N.Y. 440 pp.

Schmidt, R.L. 1978. Copper in the marine environment. Part 1. *Crit. Rev. Environ. Control* 8:101–152.

Schmitz, K. 1981. Translocation. In C.S. Lobban & M.J. Wynne (eds.), *The Biology of Seaweeds*, pp. 534–558. Blackwell Scientific Publications, Oxford.

Schmitz, K., & W. Riffarth. 1980. Carrier-mediated uptake of L-leucine by the brown alga *Giffordia mitchelliae*. *Z. Pflanzenphysiol*. 67:311–324.

Schmitz, K., & L.M. Srivastava. 1974. Fine structure and development of sieve tubes in *Laminaria groenlandica* Rosenv. *Cytobiol*. 10:66–87.

– 1979. Long distance transport in *Macrocystis integrifolia*. II. Tracer experiments with ^{14}C and ^{32}P. *Plant Physiol*. 63:1003–1009.

Schneider, H. 1980, Chlorophyll biosynthesis. Enzymes and regulation of enzyme activities. In F.-C. Czygan (ed.), *Pigments in Plants*, 2nd ed., pp. 237–307. Gustav Fischer Verlag, Stuttgart.

Schonbeck, M.W., & T.A. Norton. 1978. Factors controlling the upper limits of fucoid algae on the shore. *J. Exp. Mar. Biol. Ecol*. 31:303–313.

– 1979a. An investigation of drought avoidance in intertidal fucoid algae. *Bot. Mar*. 22:133–144.

– 1979b. Drought-hardening in the upper shore seaweeds *Fucus spiralis* and *Pelvetia canaliculata*. *J. Ecol*. 67:687–696.

– 1979c. The effects of diatoms on the growth of *Fucus spiralis* germlings in culture. *Bot. Mar*. 22:233–236.

– 1980. The effects on intertidal fucoids of exposure to air under various conditions. *Bot. Mar*. 23:141–147.

Schramm, W. 1972. The effects of oil pollution on gas exchange in *Porphyra umbilicalis* when exposed to air. *Proc. Intl. Seaweed Symp*. 7:309–315.

Schröter, K. 1978. Asymmetrical jelly secretion of zygotes of *Pelvetia* and *Fucus*: an early polarization event. *Planta* 140:69–73.

Schwenke, H. 1971. Water movement: plants. In O. Kinne (ed.), *Marine Ecology*, vol. 1, pt. 2, pp. 1091–1121. Wiley, New York.

Sears, J.R., & R.T. Wilce. 1975. Sublittoral, benthic marine algae of southern Cape Cod and adjacent islands: seasonal periodicity, associations, diversity, and floristic composition. *Ecol. Monogr*. 45:337–365.

Seely, G.R., M.J. Duncan, & W.E. Vidaver. 1972. Preparative and analytical extraction of pigments from brown algae with dimethyl sulfoxide. *Mar. Biol*. 12:184–188.

Senger, H. (ed.) 1980. *The Blue Light Syndrome*. Springer-Verlag, New York, 665 pp.

Šesták, Z., P.G. Jarvis, & J. Čatský. 1971. Criteria for the selection of suitable methods. In Z. Šesták, J. Čatský, & P.G. Jarvis (eds.), *Plant Photosynthetic Production: Manual of Methods*, pp. 1–48. W. Junk, The Hague.

Setchell, W.A. 1915. The law of temperature connected with the distribution of the marine algae. *Ann. Mo. Bot. Gard*. 2:287–305.

– 1920. Stenothermy and zone-invasion. *Amer. Nat*. 54:385–397.

Sharp, J.H., & C.H. Culberson. 1982. The physical definition of salinity: a chemical evaluation. *Limnol. Oceanogr*. 27:385–387.

Shepherd, S.A., & H.B.S. Womersley. 1970. The sublittoral ecology of West Island, South Australia: I. Environmental features and algal ecology. *Trans. Roy. Soc. S. Austr*. 94:105–137.

– 1981. The algal and seagrass ecology of Waterloo Bay, South Australia. *Aquat. Bot*. 11:305–371.

Shiels, W.E., J.J. Goering, & D.W. Hood. 1973. Crude oil phytotoxicity studies. In D.W. Hood, W.E. Shiels, & E.J. Kelley (eds.), *Environmental Studies of Port Valdez*, pp. 413–446. Inst. Mar. Sci., Univ. Alaska, Fairbanks, Occ. Publ. No. 3.

Shih, M.L., J.-Y. Floc'h, & L.M. Srivastava. 1983. Localization of ^{14}C-labeled assimilates in sieve elements of *Macrocystis integrifolia* by histoautoradiography. *Can. J. Bot*. 61:157–163.

Sieburth, J. McN. 1969. Studies on algal substances in the sea. III. The production of extracellular organic matter by littoral marine algae. *J. Exp. Mar. Biol. Ecol*. 3:290–309.

Sieburth, J.McN. & J.T. Conover. 1965. *Sargassum* tannin, an antibiotic which retards fouling. *Nature* 208:52–53.

Sieburth, J.McN., & J.L. Tootle. 1981. Seasonality of microbial fouling on *Ascophyllum nodosum* (L.) LeJol., *Fucus vesiculosus* L., *Polysiphonia lanosa* (L.) Tandy and *Chondrus crispus* Stackh. *J. Phycol*. 17:57–64.

Sikes, C.S. 1978. Calcification and cation sorption of *Cladophora glomerata* (Chlorophyta). *J. Phycol*. 14:325–329.

Silva, P.C. 1979. The benthic algal flora of central San Francisco Bay. In *San Francisco Bay: The Urbanized Estuary*, pp. 287–345. California Academy of Science, San Francisco, Cal.

Silverberg, B.A., P.M. Stokes, & L.B. Ferstenberg. 1976. Intranuclear complexes in copper tolerant green algae. *J. Cell Biol*. 69:210–214.

Simpson, W.R. 1981. A critical review of cadmium in the marine environment. *Prog. Oceanogr*. 10:1–70.

Sinclair, J., S. Garland, T. Arnason, P. Hope, & M. Granville. 1977. Polychlorinated biphenyls and their effects on photosynthesis and respiration. *Can. J. Bot*. 55:2679–2684.

Skipnes, O., T. Roald, & A. Haug. 1975. Uptake of zinc and strontium by brown algae. *Physiol. Plant*. 34:314–320.

Slocum, C.J. 1980. Differential susceptibility to grazers in two phases of an intertidal alga: advantages of heteromorphic generations. *J. Exp. Mar. Biol. Ecol*. 46:99–110.

Smetacek, V., B. von Bodungen, K. von Brödsel, & B. Zeitzschel. 1976. The plankton tower. II. Release of nutrients from sediments due to changes in the density of bottom water. *Mar. Biol*. 34:373–378.

Smith, F.A., & N.A. Walker. 1980. Photosynthesis by aquatic plants: effects of unstirred layers in relation to assimilation of CO_2 and HCO_3^- and to carbon isotopic discrimination. *New Phytol.* 86:245–259.

Smith, G.M. 1947. On the reproduction of some Pacific coast species of *Ulva*. *Amer. J. Bot.* 34:80–87.

Smith, R.C. 1969. An underwater spectral irradiance collector. *J. Mar. Res.* 27:111–120.

– 1974. Structure of solar radiation in the upper layers of the sea. In N.G. Jerlov & E. Steemann Nielsen (eds.), *Optical Aspects of Oceanography*, pp. 95–119. Academic Press, London.

Smith, R.C., & K.S. Baker. 1978. Optical classification of natural waters. *Limnol. Oceanogr.* 23:260–267.

Smith, R.C., & J.E. Tyler. 1976. Transmission of solar radiation into natural waters. *Photochem. Photobiol. Rev.* 1:117–155.

Smith, R.G., W.N. Wheeler, & L.M. Srivastava. 1983. Seasonal photosynthetic performance of *Macrocystis integrifolia* (Phaeophyceae). *J. Phycol.* 19:352–359.

Soeder, C., & E. Stengel. 1974. Physico-chemical factors affecting metabolism and growth rate. In W.D.P. Stewart (ed.), *Algal Physiology and Biochemistry*, pp. 714–740. Blackwell Scientific Publications, Oxford.

Solomonson, L.P., & A.M. Spehar. 1977. Model for the regulation of nitrate assimilation. *Nature* 265:373–375.

Somero, G.N. 1981. pH-temperature interactions on proteins: principles of optimal pH and buffer system design. *Mar. Biol. Lett.* 2:163–178.

Sorentino, C. 1979. The effects of heavy metals on phytoplankton – a review. *Phykos* 18:149–161.

Sousa, W.P. 1979. Experimental investigations of disturbance and ecological succession in a rocky intertidal algal community. *Ecol. Monogr.* 49:227–254.

Sousa, W.P., S.C. Schroeter, & S.D. Gaines. 1981. Latitudinal variation in intertidal algal community structure: the influence of grazing and vegetative propagation. *Oecologia* 48:297–307.

South, G.R., & R.D. Hill. 1970. Studies on marine algae of Newfoundland. I. Occurrence and distribution of free-living *Ascophyllum nodosum* in Newfoundland. *Can. J. Bot.* 48:1697–1701.

Spiller, H., E. Dietsch, & E. Kessler. 1976. Intracellular appearance of nitrite and nitrate in nitrogen-starved cells of *Ankistrodesmus braunii*. *Planta* 129:175–181.

Stafford, S., P. Berwick, D.E. Hughes, & D.A. Stafford. 1982. Oil degradation in hydrocarbon- and oil-stressed environments. In R.G. Burns & J.H. Slater (eds.), *Experimental Microbial Ecology*, pp. 591–612. Blackwell Scientific Publications, Oxford.

Stebbing, A.R.D. 1979. An experimental approach to the determinants of biological water quality. *Phil. Trans. Roy. Soc. Lond.* B 286:465–482.

Stebbing, A.R.D., B. Akesson, A. Calabrese, J.H. Gentile, A. Jensen, & R. Lloyd. 1980. The role of bioassays in marine pollution monitoring. *Rapp. P.-v. Réun. Cons. Intl. Explor. Mer* 179:322–332.

Steele, R.L., & M.D. Hanisak. 1979. Sensitivity of some brown algal reproductive stages to oil pollution. *Proc. Intl. Seaweed Symp.* 9:181–191.

Steinberg, P.D. 1984. Algal chemical defense against herbivores: allocation of phenolic compounds in the kelp *Alaria marginata*. *Science* 223:405–407.

Steinbiss, H.H., & K. Schmitz. 1973. CO_2-Fixierung und Stofftransport in benthischen marinen Algen. V. Zur autoradiographischen Lokalisation der Assimilattransportbahnen im Thallus von *Laminaria hyperborea*. *Planta* 112:253–263.

Steneck, R.S. 1982. A limpet-coralline alga association: adaptations and defenses between a selective herbivore and its prey. *Ecology* 63: 507–522.

Steneck, R.S., & L. Watling. 1982. Feeding capabilities and limitation of herbivorous molluscs: a functional approach. *Mar. Biol.* 68:299–312.

Stephenson, T.A., & A. Stephenson. 1949. The universal features of zonation between tide-marks on rocky coasts. *J. Ecol.* 38:289–305.

– 1972. *Life Between Tidemarks on Rocky Shores*. Freeman, San Francisco. 425 pp.

Stewart, J.G. 1977. Effects of lead on the growth of four species of red algae. *Phycologia* 16:31–36.

Stewart, J., & M. Schulz-Baldes. 1976. Long-term lead accumulation in abalone (*Haliotis* spp.) fed on lead-treated brown algae (*Egregia laevigata*). *Mar. Biol.* 36:19–24.

Stewart, K.D., K.R. Mattox, & G.L. Floyd. 1973. Mitosis, cytokinesis, the distribution of plasmodesmata, and other cytological characteristics in the Ulotrichales, Ulvales and Chaetophorales: phylogenetic and taxonomic considerations. *J. Phycol.* 9:128–141.

Stocker, O., & W. Holdheide. 1937. Die Assimilation Helgoländer Gezeitenalgen während der Ebbezeit. *Z. Bot.* 32:1–59.

Stockner, J.G., & D.D. Cliff. 1976. Effects of pulpmill effluent on phytoplankton production in coastal marine waters of British Columbia. *J. Fish. Res. Bd. Canada* 33:2433–2442.

Stockner, J.G., & A.C. Costello. 1976. Marine phytoplankton growth in high concentrations of pulpmill effluent. *J. Fish. Res. Bd. Canada* 33:2758–2765.

Streeter, V.L. 1980. Mechanics, fluid. In *The New Encyclopaedia Britannica, Macropaedia* vol. 11, pp. 779–793.

Strickland, J.D.H. 1960. *Measuring the Production of Marine Phytoplankton*. Fish Res. Bd. Canada Bull. No. 122. 172 pp.

Strickland, J.D.H., & T.R. Parsons. 1972. *A Practical Handbook of Seawater Analysis*, 2nd ed. Fish Res. Bd. Canada Bull. No. 167. 310 pp.

Strömgren, T. 1977. Short-term effects of temperature upon the growth of intertidal Fucales. *J. Exp. Mar. Biol. Ecol.* 29:181–193.

– 1979. The effect of zinc on the increase in length of five species of intertidal Fucales. *J. Exp. Mar. Biol. Ecol.* 40:95–102.

– 1980a. The effect of dissolved copper on the increase in length of four species of intertidal fucoid algae. *Mar. Environ. Res.* 3:5–13.

– 1980b. The effect of lead, cadmium, and mercury on the increase in length of five intertidal Fucales. *J. Exp. Mar. Biol. Ecol.* 43:107–119.

– 1980c. Combined effects of Cu, Zn, and Hg on the increase in length of *Ascophyllum nodosum* (L.) Le Jolis. *J. Exp. Mar. Biol. Ecol.* 48:225–231.

– 1983. Temperature-length growth strategies in the littoral alga *Ascophyllum nodosum* (L.). *Limnol. Oceanogr.* 28:516–521.

Sueur, S., C.M.G. van den Berg, & J.P. Riley. 1982. Measurement of the metal complexing ability of exudates of marine macroalgae. *Limnol. Oceanogr.* 27:536–543.

Sugimura, Y., Y. Suzuki, & Y. Miyake. 1978. The dissolved organic iron in seawater. *Deep-Sea Res.* 25:309–314.

Sun, H.H., V.J. Paul, & W. Fenical. 1983. Avrainvilleol, a brominated diphenylmethane derivative with feeding deterrant properties from the tropical green alga *Avrainvillea longicaulis*. *Phytochem.* 22:743–745.

Sunda, W.G., & R.R.L. Guillard. 1976. The relationship between cupric ion activity and the toxicity of copper to phytoplankton. *J. Mar. Res.* 34:511–529.

Sunda, W.G., R.T. Barber, & S.A. Huntsman. 1981. Phytoplankton growth in nutrient rich seawater: importance of copper-manganese cellular interactions. *J. Mar. Res.* 39:567–586.

Suto, S. 1950. Studies on shedding, swimming and fixing of the spores of seaweeds. *Bull. Jap. Soc. Sci. Fish.* 16:1–9.

Sverdrup, H.U., M.W. Johnson, & R.H. Fleming. 1942. *The Oceans: Their Physics, Chemistry and General Biology.* Prentice-Hall, Englewood Cliffs, N.J. 1087 pp.

Sweeney, B.M. 1974. A physiological model for circadian rhythms derived from the *Acetabularia* rhythm paradoxes. *Int. J. Chronobiol.* 2:25–33.

Sweeney, B.M., & B.B. Prézelin. 1978. Circadian rhythms. *Photochem. Photobiol.* 27:841–847.

Swift, D.G. 1980. Vitamins and phytoplankton growth. In I. Morris (ed.), *The Physiological Ecology of Phytoplankton*, pp. 329–368. Blackwell Scientific Publications, Oxford.

Swinbanks, D.D. 1982. Intertidal exposure zones: a way to subdivide the shore. *J. Exp. Mar. Biol. Ecol.* 62:69–86.

Syrett, P.J. 1956. The assimilation of ammonia and nitrate by nitrogen-starved cells of *Chlorella vulgaris*. II. The assimilation of large quantities of nitrogen. *Physiol. Plant.* 9:19–27.

– 1981. Nitrogen metabolism of microalgae. In T. Platt (ed.), *Physiological Bases of Phytoplankton Ecology.* Can. Bull. Fish. Aquat. Sci. No. 210, pp. 182–210.

Szeicz, G. 1974. Solar radiation for plant growth. *J. Appl. Ecol.* 11:617–636.

Tanner, C.E. 1981. Chlorophyta: life histories. In C.S. Lobban & M.J. Wynne (eds.), *The Biology of Seaweeds*, pp. 218–247. Blackwell Scientific Publications, Oxford.

Tatewaki, M. 1970. Culture studies on the life history of some species of the genus *Monostroma*. *Sci. Pap. Inst. Algol. Res., Fac. Sci., Hokkaido U.* 6(1):1–56.

Tatewaki, M., & L. Provasoli. 1977. Phylogenetic affinities in *Monostroma* and related genera in axenic culture. *J. Phycol.* 13 (Suppl.):67.

Tatewaki, M., L. Provasoli, & I.J. Pintner. 1983. Morphogenesis of *Monostroma oxyspermum* (Kütz.) Doty (Chlorophyceae) in axenic culture, especially in bialgal culture. *J. Phycol.* 19:409–416.

Taylor, P.R., & M.M. Littler. 1982. The roles of compensatory mortality, physical disturbance, and substrate retention in the development and organization of a sand-influenced rocky-intertidal community. *Ecology* 63:135–146.

Tegner, M. 1980. Multispecies considerations of resource management in southern California kelp beds. In J.D. Pringle, G.J. Sharp, & J.F. Caddy (eds.), *Proceedings of the Workshop on the Relationship between Sea Urchin Grazing and Commercial Plant/Animal Harvesting*. Can. Tech. Rep. Fish. Aquat. Sci. No. 954, pp. 125–143.

Tempest, D.W., J.L. Meers, & C.M. Brown. 1970. Synthesis of glutamate in *Acrobacter aerogenes* by a hitherto unknown route. *Biochem. J.* 117:405–407.

Terumoto, I. 1964. Frost resistance in some marine algae from the winter intertidal zone. *Low. Temp. Sci.* (Ser. B) 22:19–28.

Thomas, T.E. 1983. Ecological aspects of nitrogen uptake in intertidal macrophytes. Ph.D. Thesis, University of British Columbia, Vancouver. 185 pp.

Thomas, T.E., & D.H. Turpin. 1980. Desiccation enhanced nutrient uptake rates in the intertidal alga *Fucus distichus*. *Bot. Mar.* 23:479–481.

Thomas, W.H., D.L.R. Seibert, & A.N. Dodson. 1974. Phytoplankton enrichment experiments and bioassays in natural coastal seawater and in sewage outfall receiving waters off southern California. *Estu. Cstl. Mar. Sci.* 2:191–206.

Thomas, W.H., J. Hastings, & M. Fujita. 1980. Ammonium input to the sea via large sewage outfalls – Part 2: Effects of ammonium on growth and photosynthesis of southern California phytoplankton cultures. *Mar. Environ. Res.* 3:291–296.

Thorhaug, A., & A. Katchalsky. 1972. The role of thermoosmosis in marine algae. *Proc. Intl. Seaweed Symp.* 7:279–285.

Thorhaug, A., & J.H. Marcus. 1981. The effects of temperature and light on attached forms of tropical and semi-tropical macroalgae potentially associated with OTEC (Ocean Thermal Energy Conversion) machine operation. *Bot. Mar.* 24:393–398.

Thurman, H.V. 1978. *Introductory Oceanography*, 2nd ed. Merrill, Columbus, Ohio. 506 pp.

Tilman, D., S.S. Kilham, & P. Kilham. 1982. Phytoplankton ecology: the role of limiting nutrients. *Ann. Rev. Ecol. System.* 13:349–372.

Titlyanov, E.A., & V.M. Peshekhodko. 1973. On the transport of assimilates in the thalli of fringing seaweeds. *Trudy̆ Biol. Pochv. Inst.* (N.S.) 20:137–141. (In Russian with English summary.)

Tolbert, N.E., & C.B. Osmond (eds.). 1976. *Photorespiration in Marine Plants.* (*Austr. J. Plant Physiol.* 3:1–139), CSIRO, Melbourne.

Topinka, J.A. 1978. Nitrogen uptake by *Fucus spiralis* (Phaeophyceae). *J. Phycol.* 14:241–247.

Topinka, J.A., & J.V. Robbins. 1976. Effects of nitrate and ammonium enhancement on growth and nitrogen physiology in *Fucus spiralis*. *Limnol. Oceanogr.* 21:659–664.

Trick, C.G., R.J. Andersen, N.M. Price, A. Gillam, & P.J. Harrison. 1983. Examination of hydroxamate-siderophore production by neritic eukaryotic marine phytoplankton. *Mar. Biol.* 75:9–17.

Tricker, R.A.R. 1980. Water waves. In *The New Encyclopaedia Britannica, Macropaedia*, vol. 19, pp. 654–660.

Tseng, C.K. 1981. Commercial cultivation. In C.S. Lobban & M.J. Wynne (eds.), *The Biology of Seaweeds*, pp. 680–725. Blackwell Scientific Publications, Oxford.

Turner, C.H.C., & L.V. Evans. 1977. Physiological studies on the relationship between *Ascophyllum nodosum* and *Polysiphonia lanosa*. *New Phytol.* 79:363–371.

Turpin, D.H. 1980. Processes in nutrient based phytoplankton ecology. Ph.D. Thesis, University of British Columbia, Vancouver. 131 pp.

Turpin, D.H., J.S. Parslow, & P.J. Harrison. 1981. On limiting nutrient patchiness and phytoplankton growth: a conceptual approach. *J. Plankton Res.* 3:421–431.

Turvey, J.R. 1978. Biochemistry of algal polysaccharides. *Intl. Rev. Biochem.* 16:151–177.

Tyler, J.E., & R.C. Smith. 1970. *Measurements of Spectral Irradiance Underwater.* Gordon & Breach, New York. 103 pp.

Ukeles, R. 1962. Growth of pure cultures of marine phytoplankton in the presence of toxicants. *Appl. Microbiol.* 10:532–537.

Underwood, A.J. 1978. A refutation of critical tidal levels as determinants of the structure of intertidal communities on

British shores. *J. Exp. Mar. Biol. Ecol.* 33:261–276.
– 1980. The effects of grazing by gastropods and physical factors on the upper limits of distribution of intertidal macroalgae. *Oecologia* 46:201–213.
– 1981. Techniques of analysis of variance in experimental marine biology and ecology. *Oceanogr. Mar. Biol. Ann. Rev.* 19:513–605.
Vadas, R.L. 1977. Preferential feeding: an optimization strategy in sea urchins. *Ecol. Monogr.* 47:337–371.
– 1979. Abiotic disease in seaweeds: thermal effluents as causal agents. *Experientia* 35:435–437.
Vadas, R.L., M. Keser, & B. Larson. 1978. Effects of reduced temperatures on previously stressed populations of an intertidal alga. In J.H. Thorp & J.W. Gibbons (eds.), *Energy and Environmental Stress in Aquatic Systems*, pp. 434–451. DOE Symp. Series 48 (CONF-721114). U.S. Government Printing Office, Washington, D.C.
van den Hoek, C. 1975. Phytogeographical provinces along the coasts of the northern Atlantic Ocean. *Phycologia* 14:317–330.
– 1982. Phytogeographic distribution groups of benthic marine algae in the North Atlantic Ocean. A review of experimental evidence from life history studies. *Helgol. Meeresunters.* 35:153–214.
van der Meer, J.P. 1983. The domestication of seaweeds. *BioScience* 33:172–176.
van der Meer, J.P., & N.L. Bird. 1977. Genetics of *Gracilaria* sp. (Rhodophyceae, Gigartinales). I. Mendelian inheritance of two spontaneous green variants. *Phycologia* 16:159–161.
van der Meer, J.P., & E.R. Todd. 1977. Genetics of *Gracilaria* sp. (Rhodophyceae, Gigartinales). IV. Mitotic recombination and its relationship to mixed phases in the life history. *Can. J. Bot.* 55:2810–2817.
– 1980. The life history of *Palmaria palmata* in culture. A new type for the Rhodophyta. *Can. J. Bot.* 58:1250–1256.
van der Meer, J.P., M.D. Guiry, & C.J. Bird. 1983. Sporogenesis in male plants of *Chondrus crispus* (Rhodophyta, Gigartinales). *Can. J. Bot.* 61:2261–2268.
Vandermeulen, J.H., & T.P. Ahern. 1976. Effects of petroleum hydrocarbons on algal physiology: review and progress report. In A.P.M. Lockwood (ed.), *Effects of Pollutants on Aquatic Organisms*, pp. 107–125. Cambridge University Press, Cambridge.
van der Velde, H.H., & A.M. Hemrika-Wagner. 1978. The detection of phytochrome in the red alga *Acrochaetium daviesii*. *Plant Sci. Lett.* 11:145–149.
Velthuys, B.R. 1980. Mechanisms of electron flow in Photosystem II and toward Photosystem I. *Ann. Rev. Plant Physiol.* 31:545–567.
Veroy, R.L., N. Montaño, M.L.B. de Guzman, E.C. Laserna, & G.J.B. Cajipe. 1980. Studies on the binding of heavy metals to algal polysaccharides from Philippine seaweeds. I. Carrageenan and the binding of lead and cadmium. *Bot. Mar.* 23:59–62.
Vogel, S. 1981. *Life in Moving Fluids: The Physical Biology of Flow*. Willard Grant Press, Boston. 352 pp.
Vollenweider, R.A. (ed.) 1969. *A Manual on Methods for Measuring Primary Production in Aquatic Environments*. IBP Handbook No. 12. Blackwell Scientific Publications, Oxford. 213 pp.
von Hofsten, A., B. Liljesvan, & M. Pedersén. 1977. Localization of bromine in the chloroplasts of the red alga *Lenormandia prolifera*. *Bot. Mar.* 20:267–270.
von Stosch, H.A. 1964. Wirkung von Jod und Arsenit auf Meeresalgen in Kultur. *Proc. Intl. Seaweed Symp.* 4:142–150.
Voskresenskaya, N.P. 1972. Blue light and carbon metabolism. *Ann. Rev. Plant Physiol.* 23:219–234.
Waaland, J.R. 1978. Growth and strain selection of *Gigartina exasperata* (Florideophyceae). *Proc. Intl. Seaweed Symp.* 9:241–248.
– 1981. Commercial utilization. In C.S. Lobban & M.J. Wynne (eds.), *The Biology of Seaweeds*, pp. 726–741. Blackwell Scientific Publications, Oxford.
Waaland, J.R., S.D. Waaland, & G. Bates. 1974. Chloroplast structure and pigment composition in the red alga *Griffithsia pacifica*: regulation by light intensity. *J. Phycol.* 10:193–199.
Waaland, S.D. 1975. Evidence for a species-specific cell fusion hormone in red algae. *Protoplasma* 86:253–261.
Waaland, S.D., & R. Cleland. 1974. Cell repair through cell fusion in the red alga *Griffithsia pacifica*. *Protoplasma* 79:185–196.
Waaland, S.D., & J.R. Waaland. 1975. Analysis of cell elongation in red algae by fluorescent labelling. *Planta* 126:127–138.
Waaland, S.D., J.R. Waaland, & R. Cleland. 1972. A new pattern of plant cell elongation: bipolar band growth. *J. Cell Biol.* 54:184–190.
Waite, T.D., & R. Mitchell. 1972. The effect of nutrient fertilization on the benthic alga *Ulva lactuca*. *Bot. Mar.* 15:151–156.
Waldichuk, M. 1979. Review of problems. *Phil. Trans. Roy. Soc. Lond.* B 286:399–424.
Walker, N.A., F.A. Smith, & M. Beilby. 1979. Amine uniport at the plasmalemma of charophyte cells. II. Ratio of matter to charge transported and permeability of free base. *J. Membr. Biol.* 49:283–296.
Walther, K., & L. Fries. 1976. Extracellular alkaline phosphatase in multicellular marine algae and their utilization of glycerophosphate. *Physiol. Plant.* 36:118–122.
Wassman, R., & J. Ramus. 1973. Seaweed invasion. *Natural History* 82:24–36.
Watson, B.A., & S.D. Waaland. 1983. Partial purification and characterization of a glycoprotein cell fusion hormone from *Griffithsia pacifica*, a red alga. *Plant Physiol.* 71:327–332.
Weber, A., & M. Wettern. 1980. Some remarks on usefulness of algal carotenoids as chemotaxonomic markers. In F.-C. Czygan (ed.), *Pigments in Plants*, 2nd ed., pp. 104–116. Gustav Fischer Verlag, Stuttgart.
Weidner, M., & H. Kiefer. 1981. Nitrate reduction in the marine brown alga *Giffordia mitchelliae* (Harv.) Ham. *Z. Pflanzenphysiol.* 104:341–351.
Werz, G. 1965. Determination and realization of morphogenesis in *Acetabularia*. *Brookhaven Symp. Biol.* 18:185–203.
West, E.S., W.R. Todd, H.S. Mason, & J.T. Van Bruggen. 1966. *Textbook of Biochemistry*, 4th ed. Macmillan, New York. 1595 pp.
West, J.A. 1972. Environmental regulation of reproduction in *Rhodochorton purpureum*. In I.A. Abbott & M. Kurogi (eds.), *Contributions to the Systematics of the Benthic Marine Algae of the North Pacific*, pp. 213–230. Jap. Soc. Phycol., Kobe, Japan.
West, J.A., & M.H. Hommersand. 1981. Rhodophyta: life histories. In C.S. Lobban & M.J. Wynne (eds.), *The Biology of Seaweeds*, pp. 133–193. Blackwell Scientific Publications, Oxford.
West, K.R., & M.G. Pitman. 1967. Rubidium as a tracer for

potassium in the marine algae *Ulva lactuca* L. and *Chaetomorpha darwinii* (Hooker) Kuetzing. *Nature* 214:1262–1263.

Wheeler, A.E., & J.Z. Page. 1974. The ultrastructure of *Derbesia tenuissima* (de Notaris) Crouan. I. Organization of the gametophyte protoplast, gametangium, and gametangial pore. *J. Phycol.* 10:336–352.

Wheeler, P.A. 1979. Uptake of methylamine (an ammonium analogue) by *Macrocystis pyrifera* (Phaeophyta). *J. Phycol.* 15:12–17.

Wheeler, P.A., & J.A. Hellebust. 1981. Uptake and concentration of alkylamines by a marine diatom. Effects of H^+ and K^+ and implications for the transport and accumulation of weak bases. *Plant Physiol.* 67:367–372.

Wheeler, P.A., & J.J. McCarthy. 1982. Methylammonium uptake by Chesapeake Bay phytoplankton: evaluation of the use of the ammonium analogue for field uptake measurements. *Limnol. Oceanogr.* 27:1129–1140.

Wheeler, P.A., & W.J. North. 1980. Effect of nitrogen supply on nitrogen content and growth rate of juvenile *Macrocystis pyrifera* (Phaeophyta) sporophytes. *J. Phycol.* 16:577–582.

– 1981. Nitrogen supply, tissue composition and frond growth rates for *Macrocystis pyrifera* off the coast of southern California. *Mar. Biol.* 64:59–69.

Wheeler, P.A., & G.C. Stephens. 1977. Metabolic segregation of intracellular free amino acids in *Platymonas* (Chlorophyta). *J. Phycol.* 13:193–197.

Wheeler, W.N. 1980. Effect of boundary layer transport on the fixation of carbon by the giant kelp *Macrocystis pyrifera*. *Mar. Biol.* 56:103–110.

– 1982. Nitrogen nutrition of *Macrocystis*. In L.M. Srivastava (ed.), *Synthetic and Degradative Processes in Marine Macrophytes*, pp. 121–137. Walter de Gruyter, Berlin.

Wheeler, W.N., & M. Neushul. 1981. The aquatic environment. In O.L. Lange, P.S. Nobel, C.B. Osmond, & H. Ziegler (eds.), *Physiological Plant Ecology I. Responses to the Physical Environment. Encycl. Plant Physiol.*, vol. 12A, pp. 229–247. Springer-Verlag, New York.

Wheeler, W.N., & L.M. Srivastava. 1984. Seasonal nitrate physiology of *Macrocystis integrifolia* Bory. *J. Exp. Mar. Biol. Ecol.* 76:35–50.

Wheeler, W.N., & M. Weidner. 1983. Effects of external inorganic nitrogen concentration on metabolism, growth and activities of key carbon and nitrogen assimilating enzymes of *Laminaria saccharina* (Phaeophyceae) in culture. *J. Phycol.* 19:92–96.

White, F.N., & G.N. Somero. 1982. Acid-base regulation and phospholipid adaptations to temperature: time courses and physiological significance of modifying the milieu for protein function. *Physiol. Rev.* 62:40–90.

Whitfield, P.H., & A.G. Lewis. 1976. Control of the biological availability of trace metals to a calanoid copepod in a coastal fjord. *Estu. Cstl. Mar. Sci.* 4:255–266.

Widdowson, T.B. 1965. A survey of the distribution of intertidal algae along a coast transitional in respect to salinity and tidal factors. *J. Fish. Res. Bd. Canada* 22:1425–1454.

Wiencke, C., & A. Läuchli. 1981. Inorganic ions and floridoside as osmotic solutes in *Porphyra umbilicalis*. *Z. Pflanzenphysiol.* 103:247–258.

Wilber, C.G. 1971. Turbidity – general introduction. In O. Kinne (ed.), *Marine Ecology*, vol. 1, pt. 2, pp. 1157–1165. Wiley, New York.

Wilce, R.T. 1967. Heterotrophy in Arctic sublittoral seaweeds: an hypothesis. *Bot. Mar.* 10:185–197.

Wilce, R.T., C.W. Schneider, A.V. Quinlan, & K. vanden Bosch. 1982. The life history and morphology of free-living *Pilayella littoralis* (L.) Kjellm. (Ectocarpaceae, Ectocarpales) in Nahant Bay, Massachussetts. *Phycologia* 21:336–354.

Wiltens, J., U. Schreiber, & W. Vidaver. 1978. Chlorophyll fluorescence induction: an indicator of photosynthetic activity in marine algae undergoing desiccation. *Can. J. Bot.* 56:2787–2794.

Withrow, R.B., & A.P. Withrow. 1956. Generation, control and measurement of visible and near-visible radiant energy. In A. Hollaender (ed.), *Radiation Biology*, vol. 3, pp. 125–258. McGraw-Hill, New York.

Wong, K.F., & J.S. Craigie. 1978. Sulfohydrolase activity and carrageenan biosynthesis in *Chondrus crispus* (Rhodophyceae). *Plant Physiol.* 61:663–666.

Wong, P.T.S., Y.K. Chau, O. Kramar, & G.A. Bengert. 1982. Structure-toxicity relationship of tin compounds on algae. *Can. J. Fish. Aquat. Sci.* 39:483–488.

Wood, E.J.F., & J.C. Zieman. 1969. The effects of temperature on estuarine plant communities. *Chesapeake Sci.* 10:172–174.

Wood, J.M. 1974. Biological cycles for toxic elements in the environment. *Science* 183:1049–1052.

Woolery, M.L., & R.A. Lewin. 1973. Influence of iodine on growth and development of the brown alga *Ectocarpus siliculosus* in axenic culture. *Phycologia* 12:131–138.

Wright, S.W., & B.R. Grant. 1978. Properties of chloroplasts isolated from siphonous algae. Effects of osmotic shock and detergent treatment on intactness. *Plant Physiol.* 61:768–771.

Wurster, C.F. 1968. DDT reduces photosynthesis by marine phytoplankton. *Science* 159:1474–1475.

Wynne, M.J., & G.T. Kraft. 1981. Classification summary. In C.S. Lobban & M.J. Wynne (eds.), *The Biology of Seaweeds*, pp. 743–750. Blackwell Scientific Publications, Oxford.

Wynne, M.J., & S. Loiseaux. 1976. Recent advances in life history studies of the Phaeophyta. *Phycologia* 15:435–452.

Yamada, T., K. Ikawa, & K. Nisizawa. 1979. Circadian rhythm of the enzymes participating in the CO_2-photoassimilation of a brown alga, *Spatoglossum pacificum*. *Bot. Mar.* 22:203–209.

Yarish, C., P. Edwards, & S. Casey. 1979. A culture study of salinity responses in ecotypes of two estuarine red algae. *J. Phycol.* 15:341–346.

– 1980. The effects of salinity, and calcium and potassium variations on the growth of two estuarine red algae. *J. Exp. Mar. Biol. Ecol.* 47:235–249.

Yokohama, Y. 1972. Photosynthesis-temperature relationships in several benthic marine algae. *Proc. Intl. Seaweed Symp.* 7:286–291.

– 1973. A comparative study on photosynthesis-temperature relationships and their seasonal changes in marine benthic algae. *Int. Rev. Ges. Hydrobiol.* 58:463–472.

– 1981. Distribution of green light-absorbing pigments siphonaxanthin and siphonein in marine green algae. *Bot. Mar.* 24:637–640.

Yokum, C.S., & L.R. Blinks. 1958. Photosynthetic efficiency of marine plants. *J. Gen. Physiol.* 38:1–16.

Young, D.N., B.M. Howard, & W. Fenical. 1980. Subcellular localization of brominated secondary metabolites in the red alga *Laurencia synderae* (Rhodophyta). *J. Phycol.* 16:182–185.

Young, M.L. 1975. The transfer of ^{65}Zn and ^{59}Fe along a *Fucus serratus* (L.) → *Littorina obtusata* (L.) food chain. *J. Mar. Biol. Ass. U.K.* 55:583–610.

Index